R00172 76248

REF
TA710
.K4913
cop.1
FORM 125 M
BUSINESS AND INDUSTRY DIVISION
The Chicago Public Library
Received MAR 30 1976

DISCARD

Handbook of Soil Mechanics

Volume 1

Handbook of Soil Mechanics

Volume 1

Soil Physics

by

Árpád Kézdi

Corresponding Member of the Hungarian Academy of Sciences
Professor of Civil Engineering
Technical University of Budapest

Elsevier Scientific Publishing Company
Amsterdam · London · New York 1974

Handbook of Soil Mechanics

Vol. 1. Soil Physics
Vol. 2. Soil Mechanics of Earthworks. Foundations and Highway Engineering
Vol. 3. Soil Testing in the Laboratory and in the Field
Vol. 4. Application of Soil Mechanics in Practice. Examples and Case Histories

The original: Handbuch der Bodenmechanik. Band 1: Bodenphysik was published by Akadémiai Kiadó, Budapest in co-edition with VEB Verlag für Bauwesen, Berlin (GDR)

Translated by I. Lazányi

REF
TA
710
.K4913
VOL. 1

cop. 1

Distribution of this book is being handled by the following publishers

for the U.S.A. and Canada
American Elsevier Publishing Company, Inc.
52 Vanderbilt Avenue
New York, New York 10017

for the East European Countries, China, Northern Korea, Cuba, Northern Vietnam and Mongolia
Akadémiai Kiadó, Budapest

for all remaining areas
Elsevier Scientific Publishing Company
335 Jan van Galenstraat
P.O. Box 211, Amsterdam, The Netherlands

ISBN 0-444-99890-X
Library of Congress Card Number 73-85223
With 437 Illustrations and 36 Tables

© Akadémiai Kiadó, Budapest 1974. Á. Kézdi

Joint edition published by Elsevier Scientific Publishing Company, Amsterdam and Akadémiai Kiadó, Budapest

Printed in Hungary

BUSINESS & INDUSTRY DIVISION

MAR 30 1976

Preface

The publication of the English version of my "Handbook of Soil Mechanics" has been preceded by three Hungarian and two German editions. The first volume of the English version follows, rather closely, the text of the first German "Handbook" volume, with some additions which were necessitated by recent developments. I am glad that the book in its English version may serve a much wider circle in the field of civil engineering.

The "Handbook" consists of four volumes; according to the following scheme:

Vol. 1. The physical properties of soils
Vol. 2. Soil mechanics in foundation engineering, earthworks and road construction
Vol. 3. Soil testing in the laboratory and in the field
Vol. 4. Practical applications; case histories.

In this first volume, I have made an effort to present soil physics—as far as is possible today—on a basis which is broader and more homogeneous than usual; as a physics of granular material which can be interpreted as a particulate state of matter. The knowledge of the laws of soil physics helps the civil engineer to understand the behaviour of soils and to judge the mutual interactions between soil and structures. This volume serves as a basis and as an introduction to the applications, treated in Vol. 2, the testing procedures in Vol. 3 and the examples in Vol. 4. It is my sincere wish to provide the reader with some tools in his struggle against Nature's forces for a better and a more economical construction.

I would like to express my sincere thanks to my friend and associate I. Lazányi, for his meticulous and tiresome work in the translation of the book; a task he accomplished with great skill. Also, the care of the publishers both in Amsterdam and in Budapest helped to give the volume its present from; I appreciate all their efforts.

Árpád Kézdi

Contents

Introduction

Scope, aims and methods of soil mechanics

In a technical sense the term soil is applied to that outermost layer of the earth's crust, which supports all civil engineering structures and which in turn reacts upon their foundations and influences their behaviour. Soil is also used as the construction material of earthworks. This outer layer consists, in the overwhelming majority of cases, of what the geologist calls loose sediments, i.e. deposits made up of disintegrated fragments and decomposed particles of solid rock. Every civil engineering object, whether it be a building, a bridge, an embankment, road pavement or railway, ultimately transmits its dead and live loads onto these sediments, which form the subsoil, and its stability and permanence depend greatly on a successful foundation, i.e. on whether proper connection between subsoil and superstructure has been established. The structure thus inevitably comes into contact with the subsoil and induces strains and stresses in it. Conversely, soil behaviour reacts upon the structure and may affect its stability; structure and soil thus act on each other.

This soil–structure interaction is extremely manifold and complex in nature. To build safely and economically, to transform nature and harness her forces to man's service, constitute tasks we can only cope with if we understand the nature of these interactions and make a thorough investigation of soil behaviour. The science which comes to our aid by studying and investigating the interrelations between soil and objects erected upon or connected with it, is called *soil mechanics*. As an engineering i.e. applied science, it rests upon a theoretical basis furnished by physics—mechanics and hydraulics—and chemistry. It deals with the study of the physical properties of soils; on the basis of simplifying assumptions it works out new theories and it applies the laws thus obtained to solve civil engineering problems associated with soil. It assists the engineer to attain safety and economy in the design of foundations and earthworks, and equips him with scientific tools enabling him to observe the behaviour of the subsoil and to make use of his findings, to the improvement of the structure.

In connection with civil engineering construction the role of soil is twofold. As has been mentioned before, it either constitutes the foundation of a structure, or is used as construction material for it, e.g. for the construction of earthworks, dams, dikes or cuts. In both cases soil is regarded as a dead material whose bacteriological and biological properties are normally of no importance to us. Neither shall we be concerned with the properties of soil in relation to agricultural cultivation, and its chemical properties will also receive relatively little consideration.

These questions belong to the domain of soil science, whose interest is focussed on the interrelation between soil and living organisms. The field of soil mechanics should also be demarcated from that of geology, which is primarily concerned with the descriptive study of rocks and minerals, the study of events of geological history and the investigation of agents responsible for the formation of the earth's surface. Nevertheless, soil science and geology, which have many analytical methods in common with soil mechanics, may give useful assistance in solving a number of civil engineering problems.

In the design of a structure the civil engineer usually sets out from certain physical properties of various construction materials. The knowledge of only a few data such as modulus of elasticity, specific gravity, yield point and ultimate tensile strength of a particular structural steel suffices to enable the civil engineer to predict, by applying the laws of mechanics and strength of materials, the behaviour of the structure under any loading condition. Should the properties of the material actually delivered differ in any way from those stipulated in the specifications, he may reject it and insist on having the structure constructed of materials exactly as specified.

Where soil is concerned, however, the situation is a great deal more complex. The choice of the site for a new structure is governed, in most cases, by factors other than those dependent on the subsoil, and soil conditions must usually be accepted as they are. The possibility of improving soil properties is rarely available. Besides, these properties are subject to great variations in both space and time. This not only calls for the deter-

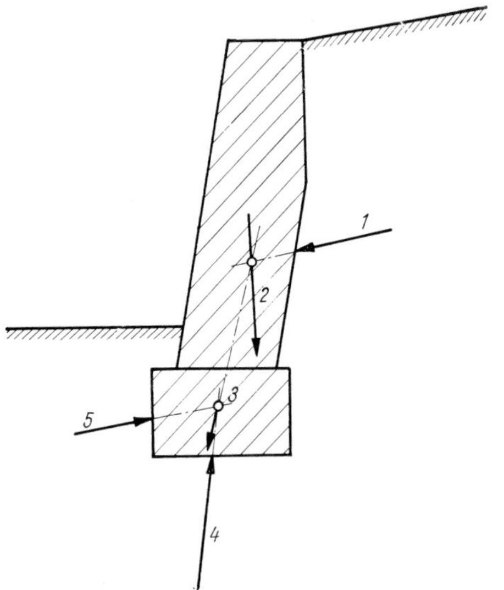

Fig. 1. Interaction between retaining wall and earth:
1 — earth pressure; 2 — weight of the wall; 3 — resulting force; 4 and 5 — reactions

mination of many more physical properties than those normally required when dealing with other construction materials; it also must be born in mind that these properties may vary from point to point under the same object, and may not even remain constant in time, due to the effect of erosion, wetting or drying out etc. Heterogeneity and anisotropy of soil should not, however, be looked upon as unsurmountable obstacles in soil testing; they only make it more difficult.

Soil mechanics, if it is to answer all the questions raised by other branches of civil engineering science, must possess first of all a knowledge of the properties necessary for the classification and identification of soils. Further, it should provide methods for the study of soil response to various effects resulting from man's engineering activity. This first group of questions, i.e. physical properties and testing of soils, is dealt with by soil physics. The knowledge of the physical properties of the soil permits the construction, on simplifying assumptions, of models which depict the behaviour of soil masses and the interaction between soil and superstructure. The applicability of theoretical results to real soils is, of course, limited; when using them, practical experience, site observations and measurements must also be taken into consideration. Problems of this kind constitute the subject matter of applied soil mechanics.

In order to familiarize the reader with the kinds and nature of soil–structure interactions outlined in the introductory part of this chapter, a few examples of them are given here.

1. If a bank of soil which otherwise would fail to remain stable is supported by a retaining wall (*Fig. 1*), then the soil exerts a lateral pressure on the wall and causes stresses in it. The earth pressure on the wall gives a resultant force, which in turn imposes an eccentric load at the base upon the soil. The interaction is two-fold; on the one hand the retaining wall is acted upon by a lateral pressure applied by the supported soil; on the other, the wall exerts at its base a pressure on the soil. The forces produce strains and stresses and we can design the wall economically only if these forces and stresses are known.

2. The safe foundation of a column of a building—e.g. a framed structure—requires a footing with a sufficiently large area of base so that the pressure p is transmitted onto the soil without causing detrimental displacements and deformations in it (*Fig. 2*). The applied load is distributed within the soil resulting in compression of the

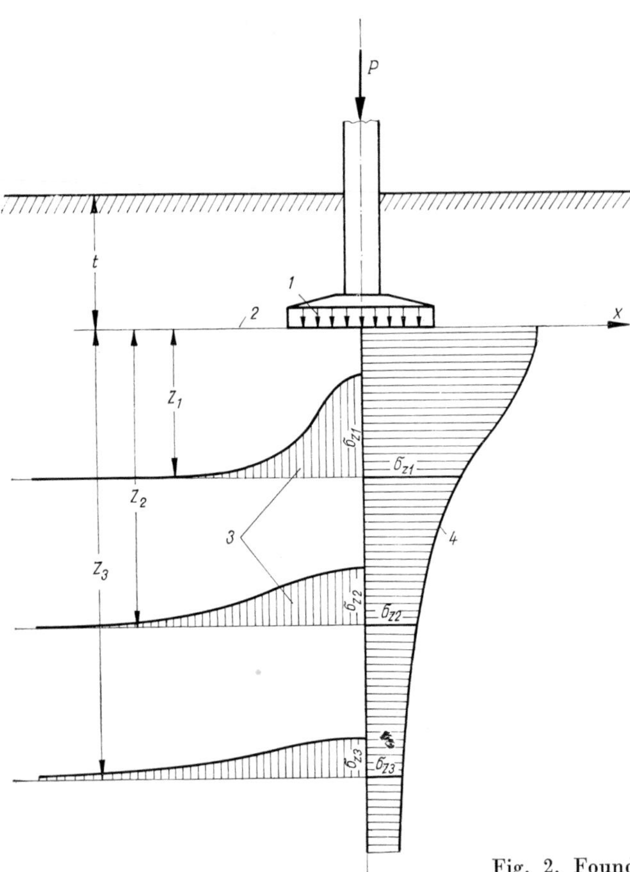

Fig. 2. Foundation of a column:
1 — contact pressures; 2 — base of foundation; 3 — vertical stresses; 4 — vertical stresses acting in the axis of the square foundation

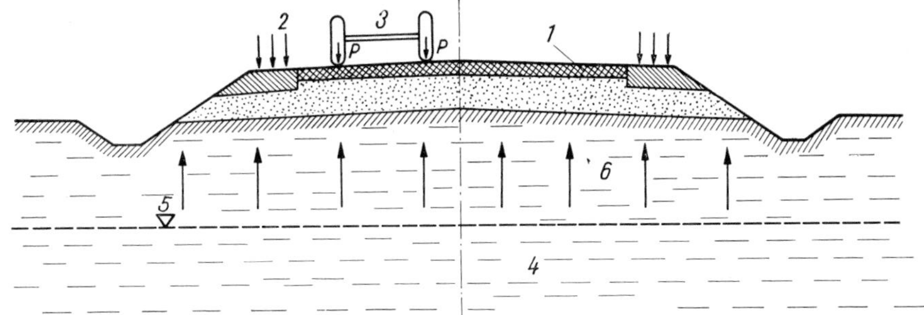

Fig. 3. Mutual effects between road pavement and subsoil:
1 – pavement; *2* – precipitation; *3* – load; *4* – subsoil; *5* – ground water; *6* – capillary rise of water

soil and settlement of the building. On the other hand, settlements may have a damaging effect to the building. This again reveals the twofold character of the interaction between soil and structure.

3. *Figure 3* shows a road pavement laid directly on the soil. The wheel load of passing vehicles is transmitted by way of the pavement onto the soil, producing there, as has been shown in previous examples, stresses and compressive deformation. On the other hand, the pavement is exposed, too, to various other soil effects. Rainwater infiltrating into the soil may cause it to swell, whereas drying out makes it shrink so that the overlying pavement may loose its support; in winter, capillary water, which may rise as high as the underside of the pavement, may freeze in the voids, ice lenses be formed and frost heave may result—another striking example of multiple interaction between soil and supported structure.

Historical background

The art of using earth as a construction material can be traced back to prehistoric times. Tens of thousands of years ago, early man of the neolithic age erected earthen mounds to mark places of religious or other significance; such earthworks are found at the beginning of many cultures. On marshy ground he built his house on piles. Later, with the development of slavery, it became possible to use massed human labour for the construction of large monuments. But the ancient builders, as revealed by archeological exploration, must have lacked a sound knowledge of the art of foundations—building. If a building settled excessively, it was simply pulled down and the debris used for the construction of a new one.

Making a proper foundation has been a major problem for builders in every period of history. The foundation of many magnificent buildings in the metropolises of ancient civilizations must have posed the architects with problems similar in many aspects to those encountered in our modern practice. The pyramids of Egypt or the Temples of Babylon, the Great Wall of China or the roads and aqueducts of the Roman empire all presented specific problems of foundation. Considering the varying and complex nature of soil it is fair to say that ever since prehistoric times, few structural problems required as much attention, ingenuity and invention to solve them as those associated with soil and foundation.

Building practice until modern times pursued an empirical approach to the solution of foundation problems. Success relied primarily upon intuition, a human gift which is unfortunately not a hereditary one. Progress was therefore inevitably retarded and intermittent, as men failed to give due consideration to relations of cause and effect. In the Middle Ages the foundation was not even recognized as a separate part of the structure. The lowermost part of every load-bearing wall was, as a matter of routine, somewhat enlarged and the base so formed brought down in trenches to a depth of one to two meters below ground level. If the soil found at the bottom was judged, by inspection, treacherous, wooden piles were driven into it to give a safer foundation.

But a few perceptive observers made some remarkably early discoveries. In the works of ancient and mediaeval philosophers one may come upon sporadic thoughts that have a bearing on the mechanics of soils, and are quite modern in that they are based on practical observations. Many fundamental concepts of foundation engineering emerged for the first time in the 15th and 16th century, bred in the atmosphere of the Renaissance, amidst the renewal of ideas and the ferment of human society. For the Renaissance meant a revival in a broad sense, not only that of the fine arts. LEONARDO DA VINCI, the "Uomo universale", is honoured by posterity as above all the creator of the "Mona Lisa" and the "Last Supper", but his works, particularly some 1700 drawings in the "*Codice Atlantico*"—preserved in the Ambrosiana of Milan—also portray him as a technical genius; almost every great achievement of our modern techniques took shape first in his plans. Of these works we can learn, to mention only a few fields of interest to us, what concept the Maestro formed of the science of mechanics and which system of fortification or type of retaining wall he held best. We can get an insight into his ideas on road construction and canal-making, and find original sketches of hoisting and conveying machines and excavators. With his innumerable elaborately detailed drawings he was centuries ahead of his time.

The 18th century, which is generally recognized as the mother of the construction engineering sciences, brought progress in the field of soil engineering too. It was then that the first clear concept of earth pressure, one of the main topics of soil mechanics ever since, was formulated and a solution devised which, subject to certain limitations, is still considered sound. This achievement is due to the great French physicist COULOMB. For a long time, indeed until the beginning of the 20th century, the theory of earth pressure remained the central problem of the scientific investigation of soils. In these studies soil was treated as an ideal material and characterized by assumed physical coefficients instead of by real properties obtained experimentally. On such a basis theoretical solutions were devised, and in order to make the problems amenable to mathematical treatment, investigators resorted to further assumptions, which deviated more or less from reality. Subsequent research work was only guided by the requirement of arriving at a mathematically "absolute" solution. New theories were developed and extremely complicated mathematical methods applied, but the benefit gained from the mathematical refinements was partly lost because of the errors due to the oversimplifying assumptions.

Similar trends and attitudes are characteristic in general of the research of the 19th century at the time when science began to get a foothold in engineering too. Research was extended to a good many problems of soil mechanics. In Russia, FUSS, KARLOVICH, KURDYOUMOV and others investigated the problems of stress distribution within the soil, slabs on elastic foundation and the traction of a rigid wheel on soft ground. In France, following the line marked by the scientific achievements of PONCELET, POISSON, DARCY and BOUSSINESQ, questions of earth pressure, elasticity and hydraulics of ground water were developed most; in England, RANKINE devised a fundamentally new theory of earth pressure. In Germany, REBHANN, MÜLLER–BRESLAU, ZIMMERMANN and others worked in similar fields. Alongside the theoretical investigators, engineers and practical men also acquired valuable knowledge by experience. But it was not possible to apply the experience amassed to other construction works, since no methods of identification and quantitative testing of soils were available yet. A review of literature of the period from 1850 to 1900 reveals that many ideas of soil mechanics were already known at that time, but all that valuable knowledge was scattered in notebooks, technical periodicals, contractor's records etc., or existed only in the practical experience of engineers. To compile all that knowledge, to explain phenomena on the basis of physics and mechanics, to determine the physical properties of soils and to devise adequate testing methods, was a job left to the researchers of the present century. In 1912 a committee was formed under the Board of Swedish Railways to carry out investigations on numerous landslides that had occurred on the lines of the railways, and to determine the internal resistance of the soil and its variation with consistency. During its ten years of existence, the committee made great contributions to our knowledge of soils.

In the same period, the Norwegian soil scientist ATTERBERG investigated the consistency of various soils and devised a new system of soil classification. In Russia academician K. K. GEDROITZ applied the laws of colloid chemistry to soils. In the United States of America a committee was set up to study the physical properties and the bearing capacity of soils.

Soil mechanics in its present sense is closely associated with the name of professor K. v. TERZAGHI (1883–1963). His book (*Erdbaumechanik auf bodenphysikalischer Grundlage*), which was published in 1925, initiated worldwide research into the subject and laid the foundation of a new civil engineering science, soil mechanics, in its modern sense as interpreted in the introduction to this chapter. His lifework, spanning well over half a century, and devoted to the development and practical application of a science initiated by himself, exerted a decisive influence throughout this branch of civil engineering activity. The initial period was marked by increased research into the physical properties of soils and by the rapid development of laboratory testing methods. Simultaneously, theoretical investigations were now carried out on the basis of soil properties obtained from laboratory tests, and therefore became efficient scientific tools in the hand of designers and construction engineers alike. In the first stage of development, papers from all over the world were published, in vast numbers, but often advocating conflicting views, or theories based on erroneous premisses. Yet this period of turmoil, this ardent experimentation and keen research were not all in vain: they helped greatly to clarify ideas and to correct the unjustifiably high expectations often attached to the role of soil mechanics. Besides, as soon as experience derived from the performance records of the first structures ever built on the basis of modern soil investigations became available, it was possible to adjust or revise the assumptions on which former theoretical methods had been based. Soil mechanics reached this stage of development by the 1940s and this marked the end of the early, pioneering period. Research today is not likely to make fundamentally new discoveries that would bring about radical changes in laboratory testing techniques.

Soil mechanics, as it stands today, is engaged in three main fields of activity: these are first, laboratory testing of soils to provide preliminary data for the purpose of safe and economic design of structures; second, field investigations and site measurements to be carried out in the course of construction works and, finally, performance records of completed structures. This latter makes it possible to assess to what extent the assumptions made in the preliminary stage of the work have proved correct, and to derive useful information for future use. It is in this threefold approach to practical problems that the semi-empirical character of soil mechanics is most evident.

In Hungary, soil mechanics research made an early start. In 1928 the illustrious scientist, Professor J. JÁKY, (Kossuth-prize winner, 1948) established the first soil mechanics laboratory at the Technical University of Budapest, years ahead of many other countries in Europe. His theoretical results, particularly his outstanding achievements in earth pressure theory, are of fundamental significance. For a long time his laboratory remained the only consulting engineering institution in this field, and with his professional reports on foundation problems he made a great contribution to the implementation of almost every major civil engineering scheme in this country. In the post-war period, he had his full share in the building program of the socialist economy, and the soil investigations of several large-scale projects, such as the Danube Steelworks and the new underground railways of Budapest, were started under his personal guidance. All those who have been actively engaged in soil mechanics research in this country are, directly or indirectly, his disciples. His name and works will be frequently referred to later in this book.

In the Soviet Union, earth sciences gained fresh momentum after the October Revolution. This especially applies to soil science, a Russian science by tradition, for which the possibilities offered by a mechanized socialist agriculture opened up new vistas. The leading personality of this period was academician VILYAMS (1863—1939), who continued and further developed the scientific work of the great Russian geologist, DOKUCHAIEV (1846—1903).

The immense construction program of consecutive Five Year Plans led to the rapid development of foundation engineering. Important advances in this field were made by GERSEVANOV (1879–1950) and his followers working at various research centres (*VIOS*, *VODGEO*) in the 1930s. Their investigations covered a wide area including various problems of shallow and pile foundations, distribution of stresses and computation of the settlement of multi-layered systems on the theoretical basis of elasticity and hydrodynamics. The process of consolidation of both saturated and partially saturated soils was studied by FLORIN. *Gersevanov*'s line of research has been adopted and further developed by the Institute of Mechanics of the Academy of Sciences, USSR, a prominent institution of the Soviet school in Mathematical Physics.

Encouraged by the inspiring achievements of Terzaghi, research was soon started on a worldwide basis, and a number of countries have contributed, to greater or lesser extent, to the advancement of soil mechanics. Our knowledge of soils, derived from laboratory and field investigations and from re-evaluation of theoretical methods in the light of practical experiences, has been accumulated over the years at a steadily increasing rate. This fact and the ever growing importance of soil mechanics have been acknowledged in a series of International Conferences on Soil Mechanics and Foundation Engineering (*Cambridge*, USA, 1936; *Rotterdam*, 1948; *Zürich*, 1935; *London*, 1957; *Paris*, 1961; *Montreal*, 1965; *Mexico City*, 1969). The number of participants and the high standard of the submitted papers have clearly testified that in modern practice, planning and construction of foundations and other underground structures without the constructive use of soil mechanics is simply inconceivable.

After the rising trend of development of soil mechanics research, a break appears to have occurred about the mid-1960s. While the important role of soil mechanics in road construction, design of earth dams, soil stabilization etc. was generally recognized, there were signs of growing scepticism concerning on the one hand the reliability of the testing methods used for the determination of the physical properties of soils, and on the other the validity of certain theoretical methods. Some fundamental assumptions have not proved true and certain simplifications turned out to be unjustified. To rectify such discrepancies by discarding false theories or by fresh analysis of misinterpreted experiences seemed to be a relatively simple matter; the real problem rather lays in an increasing need for a fundamentally new approach, on a uniform basis, to soil behaviour and soil properties. This "crisis" can only be resolved by the realization that soil as a granular medium represents a *distinct physical state*, one between the solid state and the liquid state, and that soil behaviour can be described best by the physical laws governing the behaviour of highly viscous fluids. Research that was started on these lines only a few years ago has already made some promising advances, which seem to warrant expectations that eventually a new physical concept of soil will be developed, which may well be followed by a fresh advance in theoretical research and practical applications. Some recent results and experimental evidence are presented in appropriate places in the book.

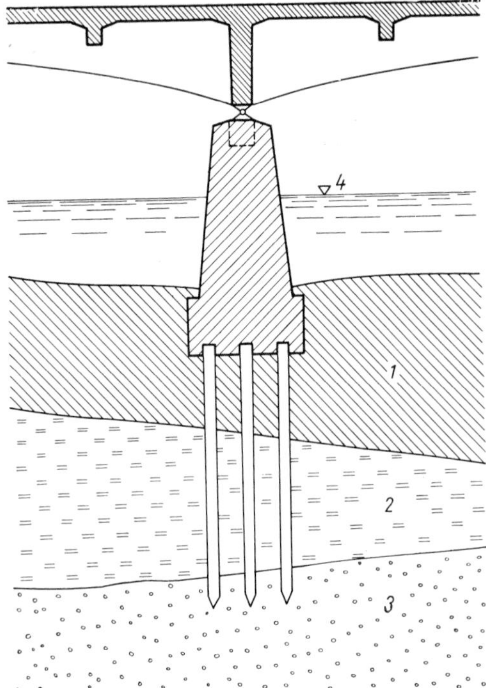

Fig. 4. Foundation of a bridge pier with piles:
1 — mud; *2* — peat; *3* — load-bearing gravel; *4* — water level

The main problems of soil mechanics

Having briefly reviewed the historical development, let us now turn to a closer study of the main problems which soil mechanics may be called upon to solve. Such problems have been mentioned previously as examples for soi–lstructure interaction; to some of them soil mechanics can offer a positive answer, for others the solution still remains to be found. The main problems are as follows.

1. *Bearing capacity and settlement of foundations.* In the design of the foundation of a building a number of questions may arise. First of all the footings must be of sufficient dimensions to withstand the superimposed loads with the required safety factor against ground failure. We must also consider the stresses due to the external loads, the manner in which the load applied to the soil is distributed over the base area of the footing, the depth, called limit depth, at which stresses induced by the external loads practically diminish to zero, and, finally, the intensity of the stresses at various points within the affected soil mass. If the soil profile and the main physical characteristics of the subsurface material are known from preliminary soil explorations, then we can estimate, on the basis of real physical properties, the bearing capacity of the soil, and, having determined the distribution of stresses in the soil mass (Fig. 2), we can judge whether the stress at a particular point of the soil is allowable. Finally, making use of the results of laboratory tests on the compressibility of the soil, we can estimate the magnitude of the settlement the building is likely to suffer.

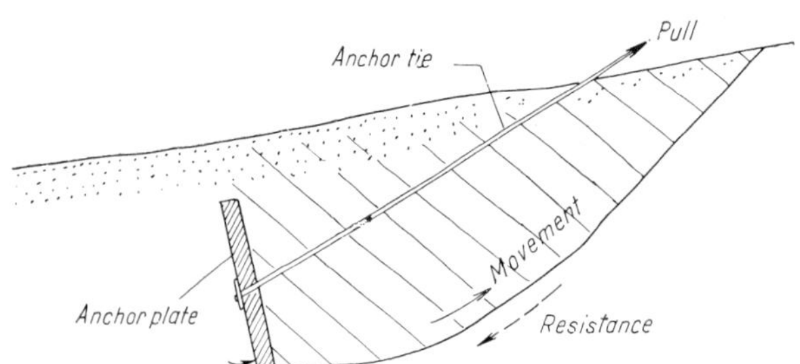

Fig. 5. Passive earth pressure prevents the breaking-out of the anchor

2. *Pile foundations.* If a solid bearing stratum is found only at a lower depth and, therefore, the loads of the superstructure are transmitted on the supporting stratum by means of piles, new problems arise. The safe load carried by a single pile, or by a group of piles, must be determined first. Hence, if the total load of the structure is given, the necessary number of piles can be calculated. An example for a pile foundation is shown in *Fig. 4.*

3. *Stability of slopes.* When we construct a cutting or an embankment, i.e. an artificial soil body bounded by free surfaces with no lateral support, then the safe angle of slope must be determined first. That angle depends on the type and the condition of the soil. Should the slope be to steep, a large mass of soil adjoining the slope may get into a critical state of equilibrium and a slide may occur. Therefore, it is necessary to study the relationship between the safe height and the angle of the slope.

This topic includes a number of similar problems, such as stability analysis of natural hillsides, case studies of various types of landslides and design considerations concerning precautions and remedial measures in connection with slides.

4. *Earth pressure.* The concept of earth pressure has already been introduced in connection with Fig. 1. That type of earth pressure occurs when a vertical bank of soil, otherwise unstable, is laterally supported by a retaining wall. Another, though somewhat different, example is presented by an anchorage (*Fig. 5*), where the question is whether the soil can sustain sufficient resistance to an inclined force transmitted by an anchor-plate, in other words, whether there is enough safety margin against failure by virtue of the passive resistance of the earth. In a broader sense, the term earth pressure can be extended to such seemingly remote problems as the forces acting against the bracing of open cuts, the pressure exerted by granular materials on the vertical walls of silo cells, or the pressure acting on flexible sheet pile

Fig. 6. Different cases of earth pressure action:
a — braced excavation; *b* — silo; *c* — sheet pile wall; *d* — tunnel

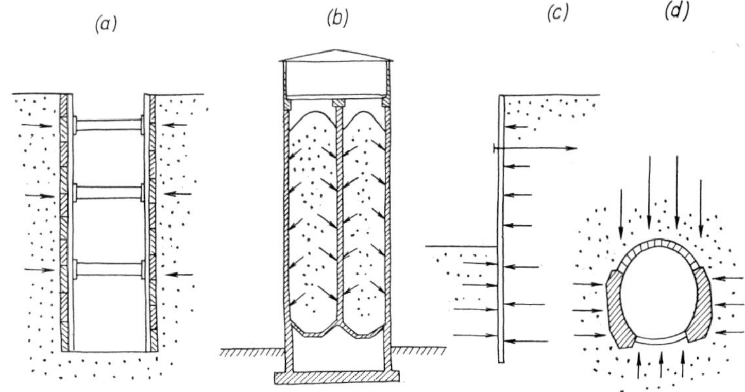

walls, tunnel linings, buried pipes and other underground structures (*Fig. 6*). Whichever is the case, the earth pressure depends basically on two things, first, on the physical properties and internal resistance of the soil, secondly, on the type and stiffness of the wall and the manner of its movement relative to the adjoining soil. Earth pressure may also change in time.

5. *Construction of earthworks and compaction of soils.* Only a few decades back, the construction of earthworks presented no particular problem. The technical requirements for railways carrying low-speed traffic or for road pavements flexible enough to follow the settlement of the fill without damage, coupled with the slow rate of construction customary at that time, did not call for any special effort to compact the earth. It is no longer so in modern practice, when rigid pavements or railway trucks designed to carry heavy and high-speed traffic safely are placed onto the fresh fill as soon as it has been completed, and any subsequent slump or settlement of the embankment which may be detrimental to the superstructure must be kept at a minimum. The selection of suitable materials for, and the proper compaction of, the embankment has therefore become the primary concern of the civil engineer working in this field. In the solution of such problems he is aided by modern methods of soil mechanics.

6. *Problems of dams.* The construction of earth dams and levees raises further problems, namely those of seepage, hydrostatic pressure and uplift. We may be called upon to answer, on the basis of our knowledge of soil properties, such questions as what sort of materials must be selected and what dimensions adopted for a dam to ensure that it withstands immense hydrostatic pressure, or to estimate the loss of stored water due to seepage through the dam (*Fig. 7*). Likewise, in the case of rockfill and concrete dams, the stability analysis of the dam, the design of the foundation and preventive measures to reduce seepage require a thorough knowledge of the laws of soil mechanics. Further seepage problems are encountered in conjunction with the drainage and lateral enclosure of open excavations.

7. *Soil mechanics in road construction.* Recent rapid growth of highway traffic has led to increasingly stringent requirements in the design of road pavements. The durability of both rigid and flexible pavements depends primarily on the behaviour of the underlying soil, and the problems relating to the interaction between pavement and subgrade need special considerations. The pavement is laid directly upon the uppermost layer of the subsoil, where soil properties may be, under unfavourable conditions, greatly affected by seasonal variations (drying out alternating with wetting or frost action) and by the agency of root system of plants, and varying soil conditions may influence the behaviour of the pavement. These problems are even more accentuated in the construction of airfields and runways for heavy aeroplanes. Current methods of pavement design must, therefore, be based on thorough soil investigations.

We must also mention here the important role of soil mechanics in the design and construction of stabilized earth roads. With the development of soil mechanics it has become possible to improve the properties of soils by artificial means. By mixing natural soils together, or by admixing certain additives to the soil, new materials can be obtained, which are suitable for the construction of all-weather roads.

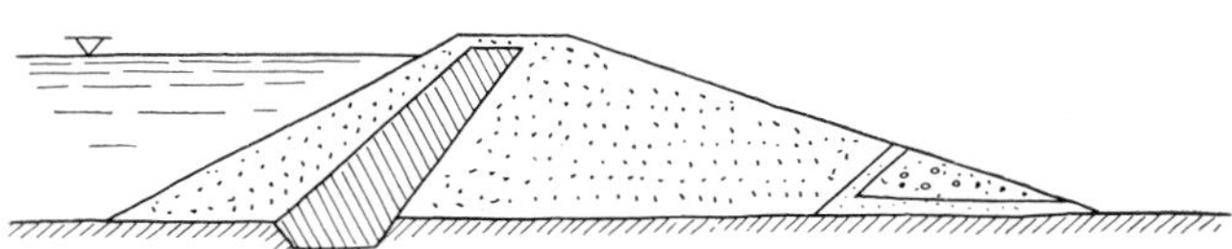

Fig. 7. Cross-section of an earth dam

8. *Soil mechanics problems of large open excavations.* In modern coal mining practice, traditional subsurface methods of exploitation have been gradually replaced, where favourable conditions prevail, by large open excavations. This means that access to a coal seam is secured by the removal of the entire overburden over an enormously large area and, usually, to considerable depth. The coal layer is often sandwiched between waterlogged strata, and the relief of hydrostatic pressure in the underlying soil, the drainage of the overburden prior to the excavation, together with the stability of the slopes of the excavated pit and the hazard of ground failure under the redeposited spoil, represent a wide range of soil mechanics problems.

Construction problems like those mentioned in the previous examples can only be coped with successfully if soil conditions have been adequately explored within a depth likely to be affected by any interference from construction activities. To this end, we take soil samples from every affected layer and subject them to laboratory tests. Such investigations, however, cannot be expected to supply reliable information unless performed on soil samples that are truly representative of the original *in situ* conditions prevailing in the respective layers. The need to preserve original soil conditions intact has led to the development of various sampling methods by means of which soil samples in a relatively undisturbed state can be recovered from boreholes, which are often of considerable depth.

The problems encountered in various fields of application of soil mechanics may be grouped into three categories. Those in the first category concern the conditions under which failure may occur in large soil masses: problems of *stability*. When a mass of soil is subjected to loads smaller than those causing the failure of the soil, the determination of the deformations may be of interest; such questions i.e. the problems of *deformation*, belong to the second category. Finally, problems in the third category relate to the *permeability* of the soil and the flow of water in soil. A brief summary of the main problems of soil mechanics is given in *Table 1.*

In practice, we often have to deal with a combination of these problems, e.g. when designing a sheet pile wall exposed to the combined effect of earth pressure and of the seepage force of the percolating water. Theoretical methods have made a fair advance in the solution of the problems of stability, while the problems of deformation, e.g. the prediction of the settlement of a building, appear to be a great deal more complex.

In the foregoing, we have seen a brief review of the main tasks soil mechanics is confronted with. Before we should set about discussing these problems methodically, we must get acquainted with those soil properties which enable us to distinguish between different soils, to describe the state of the soil adequately and to predict the response of the soil to various external effects. Then we can turn to the study of theoretical fundamentals.

These requirements determine also the scheme of the book. First, it discusses the physical properties of soils, then it investigates the stresses in soils in Ch. 5. Chapter 6 deals with the laws of water movement, Ch. 7 treats the strength problem, Ch. 8 deformation. These basic principles enable us to discuss the failure states in soil masses; this is given in Ch. 9. The second volume of the book is concerned with the different fields of application. Volume 3 is devoted to the testing procedures in the laboratory and in the field; finally, vol. 4 presents examples and case histories.

Table 1. Principal problems of soil mechanics

	Soil properties	
	Strength characteristics	Hydraulic characteristics
Stability problems	Slope stability: critical height of dams and cuts Earth pressure; value and distribution of pressures acting on retaining walls, bracings and other structures Bearing capacity; allowable bearing value of spread foundations and piles	Seepage forces: effect of pore water pressures on the stability of dams, levees, slopes, walls etc.
Deformation problems	Settlement calculations: elastic and plastic deformation beneath footings and piles. Deformation of sheet piles	Consolidation; compression of soils due to external or capillary forces
Phase movement in soils		Permeability of soils; water movement through dams or beneath them

Recommended reading

Casagrande, A. (1965): Role of the calculated risk in earthwork and foundation engineering, J. Soil Mechanics and Foundation Division. Proc. A.S.C.E. July, p. 1. (the Terzaghi Lecture, 1964).

Feld, J. (1948): Early history and bibliography of soil mechanics, Proc. 2nd Int. Conf. on Soil Mechanics and Foundation Engineering, Rotterdam, Vol. 1, pp. 1–7.

Feld, J. (1966): The factor of safety in soil and rock mechanics, Proc. 6th Int. Conf. on Soil Mechanics and Foundation Engineering, Vol. 3, pp. 185–197. University of Toronto Press.

Kézdi, Á. (1955): Soil Mechanics, Applied Mechanics Reviews. Vol. 8, No. 9, pp. 357–363.

Kézdi, Á. (1966): Grundlagen einer allgemeinen Bodenphysik, VDI-Zeitschrift, Düsseldorf, Bd. 108, No. 5.

Kézdi, Á. (1970): Szemcsés közegek fizikájának szerepe az építőmérnöki mechanikában (Role of the physics of granular median in civil engineering mechanics). Budapesti Műszaki Egyetem, Budapest.

Kézdi, Á. (1973): Die Bedeutung der bodenphysikalischen Forschung für den Erdbau. Mitteilungen des Institutes für Geotechnik und Verkehrsbau. Hochschule für Bodenkultur, Wien.

Leonards, G. A. ed. (1962): Foundation Engineering. McGraw-Hill, New York.

Lambe, T. W.–Whitman, E. V. (1969): Soil Mechanics. Part 1. Chs 1 and 2. J. Wiley & Sons, Inc. New York, London, Sydney, Toronto.

Peck, R. B. (1969): Advantages and limitations of the observational method in applied soil mechanics. Géotechnique, Vol. 19.

Proceedings, 7th Int. Conf. on Soil Mechanics and Foundation Engineering. Mexico City, 1969. State of the Art-volume.

Terzaghi, K. v. (1939): Soil Mechanics — a new chapter in engineering science. Journal of the Institution of Civil Engineers. Extra Meeting, 2 May 1939.

Terzaghi, K. v. (1944): Ends and means in soil mechanics. Journal of the Engineering Institute of Soil Mechanics.

Terzaghi, K. v. (1951): Building foundations in theory and practice, Building Research Congress, 1951. Papers presented in Division I. London.

Terzaghi, K. v. (1953): Fifty years of subsoil exploration, Proc. 3rd Int. Conf. on Soil Mechanics and Foundation Engineering, Zürich, Vol. 3, pp. 227–237.

Terzaghi, K. v. (1960): Selections from the writings of—. Prepared by *L. Bjerrum*; *A. Casagrande*, *R. B. Peck* and *A. W. Skempton*. J. Wiley & Sons, New York, London.

Tsytovich, N. A. (1955): Műtárgyak építésével kapcsolatos talajmechanikai kérdések (Soil mechanics problems related to civil engineering structures). MTA Műsz. Tud. Osztály Közleményei, Vol. XIX, 1–3, pp. 51—69.

Chapter 1

Origin of soils. Soil properties

1.1 Origin and formation of soils

All soils and sedimentary rocks located at or near the earth's surface have been formed by the weathering of solid rocks. Weathering is the result of physical and chemical processes that tend to break down solid rocks. Alternate thermal expansion and contraction; the expansive forces of ice, salt crystals and the roots of plants; the erosion by water, ice and wind charged with solid grains—these are the main factors that bring about the physical disintegration of rocks into small fragments. Oxidation, carbonization (destructive distillation of organic matter in the absence of air) and other chemical processes are the main agents in the decomposition of the constituent minerals. The loose material resulting from rock decay is carried away from its place of origin by wind, rainwash, rivers and glaciers, and is deposited elsewhere. The processes of disintegration and decomposition continue even in the course of transportation. The fresh sediments may be later exposed to various effects such as heat and pressure, which again will influence their properties. The character and the extent of weathering depend on the chemical composition, texture and strength of the solid rock, but are also affected by the environment and the prevailing climatic conditions. Products of weathering which remain at their place of origin form residual soils. The gradual downward transition from a decomposed, earthy top soil to the solid inert parent rock below is characteristic of such a residual soil. Loosened rock material which has been transported from its source of origin and laid down elsewhere forms sedimentary rocks.

Most soils that occur in civil engineering practice are water-borne sediments. As a result of rock weathering, loose, sloping deposits of debris are formed on the hillsides, near the upper course of rivers. Such deposits are composed of grains of widely varying size. Fine particles are washed by rain down the hill into the rivers, whereas larger boulders and coarse grains begin, under the action of periodical expansion and contraction due to varying temperature, a very slow, drifting movement towards the bottom of the valley. Such rock creep may finally reach the sphere of action of torrent streams, and may continually supply fresh material to be carried away by rivers. The profile of a valley side is shown in *Fig. 8*. Loosened blocks of rock fall, under the combined action of gravity and the erosion of water, from the hillside while fine material is washed down by rain, so that in the end the entire floor of the valley is covered by sediment, which will then be further transported by wind or water. In the course of transportation, angular rock fragments gradually become worn and rounded; finer particles held in suspension are moved away with a relatively greater velocity, and thus the sorting and grading of the sediment begins to take place.

As the river approaches the sea, the velocity of flow drops and, consequently, its sediment-carry-

Fig. 8. Soil forming process on a hill-side. The terrain becomes steeper and steeper; at the top there is an outcrop. The slopes and the bottom of the valley are covered with detritus which is being transported by water and wind: *1* — valley bottom; *2* — alluvium; *3* — run off water; *4* — sound rock

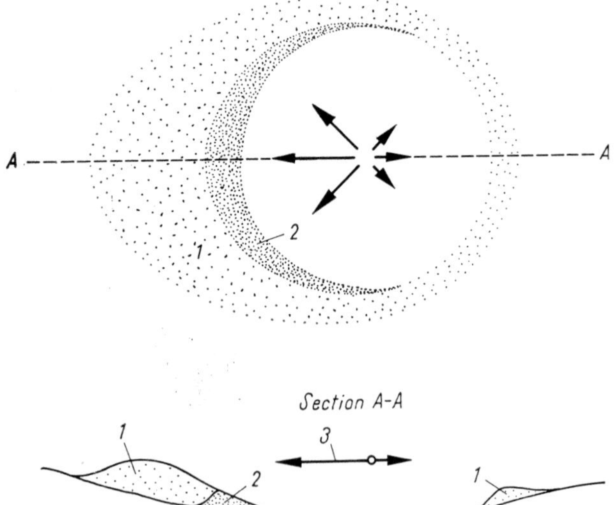

Fig. 9. Sedimentation of wind-blown particles:
1 — loess; *2* — sand; *3* — wind direction

ing power is gradually reduced. Thus, the coarse material rolled along the river bed is laid down first, then the suspended finer particles begin to settle. Coarse grains of gravel and sand are deposited first, followed later by minute particles of silt and clay so that in the end colloidal particles of clay remain the only ones still held in suspension. The solid particles in a colloidal suspension have electrical charges of the same sign; and there is,

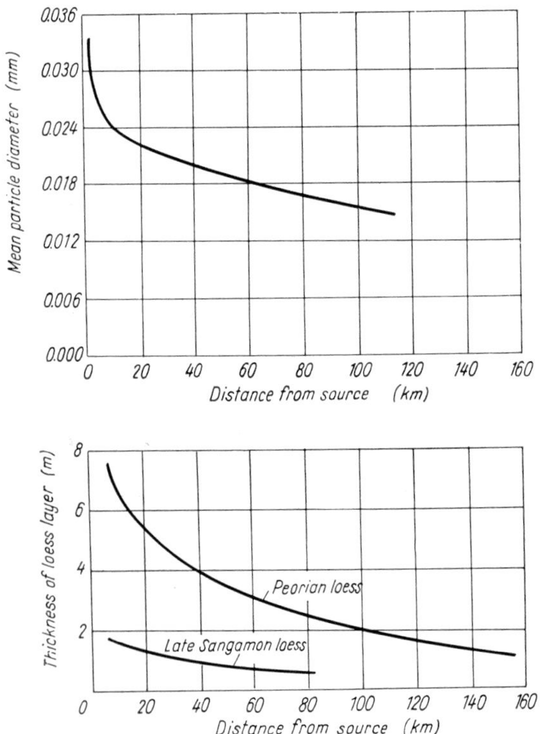

Fig. 10. Thickness of wind-blown sediments as a function of the distance from place of origin

therefore, mutual repulsion in the suspension. These electrical charges are neutralized when the colloidal suspension comes into contact with sea water which acts as an electrolyte causing the fine particles to adhere and aggregate into flocks large enough to settle. This phenomenon, known as flocculation, accounts for the fact that rivers entering the sea drop all their load of sediment and build up a delta, i.e. a deposit of varying thickness, made up of fine material having a characteristic flocculated or cellular structure.

Likewise, flocculation may occur due to exposure to solar and ultraviolet radiation when rivers pass, in their course, through fresh-water lakes.

On its way towards the sea, the sediment transported by rivers undergoes a definite process of sorting: coarse materials, gravel and sand, are laid down in the upper course of the river, fine-grained sand predominates in the plains, and silt and colloidal clay particles settle in the estuaries where the river discharges into the sea or a lake. A common feature of all water-deposited materials is that their disintegration is partly due to the transportation by water and the resulting breakdown of the larger grains. Fresh fracture surfaces immediately get into contact with water, which will then be adsorbed on the surface. Another characteristic of these fine-grained soils is that their voids are filled with water and remain saturated even if a soil layer is later elevated by geological processes above the level of the ground water.

Air-blown or Aeolian deposits have been formed under entirely different conditions. In desert regions, strong winds sweep away the fine material of rock decay. Sand grains driven by the wind against the exposed surfaces cause further abrasion of the rocks. Large grains are soon laid down in the form of sand dunes. Similar dunes occur in coastal regions where the wind sweeps away and deposits along the shore the loose debris derived from the weathering of rocks due to the tidal movements of the sea.

Very fine dust may be lifted by air currents to high altitudes and may be transported then for hundreds of kilometers. These deposits, which possess very peculiar characteristics, are usually termed loess. The manner of their formation is schematically shown in *Fig. 9*. They originate from the arid regions of deserts. The thickness of the loess layer and the average size of the deposited grains are governed by the distance from the place of blowing-out, as evidenced by observations relating to the loess deposits of the Mississippi valley (*Fig. 10*). Loess soils exhibit no distinct stratification. Their pores are mainly filled with air. Since the deposition of loess takes place over an area covered by vegetation, plant growth must keep pace with the rate at which the fine dust settles, and the roots are continually forced to move upwards, leaving behind a network of fine channels often penetrating to great depths below the ground surface. Formed under arid climatic conditions, loess normally contains a high proportion of lime, unlike water-laid sediments which have undergone a process of leaching-out of soluble salts in the course of transportation.

Besides water and wind, glaciers have been the main agents in the formation of the present-day soils. During the Ice Age, the glaciers of the Alps extended over much greater areas than they do today. Glaciers, representing a considerable load, compressed the underlying soil into a firm mass and at the same time dragged along a large amount of soil towards the valley. Rock debris dropped in the lower part of the glaciers has formed sole-moraines—heaps of irregularly deposited and poorly graded material which, as a rule, is very dense owing to the heavy load of the ice. Since glacial deposits have not been subjected to solution process, they usually contain lime. Streams emerging from the glaciers wash the fine particles out of the moraines and carry them to the valley. Thus, moraines consist mainly of a mixture of coarse-grained and irregularly graded materials. Among the soils deposited in glacier—streams, silts and clays are the most common, though in general they are not as fine-grained as the marine clays.

The mineral associations found in solid and loose rocks and in derived soil systems are functions of the elementary compositions, and the temperature, pressure, and chemical history of the system. There are many factors that influence the formation of soils, these can be listed as follows.

Soil formation will be affected, first, by the physical effects due to parent material; its chemical and mineralogical composition and also by its own physical condition and the physical properties of its primary weathering products. This effect is concerned especially with the permeability of the parent material and the depth of the weathering due to percolating waters. Residual effects of the parent material depend mainly on its chemical composition. Sometimes a difference in heat conductivity has a determinant effect.

The second important factor is the climate. This term refers to the temperature, moisture, air movement and radiant energy conditions of a system and to the changes in these conditions over an extended period of time. The thermal and moisture regimes within the soil are additionally influenced by the presence of plant and animal life.

The effect of topography on soil formation is more complex than would appear at first consideration. Topography, by controlling the rate of removal of surface particles under the action of gravity, wind, water, or ice flow, determines the period of time during which the material at a particular site is exposed to the action of a particular climate. The topography will also largely decide the particle size of the deposits. The velocity of the water running along the slope depends on the slope angle. This velocity together with the particle size of the material determines whether the particles will be washed away or a sedimentation will occur. The topography of the ocean floor greatly influences the deposits along the beaches.

The next agent is time. Soils are often considered as young, immature, mature, or even senile. Such expressions indicate the effect of time as a factor. The actual time required for the formation of a typical soil depends also to a great extent on the physical condition of the parent materials.

Finally, the influence of human activity has always been a factor in soil formation and modification. This activity, greatly escalated in modern times, pertain to all soil-forming factors that have been previously listed.

In the foregoing we have discussed the most important modes of soil formation and the many factors influencing it. But soil formation must not be seen as a once and for all process. Subsequent events in the course of the long geological history of soils may bring about substantial changes in their primary properties. For example, the load of fresh sediments exerts excess pressure on the underlying strata under which they tend to consolidate; tectonic events may produce immense heat and pressure. Long exposure to the atmosphere results in the hardening of the superficial strata due to desiccation and the formation of a solid crust. On the other hand, in humid regions the leaching-out of soluble salts may cause changes in soil properties. Atmospheric agents and organic matter produced by plant life are instrumental in the secondary weathering of the top soil.

Some of the effects caused by secondary weathering processes are illustrated in a quantitative manner in some examples. *Figure 11* shows the average clay content of soils derived from igneous rocks—

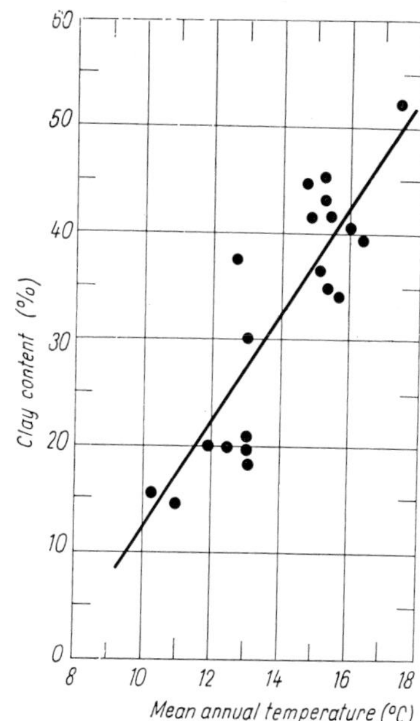

Fig. 11. Clay content of soils formed by weathering of igneous rocks, as a function of the mean annual temperature (Jenny, 1950)

mainly gabbro and diorite—, as a function of mean annual temperature. It is seen that in hot climate the process of weathering is accelerated. The variation of the lime content of virgin-land soils with mean annual rainfall is shown in *Fig. 12.* In arid regions the leaching effect is insignificant and the lime content of the soils is normally high. The advance of leaching with time is shown in *Fig. 13*, for two typical soil formations: sand dunes of Britain, and the polders of Holland. In the cold and humid climate of these countries, it takes about 300 years before all lime is completely leached out of the superficial soil layer. Changes in the chemical composition of the soil will in turn affect its mechanical properties.

If a fine-grained sediment e.g. a marine clay became covered by a moraine during the Ice Age, it would stay under an increased overburden pressure for thousands of years after. Such an excess load causes the compression of the soil; its grains tend to assume a more closely packed position, while the surplus water is being squeezed out of the pores. Should the overburden pressure be relieved later, e.g. by erosion, then the soil emerging from the process will exhibit physical properties entirely different from those which the same soil would possess if it had not undergone preconsolidation.

If a soil stratum, originally deposited in water, is later raised above the ground water level, then the pore water begins to evaporate and the upper parts of the layer gradually dry out. Seasonal rains may temporarily interrupt the process of desiccation, and the total thickness of the dried-out zone seldom exceeds 8 to 10 meters. In contrast to the general assumption that soil strength increases with depth, a desiccated soil becomes softer at greater depths.

In addition to purely physical processes various chemical actions also take part in the formation of soils, the carbon dioxide content of the air and

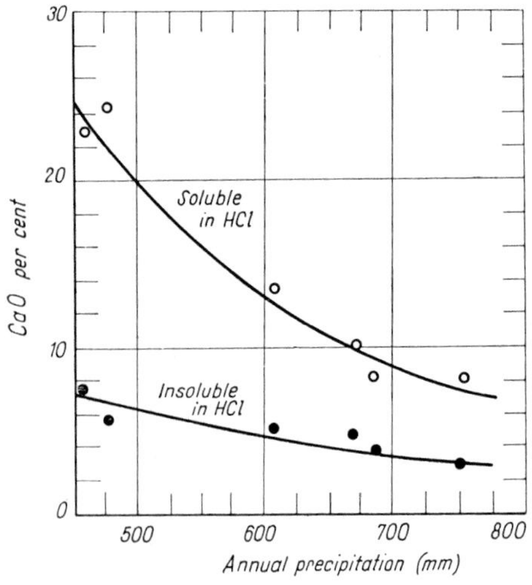

Fig. 12. Lime content of virgin soils as a function of the amount of rainfall (JENNY, 1950)

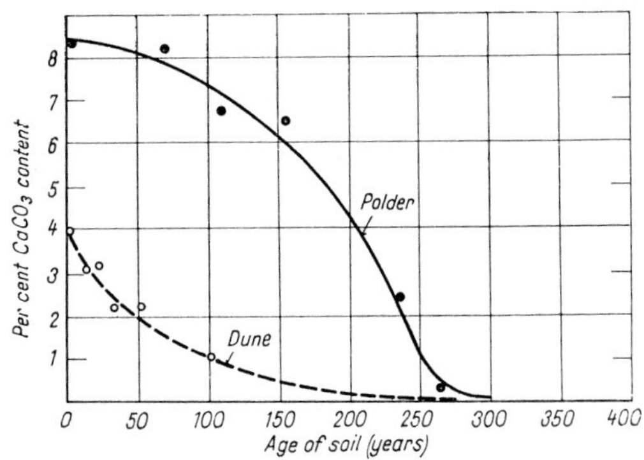

Fig. 13. Leaching of lime with time (JENNY, 1950)

various salts and organic acids produced by the vegetation being the main agents. One of the most conspicuous signs of chemical changes is discoloration which, however, does not necessarily indicate the alteration of the physical properties as well. For example, gray and blue clays, located near the ground surface may become stained in time due to oxidation, and when exposed to the atmosphere their colour may turn yellow or red. The colour of soils found at greater depths is governed, besides mineralogical composition, by the climatic conditions prevailing at the time of their formation.

Chemical actions may bring about substantial changes in soil properties because of the exchange of ions adsorbed on the surface of the soil grains. As a result, the thickness of the water film surrounding the grains is altered. Changes in the adsorbed water layer may act in two ways: the affected soil becomes more susceptible to volume changes and its strength characteristics are impaired; on the other hand, the opposite effect i.e. the strengthening of the soil may also occur.

Though less significant than the natural sediments, artificial deposits such as refuse damps, landfills made of industrial waste or rubble, slag-dumps, mining spoilbanks and the likes, deserve increasing attention. Such deposits have over the past encroached upon increasingly large areas particularly in the big cities, around industrial plants and mining works, and with the rapid growth of population and high rate of industrialization there has been an increasing demand for the reclamation of such waste-lands for housing or industrial development projects. Building on refuse deposits may always present unforeseeable diffculties, their properties being very much dependent on local conditions. Yet, with economy in view, an attitude that such deposits are *a priori* unsuitable for any foundation purposes would be unjustified. Based on thorough field exploration and soil investigations, ways and means may be found to make use of the load-bearing capacity, however limited, of these deposits.

It will have been seen from the foregoing that the knowledge of the geological conditions of a particular area chosen for constructional purposes, and the information that we can obtain from the study of the history of soil formation, are indispensable to the better understanding of soil characteristics and soil behaviour.

1.2 Physical properties of soils

The term soil, as defined at the beginning of this book refers, in reality, to a diversity of materials produced, as has been briefly described in the preceding chapter, in a sequence of natural processes. If we are to deal with a medium so complex in nature and to solve civil engineering problems associated with soils, then it is necessary to introduce a number of new physical characteristics and to devise appropriate testing methods.

Soils are, in general, made up of three distinct constituent parts: solid particles, water and air, often referred to as phases that form together a dispersed system. The properties of the soil aggregate are governed by the shape and size of the solid particles, their distribution by size within the whole mass, and the interactions between the phases forming the soil. In order to understand fully the behaviour of soil aggregate, it is necessary to acquaint ourselves with the characteristics of each separate phase and with the nature of the manifold mutual relationships existing between them. These interactions are responsible for, among others, the formation of specific soil structures.

The physical properties of soils can be divided into three main groups. Those in the *first group* are related to, or can be derived from, the mineralogical composition of soils and the size and distribution of the solid particles. They serve as a means of soil classification, i.e. distinguishing between different soils and identification of soils. Such properties are given, as a rule, in numerical form, thereby making it possible to describe soils adequately excluding any subjective element. The need for true and objective description of soils will be appreciated if one considers that no practical experience with soils acquired on a particular construction work can be applied to other localities, unless the physical properties of the respective soils were found to be identical beyond doubt. Two seemingly like soils, e.g. two "yellow clays", may exhibit greatly differing characteristics as regards compressibility, strength or swelling. Soil properties in the first group are often referred to as petrographical properties or index properties. For their actual determination, disturbed soil samples will suffice.

Physical properties falling into the *second group* provide numerical data concerning the state of the soil. On the basis of these data soils can be specified for various civil engineering uses. Such properties are, for instance, the density of sands or the consistency, i.e. the measure of adhesion between the particles, of clay. For the practical determination of soil properties in this group, undisturbed soil samples are needed.

Soil properties grouped into the first two categories are sometimes referred to as *static* properties, for they concern the soil at rest and on its own, irrespective of any external influence. They are usually given in the form of numerical data.

The *third group* comprises what may be described as *dynamic* soil properties. They are related to changes, qualitative or quantitative, in soil behaviour in response to various external effects. In the study of these changes the time factor too needs to be taken into consideration. One of the most important factors affecting soils is loading, which induces strains and stresses within a soil mass. Others to be dealt with are flow of water through soil, capillary phenomena, moisture movement due to variation of temperature or under the influence of electric current, chemical changes etc. Test results are normally given in the form of empirical relationships, i.e. plotted in graphs. Strength, deformation and hydraulic properties constitute the majority of those in the third group; their many uses and applications are shown in *Table 2*.

Table 2. Groups of soil properties

Group number	Physical properties	Engineering use
I	Size of particles, size distribution; specific gravity, organic constituents, chemical properties	Classification and identification; direct use in empirical relationships; practical rules
II	Phase composition of soils; water content, void ratio, saturation, degree of compaction, consistency	Numerical evaluation of soil condition
IIIa	Strength characteristics (compression and shear strength)	Stability problems, earth pressure, slope stability
IIIb	Deformation characteristics (compression, shrinking, swelling, collapse of structure)	Deformation problems, settlement calculation
IIIc	Hydraulic characteristics (water movement, gravitational movement, capillarity, thermal movement; permeability etc.)	Consolidation; seepage through dams and levees, seepage forces, frost problems, dewatering

Recommended reading

Application of Geology to Engineering Practice, "Berkey Volume", Geological Society of America, New York, 1950.
Bendel, L. (1949): Ingenieurgeologie. Bd. I, Springer, Wien.
Denisov, N. Sa. (1957): Inzhenernaya geologiya i gidrogeologiya. Moscow.
Emmons, W. H. et al. (1949): Geology: Principles and Processes. 3rd ed. McGraw-Hill Book Co. Inc., New York.
Holms, A. (1946): Principles of Physical Geology. London.
Krynine, D. P.–Judd, W. R. (1957): Principles of Engineering Geology and Geotechnics. McGraw-Hill Book Co. Inc., New York.
Maslov, N. N. (1941): Inzhenernaya geologiya. Gosstroiizdat Moscow.
Moos, A. v–Bjerrum, L. (1948): Soil mechanics and geology, Proc. 2nd Int. Conf. on Soil Mechanics and Foundation Engineering. Rotterdam.
Moos, A. v.–De Quervain, F. (1948): Technische Gesteinskunde. Basel, Birkhauser Verlag.
Mosonyi E.–Papp F. (1959): Műszaki földtan (Technical Geology) Műszaki Könyvkiadó, Budapest.
Pettijohn, F. J.–Potter, P. E. (1964): Atlas and Glossary of Primary Sedimentary Structures. Springer Verlag, Berlin.
Redlich, K. A.–Terzaghi, K. v.–Kampe, R. (1929): Ingenieurgeologie. Springer, Wien u. Berlin.
Terzaghi, K. v. (1955): Influence of Geological Factors on the Engineering Properties of Sediments. Economic Geology Fiftieth Anniversary Volume.
Trask, P. D. (1950): Applied Sedimentation. A Symposium. J. Wiley, New York.
Scheidegger, A. E. (1970) Theoretical Geomorphology. Springer-Verlag, Berlin–Heidelberg–New York.
Záruba, Q.–Mencl, V. (1961): Ingenieurgeologie. Akademie-Verlag, Berlin.

Chapter 2

Constituent parts of soils

2.1 The solid part

2.1.1 Size and shape of grains

The solid part of soils consists of grains of varying size surrounded by a network of voids or pores. The pores are filled with water or air or both. The solid particles are the product of physical and chemical weathering processes. The grains may vary in size within very wide limits from colloidal clay particles to large boulders. The size of the grains and the percentage, by weight, of the grains in various size ranges, called fractions, exert a decisive influence on the behaviour of soils subjected to various external actions. For example, the manner in which a coarse-grained gravel or sand responds when acted upon by static load or by dynamic compactive effort or when exposed to the action of water will be quite different from the behaviour of a fine-grained soil, silt or clay, under similar conditions. For the purpose of soil classification it is necessary to determine the size of the grains which constitute a particular soil and the percentage of the total weight represented by the various grain fractions.

The significance of the size and shape of the grains is clearly demonstrated by a comparison of the main characteristics of two extreme soil types, sand and clay. What they have in common is that both consist of discrete particles. But the size and especially the shape of their representative particles are completely different, and thus they react to various external actions in different, often opposite, ways although the differences in mineralogical composition are not necessarily great enough to account for the differing behaviour. The properties of the two typical soils are compared in *Table 3*.

Grains larger than 0.1 mm in diameter can be inspected with the unaided eye or with the aid of a hand lens. These particles constitute the coarse fraction of soils. Grains within the size range 0.1 mm to 2μ are referred to as the fine fraction; their size and shape can be viewed only under a microscope. Particles smaller than 2μ represent the very fine fractions. They can be examined only by special methods (electronmicroscope, X-ray analysis). In most natural soils there are more than one fraction present, but the properties of soils are primarily governed by the predominating

Table 3. Properties of sand and of clay

Physical property	Sandy	Clay
Plasticity	Not plastic	Plastic
Volume change	Very slight, practically negligible	High volume change, swells and shrinks
Permeability	Permeable	Practically impervious
Drainage	Drains quickly by gravity. Suitable for filter layers	No drainage through gravity. Variation in moisture content together with volume change only
Capillarity	Rate of capillary rise rather quick; the height of capillary rise a few feet only	Rate of capillary rise very slow, to great heights
Compressibility	Compressible in loose state, compression occurs very quickly. Compressibility in dense state very low	Very high in soft state, the rate of compression is very slow
Shear strength	Friction only; neutral stresses have an effect in closed systems only	Friction and cohesion; hydrodynamic stresses have substantial effects
Behaviour on vibration	Quick compaction	Almost no effect

influence of the *fine fraction*. The very coarse fraction, called gravel, consists of rock fragments, which in turn are composed of one or more minerals. The grains may be angular, subangular, rounded or flat in shape. The coarse fraction, or sand, consists mainly of quartz grains; angular, subangular or rounded shapes being the most common. The individual grains of the fine and very fine fractions are usually composed of a single mineral. Angular, oblong or needle-shaped grains are common, but usually the flake-shaped ones predominate as a result of chemical weathering.

Grain size can conveniently be specified with the *diameter* of the grain. Regular geometrical solids can be perfectly described in terms of their principal dimensions, based on which their volume, weight and surface area are readily computed. The problem becomes a great deal more difficult when the bodies in question are irregular, as are the grains of the soil. No method by means of which an irregular particle could be described geometrically has been devised so far. Neither the length of a grain, nor its width, nor its thickness can be interpreted in such a way as to serve as a basis for the computation of the volume.

The difficulties in the calculation of volume and surface area are even further increased if we deal with assemblies of irregular grains. In such case we must resort to statistical methods. These methods are based on the measurement of a single dimension, called, for the sake of convenience, the *diameter* of the grain. If numerical values of the diameter as defined above are available for a large number of grains, several types of averages can be derived. From the *mean diameter* thus obtained, the total number or the total weight of the grains, contained within a specified volume, can be estimated, and the total surface area of, or the total volume occupied by, the accumulation of solid grains can be computed.

Any one irregular particle may be replaced by a so-called *equivalent grain*, geometrically regular in shape, on the condition that it should possess the same volume or the same surface area as the irregular particle in question does. For statistical purposes the concept of *equivalent sphere* is in general use. Its diameter is defined by the formula

$$d_n = \sqrt[3]{\frac{6V}{\pi}}, \tag{1}$$

wherein V denotes the volume of the irregular particle.

If the number of grains in a soil mass weighing 1 N be n, and the unit weight of the grains is γ_s, then, d_n is given as

$$d_n = \sqrt[3]{\frac{6}{\pi \gamma_s n}}. \tag{2}$$

The term "equivalent sphere", as mentioned before, is in widespread use, but different concepts of an imaginary "diameter" characteristic of an irregular grain are equally justified. For example, the plane area covered by a grain can be determined by projecting the microscopic image of the grain onto a screen; then the diameter of the circle having the same area can be regarded as the diameter of the equivalent grain; or alternatively the arithmetic or geometric means of the longest and shortest dimensions of the image may be used. But it must be borne in mind that while an irregular grain does possess a definite volume or definite surface area, it has no well-defined diameter.

That the concept of the equivalent sphere may be very misleading as regards the shape of real soil grains is strikingly demonstrated in *Fig. 14*. In the case of flat, flaky or plate-shaped particles, the diameter of the equivalent sphere has no bearing on the behaviour of real soil grains.

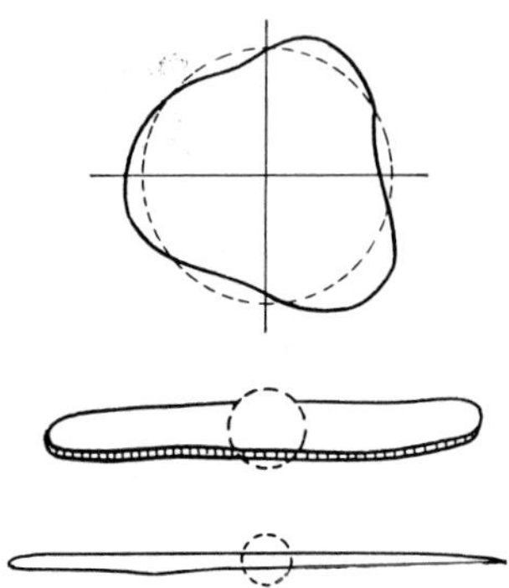

Fig. 14. The equivalent sphere gives a false picture of needle-shaped or flat particles

For the purpose of laboratory testing of soils, grain size is usually defined in two ways; for grains larger than 0.1 mm it is equivalent to the width of that circular or square aperture which the grain would just pass; for fine particles, grain size is obtained as the diameter of a spherical grain, which would sink in still water with the same velocity as the irregular grain.

Grains of gravel and sand are larger than 0.1 mm and they can be distinguished even with the unaided eye. Their shape may be either irregular with sharp corners and edges, or worn down and rounded to some degree. Grains of residual soils are chiefly of the first type, i.e. angular and irregular, whereas grains of wind-blown or water-deposited soils tend to become more and more worn down during transport. This rounding effect is the more intense, the larger the grains, the lower their hardness and the further they have been transported. The most commonly occurring grain shapes are illustrated in *Fig. 15*. In general, the grains of wind-blown sands are more apt to become worn down. In the case of water-laid sediments the degree of roundedness of the grains varies along the course of the river: in gravels it increases rapidly first,—and then at a slower rate; in sands, it decreases at a low but steady rate (Pettijohn, 1948). Some researchers attempted to explain it by the progressive breakdown of the particles, others by the sorting, according to size, of the transported sediment.

The laws of the abrasion of fluvial sediments were investigated by Stelczer (1967). He divided the factors affecting the amount of abrasion into the following groups:

mineralogical and petrographical composition;

abrasive force, which depends on the size, roughness and shape of the grains;

impact force, which depends on the velocity and weight of the colliding particles and on the duration of the impact;

specific yield and grain-size distribution of the sediment.

On the basis of mathematical models and laboratory experiments, he developed a general law of abrasion and found numerical values for the abrasion coefficients. He also suggested formulae for the decrease, with time, of the mean diameter of the sediment rolled along in an intermittent motion. In the case of continuous motion, the relationship assumes the form

$$\delta = \frac{\Delta d}{d_0} = 1 - \sqrt[3]{e^{-\alpha t}}, \tag{3}$$

where d_0 is the initial mean diameter of the sediment, Δd is the mean decrement in diameter, t is the duration of the motion and α is a coefficient dependent on mineralogical composition.

Numerical values calculated on the basis of the law of abrasion were found to be in fair agreement with those obtained by field measurements.

The loss in weight of a grain can be given by

$$\frac{\Delta W}{W_0} = \beta d_0 s, \tag{4}$$

in which W_0 is the initial weight, β is a constant, dependent on the kind of rock, and s is the distance covered by the grain.

There are methods of describing the shape of coarse grains quantitatively by means of certain coefficients; but—for civil engineering uses—visual inspection will normally suffice. But even the qualitative examination with the aid of a simple hand lense or a microscope can provide valuable information on the origin of sands—a knowledge that often proves useful from the engineering point of view. Grain shape characteristics of sands, both fluvial and aeolian, from the Great Hungarian Plain were investigated by Miháltz and Ungár (1954). They found that sand grains in the size range of 0.1 mm to 0.2 mm can be divided, according to their shape, into three categories. Grains of type I are angular or flaky with sound surfaces of fracture; they predominate in the fluvial sands of this area. Type II shows a transitional character; the grains are crudely rounded with no marks of splitting. Type III is a typical representative of dune sand; the grains are well-rounded with polished, frosted surfaces. The three types are shown in *Fig. 16*. By counting the grains of each type within a given soil sample,

I

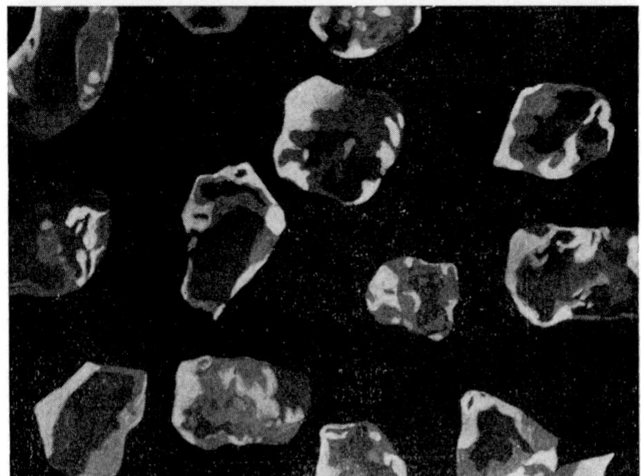
II

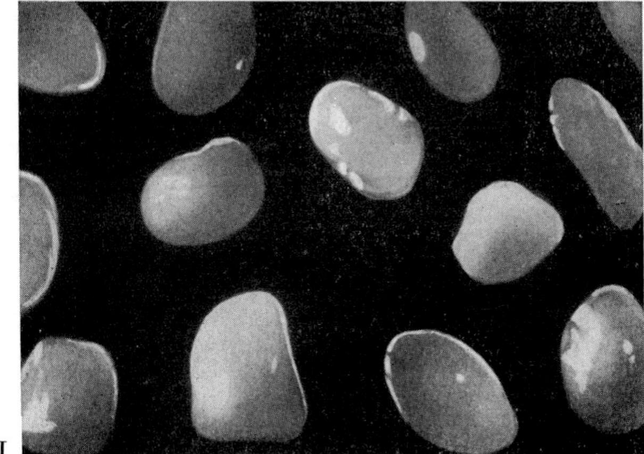
III

Fig. 16. Categories of particles shape (Mih//ltz, 1954): I — sharp edges, sound breaking surfaces; II — intermediate, partly rounded; III — rounded particles, polished surfaces

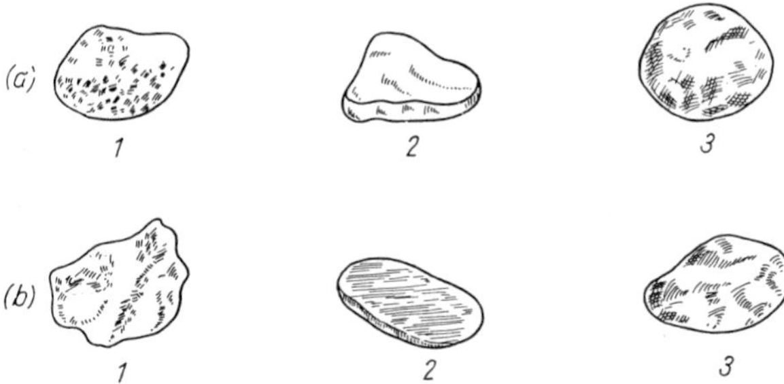

Fig. 15. Characteristic shapes of gravel- and sand particles

percentages can be plotted in a triangular diagram. River deposits and wind-blown sands can be clearly distinguished in this diagram (MIHÁLTZ and UNGÁR, 1953), and sands of unknown origin can thus be identified.

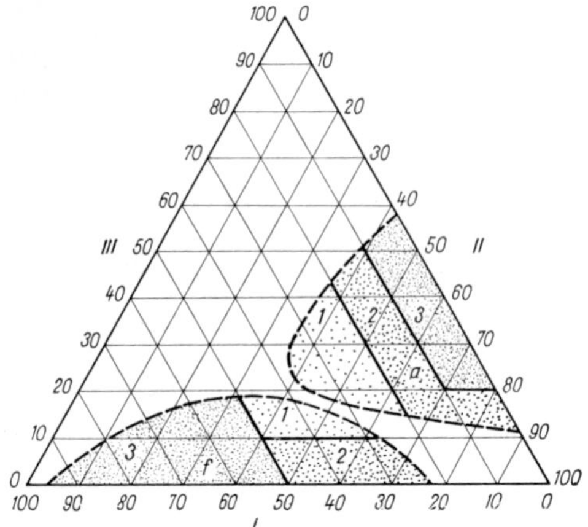

Fig. 17. Degrees in characteristics of sands. The character increases as 1 – 2 – 3. Group *a* — wind-blown sediment; Group *f* — river transported sediment

Investigations by H. GREEN (1927) and G. MARTIN (1923) revealed that for any accumulation of irregular grains the expression V/d_v^3 is a constant. V denotes the volume of a grain, d_v is a statistical diameter. If the grains are spheres of varying size, the constant is equivalent to $\pi/6$. For real soils its value will be less than $\pi/6$. A similar factor can be defined for the surface area. Let F denote the surface area of a grain of average diameter d_f, then $F/d_f^2 =$ constant. For irregular grains its value will be less than π.

Hence

$$V = \alpha_v d_v^3 , \tag{5}$$

$$F = \alpha_f d_f^2 , \tag{6}$$

where d_v is the volumetric shape factor and α_f is the superficial shape factor. The quotient of the two was found to be characteristic of the shape of the grains. For spherical grains α_f/α_v is 6, for rounded grains 6.1, for partly rounded grains 6.4, for subangular grains 7.0 and for angular grains 7.7.

For the investigation of grain shape F. SCHIEL (1948) suggests that the coarse mechanical analysis of a granular material should be performed by double sieving of the same sample, once through a circular-opening sieve, then through a square-opening sieve. In the first case the aperture size at which a grain is retained depends only on the "width" of that grain, whereas when the grains are sieved through square openings, the "thickness" of the grains must also be taken into consideration. That is why no correspondence between a particular square aperture size and an equivalent circular aperture size can be set up; there will always be a difference in the amount of material passing, depending on the shape of the grains. This discrepancy can be used, as *Schiel* suggested, to characterize the shape of the grains. Let l and m denote the size of those circular and square apertures, respectively, at which the same "percentage passing" for a given sample is obtained; the ratio l/m can then be regarded as a numerical shape factor. For equal spheres $l/m = 1$. For very thin plates whose thickness is negligible relative to their width b, $l = b$ and $m = b\sqrt{2}$, since such plates may pass a square aperture diagonally. Hence $l/m = 1/\sqrt{2}$.

The shape character of an assembly of grains can be suitably judged by the following procedure. The double sieving is performed as described above, and the resulting two grain-size distribution curves (see section 2.1.4) are plotted. The logarithm of the ratio $u = l_{50}/m_{50}$ can be obtained as the difference of the grain sizes corresponding to the ordinate 50%. The sphericity of the assembly is defined by

$$\zeta = 100\left(1 - \frac{\log u}{\log \sqrt{2}}\right). \tag{7}$$

For practical calculations, KOLBUSZEWSKI (1958) suggested the following expression

$$\zeta = d_c/d_i , \tag{8}$$

where d_c is the diameter of the circle whose area is equivalent to the area covered by the grain when laid on its long side, and d_i is the diameter of the circle inscribed into the same area.

A practical method for characterizing grain shape was put forward by H. B. SUTHERLAND (1968).

If a sample is separated by sieving into fractions, then these show as a rule great differences in mineral composition. Sand and fine sand fractions consist of a mixture of quartz, feldspar, calcite and mica. The proportion of the flat, platy particles to the bulky ones increases as grain size decreases.

The size of clay particles is of the order of microns; here so called *clay minerals* predominate. At the surface of clay grains, there act various surface forces of chemical and electrical nature which are primarily responsible for the peculiarities of clay behaviour (see Ch. 3). The shape of the individual grains can be discerned only under an electron microscope. Clay minerals have a sheeted crystal structure; thin plate-shaped or elongated, needle-shaped particles are very common. Flake-shaped particles of kaolinite and needles of halloysite crystals are shown in *Figs 18* and *19*. The ratio of the diameter of the plates to their thickness may vary between limits as wide as 10 to 1 and 250 to 1, depending upon the character of the clay minerals. Often there is no direct contact between the particles which, being held in their position by space forces, build up a skeleton in contrast to coarse-grained soils where the particles are arranged in a random packing. The prevalence of plate-shaped and flaky particles accounts, in part, for the high plasticity and compressibility of clays.

Individual soil grains are in general much too hard to be crushed under pressures due to the load of structures. But even sand can be highly compressible if it contains a considerable amount of fossil shells.

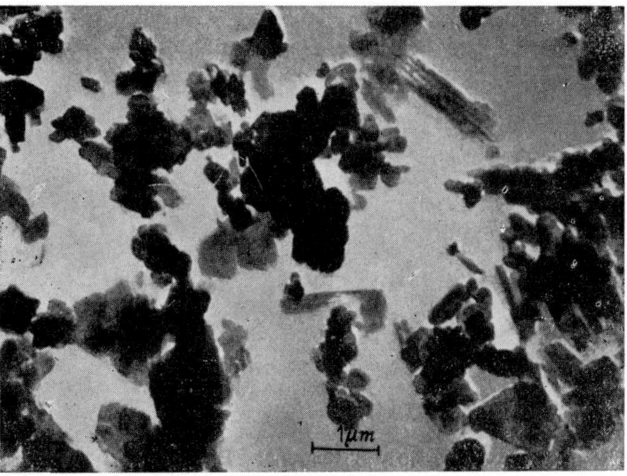

Fig. 18. The flaky particles of kaolin. (After GRIM, 1953; *Macon, Georgia*)

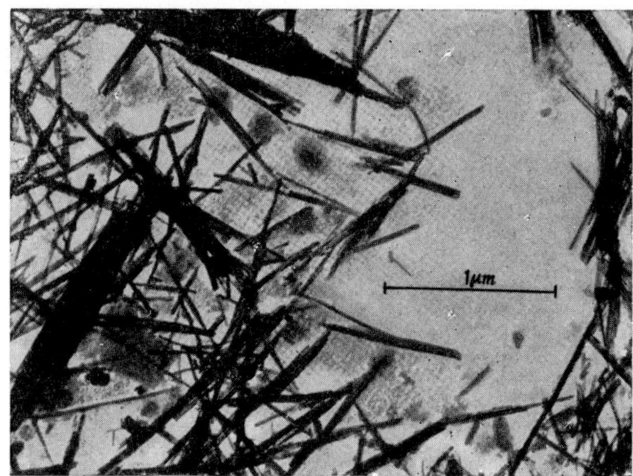

Fig. 19. Needle-shaped particles of halloysite (*Windover, Utah*; GRIM, 1953)

2.1.2 Particle-size classification

A number of systems have been put forward in the literature, which divide soil particles of widely varying sizes into certain size groups. Some of the most widely accepted conventions, together with the names adopted for each size group, are shown in *Fig. 20.* It will be seen that the names assigned to the particle-size fractions are the same as those conventionally used for natural soils. Therefore, to avoid ambiguity, the words "grains" or "particles" should be added to the name, as in the expressions "sand grains" or "clay-size particles" when it is meant to designate a grain-size fraction.

Since no unified international system of grain-size classification or nomenclature is available yet, the names of fractions should be always supplemented by the numerical values of the corresponding size limits.

It is worth noting that the size limits of *Atterberg*'s system (see Fig. 20), which forms the com-

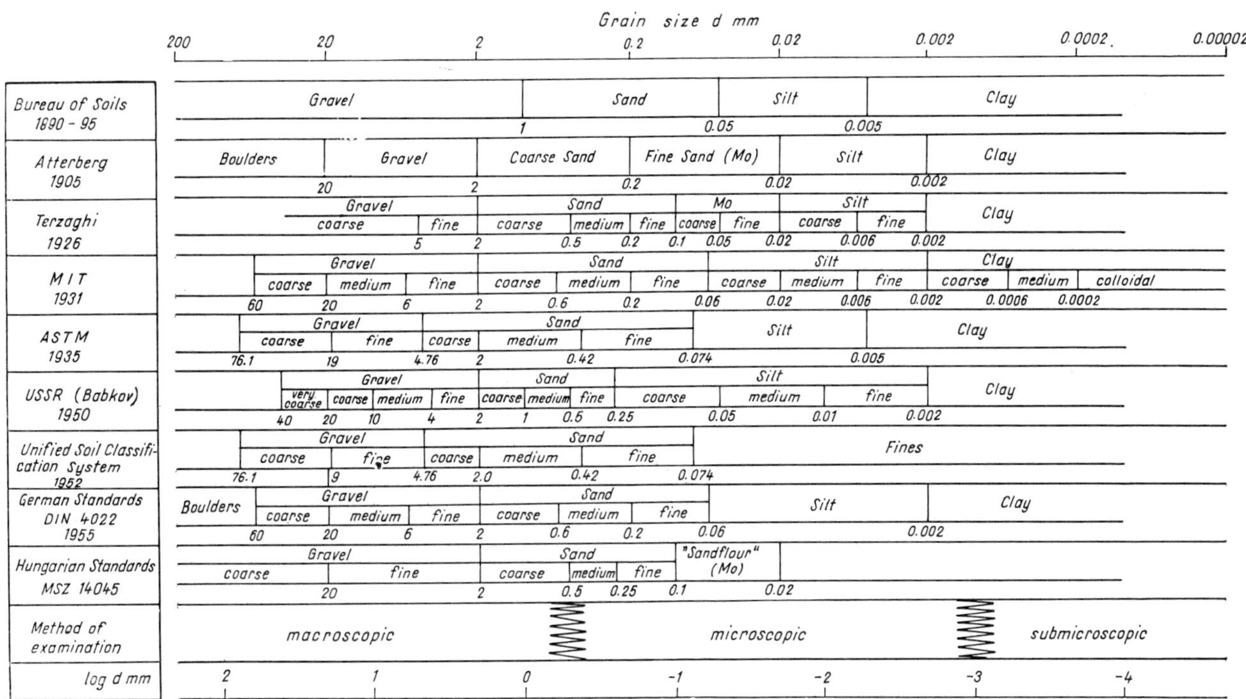

Fig. 20. Names for soil fractions

mon basis of all later systems, were determined on the basis of soil behaviour under the action of water. The limit between the grain-size fractions of gravel and sand was set at 2 mm because water can flow through a medium consisting of grains larger than 2 mm without appreciable retardation. If the grain size is less than 2 mm, the soil, though still permeable, exhibits considerable resistance to the flow of water. With a grain size of approximately 0.02 mm the rate of capillary rise of the ground water is at its maximum. Root hairs can just penetrate crevices of this size.

The grain-size limit of 0.002 mm is associated with bacteriological and physical properties; bacteria cannot move in the pores between grains smaller than this limit. Clay particles finer than 2 μ do not settle from suspension but remain in a permanent state of Brownian movement. Particles below the size limit of 200 mμ are colloidal, and their behaviour is governed by entirely different physical laws; surface forces become predominant over gravity forces. The particle size of colloidal clay ranges from 200 mμ to 20 mμ. Smoke and fog consist of particles between 20 mμ and 2 mμ in size. Gas molecules are smaller than 2 mμ.

2.1.3 Unit weight of solid particles

Unit weight is defined as the weight of solid matter per unit volume. It is denoted by γ_s and expressed, in the metric system, in units of N/cm^3 or kN/m^3. For soils, unit weight can be interpreted as the average unit weight of the constituent minerals.

Unit weights of natural soils as determined by laboratory tests are not very different from the average numerical values given in *Table 4*. Considering the slight variations in the numerical values of the unit weight, it is quite satisfactory to use these average values in most soil mechanics calculations.

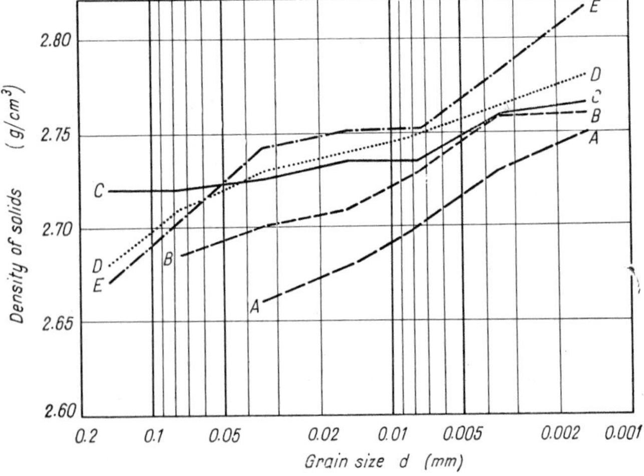

Fig. 21. Specific gravity of soil fractions:
A — clay with humus, *Szeged*; *B* — loess, *Szeged*; *C* — oligocene clay, *Óbuda*; *D* — loess, *Nagykőrös*; *E* — silt, with high lime content, *Szatymaz*

Table 4. Average values for specific gravity

Soil	Specific gravity γ_s/γ_w
Gravel, sand	2.65
Loess, rock flour, sandy silt	2.67
Silt	2.70
Lean clay	2.75
Clay	2.80

The relatively high unit weight of clay is caused by its mineral composition. Of the fractions of sedimentary soils which were formed by the breakdown of rocks, the finest fraction has the highest unit weight; with increasing grain size the unit weight gradually decreases. Contents of Fe, Ca, Mg, Al show a similar trend, decreasing from the fine fractions towards the coarse ones, and the average unit weight of a mixture of fractions is determined by the percentage of these minerals. The variation of unit weight with grain size was evidenced by the investigations by Mih́áltz (1938). Results for typical soils are shown in *Fig. 21.*

The unit weight of the solid particles is determined by measuring the volume of water displaced when a soil sample is submerged in water (pycnometer method). The volume of the adsorbed water which forms a viscous film on the surface of the grains is considered as part of the volume of the grain. The error caused by this approximation is negligible for practical purposes.

Let

W_d = weight of oven-dried sample,
W_w = weight of pycnometer filled with distilled water,
W_p = weight of pycnometer filled with water and solids,
γ_w = unit weight of water.

Hence the volume of the water displaced i.e. the volume of the solid grains is

$$V_s = \frac{W_d + W_w - W_p}{\gamma_w} \tag{9}$$

and the unit weight of the solid grains

$$\gamma_s = \frac{W_d}{V_s} = \gamma_w \frac{W_d}{W_d + W_w - W_p}. \tag{10}$$

In precise measurements a correction for temperature must be applied. Unit weight has its uses in computations related to other physical properties of soils.

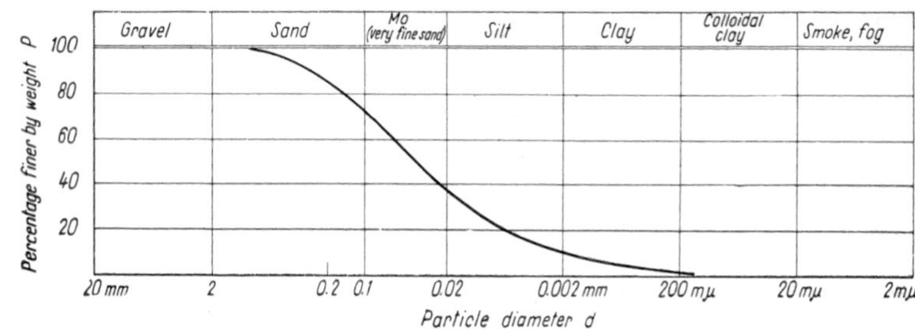

Fig. 22. Grain-size distribution curve; fractions

2.1.4 Particle-size distribution of soils

Particle-size distribution of a soil means the percentage of the total weight represented by the grains in various size ranges. Grain-size characteristics are determined by a laboratory test known as mechanical analysis. This test furnishes useful data for the identification and classification of soils. In addition, it has many practical applications, since certain other physical properties of soils such as compactibility, permeability, susceptibility to frost damage, suitability for stabilization are directly related to particle size distribution.

Grain size as obtained from the mechanical analysis must be understood as the *equivalent diameter* in the sense defined in section 2.1.1. In mechanical analysis, an assembly of particles is separated into various grain-size groups or fractions. For coarse-grained soils it is performed by sieving a sample on a set of standard sieves; by weighing the material retained at each sieve, the percentage and the frequency distribution of the fractions can be evaluated. For fine-grained soils, sieving is not practicable simply because the finest mesh readily available (No. 200 B.S. sieve) has an aperture of approx. 0.08 mm. Therefore, grain size in the fine fractions is determined by measuring the velocity with which a particle sinks in a suspension.

The results of the mechanical analysis are presented in a plot termed *particle-size distribution curve.* One ordinate of the curve represents the percentage P, by weight, of grains finer than the diameter denoted by the abscissa. The plot is thus a cumulative representation i.e. an integral curve. Considering that grain size may vary between extremely wide limits, e.g. from 20 mm to 0.001 mm for a clayey gravel, it would not be practical to plot the values of grain size to the conventional arithmetic scale. For however long (within reasonable limits) the distance representing unity was chosen, the grain-size values smaller by several orders of magnitude than unity would appear within a very short distance on the scale, and this would give a completely distorted representation of grain-size distribution. For that reason, a semi-logarithmic plot is used wherein the abscissa represents grain size on a logarithmic scale and the ordinate denotes percentage by weight of fractions finer than the corresponding grain size.

A particle-size distribution curve is shown in *Fig. 22.* The initial portion of the curve is tangential to the horizontal line representing $P = 100\%$ and the tangent point indicates the maximum grain size in the assemblage. Then the diagram continues with a downward slope, passes through a point of inflexion and approaches the horizontal line $P = 0\%$ asymptotically since, theoretically, also grains of infinitely small size may occur in the soil. The standard names of the fractions and also the corresponding grain-size limits are noted in Fig. 22.

Sieve analysis is restricted to the testing of soils whose particles are larger than 0.05 to 1 mm i.e. sands and gravels, for two reasons: first, finer particles tend to cohere when dried out and, secondly, the finest mesh available has a width of about this size.

The sieves used for sieve analysis are fitted with perforated plates for coarse material or with woven wire mesh for finer material, and are mounted in metal or wooden frames. The individual units of the set are stacked in descending order, with the largest at the top, so that each sieve has an aperture equal to half the aperture of the preceding sieve in the set. In this way it can be ensured that the points representing the results of the sieving will be equally spaced in a semi-logarithmic plot so that the diagram can be readily drawn.

The quantity of the soil required for sieving is 1 to 2 N for sand and about 5 N for gravel. The sample is oven-dried at 105 °C to constant weight, poured onto the sieve of largest size and sifted through the whole set of sieves. On completion of the sieving the material retained on each sieve is weighed. The sum of the "weights retained", $W_1, W_2, \ldots, W_i$ plus the weight of the material passing the finest sieve and collected in a receiver, make up the total weight W_0 of the sample. By definition a point of the particle-size distribution curve indicates the amount, in percentage of the total weight, of the material finer than the corresponding grain size. Hence, the ordinate pertaining to a particular sieve size d_i is obtained by subtracting the sum of the weights retained on the sieves of mesh width $d_1, d_2, \ldots, d_i$ inclusive, from the total weight W_0 and expressing the difference in percentage of the total weight. This will be referred to as "percentage passing" or "percentage finer":

$$P_i = \frac{W_0 - \sum_{n=1}^{i} W_i}{W_0} \times 100\% . \qquad (9a)$$

By plotting the percentages P_i against the corresponding diameter d_i, the grain-size distribution curve is obtained (*Fig. 23*).

Particles finer than about $d = 0.05$ to 0.1 mm cannot be separated by sieving, and the determination of the proportions of the fine fractions is carried out by a *sedimentation process* also called wet analysis. There are various methods of sedimentation all of which are based on *Stokes*' law. In a method known as elutrition the soil grains of varying sizes are separated from a suspension by the application of a current of water rising upwards at gradually increasing velocity. Grain size may also be determined on the basis of the time required for a particular grain to settle through a specified distance (*hydrometer method*). The finer the grains the more slowly they settle from a suspension, since the relative viscous resistance is greater.

Stokes' law gives the settling velocity of a single spherical grain of diameter d(cm) as

$$v(\text{cm/s}) = \frac{\gamma_s - \gamma_w}{18\eta} d^2 = Cd^2 , \qquad (10a)$$

where

γ_s = unit weight of the solid particles (N/cm³),
γ_w = unit weight of the liquid (usually water) (N/cm³),
η = coefficient of viscosity of the liquid (Ns/cm).

On the other hand, velocity can be expressed as the distance h through which the grain has fallen divided by the time required, t:

$$v = \frac{h}{t} ,$$

whence (11)

$$d = \sqrt{\frac{1}{C}\frac{h}{t}} .$$

The numerical value of the coefficient C depends on the unit weight of the grains and the viscosity of the liquid. The latter, furthermore, depends on temperature.

In the hydrometer method, a sample of 0.20 to 0.40 N for clay, and 0.50 to 1.00 N for sand, is thoroughly mixed with water, stirred up and poured into a 1000 cm³ measuring cylinder. The density of the suspension is then measured at various time intervals after the commencement of the test. Since the density of the soil–water mixture is higher than that of clean water, the depth of immersion of the hydrometer is less in the suspension than in clean water. The density of the suspension decreases as the coarser grains settle, and the hydrometer will be immersed to gradually increasing depths. From the hydrometer readings the "diameter" of the largest grain remaining in suspension at the level of the centre of the hydrometer bulb, and the concentration of particles finer than this diameter, i.e. the two co-ordinates of a point on the grain-size distribution curve, can be calculated.

The formulae of the calculation can be obtained as follows.

First we have to determine the unit weight of the suspension at a depth z below the surface after a time t from the beginning of the sedimentation. It is assumed that the solid grains are uniformly dispersed in the suspension at the beginning of the test. If the total volume of the suspension is denoted by V, and the weight of the solids dispersed by W_s, then W_s/V is the weight, and $W_s/V\gamma_s$ is the volume, of solids per unit volume. Hence, $1 - W_s/V\gamma_s$ is the volume, and $\gamma_w(1 - W_s/V\gamma_s)$ is the weight, of the

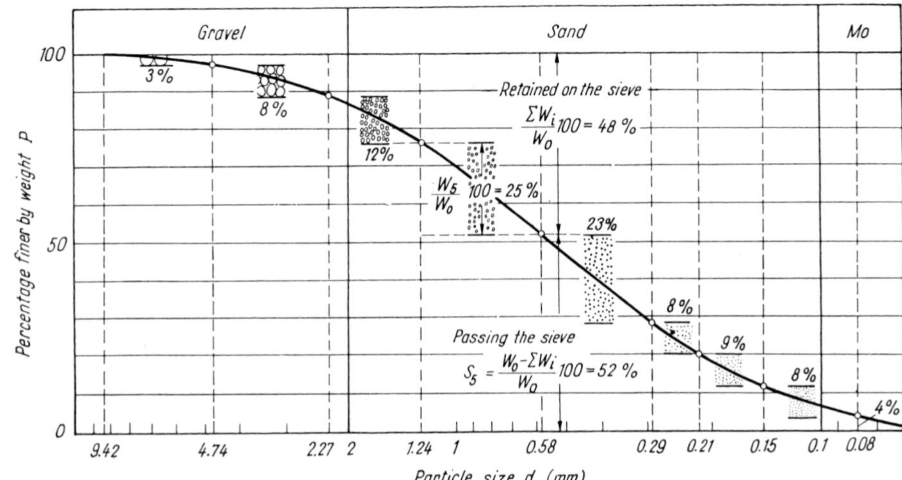

Fig. 23. Construction of grain-size distribution curve

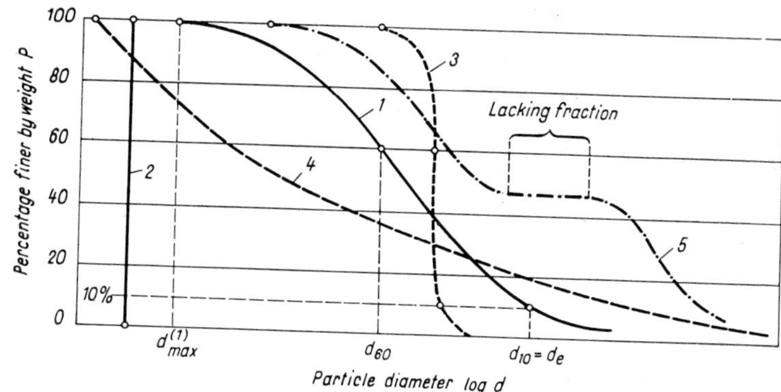

Fig. 24. Characteristic shapes of the grain-size distribution curve

liquid per unit volume of the suspension. The initial uniform unit weight of the suspension is, therefore, given by

$$\gamma_i = \frac{W_s}{V} + \gamma_w\left(1 - \frac{W_s}{V\gamma_s}\right) = \gamma_w + \frac{\gamma_s - \gamma_w}{\gamma_s}\frac{W_s}{V}.$$

Consider a section at a depth z below the surface of the suspension after time t from the beginning of the sedimentation. The size d of the grain which has fallen in time t from the surface of the suspension to the depth z can be calculated from Eq. (11):

$$d = \sqrt{\frac{1}{C} \cdot \frac{z}{t}}. \tag{11a}$$

Between the surface and the depth z there will be no grains larger than this diameter left, since after time t the coarser grains will have fallen through a distance larger than z. Within an elementary slice of the column of the suspension, taken at depth z, the concentration of the grains finer than d remains unaltered, since the number of grains finer than d entering at the upper surface is the same as the number of those leaving the element at its lower surface. Thus no particle larger than d is present within the element, whereas the concentration of the grains finer than d is the same as at the beginning of the test. If P denotes the percentage of the particles finer than d, then after time t and at a depth z the weight of the solids remaining in unit volume of suspension is given as

$$\frac{P\%}{100} \cdot \frac{W_s}{V},$$

and the unit weight of the suspension

$$\gamma = \gamma_w + \frac{\gamma_s - \gamma_w}{\gamma_s} \cdot \frac{P\%}{100}\frac{W_s}{V}.$$

Hence, the "percentage finer" is obtained from

$$P\% = 100\frac{\gamma_s}{\gamma_s - \gamma_w} \cdot \frac{V}{W_s}(\gamma - \gamma_w). \tag{11b}$$

In the practical performance of the hydrometer test, a specially devised hydrometer is immersed in the suspension and readings are taken on the graduated stem of the instrument at various time intervals after the commencement of the sedimentation. These readings indicate the density γ of the suspension at the depths determined by the centre of buoyancy of the hydrometer. From the measurements, the coordinates, d and P, of the points of the grain-size distribution diagram can be calculated by using the formulae (11a) and (11b). In an actual test, temperature effects and various other sources of error must also be taken into account and the appropriate corrections must be applied (see Vol. 3).

The hydrometer method of wet analysis is widely used in present laboratory practice. It must be pointed out, however, that the procedure has many limitations due to some inherent sources of error. The most significant deviation from the theoretical assumptions lies in the fact that real soil grains, particularly those in the fine fractions, are not spherical in shape at all and rather resemble flakes, plates or needles. It can be generally stated that the finer a particle, the less apt it is to become worn down by abrasion. According to Squires (1937) the largest dimension of a plate-shaped particle is at least five times as much as the diameter of the "equivalent sphere" as determined by *Stokes'* law. Furthermore, the soil in suspension does not redily dissolve into individual particles and rather tends to aggregate into large flocs. Therefore, dispersing and stabilizing agents must be admixed to the suspension to prevent flocculation. There may be differences in the unit weight of the grains due to their different geological origins, but in the calculations only average values are used. In a dense suspension, the settling particles tend to collide and interfere with each other's free movement, thus the result of the sedimentation test is affected also by the concentration of dispersed particles in the suspension.

For these reasons, the sedimentation test can provide only rough quantitative information about the size of the grains and the percentage of the colloidal fractions. Nonetheless, a further refinement of the procedure would not be justified in soil investigations for civil engineering purposes (Ungár, 1957). But for other uses e.g. in ceramics, the manufacture of cement, geology, numerous more accurate methods have been developed (H. A. Schweyer and L. T. Work, 1941; Dallavalle, 1948).

The results of the mechanical analysis are presented —as mentioned previously—in the form of a semilogarithmic plot known as grain-size distribution curve (also called a grading curve). Three characteristics of the curve are of major practical importance, namely the maximum grain size in the assemblage, the general slope of the curve, and the degree of grading (i.e. whether a soil is made up of one or only a few fractions, or if it contains in similar proportions a mixture of particles over a wide range of sizes. The maximum grain size d_{max} can be directly read off from the curve (*Fig. 24*). d_{max} is of importance in such questions as the suitability of a particular soil for stabilization, or the choice of a method of soil compaction; it must be considered in certain problems of permeability also.

The general slope of the curve is characterized by the *uniformity coefficient*. This is defined as the ratio

$$U = \frac{d_{60}}{d_{10}}, \tag{12}$$

where d_{60} = grain size corresponding to the percentage $P = 60\%$,
d_{10} = grain size at percentage $P = 10\%$.

d_{10} is often referred to as the effective size and is denoted by d_e. In the case of a material consisting of equal spheres the grain-size distribution "curve" is a vertical straight line (curve *2* in Fig. 24), and the uniformity coefficient is unity. Thus, a very steep grading curve similar to curve *2* in Fig. 24 bears a close resemblance to the theoretical distribution curve of the equal spheres, indicating a very high degree of uniformity. Sands with a U value close to unity are usually encountered in very loose natural deposits, their grains readily move in flowing water, and large masses of the soil may liquify on vibrating. In such soil, the dewatering of an open excavation by pumping would be a hazardous process likely to cause subsurface erosion or general liquefaction and, thus, a radical reduction or complete loss of the load bearing capacity of the soil.

A soil with a flat grain-size distribution curve (like curve *4* in the figure) is said to be well-graded; the more fractions which are present in a given soil the higher is the figure U. In extreme cases, for example silty or clayey gravels, U may have values of the order of hundreds or even thousands.

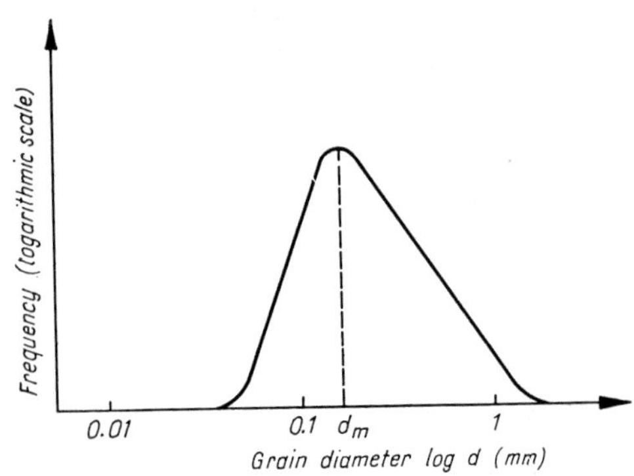

Fig. 26. Frequency curve of river sand sediments

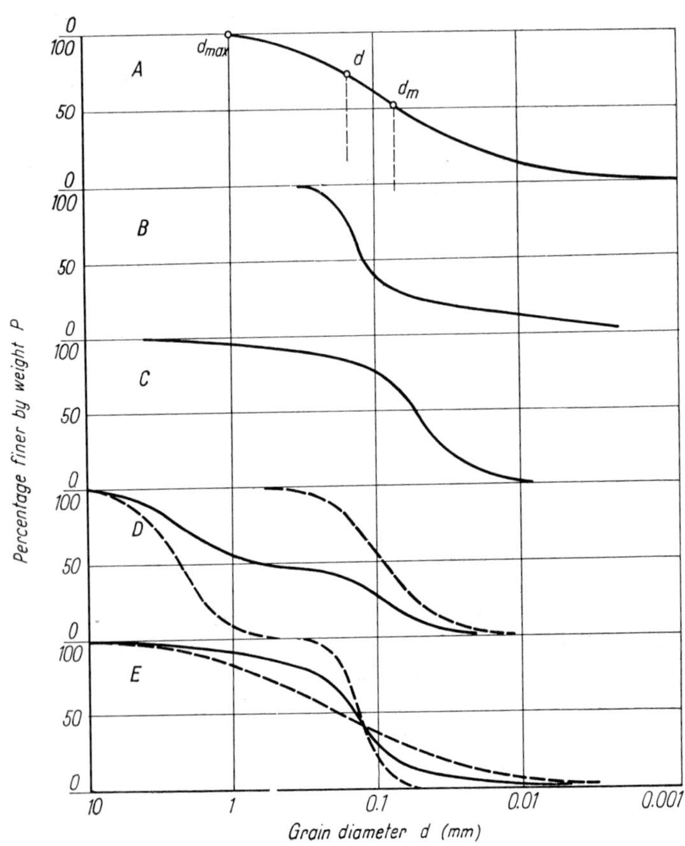

Fig. 25. Effect of the conditions of origin on the shape of the grain-size distribution curve

This means that a well-graded soil is made up of grains covering a very wide range of sizes. The large voids between the coarser grains are filled with the finer particles. This explains why well-graded soils occur normally in dense natural deposits and show a high bearing capacity. They can readily be compacted into fills and form stable slopes.

A horizontal section or step on the grain-size distribution curve (5 in Fig. 24) indicates the complete absence of grains between the size limits corresponding to the ends of the horizontal section, and the soil is said to be gap-graded or poorly graded.

From the shape of the grain-size distribution curve, much can often be learned about the geological origin of the soils. Riverbed deposits are as a rule well graded, but with distribution characteristics changing irregularly from point to point. The cause of this is that the stream-flow velocity at the riverbed fluctuates continuously according to the momentaneous water level, the trend of the variation of the water level etc., resulting in alternating periods of erosion and sedimentation. The shape of curve 5 in Fig. 24 is also common, and shows the simultaneous deposition of two different fluvial sediments. Such curves were obtained for the deposits in the river *Tisza* below the mouth of the tributaries *Sajó* and *Maros* in Hungary.

Curve *A* in *Fig. 25* shows a normal cumulative frequency curve, composed of two nearly identical halves symmetrical about the axis $P = 50\%$. The coarse grains of soil *B* are fairly uniform, but the curve includes a wide-size range of the finer fractions; immature soils show a distribution similar to this. With increasing age the coarser grains trend to break down under the action of atmospheric weathering, and the curve becomes more similar to the normal distribution curve. The grain-size distribution of a mature soil is shown in curve *C*, which has a characteristic flat section in the coarser size ranges. Curves similar to those shown in *B* and *C* are common in the case of soils of glacial or fluvioglacial origin.

The soil represented by curve *D* can be described as poorly graded, since certain intermediate fractions are absent. This type of distribution is characteristic of sandy gravels deposited from swiftly flowing rivers carrying a large load of sediment.

A conspicuous break in the grading like that shown in curve *A* represents a composite soil made up of two different soils deposited simultaneously by different agents. For instance, one fraction might be washed into a glacial lake by a river and the other fraction was deposited from a melting glacier.

The correlation between the grain-size characteristics and the distribution of fluvial deposits was investigated by Shockley and Garber (1953). They determined the grain-size distribution for a large number of sand samples and plotted the results in frequency distribution curves on a log—log system. Diagrams with triangular shape were obtained (*Fig. 26*), showing three well-definable geometrical characteristics: the grain size d_m corresponding to the apex of the triangle, and the slopes to the sides (m_1 and m_2). m_1 was found to be constant for every sample, whereas m_2 varied with varying d_m and had a maximum at $d_m = 0.2$ mm. It was also possible to find a relationship between the void ratio in the loosest, natural and densest states on the one hand, and the permeability of the tested soils on the other.

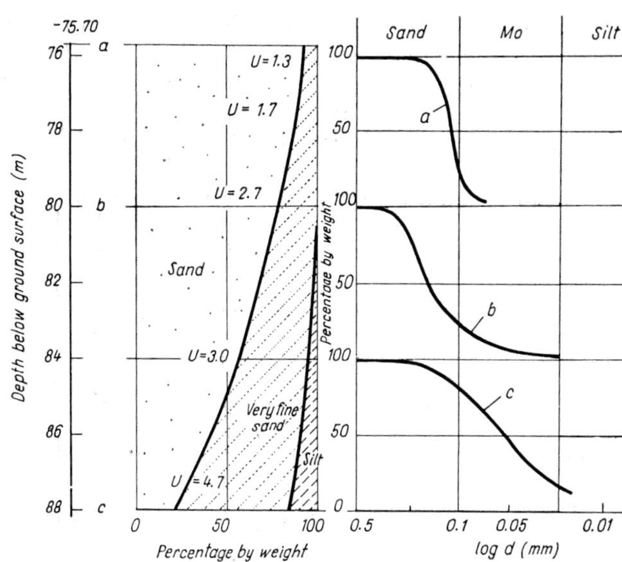

Fig. 27. Sea sediments or lake-shore sediments; sequence of layers due to the regression of the sea

The average grain-size of alluvial deposits in a river bed as a rule increases as the depth increases. This constitutes a very unfavourable condition for a dam because of the danger of seepage through the pervious layers of the subsoil.

The variation of grain-size distribution with the depth from the surface may be well demonstrated in a graph showing the percentage of each constituent fraction separately. An example is given in *Fig. 27*; here the soil becomes gradually finer as the depth increases. Such a situation may be caused by the regression of the sea.

2.1.5 Analysis of grain-size distribution curve

The grain-size distribution curve, as was shown in the preceding chapter, is an *integral curve.* The intercepts below, or above, the curve on the vertical at a particular grain size *d* indicate the percentage, by weight, of all particles finer or coarser, respectively, than that size. If the number of grains falling into an interval between two given size limits is plotted as an ordinate against the average grain size within the interval as abscissa, a line graph termed a size-frequency polygon is obtained (*Fig. 28b*). With size intervals chosen sufficiently small it is possible to obtain a smoothed curve called a size-frequency curve which mathematically is the derivative of the cumulative frequency curve (Fig. 28*a*). The mode, i.e. the grain size which occurs with the greatest frequency, is obtained as the abscissa at the maximum of the size-frequency curve. This, however, does not coincide with the point of inflexion on the integral curve because of the logarithmic representation.

The *average* diameter of a group of particles can be interpreted in various ways, the most common being the arithmetic mean, or simply the mean, d_a, the geometric mean d_g and the harmonic mean d_h. These terms are defined as follows

$$d_a = \frac{\sum_{i=1}^{N} n_i d_i}{\sum_{i=i}^{N} n_i}, \tag{13}$$

$$d_g = \sqrt[N]{d_1 d_2 d_3 \ldots d_N}, \tag{14}$$

$$\frac{1}{d_h} = \frac{1}{\sum_{i=1}^{N} n_i} \sum_{i=1}^{N} \frac{n_i}{d_i}, \tag{15}$$

where n_i denotes the number of particles of diameter d_i. An example of the calculation of the

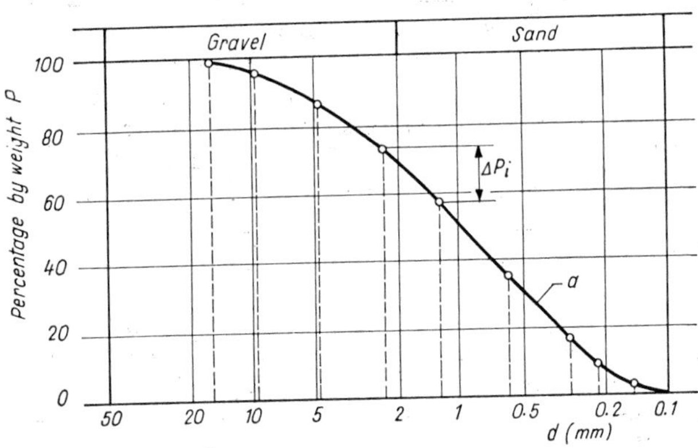

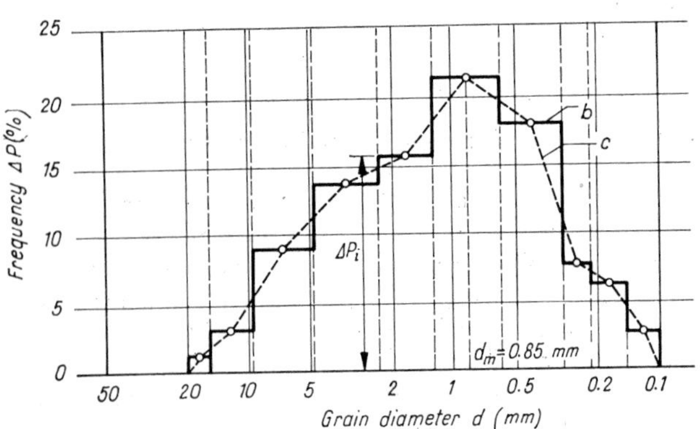

Fig. 28. Grain-size distribution curve and frequency curve

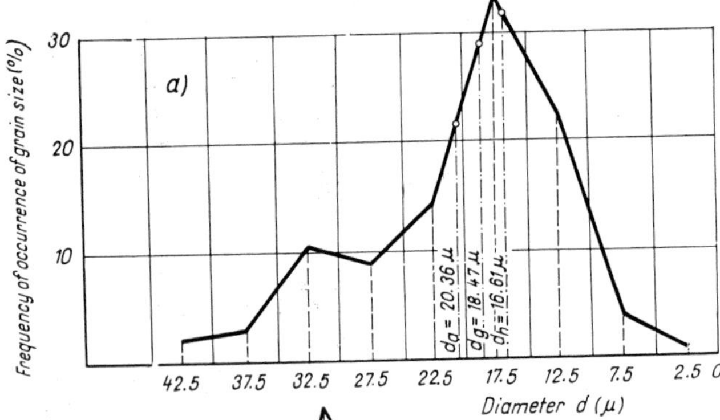

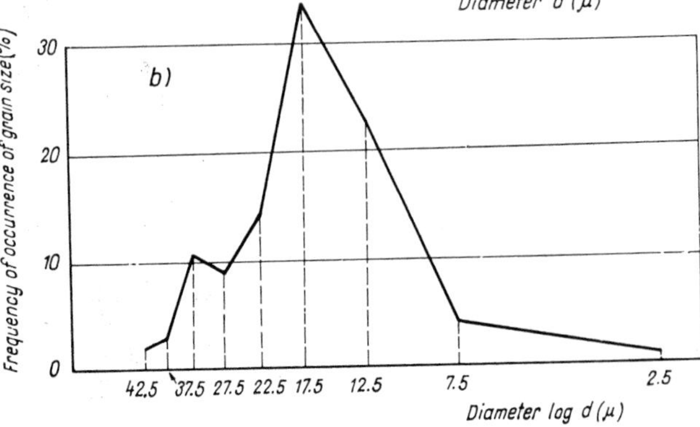

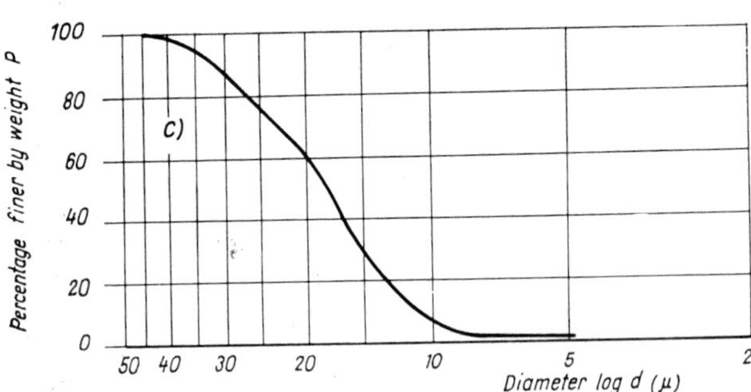

Fig. 29

a — Frequency curve; *b* — frequency curve in semi-logarithmic plot; *c* — grain-size distribution curve

average diameters is given in *Table 5*. The corresponding size-frequency curves that indicate the average diameters are also shown in *Fig. 29*.

The harmonic mean defined by Eq. (15) is related to the specific surface. If n denotes the number per unit weight of the particles (assumed to be spherical), d the mean diameter of the particles and γ_s their unit weight, then

$$\frac{\pi}{6} d^3 n \gamma_s = 1;$$

and since the specific surface (i.e. the total surface area of the particles in a unit weight) is $F = n\pi d^2$ hence

$$F = \frac{6}{\gamma_s} \frac{1}{d_h}. \tag{16}$$

Table 5. Calculation of mean particle diameters

Limits of particle size, μ	Average size d, μ	Number of particles n	nd	n log d	$\frac{n}{d}$	Mean values
0– 5	2.5	2	5.0	0.8	0.80	
5–10	7.5	10	75.0	8.7	1.33	
10–15	12.5	56	760.0	61.4	4.48	$d_a = 20.36\,\mu$
15–20	17.5	82	1435.0	102.0	4.68	
20–25	22.5	35	787.5	47.3	1.55	
25–30	27.5	22	605.0	31.7	0.80	$d_g = 18.47\,\mu$
30–35	32.5	26	845.0	39.3	0.80	
35–40	37.5	7	262.5	11.0	0.19	$d_h = 16.61\,\mu$
40–45	42.5	5	212.5	8.1	0.12	
		245	4987.5	310.3	14.75	

Assuming a continuous distribution $[P = f(d)]$, the harmonic means is obtained from

$$\frac{1}{d_h} = \int_0^1 \frac{dP}{d}. \qquad (17)$$

On the basis of Eq. (17), the harmonic mean diameter may be determined by a simple graphical method developed by KOŽENY (1931). The procedure is based on the data obtained from the standard grain-size distribution test, but the percentages P, in contrast to the usual semi-logarithmic representation, are plotted against the values of $1/d$. (Both curves are shown in *Fig. 30.*) The area below the curve is then converted into a rectangle of height $P = 100\%$ (it means that the shaded areas are equal). According to Eq. (17), the basis of the rectangle thus obtained is $1/d_h$. Hence d_h can be calculated, or obtained by a simple projection as shown in the figure.

The harmonic mean diameter is of importance in the study of soil permeability since permeability depends greatly on the specific surface of the soil because of the viscous effects associated with the surface characteristics of the grains. JÁRAY (1955) has shown that specific surface is also related to many other physical properties of soils.

It has already been mentioned in connection with Fig. 14 that for soil particles the term *diameter* can only be defined on the basis of some rather arbitrary assumptions. This implies that the *grain-size* distribution curve, in fact, reflects, though not in a numerically expressible manner, the shape characteristics of the grains too. It appears, therefore, that much more reliable information might be obtained by measuring the volume, and not some arbitrarily defined diameter, of the grains and examining the distribution of the grains in terms of their volume. Some examples for the application of this principle are shown in *Fig. 31.* Unfortunately, the experimental determination of such graphs is feasible in the case of coarse-grained gravel or crushed stone only, and this fact sets limitations to the method which might otherwise be proven a useful aid to the study of the problems of filtration, migration of particles etc.

Besides grain volume distribution, another important factor governing the properties of soils is the specific surface

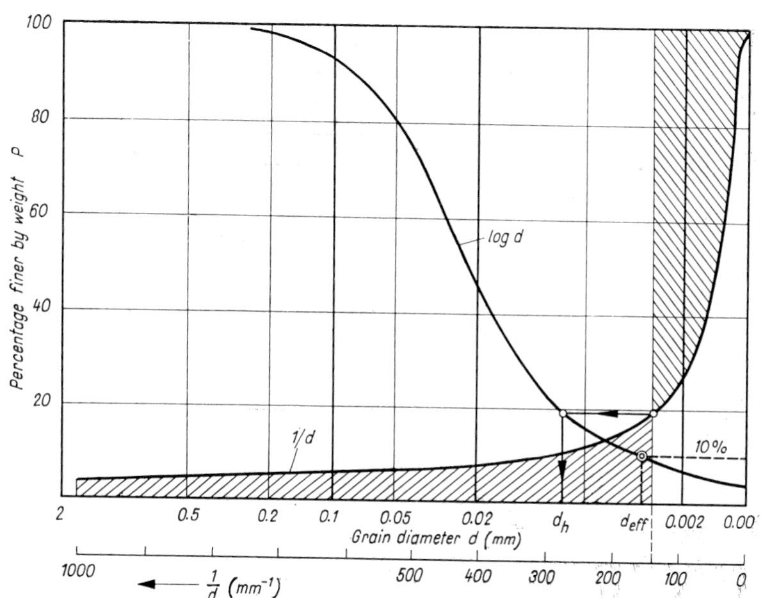

Fig. 30. Determination of the harmonic mean diameter; effective grain size in KOŽENY's definition

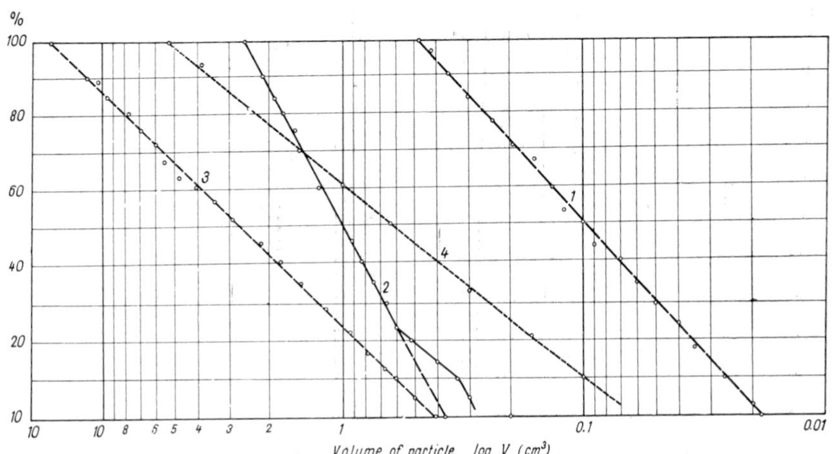

Fig. 31. Volume-size distribution curves:
1 — crushed basalt, 15–2.27 mm; *2* — crushed basalt, 25–5 mm; *3* — Danube gravel; *4* — pit gravel

area, which is considered by *Járay* as the foremost soil characteristic. The specific surface, i.e. the total surface of the particles per unit weight [Eq.(16)], is of particular importance in the study of fine-grained soils, since the electrical forces responsible for interparticle effects are proportional to the specific surface. (See also Ch. 4.) Specific surface increases very rapidly as the grain size decreases, as is well demonstrated by the following data: a cube with sides 1 cm long has a volume of 1 cm^3 and a surface area of 6 cm^2. If the same cube is subdivided into smaller ones with sides of 1 μm, then the number of the small cubes will be 10^{12} and their total surface area 6 m^2! Grain shape is another important factor: considering various solids all having a constant volume of $V = 1$ cm^3, the surface area is 4.83 cm^2 for the sphere; 6 cm^2 for the cube, 21.1 cm^2 for a disc 1 mm thick, and 2 m^2 for a plate 1 μm thick.

To demonstrate the significance of the specific surface in the case of natural soils, the grain-size distribution curve of a silty sand is shown in *Fig. 32a*, together with a graph (*b*) in which the ordinates represent the surface area of the fractions defined by the grain-size divisions on the horizontal axis. It is seen that the surface area of the particles finer than 0.02 mm amounts to about 30% of the total surface area of the whole mass, while their percentage by weight is only 7%.

In order to facilitate the determination of the average grain-size distribution of a given natural soil deposit it is convenient to divide the grains into major groups according to their size. The following division has proved particularly suited to the study of sands and fine sands—soils occurring commonly in Hungary:

coarse grains: $d \geq 2$ mm,
medium grains: $2 > d \geq 0.1$ mm,
fine grains: $d < 0.1$ mm.

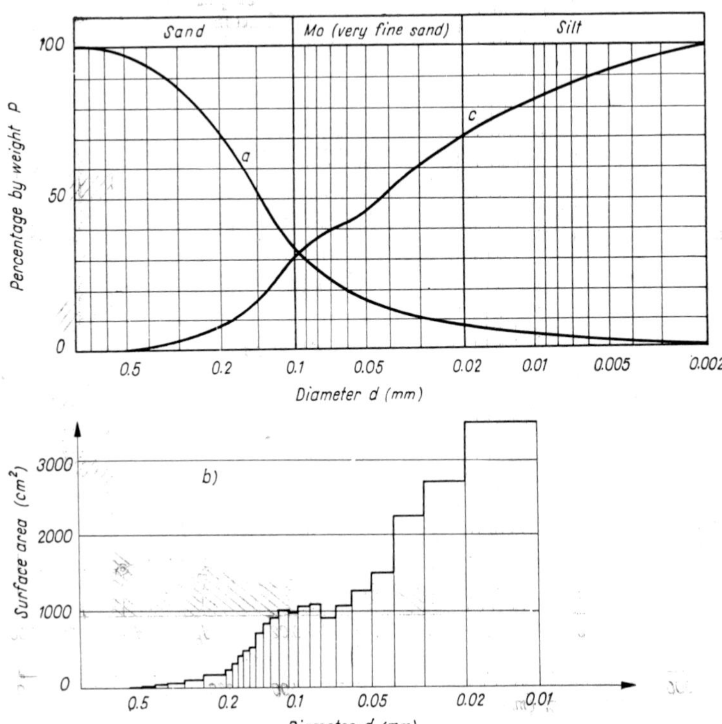

Fig. 32. Grain-size distribution curve of silt (*a*); specific surface of fractions (*b*); and total particle surface (*c*)

The composition of a given soil according to this grain-size division can be visualized with the aid of a triangular diagram, as shown in *Fig. 33*. If a soil contains 100% of a single fraction, then the point representing the soil coincides with one of the vertices. Points falling into one of the small triangles *a*, *b*, *c*, cut out by the $P = 50\%$ lines, represent, in this order, coarse, medium or fine-grained soils. Points within the central shaded triangle belong to mixed-grained soils.

Another method for the statistical evaluation of the grain-size distribution of soils in a given region is shown in *Fig. 34*. Having plotted the grading curves, their frequency distribution at selected grain sizes can be examined and, hence, their range of scatter and the "average" grain-size distribution curve obtained. The example shown was taken from the investigation of the overburden strata of an open-cast coal-pit in Hungary.

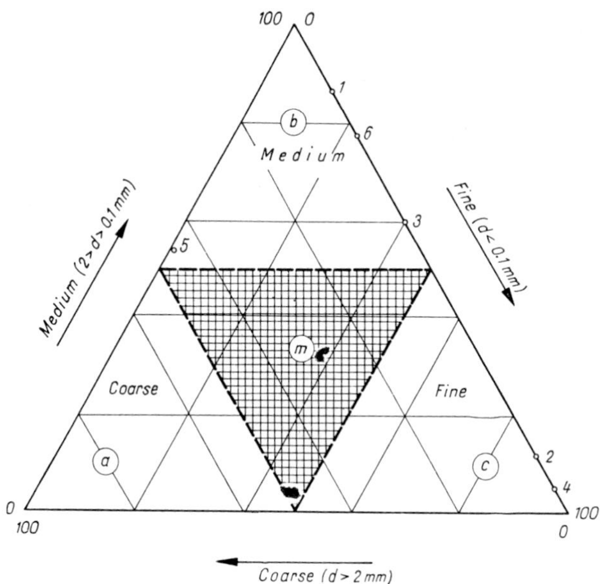

Fig. 33. Triangular diagram; characterizing the grain-size distribution

2.1.6 Typical grain-size distribution curves

Figures 35 and *36* present a selection of typical grain-size distribution curves originating from both Hungary and abroad. The grading curves *1* and *2* give the composition of the *Danube* gravel. The curve obtained for *Vienna* gravel has its full length in the gravel fraction, while the *Budapest* gravel contains also a considerable amount of sand. It is interesting to compare the grading curve of Danube gravel with the grading limits for suitable concrete aggregates (hatched area bounded by dotted lines): the sandy gravel at Budapest would make a poor aggregate owing to its high sand content and its lack of uniformity.

The sediment of the Danube section near *Győr*, below the mouth of the tributaries *Rába* and *Rábca*, is shown in curve *3*. Here the coarse gravel is mixed with fine sand and the resulting curve is gap-graded. *Sajó* gravel (*4*) contains a high percentage of sand. The sediment from the upper reaches of *Dráva* (*5*) is a poorly graded gravel—sand mixture with no fines. A very uniform coastal gravel from *Brighton* is shown in curve *6*.

"*Sable de Fontainebleau*", the characteristic fine sand of the environs of Paris (curve *7*), is a highly uniform soil with $U = 1.2$, whose particles are practically all of the same size. Fine sands of the *Tisza* valley often have a grain-size distribution curve similar to that shown in curve *8*; fine sand is the dominant fraction with some fines (particles smaller than 0.1 mm), and the uniformity coefficient is as a rule very close to unity. Such soils can readily liquify under the action of percolating water which means that the soil behaves like a viscous fluid and completely looses its former internal resistance. Trenches excavated in these soils cave in and footings may sink into the ground.

Curve *9* shows the typical soil of the southern shore of Lake *Balaton*; it consists mainly of fine sand and silt particles. Because of the uniform particles and the predominance of micaceous fine sand and silt fractions, this soil has a very smooth

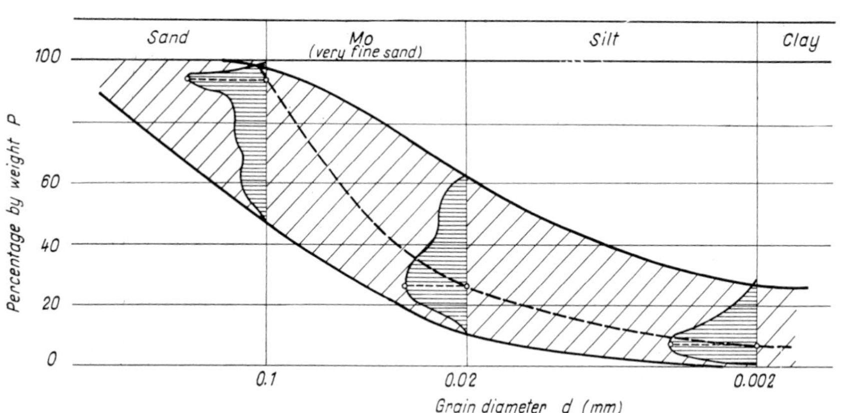

Fig. 34. Frequency of grain-size distribution curves; samples from overburden layers of open-pit coal mines

velvety touch, and the relatively low U value causes the grains to roll readily upon one another.

Curve *10* is for a dune sand from the Great Hungarian Plain, which contains predominantly fine sand with some silt. Curve *11* shows the coarse red sand of the *Sahara.*

In Western Hungary along a belt where plain and highland meet, the coincidence of special geological and geomorphological conditions (intensive weathering, sudden change in the gradient of water courses etc.) resulted in the formation of extremely mixed-grained soils. Typical examples from the regions of *Csorna* and *Szombathely* are shown in curves *12* and *13*; from $d = 40$ mm to $d = 0.002$ mm, all fractions are represented in a uniform distribution. Such well-graded soils covering a wide range of grain sizes are excellently suitable for the construction of all-weather earth roads. The coarse grains form a skeleton structure that remains stable in wet periods, whereas the fine particles act as filler and binder, and make the wearing surface of the earth road usable in dry weather.

Curve *14* represents a loess from the bank of the Danube at *Dunaújváros.* Curve *15*, which runs close to the curve of the loess, shows the grain-size distribution of the standard Grade 500 Portland cement. According to soil mechanics nomenclature, cement is composed of fine sand-sized and silt-sized grains. Curve *16* is for a highly plastic clay from *Szolnok*, and curve *17* is for a colloidal clay from the *Nile* delta. "*Kiscell*" clay, a typical representative of the subsoil of Budapest, is shown in curve *18.* On the basis if its grain-size distribution, this soil would be classified as a silt. Curve *19* is for the fine-grained fissured *London* clay and the last example, curve *20*, is for a bentonite from *Wyoming, U.S.A.*

As has been demonstrated by these examples, the grain-size distribution curve may provide very useful information, leading to a better understanding of soil properties and soil behaviour. This conclusion, however, applies chiefly to coarse-grained soils, gravels and sands, since the physical behaviour of cohesive soils is primarily governed by other factors such as soil structure, orientation of particles and surface activity of the grains, whose influence is not reflected in the grain-size distribution curve. That is why mechanical analysis, which is an indispensable part of soil investigations concerning coarse-grained soils, is of little use in the testing of cohesive soils.

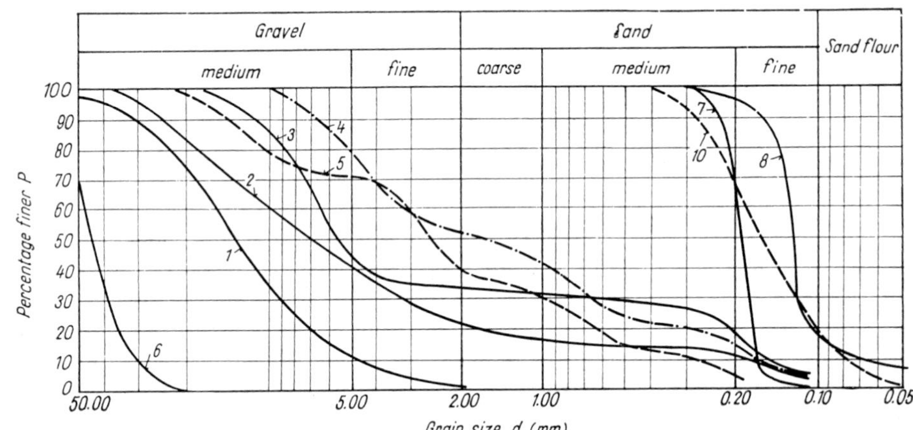

Figs 35–36. Characteristic grain-size distribution curves:

1 — Danube-gravel in *Vienna; 2* — Danube-gravel in *Budapest; 3* — Danube-gravel in *Győr; 4* — *Sajó*-gravel; *5* — *Dráva*-gravel; *6* — seashore gravel *(Brighton)*; *7* — sand from the *Paris* area; *8* — sand from the *Tisza*-valley; *9* — fine sand from Lake *Balaton*; *10* — running sand from the Great Hungarian Plane; *11* — sand of *Sahara*; *12–13* — soils suitable for earth roads; *14* — loess, *Dunaújváros*; *15* — cement; *16* — clay, *Szolnok*; *17* — clay of *Nile* delta; *18* — *Budapest* clay; *19* — *London* clay; *20* — *Wyoming* bentonite

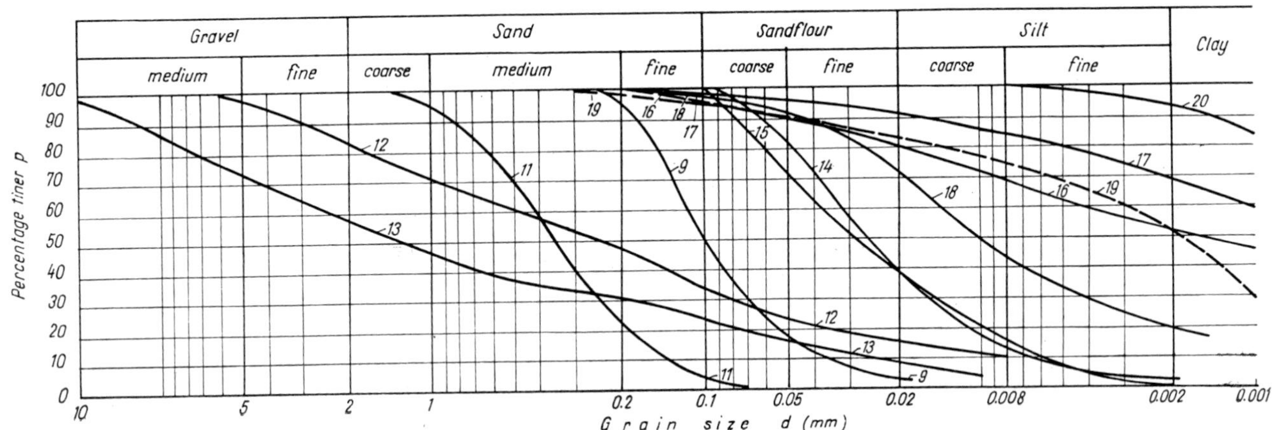

2.2 Water and air in soil

Among the factors influencing the interactions between the solid constituent particles of soils, water has a dominant role; both its quantity and its physical state are important, particularly from the point of view of the physical properties of cohesive soils. Therefore, some relevant properties of water need to be discussed in more detail.

Water is a unique substance; at normal temperature and pressure, by analogy with compounds like H_2Se and H_2Te, it would be expected to be a gas, yet it is a liquid and as a liquid it exhibits, in certain tests, signs of molecular orientation normally associated with the crystal structure of solids, even though this orientation is restricted to relatively few molecules. An examination of ice crystals shows their atomic structure to be a lattice in which each oxygen atom O is surrounded by four other O atoms in a tetrahedral arrangement. The distance between them is 2.76 Å (Å is an Ångstrom unit). Between every two oxygen atoms a hydrogen atom *H* is asymmetrically positioned, which is linked to both adjacent O atoms, but with a stronger bond to the nearer, 1 Å distant, than to the other, 1.76 Å distant (*Fig. 37*). In the liquid state of water, the distance between the O atoms is increased to 2.9 Å. When ice melts, some of the hydrogen bonds are broken allowing a closer packing of molecules and the formation of a quartz type arrangement which has a higher density than ice. This accounts for the increase in the density of water upon thawing. As the temperature increases, both the distance between the O atoms and the number of free water molecules increases. The variation of the density of water and ice with temperature is shown in *Fig. 38*.

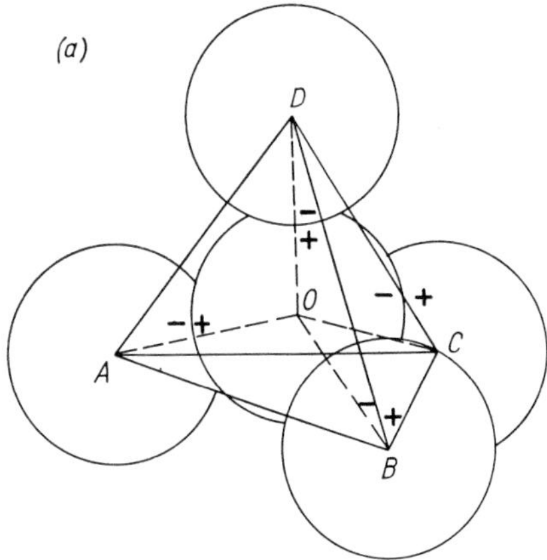

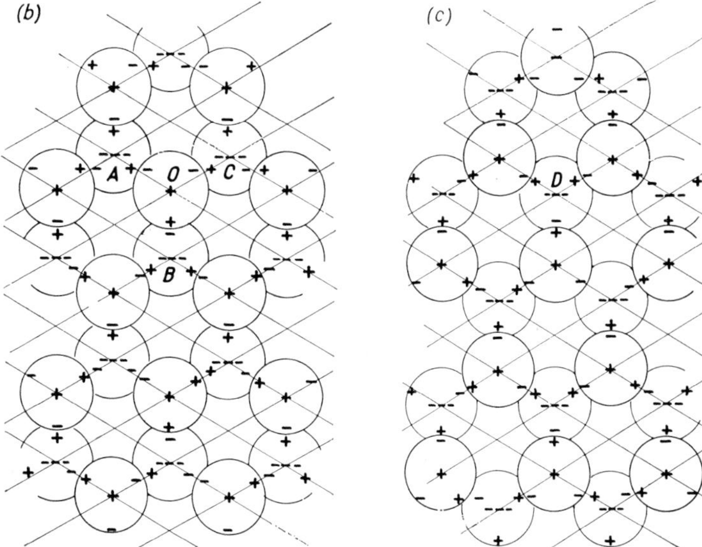

Fig. 37. Crystal structure of water:

a — tetrahedral grouping of four water molecules around a fifth; *b* and *c* — layers of water molecules in ice structure. (Layer *b* is superimposed on layer *a*, and further layers continue the structure. Spheres ABCDO correspond to those lettered in *a*)

A unique character of water molecules lies in the fact that they behave as dipoles owing to their unsymmetrical electrical fields. This structure is shown in *Fig. 39*: the molecule is neutral, but the centres of positive and negative charges are separated. Thus free ions present in the water may associate with water molecules and a shell

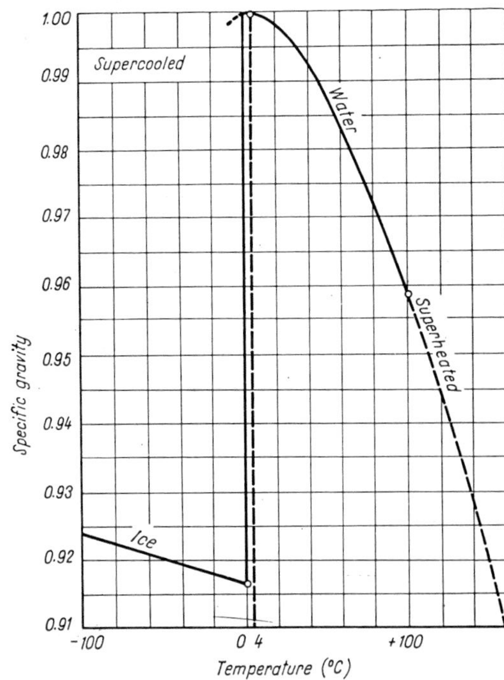

Fig. 38. Specific gravity of water and ice as function of temperature

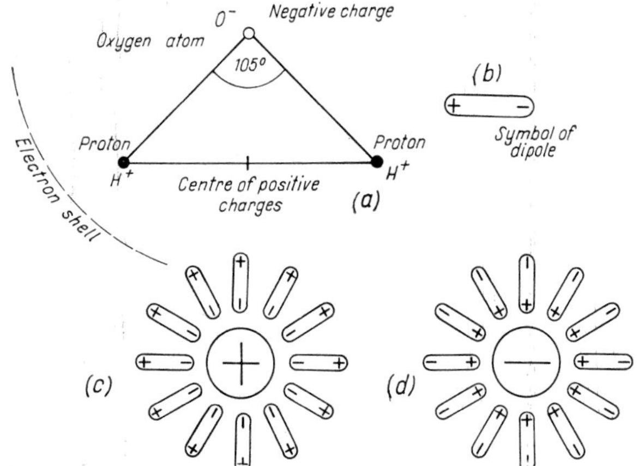

Fig. 39
a — triangular shape of the H_2O molecule due to the polarization of the oxygen ion; *b* — dipole in equilibrium condition; *c* and *d* — dipoles around ions

of dipoles may be formed round the ions as illustrated in Fig. 39*c* and *d*. This process is termed the hydration of ions.

Water dissociates to a very small extent into hydrogen ions (H^+) and hydroxyl ions (OH^-), but the H^+ ions instantaneously combine with water molecules forming H_3O^- ions. At constant temperature, the product of the numbers of H^+ and OH^- ions is also constant. Thus if in a given solution the number of H^+ ions is increased, the concentration of OH^- ions must recrease accordingly, and vice versa. The numerical value of the constant previously mentioned is 10^{-14} at a temperature of 22 °C. In neutral distilled water the H^+ ion concentration and the OH^- ion concentration are equal, 10^{-7}. This means that at 22 °C the ratio of hydrogen ions in clean water to those in normal acid or normal alkali is one to ten million. In normal acid the H^+ ion concentration is 1, and the OH^- ion concentration is 10^{-14}. The corresponding values for normal alkali are in the reversed order. The value of the hydrogen ion concentration is used as a measure of the acidity or alkalinity of a solution. For practical reasons the negative logarithm of this number, denoted by pH, is used. Hence, the pH value indicates, in terms of negative powers of 10, the mass (in *g*) of hydrogen ions H^+ per one liter solution. At pH = 7 the solution is neutral. The lower the pH the higher is the acidity of the electrolyte (*Table 6*).

Table 6. Soil reaction and pH value

Hydrogen ion concentration C_{H^+}	Hydroxyl ion concentration C_{OH^-}	pH	Concentration of corresponding acid and base
10^{-0}	10^{-14}	0	1 normal acid
10^{-1}	10^{-13}	1	0.1 normal acid
10^{-2}	10^{-12}	2	0.01 normal acid
10^{-3}	10^{-11}	3	
10^{-4}	10^{-10}	4	
10^{-5}	10^{-9}	5	
10^{-6}	10^{-8}	6	
10^{-7}	10^{-7}	7	neutrality
10^{-8}	10^{-6}	8	
10^{-9}	10^{-5}	9	
10^{-10}	10^{-4}	10	
10^{-11}	10^{-3}	11	
10^{-12}	10^{-2}	12	0.01 normal base
10^{-13}	10^{-1}	13	0.1 normal base
10^{-14}	10^{0}	14	1 normal base

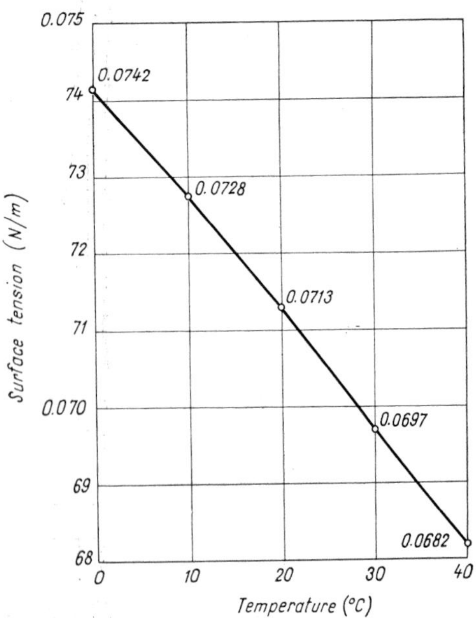

Fig. 40. Surface tension of water as function of temperature

It should be noted that the value of the product $K = C \cdot H^+ \times C \cdot OH^-$ (*C* denotes concentration) varies with the temperature. At 22 °C, as mentioned before, $K = 10^{-14}$, but at 34 °C it is four times as much as at 16 °C. ($K_{16} = 0.63 \times \times 10^{-14}$, $K_{34} = 2.51 \times 10^{-14}$)

An important physical property of water is that a film of water, several molecules thick, in the boundary zone between air and water is in a state of tension. This is known as the *surface tension* of the water. Its magnitude depends on the temperature of the water; numerical values are given in *Fig. 40*. The surface of a liquid can be visualized as a stretched elastic membrane; a thin needle or razor blade placed gently onto the surface would not sink but remain floating (*Fig. 41*).

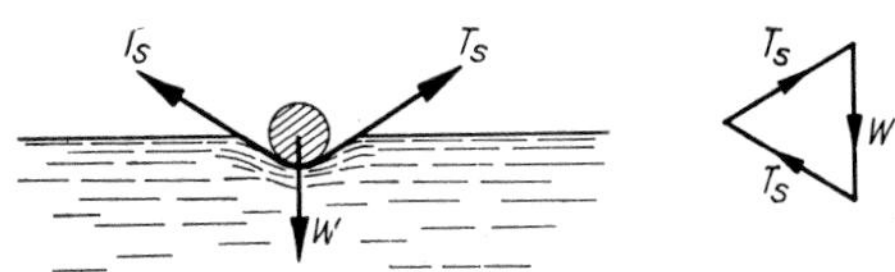

Fig. 41. The needle floats on the surface of water, due to the surface tension

A simple concept derives the surface tension from the molecular interactions in a liquid. According to this, the molecules in the interior of the liquid mutually attract each other by forces acting in every direction. At the liquid–air interface, a molecule is acted upon by these forces on one side only. If a molecule is to be moved from the interior to the surface, the molecular interaction on one side must be removed. Thus, to increase the surface of the liquid, i.e. to bring molecules to the surface, entails doing work. This is equivalent to the assumption that tensile stresses exist in a surface film of the liquid. A liquid acted upon by no forces other than surface tension tends to assume a shape in which its potential energy is a minimum. Surface tension always manifests itself at the common lines of contact of three substances: solid—liquid—air.

The thickness of the superficial water film in which tension prevails is of the order of 10^{-7} cm. A result of surface tension is that water rises in a small diameter capillary tube to a definite height and remains in that position indefinitely (*Fig. 42*). The upper surface of the rising water column assumes a curved surface, called the meniscus, that joins the wall of the tube at a definite angle α known as the contact angle. The value of this angle depends on the chemical composition of the material of the tube and on the impurities deposited on the wall surface. For a glass tube, if its surface is absolutely clean, $\alpha = 0$. If the wall surface is covered with a thin film of grease, α may assume a value greater than 90°, which means that the meniscus is convex and the surface of the water in the tube is held below the outer free water level. In general, the value of the contact angle ranges from 0 to 80°.

Let us now consider the pressure conditions at the curved surface of a liquid in the capillary tube. If a thin curved membrane of very flexible material is in a state of tension, then, for equilibrium, the pressures acting on the inner side of the membrane must be different from those acting on the outer side. This requirement applies also to curved liquid surfaces.

The cross section of a curved surface is shown in *Fig. 43*. The surface is assumed to be cylindrical with a radius of curvature r_1. An elementary surface $\mathrm{d}A$ is acted upon by the forces $p\,\mathrm{d}A$ on the outside and $(p + p_e)\,\mathrm{d}A$ on the inside, giving a resultant force $p_e\,\mathrm{d}A$ acting outward. By integrating the elementary forces over the entire area and applying equilibrium conditions, we find that the components parallel to the main chord of the arc balance each other, whereas the sum of the components normal to the chord gives a resultant force of $p\ 2r_1 \sin\alpha$. Equilibrium requires that this resultant force is balanced by the component, normal to the chord, of the surface tension T_3. Thus

$$2T_s \sin\alpha = p_e\, 2r_1 \sin\alpha$$

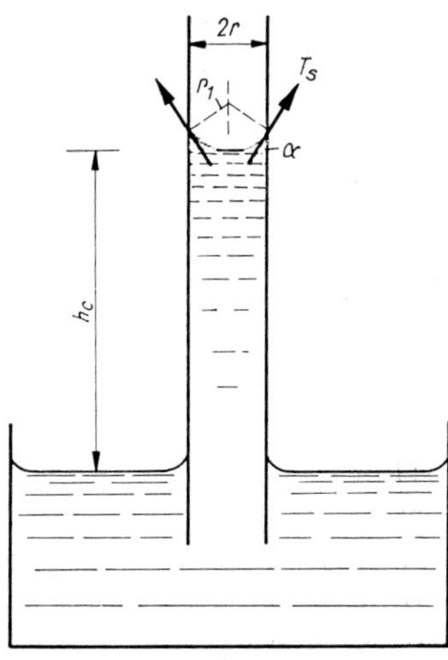

Fig. 42. Capillary rise of water in a capillary tube

and

$$p_e = \frac{T_s}{r_1}.$$

If the surface has a double curvature, the total pressure excess can be calculated as the sum of the components obtained in two principal directions:

$$p_e = T_s\left(\frac{1}{r_1} + \frac{1}{r_2}\right). \tag{18}$$

For a surface of uniform curvature ($r_1 = r_2 = r$):

$$p_e = \frac{2T_s}{r}. \tag{19}$$

Let us apply these relationships to the meniscus shown in Fig. 42. The curved surface of the water is acted upon from above by the atmospheric pressure. The pressure acting on the other side of the meniscus is less than the atmospheric pressure

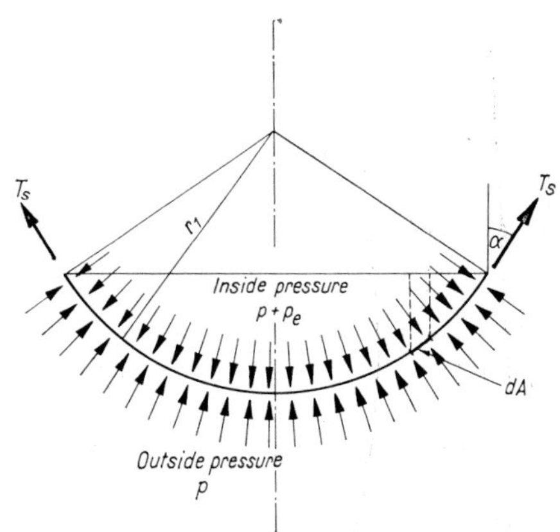

Fig. 43. Curved surface under the action of external and internal pressures

by an amount of $p_e = h_c\gamma_w$. If r_c is the radius of the capillary tube, α is the contact angle, and the meniscus is assumed to be spherical, then $r_1 = r_c/\cos\alpha$. The weight of the water column rising to the height h_c above the free water level is $h_c r_c^2 \pi\gamma_w$. This must be balanced by the force due to the pressure difference at the meniscus, defined by Eq. (19). Therefore,

$$h_c\gamma_w = \frac{2T_s \cos\alpha}{r_c}.$$

Hence

$$h_c = \frac{2T_s}{\gamma_w r_c}\cos\alpha. \tag{20}$$

At ordinary temperature, the surface tension has an average value of $T = 7.5\times10^{-4}$N/cm and $\gamma_w = 10^{-2}$ N/cm³. Substituting these values in Eq. (20), the height of capillary rise becomes

$$h_c^{(\mathrm{cm})} = \frac{0.15}{r}\cos\alpha. \tag{20a}$$

Another important property of water, considering its role in the soil, is its viscosity, since for example the flow of water through soils is greatly affected by it. The viscosity of water is a function of pressure and temperature. *Figure 44a* shows the variation of viscosity at the atmospheric pressure, as a function of temperature. The effect of temperature is very significant; in all those tests where the viscosity enters into the calculations temperature corrections are essential. Such tests are e.g. hydrometer analysis and permeability measurements. In Fig. 44*b* relative values are given, taking as a basis the viscosity at 0°C temperature and atmospheric pressure. Here the temperature was selected as a fixed parameter and the relative values were plotted as a function of the pressure. It can be seen that there is a pressure value between 1° and 10 °C where the viscosity is lowest. The temperature of the ground water is usually between these limits, so it can happen that the viscosity of the "free" water in the pores is higher than that of the outer layers of the absorbed water around the particles.

The voids between the solid particles often contain air. In respect of compressibility, air as a gas follows *Boyle–Mariotte*'s law: $p_1V_1 = p_2V_2$. It dissolves in water; the volume of air (in cm³) dissolved in 1 liter water at normal temperature is shown as a function of temperature in *Fig. 45.* It is seen that the quantity of dissolved air decreases as temperature increases. This explains why dissolved air can be extracted from water by boiling.

An air–water mixture subjected to an isothermic pressure change, i.e. one in which the temperature is constant, will suffer volume changes due to the compression of the air on the one hand, and the absorption of the air in the water, on the other. Let V denote the total volume of the voids,

Fig. 44. Viscosity of water as a function of temperature

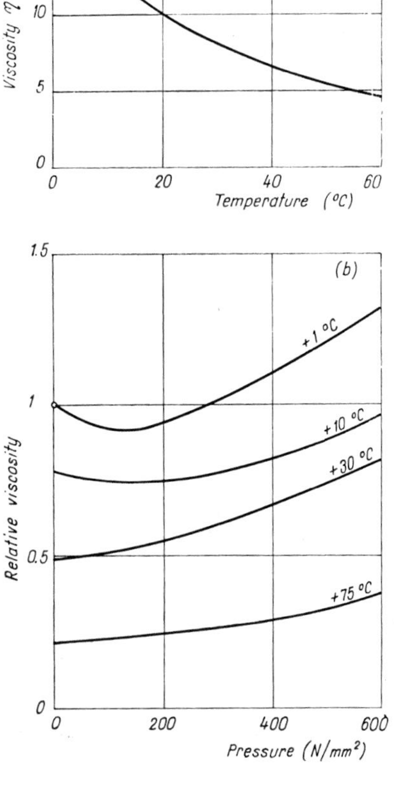

Fig. 45. Solubility of air in water

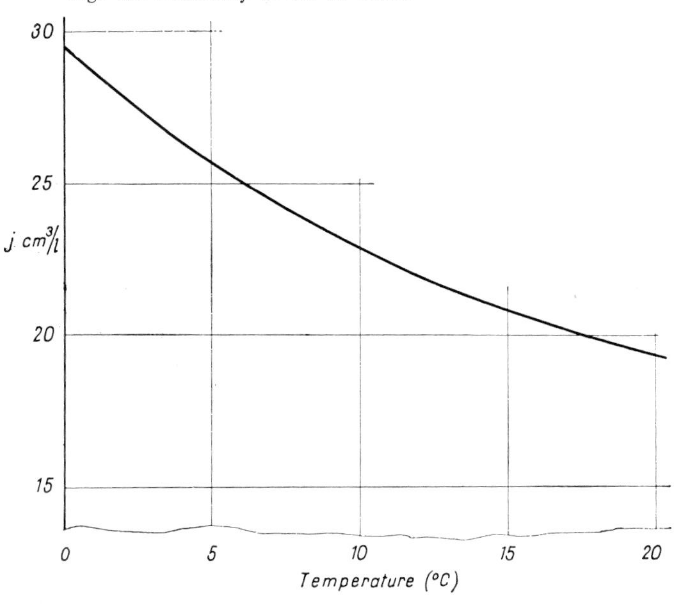

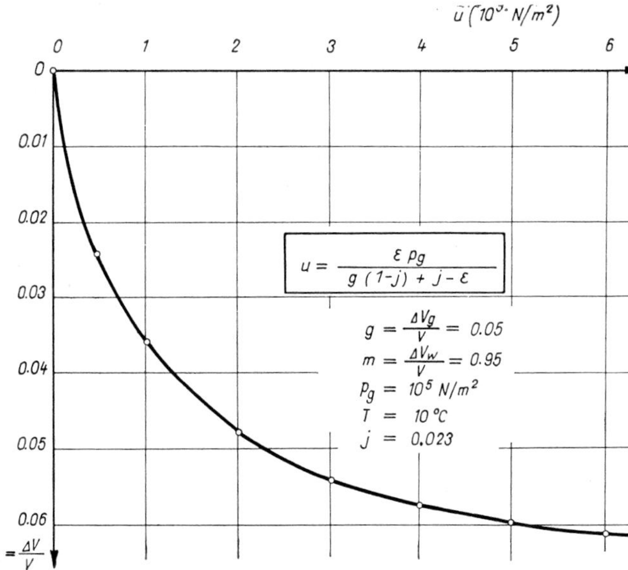

Fig. 46. Compression of air–water mixture

wherein V_g is the volume of the free air, and V_w the volume of the water in the mixture. Assume that a system, which was under an initial atmospheric pressure of $p_{a0} = p_1 = 0.1$ N/mm² $= 100$ kN/m² is subjected to an additional all-around pressure u. According to *Boyle–Mariotte*'s law

$$p_g V_g = (p_g + u)\,[V_g - (\Delta V - jV_w)]\,. \qquad (21)$$

In the equation j denotes *Henry's* coefficient of solubility. Its numerical value depends on temperature. From Eq. (21) we have

$$u = \frac{p_g\, \Delta V}{V_g + jV_w - \Delta V}\,. \qquad (22)$$

The compressibility law of the air–water system is shown diagrammatically in *Fig. 46.*

2.3 Phase composition of soils

2.3.1 Relative volumes of constituent parts

Soil is a multiphase system; its properties depend on the size and shape characteristics of the constituent parts, or phases, on their relative volumes and weights, and finally on the interactions between them. In the preceding chapter we have discussed the properties of each separate phase, and now we will discuss the characteristics which describe the relative proportions of the volumes or weights of the constituent phases. Problems relating the interactions between the phases will be dealt with, in connection with soil structure, in Ch. 3.

Figure 47 shows a soil sample of volume V. If the three phases—solid, grains water and air—are imagined separated one from the others, then the proportion of their respective volumes V_s, V_m, V_g within the whole volume V is as shown in Fig. 47*b*. Numerically, the phase composition of a soil can be characterized by the relative volumes of the three phases. This term denotes the ratio of the volume of a separate phase to the total volume of the sample:

$$s = \frac{V_s}{V}\,; \quad m = \frac{V_m}{V}\,; \quad g = \frac{V_g}{V}\,;$$

since

$$V = V_s + V_m + V_g\,;$$

hence

$$s + m + g = 1\,, \qquad (23)$$

in percentages

$$\boxed{s\% + m\% + g\% = 100\,.} \qquad (23a)$$

If we have three variables the sum of which is always constant, then these variables can be conveniently plotted in a triangular coordinate system.

The principle of this representation is as follows (*Fig. 48*).

Consider a point P inside an equilateral triangle with sides a and vertices lettered clockwise, A, B and C. If, proceeding again clockwise, straight lines are drawn from P parallel to the sides, then the sum of the intercepts a_1, a_2 and a_3 on the sides

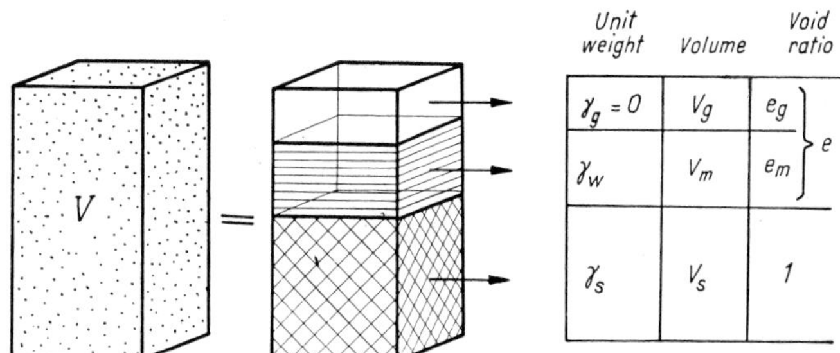

Fig. 47. Phase composition of a soil sample having volume V

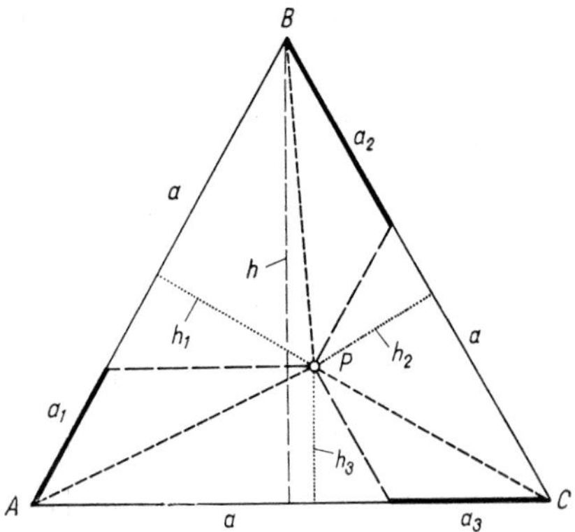

Fig. 48. Principle of triangular diagrams

is equivalent to the length of a side of the triangle. As is seen from the figure, the sum of the areas of the triangles ABP, BCP and APC is equivalent to the area of the triangle ABC, i.e.

$$\frac{ah_1}{2} + \frac{ah_2}{2} + \frac{ah_3}{2} = \frac{ah}{2} .$$

Since

$$h_i = \frac{a_i}{\cos 30°} , \quad h = \frac{a}{\cos 30°}$$

hence

$$a_1 + a_2 + a_3 = a ,$$

as stated above.

If the values of s, m and g relating to a given state of a soil are known, then this state can be represented by a point in the triangular diagram. To this end, the sides of the triangle are divided into 100 parts. Then the percentages s, m, g are plotted, proceeding say clockwise, on consecutive sides. The points thus obtained are projected next, inward and parallel to the previous side of the triangle; the point P where the projection lines meet will represent the phase composition of the given soil. The corners of the triangle correspond to the single-phase states, i.e. when either s or m or g equals 100% with the other two equal 0. The sides represent two-phase states (solids–air, solids–water, air–water), and the points falling inside the triangle the general three-phase state (*Fig. 49*).

If the volume (V), the wet weight (W) and the dried weight (W_d) of a sample and the unit weights of the solids, γ_s, and of the water, γ_w, are known, the relative volumes can be calculated from

$$s = \frac{W_d}{V\gamma_s} ; \quad m = \frac{W - W_d}{V\gamma_w} ; \quad g = 1 - s - m . \tag{24}$$

The representation of the relative volumes in the triangular diagram is a very useful aid in visualizing the variations in the state of the soil. Physical processes such as the sudden collapse of loess due to submersion, desiccation of sands, swelling of clays can be easily followed in this representation. Examples will be given in relevant chapters.

It may often be useful if an alteration of the state of a soil, as appears in the triangular diagram, is treated as a vector (*Fig. 50*). The length of the vector connecting points $P_1(s_1, m_1, g_1)$ and $P_2(s_2, m_2, g_2)$ is given by

$$P_1P_2 = \sqrt{\Delta_1^2 - \Delta_2\Delta_3} = \sqrt{\Delta_1^2 + \Delta_2^2 + \Delta_1\Delta_2} , \tag{25}$$

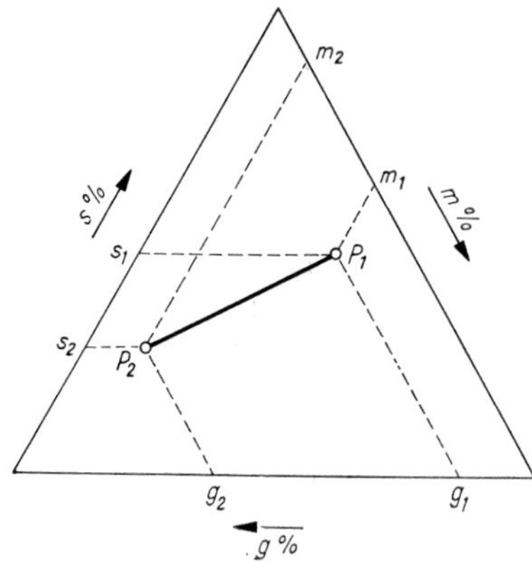

Fig. 50. Change in the phase composition represented by a vector

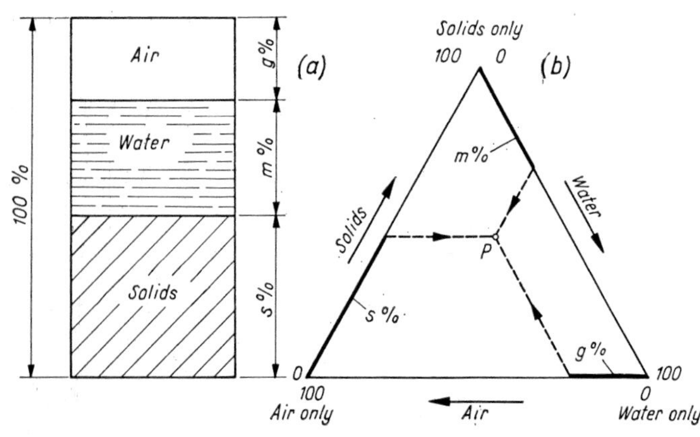

Fig. 49. Representation of the phase composition in triangular diagram

wherein Δ_1, Δ_2, Δ_3 denote the differences in the values of the relative volumes ($\Delta m = m_1 - m_2$, Δs, Δg).

The slope of the vector is obtained from

$$\tan \alpha = \frac{0.866\,\Delta s}{0.5\Delta s + \Delta m}. \tag{26}$$

In order to illustrate the phase composition of different soils, *Fig. 51* gives some examples for normal and abnormal cases. Here we see the areas where the points for loose and dense granular soils are located. Points for hard clays, plastic silts and very soft clays are also given. A very peculiar soil occurs in *Mexico City*; its volume contains 15 to 20% solid particles only. Peats, organic clays and silts also display very high values for $m\%$; they are usually saturated. The cross-hatched areas for these cases in Fig. 51 have all been determined by actual measurements.

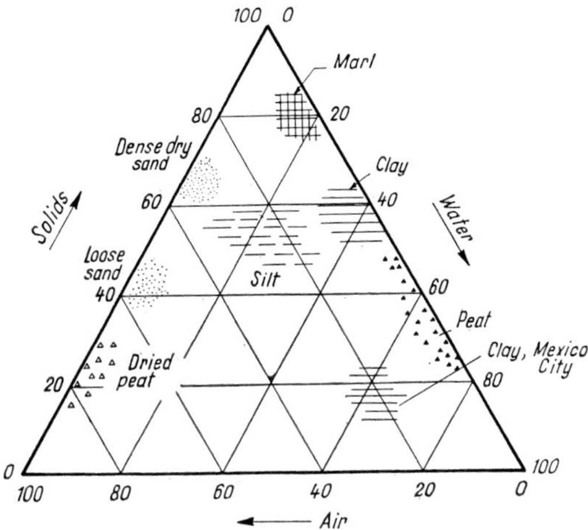

Fig. 51. Phase composition of some typical soils

The use of relative volumes for the vizualization of the changes in soil conditions in a soil profile will be shown later.

The state of a soil can be adequately described in terms of the relative volumes, s, m, g. Nevertheless, as a result of historical development, various other measures of the relations between the volumes or weights of the constituent parts have been developed, and are, in fact, in general use today. These will be discussed in the following sections.

2.3.2 Natural moisture content of soils

Water content—or moisture content—means the ratio of the weight of the water present in the soil to the weight of the dried soil. It is usually expressed as a percentage. If W denotes the weight of the wet soil, and W_d is the weight of the dried soil (i.e. weight of solids), then the water content

$$w\% = 100\frac{W - W_d}{W_d} = 100\frac{\Delta W}{W_d}. \tag{27}$$

Water content can also be expressed in terms of relative volumes as

$$w = \frac{m\gamma_w}{s\gamma_s}. \tag{28}$$

For the purpose of the measurement of the water content a small sample about 0.20 to 0.30 N weight is taken from the wet soil, and placed in an air-tight container of known weight. Pairs of numbered watch-glasses with ground rims and held together with a metal clamp, or aluminium tins with tight fitting lids will serve the purpose. The container and contents are weighed to an accuracy of 10^{-4} N. The clamps or lids are then removed, the sample put in an oven and dried at a temperature of 105 °C until no loss in weight is observed. The time required for this is about 1 to 2 hours for coarse-grained soils, and 8 to 10 hours for fine-grained soils (clays). After drying, the container with the dried sample is transferred to a desiccator (a glass vessel, with an air-tight cover, containing a highly hygroscopic material e.g. calcium chloride, which absorbs vapour from the environment), and allowed to cool down to normal room temperature. The clamp or lid is then replaced, and the container plus dried sample are weighed. The difference between the two weights gives the weight of the water evaporated.

By drying the soil at 105 °C, only the free interstitial water is removed from the pores, since the chemically bonded or adsorbed water cannot be driven off at this temperature. Besides, certain clay minerals contain, bonded in their crystal lattice, a considerable amount of water, which remains there when heated at 105 °C. The variation of water content of some typical clay minerals upon heating at increasing temperature is given in *Fig. 52*. Kaolinite, for example, looses all its lattice water rapidly at about 400 to 500 °C; at this stage the clay is said to be dehydrated.

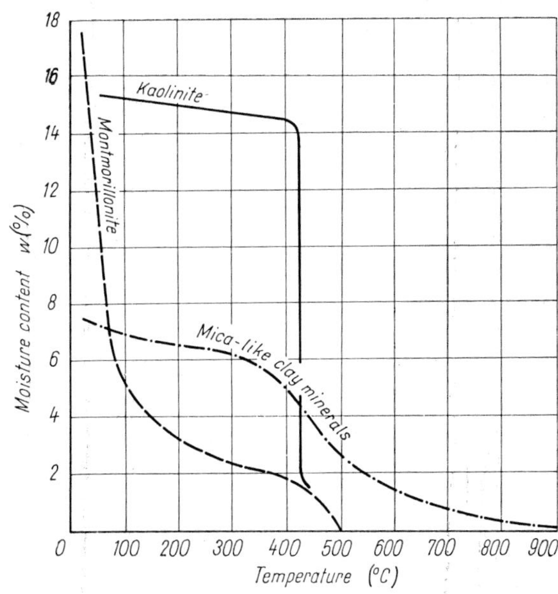

Fig. 52. Typical dehydration curves of different clay minerals

As seen in Fig. 52, no discontinuities or other characteristic marks can be distinguished on the dehydration curves at 105 °C, thus the use of this temperature in the laboratory drying method is entirely arbitrary (LAMBE, 1949). It is not the total amount of water present in various forms in the soil, but only the so-called free water and a portion of the adsorbed water that are removed at 105 °C.

The drying method has the disadvantage that it takes a relatively long time. Very often, particularly in field density measurements, quicker methods are preferred, which furnish "instant" results. There are several ways of doing this. The water content of coarse grained materials can be determined by submerging a wet sample in a pycnometer jar filled with water, and measuring the volume of the water displaced. Other rapid methods, based on the rapid removal of water by excessive heating, or on the measurement of the electrical resistance of the soil, are also in use.

A recent development has been the increasing use of rapid *in situ* methods of water content measurement. Some devices work on the principle that soil moisture can be determined by measuring the variation of the electrical resistance of nylon threads or of gypsum probes embedded in the soil. But this resistance depends also on the vapor pressure of water which in turn is a function of various factors and not only of water content. Recently, nuclear methods have been employed to determine moisture content. In these tests, a neutron source is inserted in the tip of a probe driven into the soil. The fast neutrons emitted slow down due to impacts with the hydrogen atoms of the water. The number of the decelerated neutrons intercepted by a counter is proportional to the water content of the soil. The counter is made of a material, which can be radio-activated when bombarded by slow neutrons.

These methods give, as a rule, higher values for the water content than those obtained by ordinary drying, since both the (OH^-) radicals and that portion of the adsorbed water which cannot be removed by drying at 105 °C interfere with the movement of neutrons.

Some of the workable rapid *in situ* testing methods are dealt with in Vol. 3 (Soil testing).

The value of the natural water content of the soil depends on many factors. In the first place, it is governed by the hydrological conditions in the ground i.e. by the position of the ground water table and its seasonal fluctuation. Moreover, it may be affected by climatic changes and weather, particularly in the zone near the surface. The extent of that zone where seasonal changes are likely to take place depends on the climate, whether extreme or moderate, and also on the tpye of the soil. In Hungary, the thickness of this layer seldom exceeds 1.0 meter and it is only in special cases that seasonal variations need be considered in connection with foundation work. In soil mechanics problems related to road construction and earth works seasonal changes play, of course, a much more important role.

Figure 53 shows an example, after BALLENEGGER (1938), of the annual variation of the water content of the soil at various depths near the ground surface. The variation is as much as $\Delta w = 10\%$ (extreme values for the water content being 11% and 21%) in the 30 cm thick top layer, while it is only about 5% or less at a depth of 1 m. The water content reaches its maximum between November and May, and its minimum in the dry period from August to October. A comparison of these measurements with the relevant meteorological data shows that the wetting of the soil upon the rainy autumn—winter period, as well as the desiccation following the long dry spell in the summer, take place with a certain time lag.

The average water contents of various soil types, in what could be described as the moist state, may differ considerably according to the proportion of the fine fractions in the soil, since the fine particles are capable of holding a high percentage of water owing to their high specific surface. The water content of moist sands seldom exceeds 5 to 10%. Below the ground water table the pores of sands are of course saturated and their water content may be as high as $w = 20\%$ or more.

Very fine sands, with a predominant fraction of dust-sized grains, can hold 10 to 15% of water in the moist state. This figure increases to $w = 10$ to 20% for silts and $w = 20$ to 30% for clays. In soils which are under, or in direct contact with, water much higher values may occur.

Organic soils have exceptionally high water contents, for organic silts $w = 40$ to 80%, for organic clays $w = 50$ to 100%. Peats, especially clayey peats, may contain water weighing several times the weight of the solids i.e. the water content runs to several hundred percent. Thus, an enourmously high water content is usually an indication of the presence of organic matter in the soil.

The various forms in which water exists in the soil will be discussed in detail in Ch. 3.

2.3.3 Porosity and void ratio

These two terms are used to describe numerically the same quantity, namely, the relative volume of the voids in the soil.

Porosity is the ratio of the volume of voids to the total volume of soil. It is generally expressed

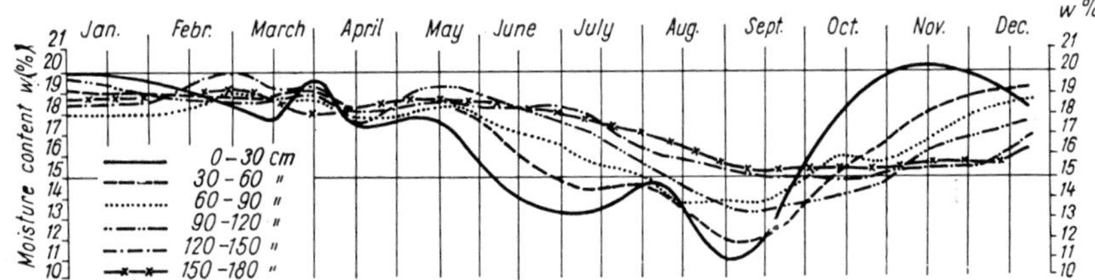

Fig. 53. Variation of moisture content at differents depths, according to BALLENEGGER (1938)

as a percentage. Using the symbols of Fig. 47, porosity is defined by the formula

$$n\% = 100\frac{V_m + V_g}{V}. \tag{29}$$

In terms of percentage relative volumes

$$n\% = 100 - s\% . \tag{29a}$$

As a geological term, porosity has long been in use. In soil mechanics, another related term, the void ratio has been introduced to indicate the relative volume of voids. By this is meant the ratio of the volume of voids to the volume of the solid mineral grains. With the symbols of Fig. 47,

$$e = \frac{V_m + V_g}{V_s}, \tag{30}$$

or, expressed in terms of relative volumes:

$$e = \frac{1 - s}{s}. \tag{30a}$$

Porosity and void ratio are interrelated quantities, according to the following formulae:

$$n\% = \frac{e}{1 + e}100, \tag{31}$$

and

$$e = \frac{n\%}{100 - n\%}. \tag{31a}$$

For the purpose of the determination of porosity and void ratio in the laboratory, a sample of volume V is taken from the soil, dried to constant weight, and weighed (W_d). If the unit weight of the solids γ_s is known, then the volume occupied by the solid grains is equal to W_d/γ_s. Hence, porosity and void ratio are calculated from

$$n\% = 100\frac{V - \frac{W_d}{\gamma_s}}{V}; \tag{32}$$

and

$$e = \frac{V - \frac{W_d}{\gamma_s}}{\frac{W_d}{\gamma_s}}. \tag{32a}$$

The volume V is either calculated from the dimensions of the samples, or, in the case of irregular samples, determined by submerging the sample in mercury and measuring the weight, and hence the volume, of the mercury displaced. Submersion in water is also used with samples coated in paraffin-way.

For the determination of e and n in the field, a hole is dug in the soil, the material removed is dried and weighed, and the volume of the hole is measured by filling it up with water or with a standard sand of known constant density. Details of testing methods will be dealt with in Vol. 3.

The porosity of soils in their natural state varies between wide limits. Porosity and unit weight for typical soils are given in *Table 7*. Porosity depends on the manner of the origin of the soil, on the degree of uniformity of the grain-size distribution, on the water content and, to a great extent, on the shape of the grains. Thus, for example, sands and gravels laid down by the swift and varying current of flooding rivers are, as a rule, very loose, whereas sands deposited from still water or slow-flowing rivers are densely packed. In the case of cohesive soils it is important whether the deposition took place in the presence of electrolytes, which tend to accelerate the sedimentation of colloidal particles, or if the process was retarded by the agency of protective colloids, which prevent flocculation. (See Ch. 3.)

In general, the more uniform the grain-size distribution, i.e. the lower the uniformity coefficient U, the greater is porosity. A well-graded soil is composed of particles varying in size in such a manner that the finer particles can readily fill the voids formed by the larger grains.

Grain size is another important factor; the finer the grains, the greater, as a rule, is porosity. The reason for this is that as grain size decreases, the number of contacts between the solid grains per unit weight increases, and, as a consequence, the ratio of the weight of an individual grain to the adhesive forces working against the tendency of the grain to shift to a more stable position decreases. This results in a looser packing, i.e. in greater porosity.

Furthermore, porosity is greatly influenced by the shape of the particles. The higher the proportion of the flaky, platy particles, the greater is the porosity in natural state, since the fine particles tend to aggregate in edge-to-edge or edge-to-face arrangement. (See Ch. 3.) The effect of grain shape on porosity can be studied experimentally on a mixture of angular sand grains with fine, platy fragments of crushed mica (TERZAGHI, 1926). If the proportion by weight of the mica is equal successively to 0.5, 10, 20 and 40%, then the porosity of the mixture when poured slowly into a container will have the following values: $n = 47$, 60, 70, 77 and 84%. The predominance of the flake-shaped and lath-shaped particles explains the high porosity of clays.

The only granular system for which porosity can be determined by relatively simple mathematical methods is a packing of uniform spheres. As stated by SLICHTER (1899), in the loosest stable arrangement of equal-sized spheres, when sphere centres form a rectangular space lattice (cubic packing, *Fig. 54*) and each sphere is in contact with six neighbouring spheres, the maximum porosity is $n = 47.6\%$; in the densest state of packing, when sphere centres form a

Table 7. Void ratio, porosity and bulk unit weight of typical soils

Soil type	State of soil	Porosity n%	Void ratio e	Bulk unit weight, kN/m^3			
				dry	natural	saturated	buoyant
Sandy gravel	loose	38–42	0.61–0.72	14–17	18–20	19–21	9–11
	dense	18–25	0.22–0.33	19–21	20–23	21–24	11–14
Coarse sand, medium sand	loose	40–45	0.67–0.82	13–15	16–19	18–19	8– 9
	dense	25–32	0.33–0.47	17–18	18–21	20–21	10–11
Uniform fine sand	loose	45–48	0.82–0.82	14–15	15–19	18–19	8– 9
	dense	33–36	0.49–0.56	17–18	18–21	20–21	10–11
Coarse silt	loose	45–55	0.82–1.22	13–15	15–19	18–19	8– 9
	dense	35–40	0.54–0.67	16–17	17–21	20–21	10–11
Silt	soft	45–50	0.82–1.00	13–15	16–20	18–20	8–10
	slightly plastic	35–40	0.54–0.67	16–17	17–21	20–21	10–11
	hard	30–35	0.43–0.49	18–19	18–19	18–22	11–12
Lean clay	soft	50–55	1.00–1.22	13–14	15–18	18–20	8–10
	slightly plastic	35–45	0.54–0.82	15–18	17–21	19–21	9–11
	hard	30–35	0.43–0.54	18–19	18–22	21–22	11–12
Fat clay	soft	60–70	1.50–2.30	9–15	12–18	14–18	4– 8
	slightly plastic	40–55	0.67–1.22	15–18	15–20	17–21	7–11
	hard	30–40	0.43–0.67	18–20	17–22	19–23	9–13

rhombohedral array ($\alpha = 60°$) and each sphere is in contact with 12 neighbouring spheres, (*Fig. 55*), porosity is $n = 25.9\%$. In intermediate cases, the centres of any 8 spheres, originally arranged in cubic packing, form the corners of a rhombohedron, and the porosity of the system depends on the acute face angle α called the angle of orientation of the rhombohedron. According to *Slichter*'s formula

$$n = \frac{\pi}{6(1 - \cos \alpha)\sqrt{1 + 2 \cos \alpha}} . \quad (33)$$

The variation of porosity with the angle of orientation is shown diagrammatically in *Fig. 56.*

Filep (1937) has proved that theoretically it is possible to construct a regular packing of uniform spheres such as to have porosity $n = 60$ to 70% and yet remain stable.

[For further discussion on the physical properties of regular packings of uniform spheres and their significance from the point of view of soil mechanics see Kézdi (1964), (1966) and (1967).]

Porosity is also influenced by existing or past loads acting on the soil. In the case of granular soils i.e. sands and gravels, the effect of "stress history" is of minor importance, since pressure is transmitted from grain to grain, and a relatively small increment in load does not usually give rise to an appreciable change of soil structure. Should, however, the pressure be increased to such intensity as to exceed the crushing strength of the individual grains, a sudden decrease of porosity will follow. In the case of cohesive soils, porosity may

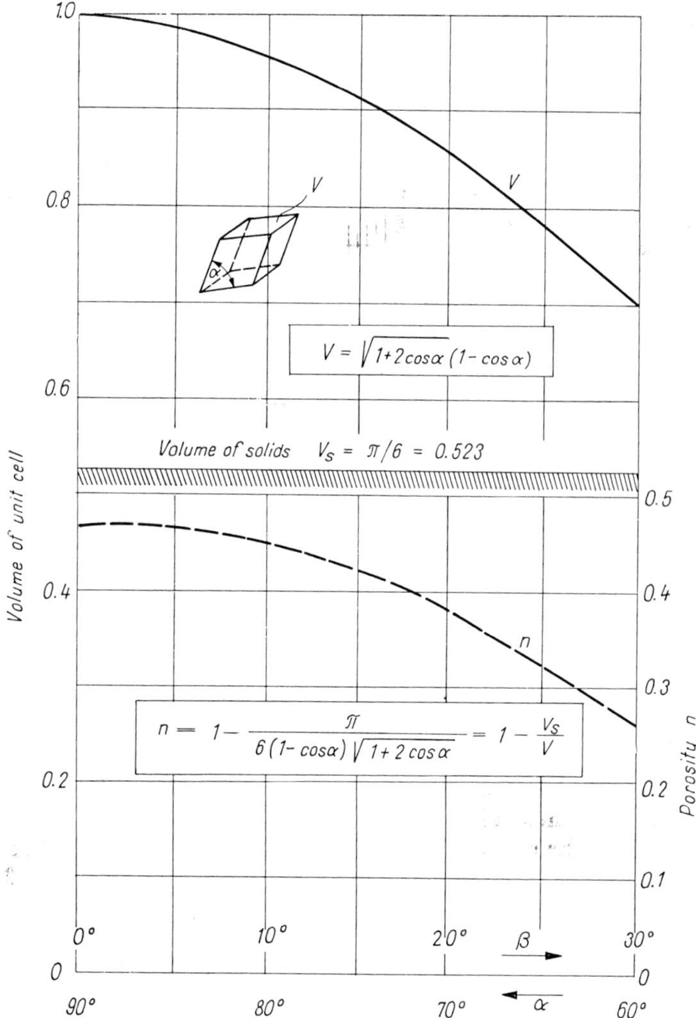

Fig. 56. Relationship between porosity and orientation angle

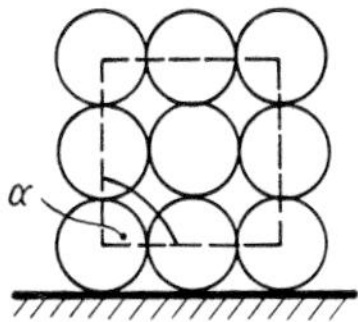

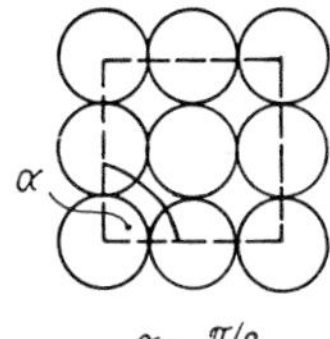

Fig. 54. Loosest state of packing of uniform spheres

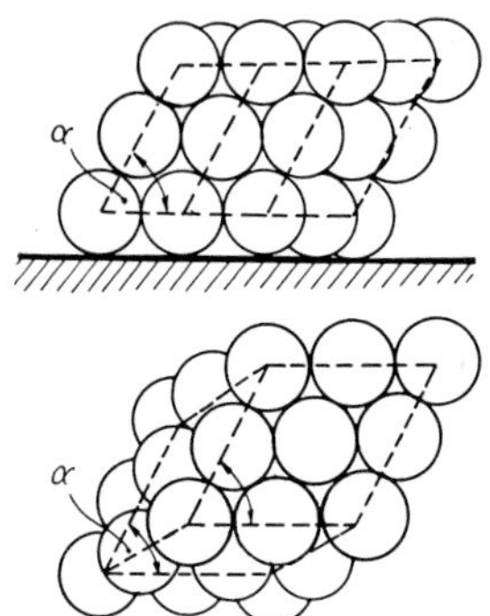

Fig. 55. Densest state of packing of uniform spheres

decrease considerably with the increase of pressure. For example the layer deposited first in the course of sedimentation will be subjected to the increasing overburden pressure of the subsequently deposited layers and will gradually be compressed. The decrease of porosity depends on the strength of the flaky particles, and on that of the bonds between them; the rate at which water drains out of the pores is also important. These questions will be discussed in detail in the chapters dealing with compression and consolidation (Chs 6 and 8).

2.3.4 Degree of saturation

The degree of saturation, S, is the ratio of the volume of water in the soil to the volume of voids. By the notations of *Fig. 57*:

$$S = \frac{e'}{e}.$$

To obtain a formula suitable for numerical calculations, let us express the water content of the model soil sample shown in Fig. 57 as the weight of water in the soil divided by the weight of the solids:

$$w = \frac{e'\gamma_w}{1\cdot\gamma_s}.$$

Hence

$$e' = w\gamma_s/\gamma_w,$$

and

$$S = \frac{w\gamma_s}{e\gamma_w}. \tag{34}$$

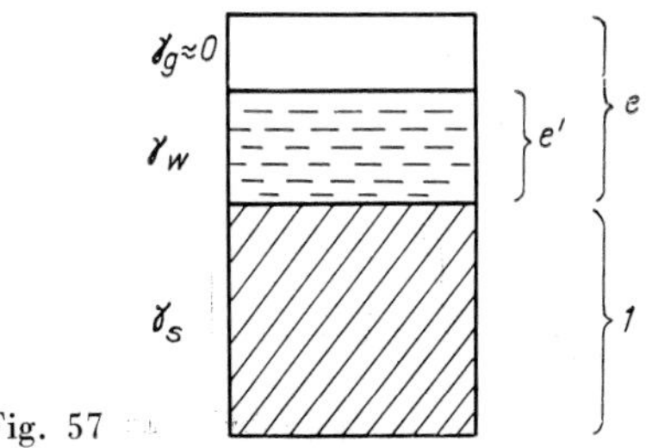

Fig. 57

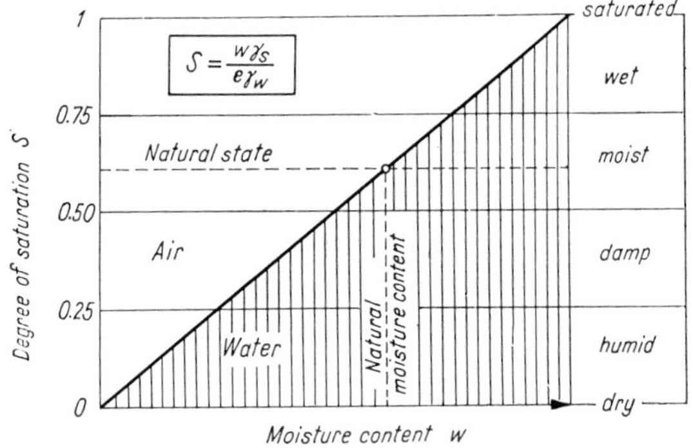

Fig. 58. Water content and degree of saturation

In the calculation of the degree of saturation, the water content and the void ratio need be determined by laboratory measurements. As regards the unit weight of the solids, γ_s, the use of the tabulated average values (see Table 4) will normally be accurate enough.

The degree of saturation of sands is commonly described by such words as dry or moist etc. Such descriptive terms and the corresponding values of S are given in *Fig. 58*. Coarse sands located above the ground water table are usually humid. Fine sands and silty sands are moist or saturated. Clays are almost always saturated, except for the layer near the surface, which is subject to seasonal variations of temperature and moisture content. If a clay contains gas, the gas is present in the form of a small entrapped bubbles dispersed throughout the soil. The bubbles may consist of air that mingled with the fresh deposit in the course of sedimentation, or of gas given off in some sort of chemical process. The gas in the pores of the clay may be under considerable pressure, which may cause an intensive swelling of the clay should the confining pressure be reduced for some reason. The determination of the gas content is extremely difficult.

2.3.5 Unit weight of soil

The determination of the unit weight of the solids (see Section 2.1.3) is only the first step to the determination of the unit weight of soil in its natural state. The unit weight γ is defined as the weight of the soil (composed, in the general case, of three phases: solids, water and gas) per unit volume. The practical determination of the unit weight is very simple: an undisturbed sample is taken from the soil and its weight and volume are measured. The unit weight is obtained by dividing the weight by the volume i.e.

$$\gamma = \frac{W}{V}.$$

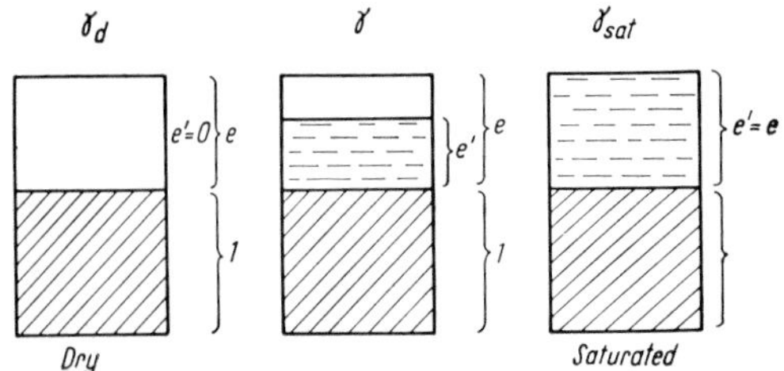

Fig. 59. Calculation of bulk unit weight

It is often necessary to calculate the unit weight from other phase relationships. The formulae required can be easily derived with the aid of the model soil sample shown in *Fig. 59*.

By using the symbols of the figure,

$$\gamma = \frac{\text{total weight}}{\text{total volume}} = \frac{1 \cdot \gamma_s + e'\gamma_w + 0}{1 + e},$$

and since

$$e' = w\gamma_s/\gamma_w,$$

hence

$$\gamma = \gamma_s \frac{1 + w}{1 + e} = \gamma_s(1 + w)(1 - n). \qquad (35)$$

Equation (35) gives the unit weight in the general case when the soil is composed of three phases. Let e.g. the water content of a given clay $w = 30\%$, its void ratio $e = 0.935$. Then, using the tabulated average value of the unit weight of the solids,

$$\gamma = 28 \frac{1.30}{1.935} = 18.8 \text{ kN/m}^3.$$

Soil strata located at greater depth below the surface often contain no air, their pores being completely saturated with water. In such condition, described as a two-phase system, the void ratio can be expressed in terms of water content w. Since $S = 1$, $e = e' = w\gamma_s/\gamma_w$. Substituting this expression for e in Eq. (35) we obtain:

$$\gamma_{sat} = \gamma_s \frac{1 + w}{1 + w\gamma_s/\gamma_w}. \qquad (36)$$

If the soil given in the previous example becomes saturated, the altered values of its water content and unit weight will be

$$w\% = 100 \frac{e\gamma_w}{\gamma_s} = 100 \frac{0.935 \times 10}{28} = 33.4\%,$$

and

$$\gamma_{sat} = 28 \frac{1.334}{1 + 2.8 \times 0.334} = 19.3 \text{ kN/m}^3.$$

Equation (36) can be easily written in terms of porosity. Denoting the total volume of the model soil sample by unity, the volume of voids is n and that of the solids is $1 - n$. The weight of water completely filling the voids is $n\gamma_w$. Hence, the unit weight is given by

$$\gamma_{sat} = \gamma_s(1 - n) + \gamma_w n \,. \tag{37}$$

Besides the saturated condition, another two-phase state is possible in the extreme case when the soil is completely dry, i.e. it consists of solids and air only. Substituting $w = 0$ in Eq. (35) the unit weight of dry soil (often referred to in practical use as the "dry density") is obtained from

$$\gamma_d = \frac{\gamma_s}{1 + e} = \gamma_s(1 - n) \,. \tag{38}$$

Let us examine for example the possible range of variation in the unit weight of a sand, considering completely dry and saturated conditions in both the loosest and the densest states of the soil.

(i) In *dry, loose state:*

Given: $n = 50\%$; $e = 1.000$; $\gamma_s = 26.5$ kN/m³

the unit weight: $\gamma_d = \dfrac{\gamma_s}{1 + e} = \dfrac{26.5}{2} = 13.2$ kN/m³

(ii) In *saturated, loose state:*

Given: $n = 50\%$; $e = 1.000$; $\gamma_s = 26.5$ kN/m³

$e = w\gamma_s/\gamma_w$; $w = \dfrac{e\gamma_w}{\gamma_s} = \dfrac{1 \times 10}{26.5} = 0.38$

$\gamma_{sat} = \gamma_s(1 - n) + n\gamma_w = 26.5 \times (1 - 0.5) + 10 \times 0.5 = 18.2$ kN/m³

(iii) *Dry, dense state:*

Given: $n = 35\%$; $e = 0.35/0.65 = 0.538$;

$\gamma_s = 26.5$ kN/m³

$\gamma_d = \dfrac{26.5}{1.538} = 17.2$ kN/m³

(iv) In *saturated, dense state:*

$\gamma_{sat} = 26.5 \times 0.65 + 10 \times 0.35 = 20.7$ kN/m³

It is seen from the figures that the difference between the values of the unit weight measured in loose and dense state of sands may amount to 3 to 4 kN per cubic meter.

The unit weight can easily be expressed in terms of the relative volumes defined in Section 2.2.1. If the soil is

$$\left.\begin{array}{ll} \text{dry} & \gamma_d = s\gamma_s \,, \\ \text{partially saturated} & \gamma = m\gamma_w + s\gamma_s \,, \\ \text{saturated} & \gamma_{sat} = (1 - s)\,\gamma_w + s\gamma_s \end{array}\right\} \tag{39}$$

The percentage relative volumes (s, m, g) and the other basic physical characteristics discussed

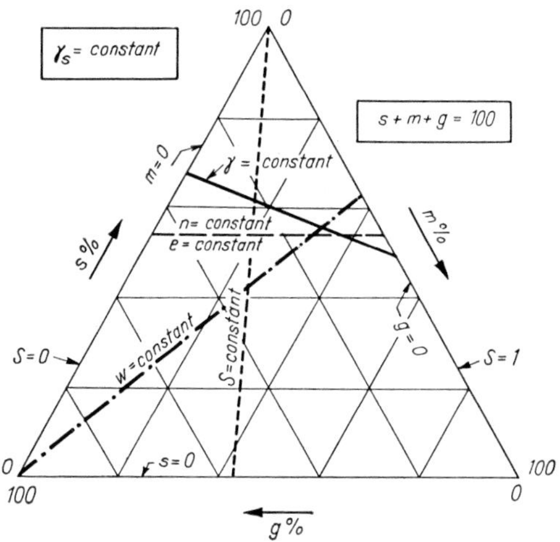

Fig. 60. Relations of index properties plotted in the triangular diagram

in Sections 2.3.2 to 2.3.5., can be related in many ways. A useful method of representing these manifold relations in a diagram is the triangular coordinate system, as in *Fig. 60.* The straight lines radiating from the point $s = 0$ correspond to w = constant, and those radiating from $m = 0$ represent constant values of the degree of saturation. The n = constant (or e = constant) condition is represented by straight lines drawn parallel to the axis g. The γ = constant lines run obliquely across the diagram; the terminals at which the lines intersect the axes s and m indicate the unit weights in the dry and saturated state of the soil, respectively. If these lines are constructed for a full set of conveniently chosen constant values of (w, S, n, γ), a universal diagram is obtained, by means of which e.g. (s, m, g) values can be determined from (w, n) values or vice versa by simple interpolation. A complete diagram detailed enough to be used for practical purposes will be given in Vol. 3.

2.3.6 Relative density and degree of compaction

The porosity of the soil is greatly influenced by the shape of the grains and the degree of uniformity of the grading, thus, whether a particular soil is loose or dense, it cannot be judged on the basis of the porosity alone. The degree of compaction can be assessed only by comparing the natural porosity of the given soil with the extreme values of porosity of the same soil in its loosest and densest states. For sands and gravel the relative density defined by the following equation has been adopted as a measure of the denseness of the soil:

$$D_{re} = \frac{e_{max} - e}{e_{max} - e_{min}} \,, \tag{40}$$

wherein e_{max} = void ratio of the soil in its loosest state,
e = void ratio in the state examined,
e_{min} = void ratio in the densest state that can be obtained in the laboratory (*Fig. 61*).

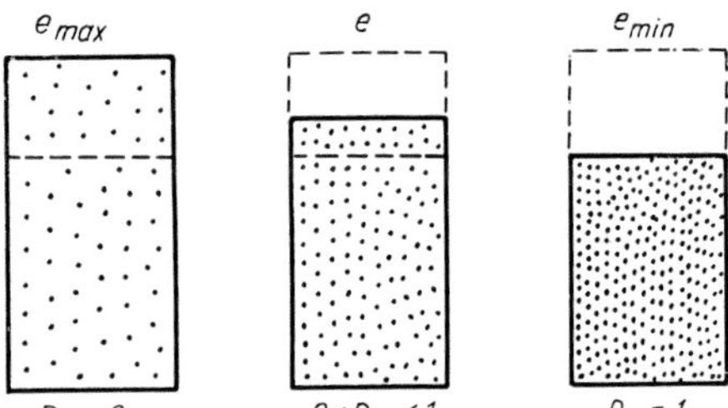

Fig. 61. Relative density

Of the quantities in Eq. (40), the determination of e has already been dealt with. e_{max} can be measured by the following test. The oven-dry, pulverized soil is slowly poured through a funnel into a cylindrical container of known volume, care being taken to keep the outlet of the funnel only a few mm clear of the surface of the sample (*Fig. 62*). Pouring is continued until the material

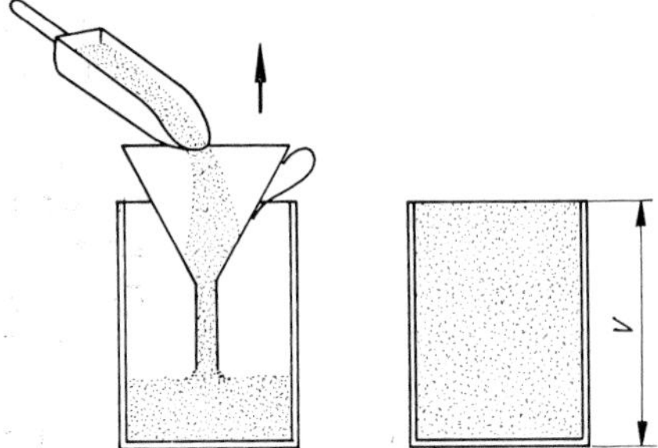

Fig. 62. Determination of the void ratio in the loosest state

overflows. The surface of the sample is then carefully struck flush with the brim of the container and the dry weight of the sample determined. The maximum void ratio e_{max} is calculated from Eq. (32a). In the case of very fine sands the loosest state can be reproduced by mixing the dry pulverized soil with enough water to make a thick slurry which is then allowed to settle. e_{max} is then obtained as the void ratio of the sediment.

The determination of the void ratio pertaining to the densest state, e_{min}, requires a somewhat tedious experiment. In principle, it would seem possible to compute e_{min} on the basis of the grain-size distribution curve, as the void ratio in the theoretically possible closest packing of the grains. Such a theoretical computation, however, would be feasible only on the assumption that the soil grains were spherical in shape, whereas the real grains are irregular, and this fact cannot be disregarded, considering the great influence of grain shape on the packing of the grains. For this reason, e_{min} is determined by a laboratory compaction test. A sample of wet soil is compacted into a mould with a volume of 2 liters. The compaction is performed by ramming the soil in three layers, each layer being compacted by 25 blows with a rammer of 25 N weight dropped from a height of 30.5 cm. The wet sample which fills the mould is weighed and its water content determined. The whole procedure is then repeated with a number of samples of different water content. Knowing the water content in each test, the weights of the solids and, using Eq. (32a), the void ratios can be calculated. If the values of e are plotted against the water content w, a characteristic curve is obtained, which has a definite minimum e_{min} (*Fig. 63*).

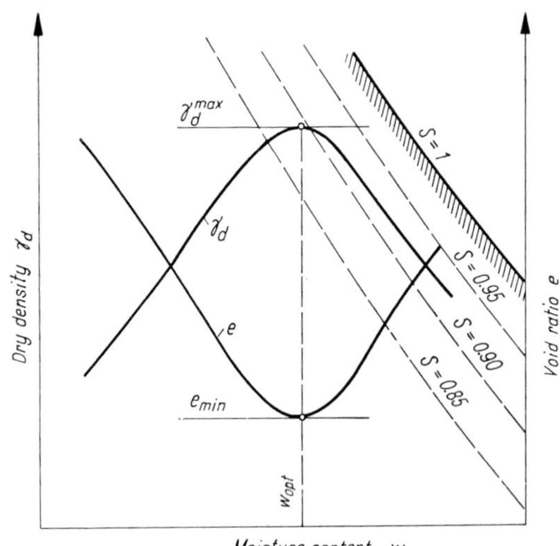

Fig. 63. Compaction curve; determination of the void ratio and dry density in the densest state, respectively

The relative density is sometimes expressed in terms of the porosity, n:

$$D_{rn} = \frac{n_{max} - n}{n_{max} - n_{min}} . \tag{41}$$

The analogous equations (40) and (41), however, do not give the same result, since at constant void ratio

$$\frac{D_{rn}}{D_{re}} = \frac{1 + e_{min}}{1 + e} \neq 1 .$$

The relative density, as defined by Eq. (40), is not applicable to cohesive soils because of the uncertainties in the laboratory determination of the void ratio in the loosest state of the soil, e_{max}. The state of compaction of cohesive soils if therefore characterized by the following ratio, called the degree of compaction:

$$D_{r\gamma} = \frac{\gamma_d}{\gamma_d^{max}}, \tag{42}$$

in which

γ_d = unit weight of dry soil in the state examined,

γ_d^{max} = maximum unit weight of dry soil obtained from standard compaction test.

(The terms "dry density" γ_d and "maximum dry density" γ_d^{max} are also in current use in many textbooks, standard specifications etc. on the subject.)

The value of γ_d^{max} depends on the general character of the grain-size distribution, on the size and shape of the grains and on chemical effects.

The degree of compaction be related to expressions previously defined for the relative density [Eqs (40) and (41)]. Since $\gamma_d = \gamma_s/(1+e)$, we have

$$D_{r\gamma} = \frac{1+e_{min}}{1+e} = \frac{D_{rn}}{D_{re}}. \tag{43}$$

In recent years, the use of the degree of compaction [Eq. (42)] has become a generally accepted practice for judging the measure of compaction of both coarse-grained and cohesive soils.

In this connection, it must be pointed out that both expressions for the relative density, Eqs (40) and (41), give values varying from zero to unity as the void ratio changes over its full range from e_{max} to e_{min}; whereas the least value of $D_{r\gamma}$ cannot be zero, since it would represent air with no solids.

In order to find γ_d^{max} for use in Eq. (42) the results of the standard compaction test are usually represented in a compaction curve (also called moisture—density curve), values of the unit weight of the dry soil (= "dry density") being plotted against the water contents on the abscissa. The peak of the curve indicates the condition in which the highest dry unit weight can be achieved by a given standard compaction effort. The value of γ_d at the peak is called the maximum dry unit weight and the corresponding value of w is the optimum moisture content (w_{opt}). Its value depends on the type of the soil and also on the method of compaction and on the intensity of the compactive effort. It will be useful to construct the saturation line (also called zero-air-voids curve), which represents the relation between dry unit weight and water content for fully saturated soil ($S = 1$). Since in saturated state $e = w\gamma_s$,

$$\gamma_d = \frac{\gamma_s}{1 + w\gamma_s}.$$

For partially saturated state ($S < 1$), the expression can be written as

$$\gamma_d = \frac{\gamma_s}{1 + \dfrac{w\gamma_s}{S}}. \tag{44}$$

The dry density/water content relations at constant values of S may also be plotted in the graph (dashed lines in Fig. 63).

The relative density (D_{re}) and the degree of compaction ($D_{r\gamma}$) are of great practical importance as means of quality control in the construction of earthworks. A fill which is loosely dumped, will experience settlement under its own weight, which may be intensified by the vibrating effect of surface traffic. Precipitation water can readily find its way into the loose fill giving rise to softening and swelling of the material, to frost damage etc., and as a result the completed earthworks will be seriously damaged. These detrimental effects can be eliminated, or at least greatly reduced, by adequate compaction of the fill. The problems of compaction are dealt with in Vol. 2.

2.3.7 Examples

Example 1. A cylindrical sample of sand, 4 cm high and 6 cm in diam., has a weight of $W = 1.362$ N in a wet state. After drying it weighs $W_d = 1.229$ N. The unit weight of the solids is $\gamma_s = 0.0265$ N/cm³. Determine the water content, the void ratio, the degree of saturation and the unit weight of the wet sample. Also calculate the percentage relative volumes.

Volume of sample:

$$V = \frac{4^2\pi}{4} \times 6 = 75.4 \text{ cm}^3.$$

Water content:

$$w = \frac{1.362 - 1.229}{1.229} = \frac{0.133}{1.229} = 10.8\% \sim \mathbf{11\%}.$$

Volume of solids:

$$V_s = \frac{W_d}{\gamma_s} = \frac{1.229}{0.0265} = 46.3 \text{ cm}^3.$$

Volume of voids:

$$V_m + V_g = V - V_s = 75.4 - 46.3 = 29.1 \text{ cm}^3.$$

Void ratio:

$$e = \frac{V_m + V_g}{V_s} = \frac{29.1}{46.3} = \mathbf{0.628}.$$

Porosity:

$$n = 100\,\frac{V_m + V_g}{V} = \frac{29.1}{75.4} \times 100 = \mathbf{38.6\%}.$$

Degree of saturation:

$$S = \frac{w\gamma_s}{e\gamma_w} = \frac{0.108 \times 2.65}{0.628} = \mathbf{0.455}.$$

Unit weight, wet:

$$\gamma = \gamma_s \frac{1+w}{1+e} = 0.0265 \times \frac{1.108}{1.628} = \frac{1.362}{75.4} =$$

$$= 0.0181 \text{ N/cm}^3 = \mathbf{18.1} \text{ kN/m}^3.$$

Relative volumes of constituent phases:

Solids: $s = \frac{W_d}{V\gamma_s} = \frac{1.229}{75.4 \times 0.0265} = \mathbf{0.615}.$

Water: $m = \frac{W - W_d}{V\gamma_w} = \frac{1.362 - 1.229}{75.4 \times 0.010} = \mathbf{0.177}.$

Air: $g = 1 - (s + m) = \mathbf{0.208}.$

Example 2. In a standard compaction test on a silt ($\gamma_s = 0.0268$ N/cm³) six cylindrical specimens were compacted by constant compactive effort, at gradually increasing water contents. The initial volume of the specimens was $V = 942$ cm³. The weights and the water contents of the wet samples were as follows:

Test sample No.	1	2	3	4	5	6
weight of wet soil, W N	17.70	18.88	19.74	19.85	19.15	18.78
average water content w %	7.5	10.2	12.4	14.7	17.0	19.2

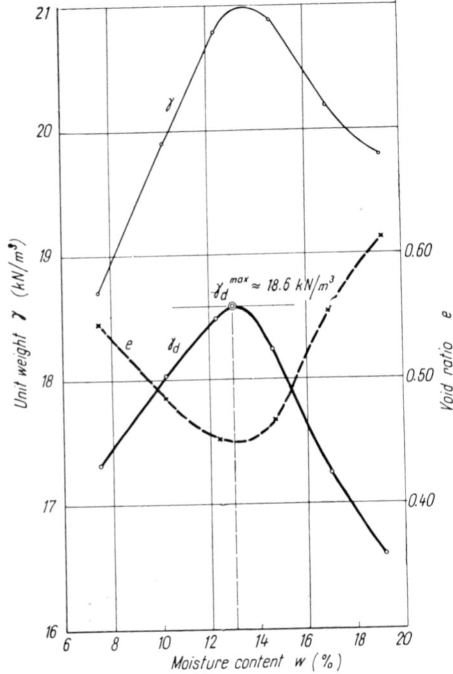

Fig. 64. Void ratio and dry density versus moulding water content

Determine the optimum moisture content, the minimum void ratio and the maximum dry unit weight. What is the degree of compaction of a fill whose average void ratio was found to be $e = 0.50$?

The calculation can be conveniently arranged in tabular form.

Test sample No.	1	2	3	4	5	6
Water content w	0.075	0.102	0.124	0.147	0.170	0.192
Weight of wet soil W N	17.70	18.88	19.74	19.85	19.15	18.78
Unit weight of wet soil γ kN/m³	18.7	19.9	20.8	20.9	20.2	19.8
Weight of dry soil W_d N	16.45	17.14	17.53	17.32	16.36	15.75
Unit weight of dry soil γ_d kN/m³	17.32	18.04	18.47	18.25	17.24	16.60
Void ratio $e = \gamma_s/\gamma_d - 1$	0.545	0.485	0.452	0.468	0.554	0.613

The results are plotted in *Fig. 64.* The optimum moisture content, read off from the figure is $w_{opt} = 13\%$, and the maximum dry unit weight $\gamma_d^{max} = 18.6$ kN/m³. If the void ratio measured in the fill was $e = 0.49$ then the dry unit weight of the fill $\gamma_d = 26.8/(1 + 0.49) = 18.0$ kN/m³ and the degree of compaction $D_{r\gamma} = \frac{18.0}{18.6} = \mathbf{0.965}.$

Example 3. The volume of sand excavated from a borrow pit was $V_1 = 2560$ m³, the unit weight of the solids being $\gamma_s = 26.8$ kN/m³, the water content $w = 12\%$ and the unit weight of the wet soil in the borrow pit $\gamma = 16.8$ kN/m³. How many cubic meter of compacted fill could be constructed of this material if the required value of the porosity in the completed fill is $n = 32\%$?

Assuming that the weight and the water content of the material do not change during construction, the change of volume can be calculated from the change of the unit weight. I.e.

$$\frac{V_1}{V_2} = \frac{W/\gamma_1}{W/\gamma_2} = \frac{\gamma_2}{\gamma_1} = \frac{1+e_1}{1+e_2},$$

whence

$$V_2 = V_1 \frac{1+e_2}{1+e_1} = V_1 \frac{1-n_1}{1-n_2}. \quad (45)$$

Void ratio in the natural state:

$$e_1 = \gamma_s \frac{1+w}{\gamma_1} - 1 = \mathbf{0.787}.$$

Void ratio of compacted fill:

$$e_2 = \frac{n_2}{1-n_2} = \frac{0.32}{0.68} = \mathbf{0.471}.$$

Volume of compacted fill:

$$V_2 = 2560 \frac{1.471}{1.787} = \mathbf{2100}\ \text{m}^3 .$$

The degree of saturation will be considerably increased in the fill:

in the natural state

$$S_1 = w\gamma_s/e_1\gamma_w = 0.12 \times 2.68/0.787 = \mathbf{0.41};$$

in the compacted fill

$$S_2 = w\gamma_s/e_2\gamma_w = 0.12 \times 2.68/0.741 = \mathbf{0.68}.$$

Example 4. A sand has the following properties in its natural state: $\gamma_1 = 18.2$ kN/m³; $w = 11\%$, $\gamma_s = 26.5$ kN/m³. The same sand when dumped loosely will experience a volume change of 20%. This phenomenon is called *bulking*. What will the porosity be in the loosened state of the sand?

The solution can be obtained by the reasoning employed in Example 3.

Consider a unit volume of the sand in its natural state ($V_1 = 1$ m³). This will be increased to $V_2 = 1.2$ m³; due to bulking.

Initial unit weight:

$$\gamma_1 = \gamma_s \frac{1+w}{1+e_1} = \gamma_s(1+w)(1-n_1) .$$

Hence

$$1 - n_1 = \frac{\gamma_1}{\gamma_s(1+w)} = \frac{18.5}{26.5 \times 1.11} = 0.628 .$$

Initial porosity: $n_1 = \mathbf{37.2\%}$.

Using Eq. (45):

$$1 - n_2 = \frac{V_1}{V_2}(1 - n_1) ;$$

whence

$$n_2 = 1 - \frac{1}{1.2} 0.628 = \mathbf{48\%} .$$

Example 5. The pavement of a road on level ground is to be laid on a base course 50 cm thick, consisting of a coarse-grained sandy gravel with good draining properties, placed evenly on the impervious subsoil. The porosity of the gravel is $n = 38\%$, and its degree of saturation $S = 0.60$. Assuming that a heavy rain occurs during the construction work and all water immediately infiltrates into the gravel, calculate the maximum precipitation, in mm, that would saturate the base course in its full thickness.

Consider a prism of sand, 50 cm thick with a base area of $A = 1$ m².

Total volume of pores in prism:

$$V_m + V_g = nV = 0.38 \times 0.5 = 0.19\ \text{m}^3 .$$

Of this, the volume filled with water

$$V_m = SnV = 0.60 \times 0.19 = 0.114\ \text{m}^3 .$$

Volume of pores to hold precipitation water:

$$V_g = 0.190 - 0.114 = 0.076\ \text{m}^3 .$$

This is equivalent to a rainfall of 76 mm. Violent summer rainstorms of this magnitude are not uncommon in this country.

Example 6. A saturated mass of soil has a volume of $V = 1$ m³ and a weight of $W = 20.5$ kN. The unit weight of the solid grains is $\gamma_s = 27.9$ kN/m³. Assuming that the pore liquid is pure water, calculate the water content and the void ratio.

Unit weight of the saturated soil:

$$\gamma_{sat} = \frac{W}{V} = 20.5\ \text{kN/m}^3 .$$

According to Eq. (35):

$$\gamma_{sat} = \gamma_s \frac{1+w}{1+w\gamma_s} .$$

Hence

$$w = \frac{1}{\gamma_{sat} - 1}\left(1 - \frac{\gamma_{sat}}{\gamma_s}\right) = \frac{1}{1.05}\left(1 - \frac{20.5}{27.9}\right) = \mathbf{25.2\%} ;$$

$$e = \frac{w\gamma_s}{\gamma_w} = 0.252 \times 2.79 = \mathbf{0.704} .$$

Now, assuming that the liquid filling the pores is salt water whose unit weight is $\gamma_w' = 10.25$ kN/m³, calculate again the void ratio, and the water content.

In this case the saturated unit weight is given by

$$\gamma_{sat} = \frac{\gamma_s + 10.25\,e}{1+e} ,$$

whence the void ratio

$$e = \frac{\gamma_s - \gamma_{sat}}{\gamma_{sat} - 10.25} = \frac{27.9 - 20.5}{20.5 - 10.25} = 0.721 ,$$

and the water content

$$w = \frac{e\gamma_w}{\gamma_s} = \frac{0.721}{2.79} = 25.8\% .$$

If the difference in the unit weight of the liquid due to the salt content is neglected, then the error in the void ratio comes to

$$\Delta = \frac{0.721 - 0.704}{0.721} 100 = \mathbf{2.5\%}.$$

Example 7. A disc-shaped sample of loess (75 mm in diam, $h = 20$ mm high), weighed $W_{w1} = 1.307$ N in its natural state and $W_d = 1.126$ N after drying. The unit weight of the solids was 0.0267 N/cm³. Subjected to a certain vertical pressure, the sample, which was laterally confined (oedometer test, Section 8.2) underwent a vertical compression of $\Delta h_1 = 0.8$ mm.

If a true loess is soaked through with water, a sudden vertical settlement occurs due to the collapse of the soil structure (Section 8.5). This can be simulated in the oedometer test, by wetting the sample from above. In the example, the additional compression of the sample was found to be $\Delta h_2 = 3.0$ mm. The soaked sample weighed $W_{w2} = 1.401$ N after the test. Calculate the relative volumes (s, m, g) of the sample in its natural state and also in the two stages of the test. Plot the results in the triangular diagram.

(*a*) In the natural state:

$$A = \frac{d^2\pi}{4} = \frac{7.5^2\pi}{4} = 44.2\ \text{cm}^2,$$

$$V_1 = Ah = 44.2 \times 2.0 = 88.4\ \text{cm}^3,$$

$$S_1 = \frac{W_d}{V_1\gamma_s} = \frac{1.126}{88.4 \times 0.0267} = \mathbf{0.477},$$

$$\Delta W_1 = W_{w1} - W_d = 1.307 - 1.126 = 0.181\ \text{N},$$

$$m_1 = \frac{\Delta W_1}{V_1\gamma_w} = \frac{0.181}{88.4 \times 0.010} = \mathbf{0.205},$$

$$g_1 = 1 - (s_1 + m_1) = \mathbf{0.318}.$$

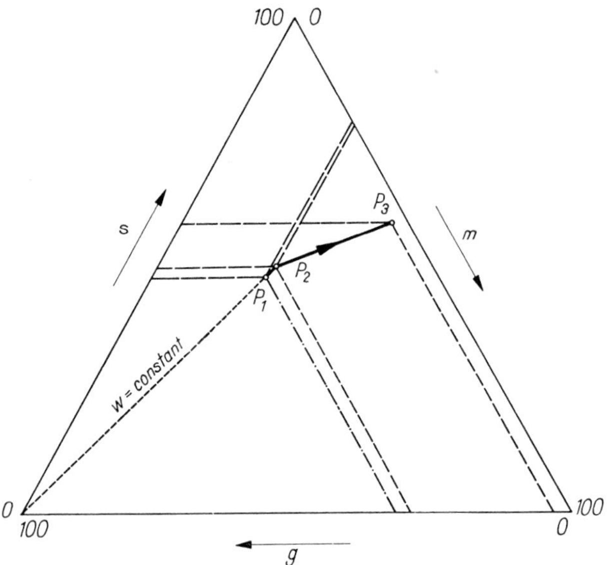

Fig. 65. Results of the numerical example, as plotted in the triangular diagram

(b) After normal compression:

$$\Delta V_1 = A\,\Delta h_1 = 44.2 \times 0.08 = 3.54 \text{ cm}^3,$$

$$V_2 = V_1 - \Delta V_1 = 88.4 - 3.54 = 84.86 \text{ cm}^3,$$

$$S_2 = \frac{W_d}{V_2 \gamma_s} = \frac{1.126}{84.86 \times 0.0267} = \mathbf{0.496},$$

$$m_2 = \frac{\Delta W_1}{V_2 \gamma_w} = \frac{0.181}{84.86 \times 0.010} = \mathbf{0.213},$$

$$g_2 = 1 - (s_2 + m_2) = \mathbf{0.291}.$$

(c) After compression with soaking:

$$\Delta V_2 = A\,\Delta h_2 = 44.2 \times 0.3 = 13.26 \text{ cm}^3,$$

$$V_3 = V_2 - \Delta V_2 = 84.86 - 13.26 = 71.60 \text{ cm}^3,$$

$$S_3 = \frac{W_d}{V_3 \gamma_s} = \frac{1.126}{71.60 \times 0.0267} = \mathbf{0.589},$$

$$W_{w2} = 1.401 \text{ N},$$

$$\Delta W_2 = W_{v2} - W_0 = 1.401 - 1.126 = 0.275 \text{ N},$$

$$m_3 = \frac{\Delta W_2}{V_3 \gamma_w} = \frac{0.275}{71.6 \times 0.010} = \mathbf{0.348},$$

$$g_3 = 1 - (s_3 + m_3) = \mathbf{0.027}.$$

The calculated results are plotted in the triangular diagram (*Fig. 65*). The processes of normal compression and sudden collapse upon soaking are represented by the vectors $\overline{P_1P_2}$ and $\overline{P_2P_3}$ respectively.

2.4 Consistency limits

2.4.1 Consistency of soils

The term consistency is used to describe the physical state, i.e. the degree of coherence between the particles, of various substances. As applied to cohesive soils, consistency can be characterized, according to the water content of the soil, by such descriptive words as soft, plastic, stiff or hard. Water content is the chief factor determining the strength and also the deformation characteristics of cohesive soils, and it is therefore necessary to examine the influence of water content on the state of consistency of soils in more detail. First of all, it must be realized that the same water content may be coupled with entirely different states of consistency for different soils; e.g. at a water content of $w = 30\%$ a silt is practically in a liquid state whereas a fat clay of the same water content may be very stiff and exhibits high strength. Thus, the water contents at which different soils pass from one state into another are characteristic of the type of soil. The boundaries, in terms of water content, between successive states of consistency can thus be regarded as soil characteristics and are called consistency limits.

A clay when mixed with a sufficient amount of water becomes a thick paste that has practically no cohesion and would flow under its own weight even on gentle slopes. If the paste is gradually dried it passes from the liquid state to the plastic state, so that it can be readily moulded. On further drying the clay becomes stiff, its colour turns into a lighter shade, its strength gradually increases until finally it becomes quite hard and rigid (*Fig. 66*). In the course of drying the clay undergoes a decrease in volume that is equivalent to the loss in volume of the water due to evaporation. At a definite water content the process of volume change ceases and beyond this stage air begins to enter the pores causing the colour of the clay to turn light, and the soil becomes gradually harder. Four characteristic consistency states can be distinguished in the course of these changes: liquid, plastic, stiff, (or semi-solid), and hard (solid). The transition from one state to another

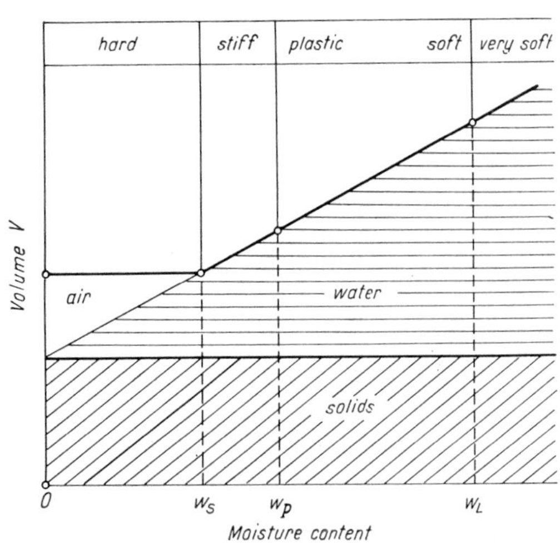

Fig. 66. Change in the consistency of clay during drying

does not occur abruptly (unlike the sudden change of fluid water into ice at the freezing point). It is a gradual transition with no well-defined water content values that would mark the boundaries between successive states of consistency. The consistency limits can be defined, therefore, only on the basis of more or less arbitrarily set criteria. The definitions currently used in soil mechanics were adopted with some modification from agricultural soil science. Three consistency limits are in general use:

— *liquid limit,*
— *plastic limit,*
— *shrinkage limit.*

The definitions of these limits, together with a brief description of their laboratory determination and notes on the practical applications are given in the succeeding chapters.

2.4.2 Liquid limit

If a cohesive soil is thoroughly mixed with increasing amounts of water, it will reach a state when the attractive forces so far holding the particles together cease to exist, and the soil possesses no cohesion, but rather behaves like a thick paste or viscous liquid. Fine sediments deposited from still water are in a similar state at the stage of sedimentation. The water content required for cohesive soils to reach this state is called the *liquid limit.* This is, however, not a clear-cut definition and the liquid limit can be determined, as discussed before, only by some arbitrarily set procedure. The need for comparability and reproducibility of tests performed at various times and various places requires the standardization of the test procedure. The liquid limit device (*Fig. 67*) proposed in 1932 by A. Casagrande, has since been internationally adopted as a standard apparatus. It consists of a circular brass dish which, by means of a handle rotating a cam, can be repeatedly dropped from a height of 1 cm. An essential part of the apparatus is a standard grooving tool.

The fractions larger than 2 mm must be removed from the soil prior to the test. The remaining material is then thoroughly mixed with as much water as is needed to obtain a water content a few percent less than the liquid limit. About 50 to 100 g of the wet soil is then placed in the cup, and its surface levelled off to a maximum thickness of 12 mm, care being taken that the paste contains no entrapped air-bubbles. Using the standard grooving tool, a trapezoid-shaped groove is cut in the paste along the cup diameter perpendicular to the axis of the driving shaft. By turning the handle at the rate of two rotations per second

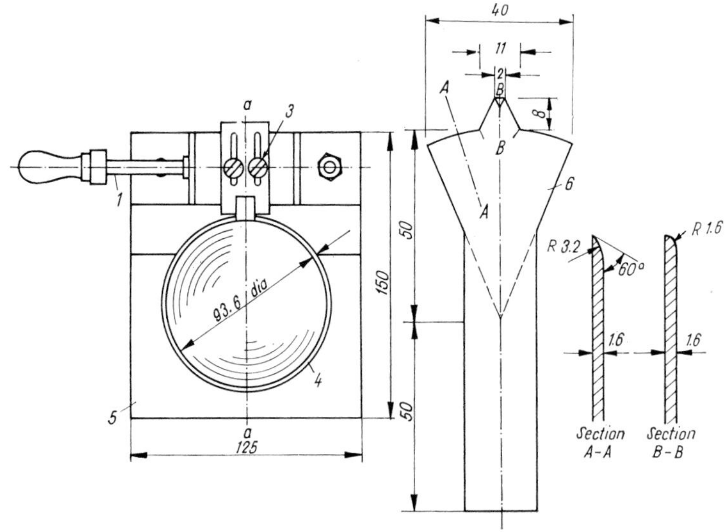

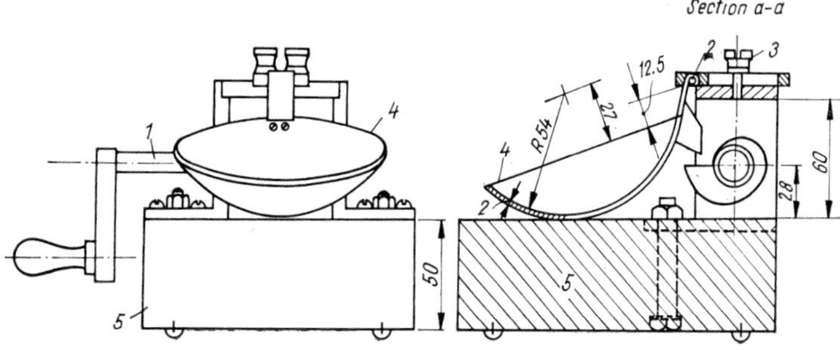

Fig. 67. Liquid limit device:

1 — handle; *2* — hinge; *3* — adjusting screws; *4* — copper cup; *5* — hard rubber base; *6* — grooving tool

the cup is repeatedly dropped until the two sides divided by the groove touch over a length of 1.2 cm. The cross-sections of the groove prior to, and after, the test are shown in *Fig. 68.* A small portion of the paste in the cup is then set aside for the determination of the water content

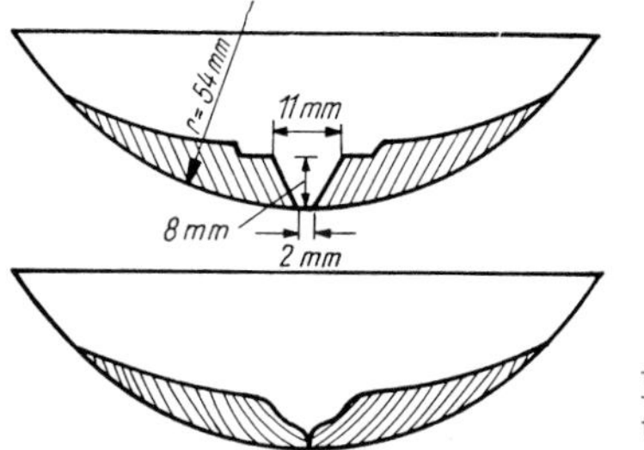

Fig. 68. Liquid limit test

and the number of blows causing the gap to close is recorded. If this number is greater than 25, more water is added to the soil and the test is repeated; if the number is less than 25, the soil is too wet and must be allowed to dry. The test is repeated at least three times and the results are presented in a semi-logarithmic graph, the water contents being plotted to an arithmetic scale against the number of blows to a logarithmic scale. In the case of carefully performed tests, the resulting points are arranged with fair approximation along a straight line, called the flow curve. The water content corresponding to 25 blows can be read off from the line. By definition this is the liquid limit w_L of the soil (*Fig. 69*).

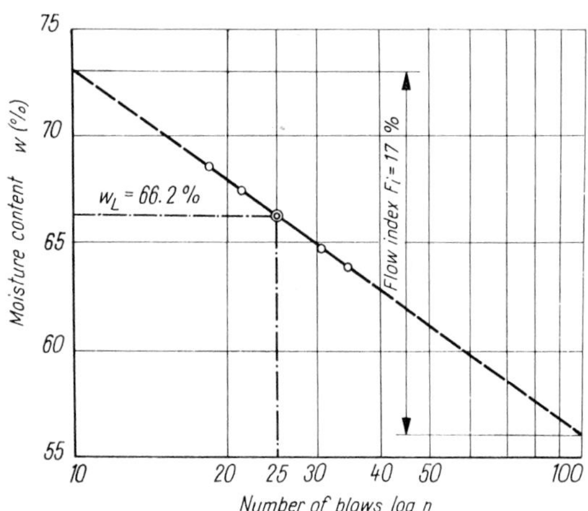

Fig. 69. Flow line; determination of the liquid limit and flow index

The slope of the flow curve can be calculated as the difference between the water contents at 1 and 10 blows (or at 10 and 100 blows), and is termed the flow index, F_i.

Further practical details and the sources of error of the liquid limit test are discussed in Vol. 3.

It should be briefly noted here that, as shown by JÁRAY (1955), the liquid limit depends greatly on the ions adsorbed on the surface of the clay particles. If the sodium ions (Na) in the adsorption complex of a montmorillonite are exchanged for calcium ions (Ca), the value of the liquid limit is reduced from 700 to 124%, and the plastic limit (see Ch. 3) from 93 to 72% (GRIM, 1948). JÁRAY attempted to correlate the specific surface, as obtained from the grain-size distribution curve, with the liquid limit of the soil.

According to LAMBE (1951), flow curves plotted in a double logarithmic system (log w, log n) give parallel straight lines, represented by the general equation

$$w_L = w\left(\frac{n}{25}\right)^{0.121}. \tag{46}$$

n denotes the number of blows, and w is the corresponding water content.

The validity of Eq. (46), which has been confirmed by the author's own tests, implies that a single test is enough for the determination of the liquid limit, furthermore, that the flow index and other indices derived from it are not independent soil properties, since they can all be derived from the liquid limit.

Solving Eq. (46) for w we obtain

$$w = 1.48\,\frac{w_L}{n^{0.121}}$$

hence

$$\frac{F_i}{w_L} = 0.273 .$$

This empirical relation between the flow index and the liquid limit together with test results are shown in *Fig. 70.*

Every soil has a definite value of the liquid limit expressed as a water content which is therefore a very suitable means of soil classification and identification. Obviously, the coarser are the grains of a soil, the smaller is the surface capable of adsorbing water, consequently, the smaller is the numerical value of the liquid limit. The least value is obtained for sands: $w_L = 15$ to 20%. For clays it is much higher since much water is needed for a soil with large specific surface area to be transformed into a paste which exhibits flow properties. The liquid limit of clays is therefore at least $w_L = 40\%$, but normally much higher, up to 60 to 80%. If the liquid limit has a value less than 35%, the soil is described as having low plasticity. For soils of medium plasticity the liquid limit ranges from 35 to 50%, and for those of high plasticity w_L is greater than 50%.

Soils of high plasticity are always fine-grained and usually contain a high proportion of adsorptive minerals. Such soils are unsuitable for many engineering purposes and should be treated with

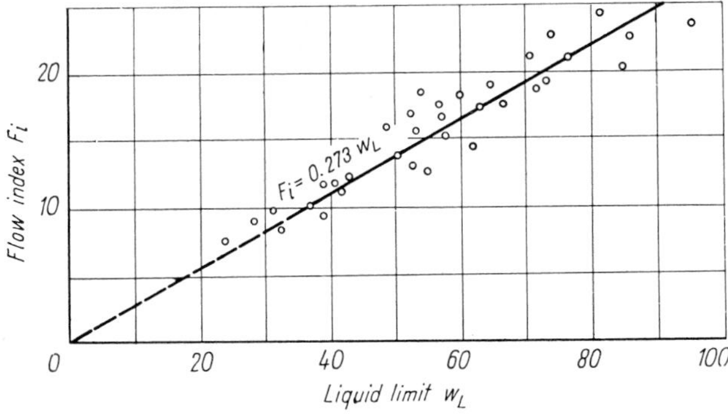

Fig. 70. Relationship between liquid limit and flow index

caution. They are highly compressible and, having very low internal friction, are likely to cause slides.

Soils with a low liquid limit are susceptible to softening on wetting, their strength being radically reduced by even relatively small increases in the water content.

If the liquid limit of a given soil is known, it is possible to estimate the danger of flow by comparing the water content in the natural state of the soil with its liquid limit. But it must be realized that natural soils cannot in actual fact be brought to the liquid state by simply increasing the water content up to the value of the liquid limit; it is also necessary to destroy the structure of the soil.

2.4.3 Plastic limit

If a pat of wet plastic soil is dried, it gradually looses its plasticity, which means that it can no longer be moulded, but begins to crumble when rolled into threads. The water content at which the soil passes from the plastic state to the semi-solid state is called the plastic limit (w_P). Its actual determination is very simple; only a glass plate and a sheet of filter paper are needed (apart from the water content test apparatus).

A small sample of the soil, from which the grains larger than 2 mm have been previously removed, is dried or wetted, as the case may be, until it can be readily moulded into a pat. About 5 g of the wet soil is then shaped into a ball and rolled into a thread on the blotting paper (*Fig. 71*). When the thread just crumbles at a diameter of 3 to 4 mm, the soil is said to be at the plastic limit. If the sample can be rolled into too thin thread, it is folded and rolled repeatedly. This process is continued until the 3-mm thread begins to crumble. At the plastic limit, the soil leaves no mark on the paper. The crumbled threads are gathered into a watchglass and their water content is determined. The result is the plastic limit w_P.

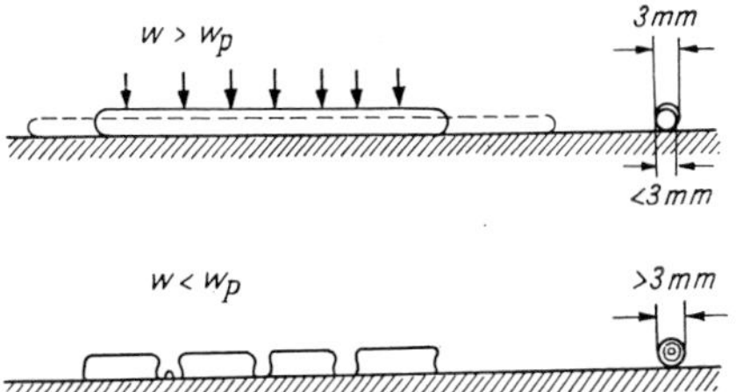

Fig. 71. Plastic limit test

The plastic limit is an important soil property from the point of view of the civil engineer. Agricultural cultivation and excavation work, by both hand and machines, can be carried out with the least effort i.e. most economically, with soils at the plastic limit. Earth roads are easily usable at this water content. Optimum conditions for soil compaction occur at water contents near the value of the plastic limit.

The range of the numerical values of the plastic limit is much narrower than that of the liquid limit. Sand has no plastic stage at all. Very fine sand (Mo) exhibits slight plasticity, its plastic limit being about 17 to 20%. For silts $w = 20$ to 25%, and for clays $w_P = 25$ to 35%.

The difference between the liquid limit and the plastic limit is called the plasticity index. In symbols

$$I_P = w_L - w_P .$$

Soils with no plastic limit have no plasticity index either. The plasticity index represents the range of water content over which a soil is plastic, i.e. it is a measure of the cohesiveness of the soils. The greater the plasticity index, the higher, in general, the attraction between the particles of the soil will be. The plasticity index is very useful both in soil classification and, as a basic soil characteristic, in various empirical rules and correlations with other soil properties. Its many uses will be referred to later in relevant chapters.

Important relations can be established between the consistency limits and the clay content of the soil. This is of great significance in relation to soil stabilization. A comprehensive treatise on this subject is given in *Stabilized earth roads* by Kézdi (1967).

2.4.4 Shrinkage limit

If a mass of soft, saturated soil gradually dries, its volume is reduced by an amount equivalent to the volume of the water lost by evaporation. This volume change is caused by capillary forces acting on the surface of the soil mass. At a certain water content the sum of the capillary forces reaches its greatest value and the volume change ceases. On further drying the water begins to withdraw into the interior of the soil, whose colour then changes from dark to light. The

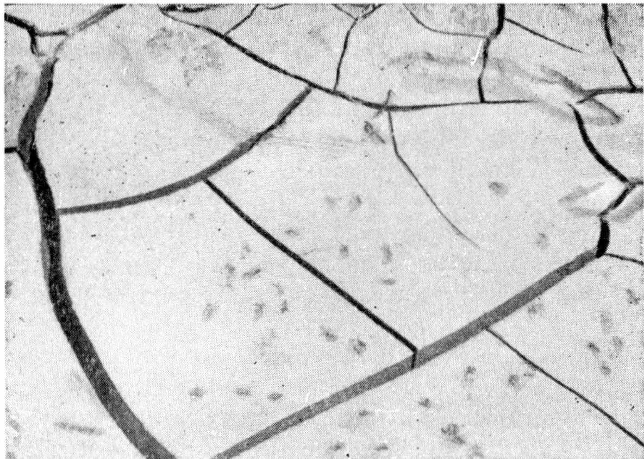

Fig. 72. Pattern of shrinkage cracks on the surface of drying clay layer

surface of the desiccating soil shows a characteristic pattern of shrinkage cracks (*Fig. 72*). The finer the particles of the soil, the greater is the amount of shrinkage.

The water content below which further loss of water by drying does not result in reduction of the volume of the soil is called the ***shrinkage limit***. It is denoted by w_S.

Working on the above definition, the shrinkage limit is determined by completely drying out a lump of soil and measuring its final volume and weight. The volume of the oven-dried sample may be assumed to be equal to its volume at the shrinkage limit, since during the drying below this limit no appreciable volume change has taken place. Since the degree of saturation at the shrinkage limit is unity (as it is throughout the process of desiccation at water contents greater than w_S), the shrinkage limit can be calculated from the minimum volume V_{min} and weight W_d of the oven-dried sample, as follows.

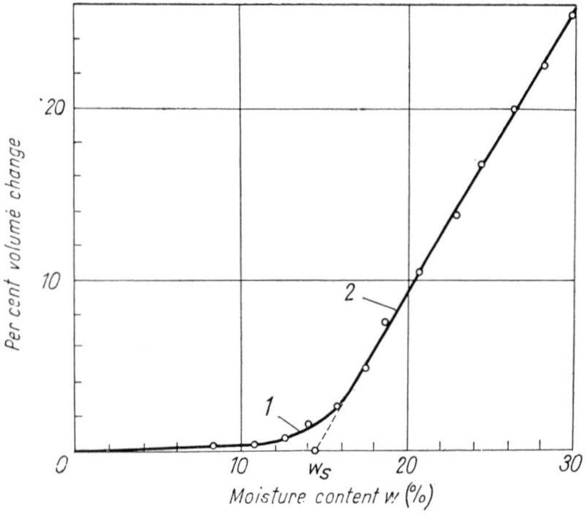

Fig. 73. Specific volume change plotted versus water content: *1* — shrinkage due to hydration forces; *2* — shrinkage due to capillary action

$$w_s = 100\,\frac{\text{weight of water filling pores completely at minimum volume of sample}}{\text{weight of oven-dried sample}}\,.$$

In symbols

$$w_s = 100\,\frac{\left(V_{\text{min}} - \dfrac{W_d}{\gamma_s}\right)\gamma_w}{W_d}\,. \tag{47}$$

A more accurate method of the determination of the shrinkage limit involves measuring the volume of the sample at various stages of the desiccation process and plotting the volume change $[(V - V_{\text{min}})/V_{\text{min}}]$ in percent, against the water content. A graph similar to that shown in *Fig. 73* results. The plotted points lie, with fair approximation, on a straight line as long as the shrinkage is solely due to capillary action. At the shrinkage limit the curve shows a conspicuous break; but this is not abruptly sharp, as some small volume change occurs even below the shrinkage limit due to hydration forces which act in the water film between the grains. The straight section of the curve is therefore prolonged to the $v = 0$ axis; the water content indicated by the intercept on this axis is the shrinkage limit.

2.4.5 Examples

Example 1. A lump of clay was completely dried out. In this state it weighed $W = 0.5835$ N. The volume of the dry lump was found, by immersion in mercury, to be $V = 28.42$ cm³. The unit weight of the solids was $\gamma_s = 0.027$ N/cm³. Calculate the value of the shrinkage limit.

The volume of the solids $V_s = 0.5835/0.027 = 21.61$ cm³. Hence the volume of the voids was $28.42 - 21.61 = 6.81$ cm³. At the shrinkage limit the voids were completely filled with water ($S = 1$). On further drying the volume remained unaltered. The weight of water filling the pores completely at the shrinkage limit was, therefore, 6.81 cm³ × 0.010 N/cm³ = 0.0681 N. Hence the shrinkage limit

$$w_s = 100\,\frac{0.0681}{0.5835} = \mathbf{12\%}\,.$$

Example 2. A clay ($\gamma_s = 0.028$ N/cm³) has the following consistency limits: $w_L = 68\%$, $w_P = 27\%$, $w_s = 14\%$. Calculate the values of the void ratio at the consistency limits.

Since the soil is assumed to be saturated at the consistency limits ($S = 1$), the void ratio is obtained from the formula $e = w\gamma_s/\gamma_w$. The results are diagrammatically shown in *Fig. 74*.

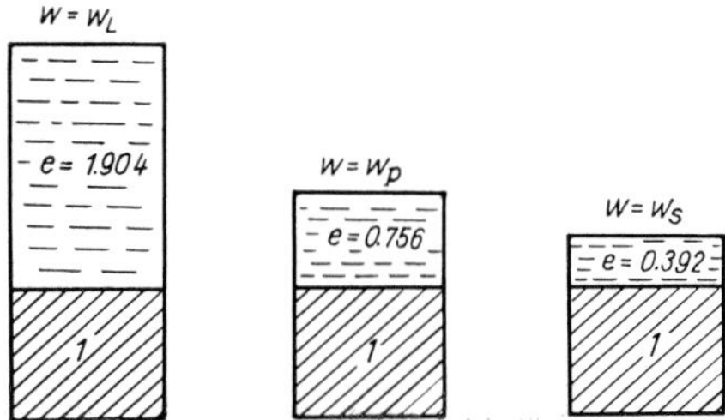

Fig. 74. Values of void ratio at the consistency limits

Average values of the consistency limits for various soils are given in Table 8.

Table 8. Consistency limits of soils

Water content	Sand	Rock flour	Silt	Clay
Liquid limit, w_L	15–20	20–30	30–40	40–150
Plastic limit, w_P	—	17–20	20–25	25– 50
Plasticity index, I_P	0	3–10	10–15	10–100
Shrinkage limit, w_S	12–18	12–20	14–25	8–35

2.4.6 Consistency index

Having defined the limits of consistency, we can now examine the varying states of a natural soil in terms of these limits. To this end a new term, the *consistency index*, is introduced which indicates the relation of the natural water content of a soil to the constant consistency limits of the same soil.

The consistency index is defined as the ratio of the difference between the liquid limit and the natural water content to the difference between the liquid and plastic limits. In symbols

$$I_c = \frac{w_L - w}{w_L - w_P} . \tag{48}$$

Figure 75 shows the variation of the consistency index with the natural water content, giving at the same time, the descriptive terms commonly used to characterize the state of consistency of soils at different values of the consistency index. As is seen, the water content alone does not indicate the state of consistency of the soils, since a water content of $w = 20\%$ may exceed the liquid limit of sands, while in the case of clays the same water content does not even reach the value of the plastic limit, and represents a hard, semi-solid state. For this reason, the natural water content must be compared with the liquid and plastic limits and soil consistency judged quantitatively on the basis of the consistency index. All the same, the use of this index is open to criticism on the grounds that combined within the same expression there are soil properties pertaining to the remolded state (w_L and w_P) and a property representing the natural state of the soil (w).

If the consistency index is equal to 1.0 the water content is at the boundary between the semi-solid and the stiff state, i.e. $w = w_P$. If $I_c = 0.5$, the water content is half-way between w_L and w_P, the soil is soft. I_c may be greater than 1.0 when w is smaller than w_P. It may also assume negative values when w is greater then w_L. The latter represents the state when the soil flows and is utterly unsuitable for foundation purposes. A significant value is $I_c = 0.5$, which indicates the boundary between the soft state and the stiff-plastic state with fair load-bearing behaviour.

Natural clay deposits which have never been subjected to preconsolidation have a consistency index near zero (Hogentogler, 1937; Skempton, 1944). If, however, a suspension of clay is allowed to settle under laboratory conditions, the consistency index of the sediment is always less than — 1. This state corresponds to an extremely high initial

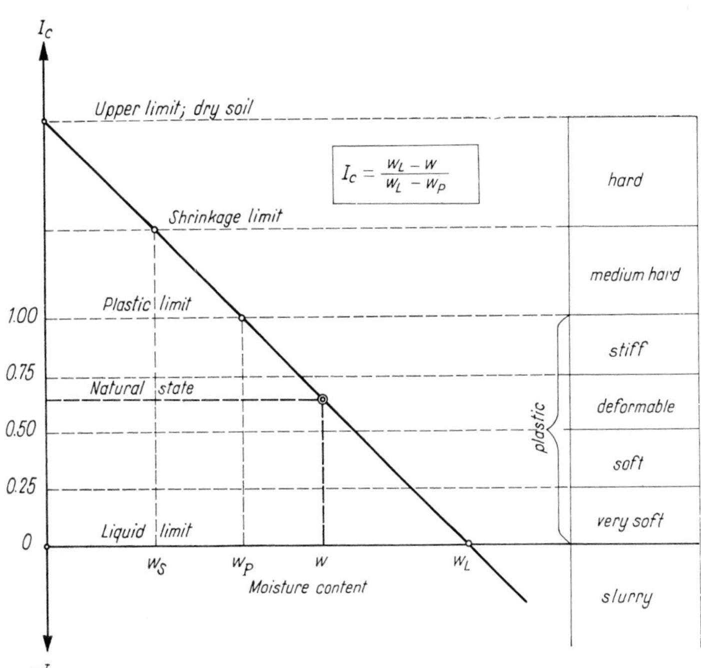

Fig. 75. Variation of the consistency index with the natural water content

porosity which, under natural conditions, is gradually reduced in the top layer of the clay to a much lower value corresponding to the liquid limit, through what is believed to be the agency of some unknown physical processes. The contraction of the clay is presumably the result of a process similar to the aging of gels (*syneresis*).

The degree of the consistency of clays may also be measured by means of the unconfined compressive strength (see Ch. 7). The designations of the various consistency states on this basis are given in *Table 9.*

Table 9. Qualitative and quantitative criteria of consistency of clays

Consistency	Filed identification	Unconfined compressive strength kN/m²
Very soft	The fist can be pressed in easily	<25
Soft	The thumb can be pressed in easily	25– 50
Medium	The thumb can be pressed in with light pressure	50–100
Stiff	The thumb leaves imprint, but can be pressed in with great effort only	100–200
Very stiff	Can be readily indented with thumb nail	200–400
Hard	Indented with difficulty by thumb nail	>400

The dry strength may also be a useful means of identifying and comparing soils. It can be determined by simply trying to crumble a small lump of air-dry soil between the fingers. The strength is said to be "medium" if the lump can be crumbled into a powder only with great effort. If the sample cannot be broken up at all, its strength is "high"; and if it disintegrates even under a gentle pressure, the strength is called "low".

The consistency index can be suitably used to show the variation of soil consistency, i.e. the degree of dryness or wetness of the strata, in the profile of an exploratory boring. This is most appropriately done in a graphical representation: a vertical section through the axis of the borehole is drawn first on a convenient scale, showing the location and thickness of the individual strata, the description of the soils encountered, and the depths from which samples were taken. A second diagram shows the variation of the consistency limits and of the natural water content with the depth below the surface. This diagram itself furnishes valuable information regarding the soil conditions, but the changes in soil consistency are even more accentuated in a third diagram showing the variation of consistency index with depth. Such a diagram may be very useful in locating the reliable soil strata with I_c values greater than 1.0, capable of supporting structures without the hazard of excessive settlements. The depth of foundation can thus be determined.

A typical boring profile is presented in *Fig. 76.*

A thin layer of clayey topsoil is underlain by a yellow-brown clay. In the upper part the clay is relatively dry, but it becomes gradually wetter as the depth approaches the water table. The value of the consistency index is accordingly reduced to about $I_c = 0.5$. Fine sand and sandy gravel lie below, with a thin layer of soft saturated silt between the pervious strata. These strata are under ground water, and as a result the clay strata underlying the gravel are also waterlogged, particularly the blue-grey, fat clay in its upper part adjacent to the gravel, and the yellow clay with streaks of fine sand. The next layer, a slightly organic clay is also of plastic consistency and favorable soil conditions are met with only at greater depths, in the light-grey calcareous clay.

The ground elevation and the water table, together with the date of observation, are also recorded in the profile of the boring.

Numerous other methods have been suggested, and are, in fact, in use, for characterizing soil consistency. Among these, the water content values introduced by Ohde (limit of pastiness, unit-water content), Polshin's squeeze test, Tsytovich's ball penetration test, and the Swedish cone penetration test are worth mentioning. Details of the laboratory procedures and the evaluation of test results are dealt with in Vol. 3.

2.4.7 Consistency and deformation of cohesive soils

From an engineering point of view the most important property of soils is their behaviour under the influence of shearing stresses. Shear deformation must be related, in some way, to the state of consistency of the soil and in the following this question will therefore be studied in some detail.

Figure 77 shows the deformation of a volume element subject to shearing stresses. Cohesive soils may respond to shear in two ways. In the first instant, the material acted upon by normal pressure will suffer a deformation under the action of shear stresses. This normally takes some time, after which the material comes again to a rest. In the second case, the initial deformations will not be followed by a state of rest, instead the material suffers continuous deformation at a slow but constant rate. In the first case the material is considered to be a solid, in the second it is a liquid. A typical stress versus strain curve for a solid material is represented by the line *OA* in *Fig. 78a.* After a relatively small and well-defined deformation, the failure takes place abruptly with no transitional, plastic state. This type of brittle failure is characteristic of cohesive soils with a water content which is below the shrinkage limit. If the water content is increased while the chemical composition of the soil remains unaltered, plastic

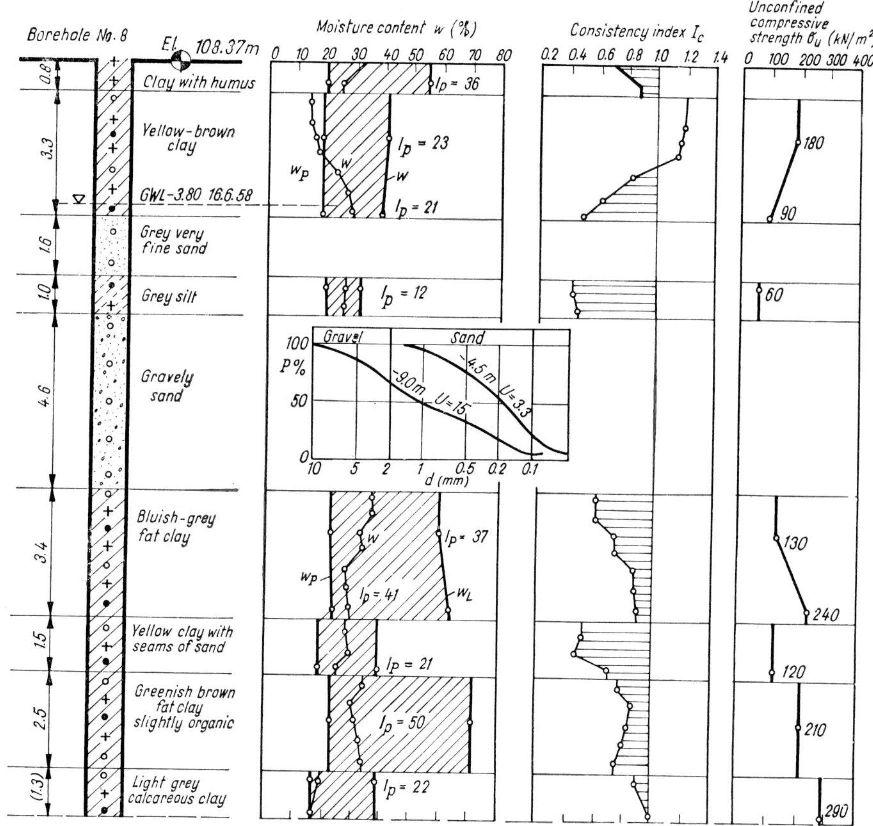

Fig. 76. Boring profiles (a) and (b) with physical characteristics

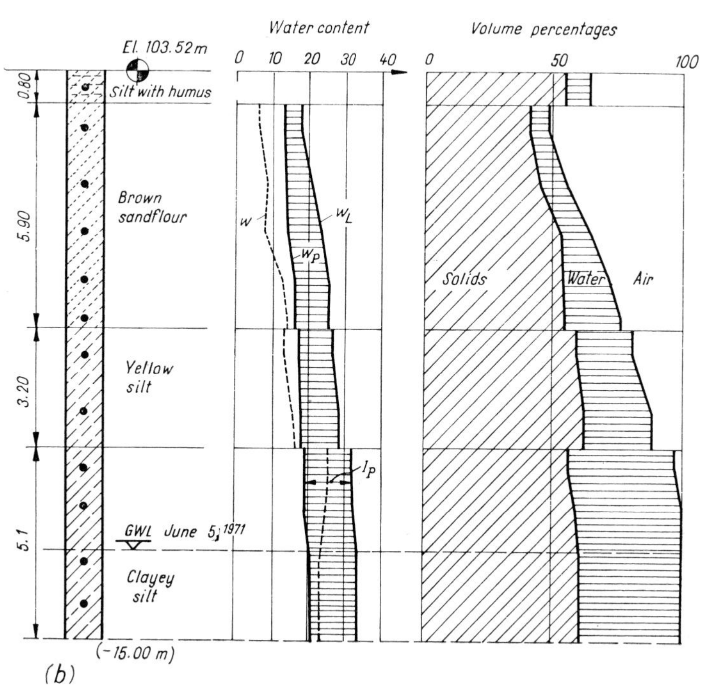

Fig. 77. Deformation of a volume element subjected to shearing stress

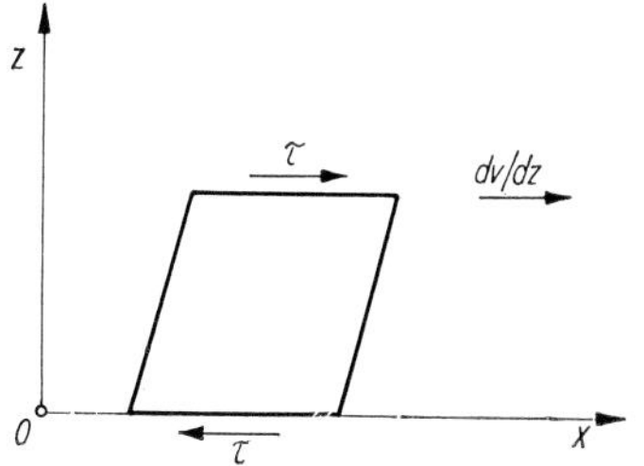

deformations gradually occur and the stress/strain relationship becomes indefinite. At this stage, the process of deformation can be described only in terms of the velocity of the deformation: the gradient of the shear deformation velocity with respect to the vertical movement (Fig. 77) is plotted against the shear stress as abscissa. For a clay at the plastic limit the diagram will be similar to curve *1* shown in Fig. 78*b*. The low and constant velocities in the region of small shear

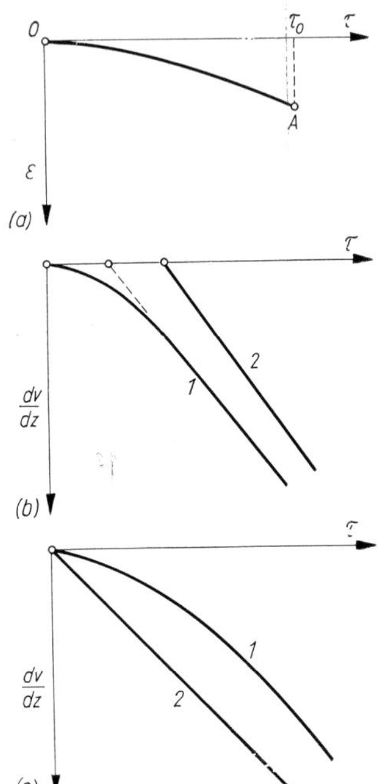

Fig. 78. Deformation diagrams:

a — solid body; *b* — velocity of deformation; Bingham solid; *c* — flow lines. *1* — clay in the plastic state; *2* — idealized flow line

stresses gradually give way to flow phenomena manifested by a much higher rate of deformation as the shear stress increases. An idealized diagram may be obtained by prolonging the straight section of the diagram to the τ-axis (dashed line); the intercept is supposed to signify the threshold value of the shear stress, below which no flow deformation is possible. An idealized deformation diagram is represented by line *2* in Fig. 78*b*. Materials exhibiting such properties are known as *Bingham* materials. Clays in the plastic state show such behaviour.

Approaching the liquid state, the initial section of the curve i.e. the threshold value of the shear stress vanishes, and the deformation diagram converts into a continuous curve right from the origin (curve *1* in Fig. 78*c*). The viscosity of such materials depends on the shear velocity; these are the so-called *non-Newtonian* fluids. The special case characterized by the equation

$$\tau = \mu \frac{dv}{dz} \tag{49}$$

is an approximation of the real behaviour, employed to facilitate the theoretical treatment of the problem. This ideal case, shown by line *2* in Fig. 78*c* represents the behaviour of the so-called *Newtonian* fluids.

As has been pointed out in connection with the *Atterberg* limits, the changes in the nature of soil behaviour do not occur abruptly but as a gradual transition. This statement holds also for the manner and character of the deformations. Therefore, the *Atterberg* limits (liquid limit, plastic limit, shrinkage limit) should in no way be interpreted as actual boundaries between different types of deformation behaviour. They are only experimental values representing approximately the water contents at, or near, which the changes in soil behaviour occur.

Recommended reading

ATTERBERG, A. (1911): Die Plastizität der Tone. Int. Mitt. Bodenkd. Oversatt fr. svenska ur Kungl. Lantbruks-akademiens Handlingar och Tidskr. 50. pp. 132–158.

BAVER, L. D. (1948) : Soil Physics. 2nd ed. Wiley & Sons, Inc. New York.

CASAGRANDE, A. (1932–33): Research on the Atterberg limits of soils. Public Roads 13 pp. 121–130.

CASAGRANDE, A. (1958): Notes on the design of the liquid limit device, Géotechnique 8. p. 84.

KÉZDI, Á. (1964): Some Properties of Packings. Highway Research Record, No. 52. Washington, D. C.

KÉZDI Á. (1966): Új eredmények a talajfizikában (New achievements in soil physics). Mélyépítéstudományi Szemle, 16, pp. 261–270.

LEIBNITZ, O. (1956): Die Schrumpfgrenze, ein bautechnischer Erdstott-Kennwert. Bauplanung u. Bautechnik, 10, p. 124.

ORR, C., DALLAVALLE, J. M. (1959): Fine particle measurement. The Macmillan Co., New York.

SEED, B. H.–WOODWARD, R. J.–LUNDGREN, R. (1964): Fundamental Aspects of the Atterberg Limits. Proceedings, Am. Soc. Civ. Engrs.; Journal of the Soil Mechanics and Foundations Division. Vol. 90, No. SM 6. Nov.

SCHULTZE, E.–MUHS, H. (1967): Bodenuntersuchungen für Ingenieurbauten. 2. Auflage. Springer-Verlag, Berlin (Heidelberg) New York.

Chapter 3

Soil structure

3.1 Introduction

One might well suppose that the engineering properties of soils are greatly affected by the mineral composition of the soil-forming grains. However, as far as coarse grained soils are concerned, this is not the case. The minerals of which the grains are composed make hardly any difference to the engineering properties of coarse-grained soils except perhaps to surface friction between the grains. In granular soils, interparticle forces other than those due to gravity or to external loads are negligible. On the other hand, the finer the grains, the more significant become the forces associated with the surface area of the grains. In this case, the chemical character of the individual grains is an essential factor in relation to the behaviour of the whole assemblage. Interparticle attraction holding the grains together also becomes increasingly important as the grain size decreases. Under the influence of various physical effects and fields of forces, the constituent parts present in different physical states are combined in what is called the structure of the soil—the dominant factor governing the behaviour of cohesive soils.

The term *soil structure* refers to the manner in which soil particles are arranged relative to each other and held in their position by forces acting between them. The concept of soil structure also implies the mineral composition of the soil, the electrical properties of the surface of the particles, the physical characteristics and ionic composition of the pore water, moreover, the interactions among the solid particles, the pore water and the adsorption complex. Besides structure, present research attributes more and more importance to the fabric, i.e. the orientation of the particles, of the soil.

This chapter is devoted to the study of those questions of soil structure which are essential to the understanding and correct application of such important engineering properties of soils as strength, deformation and permeability.

The formation of soil structure is the result of a diversity of forces. In the case of coarse grains, the main factor is the force of gravity, whereas for finer grains the influence of the forces acting at the surface of the particles becomes predominant since the specific surface (i.e. the total surface of the grains per unit weight) increases rapidly as the grain size decreases. This is the reason why the behaviour of a given soil depends chiefly on the properties of the fine fractions present in it (cf. Fig. 32). The surface forces are partly of a physical and electrical, partly of a chemical nature. Their magnitude is a function of numerous factors. Lastly, water is to be mentioned as a factor with a decisive role in the formation of soil structure.

3.2 Structure of gravels and sands

These coarse-grained deposits form a so-called single-grained structure. This term refers to a simple packing of individual grains that form separate units with no tendency to adhere to each other. Each grain is in touch with the adjacent grains at several points, but has no attraction or adhesion for them at the points of contact. The only force determining the formation of this type of structure is the force of gravity, i.e. the weights of the individual grains. These soils consist of rounded or subangular grains which bear a fairly close resemblance to the spherical shape, and, indeed, their porosity in the natural state ranges between the limits theoretically derived for the ideal packing of equal-sized spheres. I.e. their porosity seldom exceeds $n = 50\%$ or is less than 25%. Each grain is in contact with several neighbouring grains, and in the skeleton thus formed any load applied to the assemblage is transmitted from grain to grain. Therefore, the compressibility of the soil is relatively low. The formation of the single-grained structure is illustrated in *Fig. 79.* The free placement of a grain deposited from water or air is hampered by surface friction only. As long as the resultant force acting at the minute

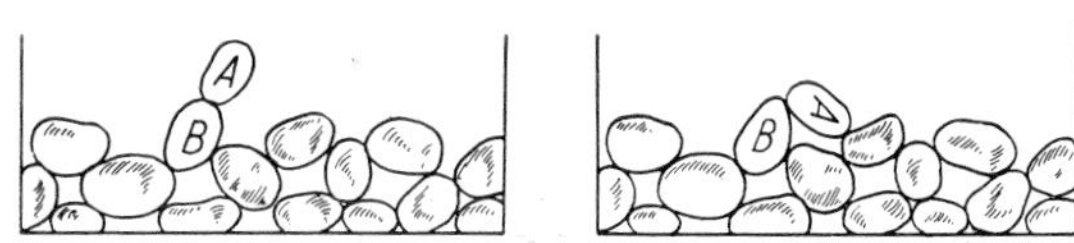

Fig. 79. Formation of single-grained structure

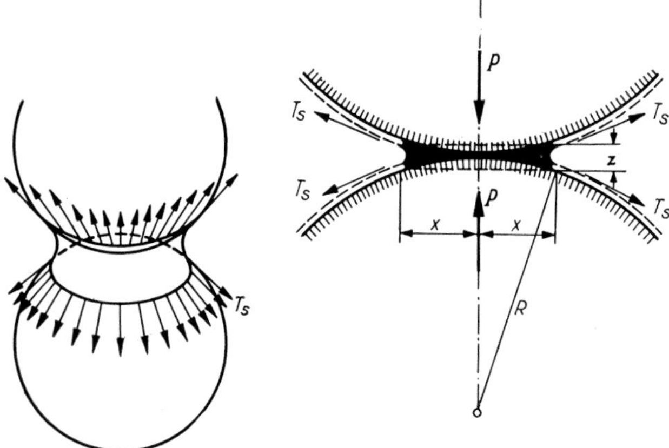

Fig. 80. Effect of surface tension between two spheres

surface of contact makes an angle with the normal to that surface exceeding the angle of surface friction, the grain continues to slip until, eventually, it comes to rest in some random position on the lower grains. The magnitude of the frictional resistance depends on the normal contact pressure, and the ratio between them remains unaltered as long as static conditions prevail. Vibration or dynamic effects may bring the grains into a pulsating movement causing them to move alternately closer to and farther from another. The total area of the momentary contact surfaces may thus be radically decreased, and the influence of the friction substantially reduced. That is the reason why granular soils can be readily compacted by vibration.

The above statements apply to dry or saturated sands and gravels. If the degree of saturation has an intermediate value between zero and unity, the influence of the surface tension of the water also comes into play.

In the three-phase state, surface tension causes the grains of a sand or gravel to adhere to one another. Let us consider two spherical grains in contact, whose contact point is surrounded with capillary water called contact moisture (*Fig. 80*). The meniscus assumes a curved surface tangential to the spherical surface of the grains; it is along the line of contact where surface tension comes into action.

Let the radius of the wetted perimeter be denoted by x, and the perpendicular distance measured between the lines of contact (i.e. the chord length of the meniscus) by z. Considering that x is small relative to r and so is z relative to x, the radius of curvature of the meniscus can be approximately obtained as

$$\frac{z}{z} \approx \varrho \approx \frac{x^2}{2r}.$$

By substituting this in Eq. (18) and taking the opposite signs of the two radii of curvature into consideration, we have

$$p = T_s\left(\frac{1}{\varrho} - \frac{1}{x}\right) \approx \frac{T_s}{\varrho} = \frac{2T_s r}{x^2}.$$

The force of adhesion between the two grains is

$$P = \pi x^2 p + 2\pi x T_s = 2\pi T_s r + 2\pi T_s x.$$

Taking the limiting value

$$\lim P = 2\pi T_s r. \tag{50}$$

This force is known as the contact pressure.

This is the force giving rise to the phenomenon called *apparent cohesion*, which means that sand may acquire compressive and tensile strength as a result of interparticle adhesion due to capillary action. This phenomenon may be observed in sands in the three-phase state only; if the pores become fully saturated with water or else the sand dries out completely, the meniscus disappears and the adhesion between the particles ceases to exist.

By using Eq. (50), let us calculate the apparent cohesion for an assemblage of grains with a uniform diameter of $d = 0.1$ mm. Assuming equal spheres arranged in the type of packing shown in Fig. 54, the number of contact points within an area of 1 cm² is $1\ \text{cm}^2/0.01^2 = 10^4$. Hence the limiting value of the adhesion per unit area is

$$A = \frac{10^4 \times 2\pi \times 0.05\ \text{cm} \times 7.5 \times 10^{-4}\ \text{N/cm}}{1\ \text{cm}^2} =$$
$$= 2.4\ \text{N/cm}^2.$$

For a uniform grain size of $d = 0.1\ \mu$ A is equivalent to 240 N/cm².

The apparent cohesion offers an explanation for the *bulking* of sands. If a mass of dry sand is moistened and then shovelled or dumped loosely into a heap, its volume increases considerably relative to the dry state. The finer the grains, the greater is the increase in volume. The volume change depends also on the water content, its maximum value being when $w = 5$ to 6%. Somewhat below complete saturation, no such bulking occurs. The difficulties caused by this phenomenon are wellknown to those working with concrete, and in modern practice the batching of aggregates for concrete mixes is therefore always done by weight and not by volume.

It was mentioned in Section 2.1.5 that the distribution of the grains according to their size cannot be adequately characterized by means of the usual grain-size distribution curve. Much better information could be obtained by determining the volume of the grains and plotting their distribution according to the volume (*Fig. 81*). The size of the pores between the solid grains, the forms in which water and air are present in the soil and the proportion in which they fill the pores, must also be considered for an understanding of soil behaviour.

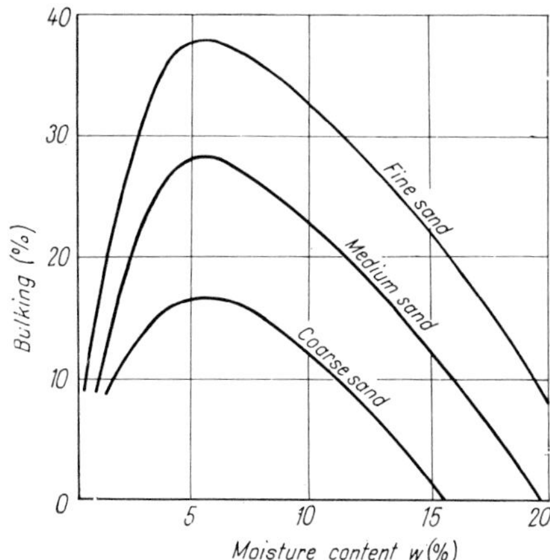

Fig. 81. Bulking of sand

Let us consider first the dry state, i.e. a two-phase system. As is well known, if two substances are mixed in the molten state then, owing to the molecular interaction of the two constituent parts, a solution is formed. The natural attraction of the molecules of like character decreases, and the temperature-dependent dispersion force increases, with the result that the melting point of the mixture is invariably lower than that of either of the pure constituent parts. Plotting the melting temperature against the mixing ratio gives the diagram shown in *Fig. 82*. Point *E* represents the composition with the lowest melting point. This is called the *eutectic point*.

Soils are made up of components present in varying proportions by weight in granular assemblages, and temperature, which is the measure of the energy of a multi-component system is also,

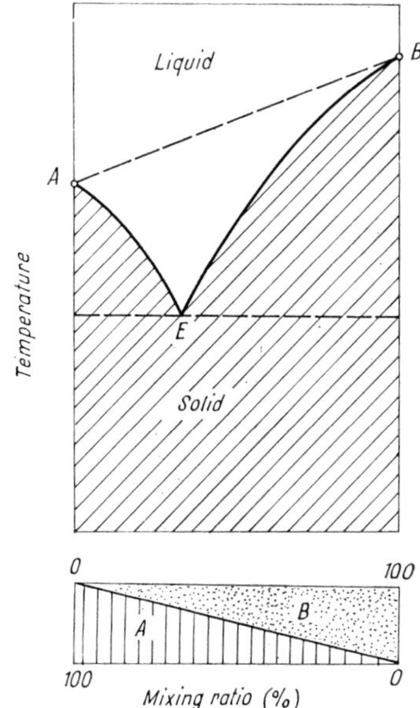

Fig. 82. Phase diagram of a mixture; melting temperature versus mixing ratio

at a given pressure, a measure of its volume (or, more exactly, a quantity in a functional relationship with it). Hence the analogy between molecular liquids and granular media—called here *macromeritic liquids* —is obvious. The greater the void ratio of the system, the greater is its free energy; thus the volume of the system is determined by the void ratio. By analogy, the void ratio plays the same part in macromeritic liquids as the temperature in molecular liquids. Conversely, the temperature of molecular liquids may be regarded as a measure of their "void ratio".

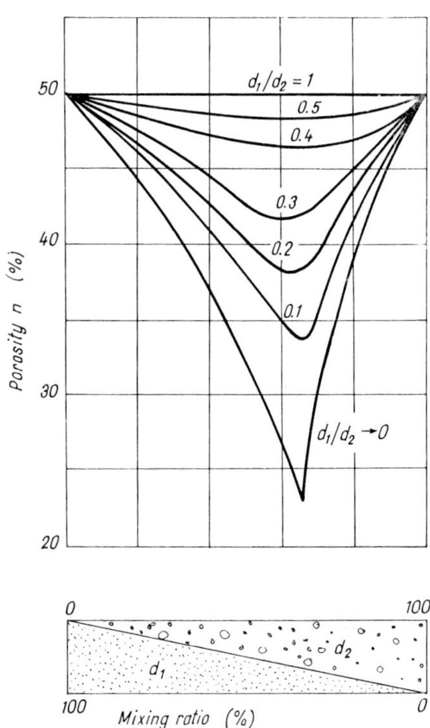

Fig. 83. Void ratio of mixtures of two components (granular soils) versus mixing ratio

There have been numerous investigations aimed at clarifying the grain-size/density relations of two-component systems. A study of the mixtures of two different grain assemblages, each by itself composed of equal-sized spheres, showed that the porosity of the mixtures, after having been subjected to constant compactive effort, varied with the mixing ratio exactly in the same manner as melting temperature varies with the concentration of the components. For a system with a given grain-size ratio, there is a certain composition that yields the lowest porosity. The shape of the diagram showing porosities against the mixing ratio is essentially the same as that of the phase diagram shown in Fig. 82.

Figure 83 shows the result of such tests, d_1 and d_2 being the diameters of the components (uniform spherical grains). The ratio $d_1/d_2 = 0$

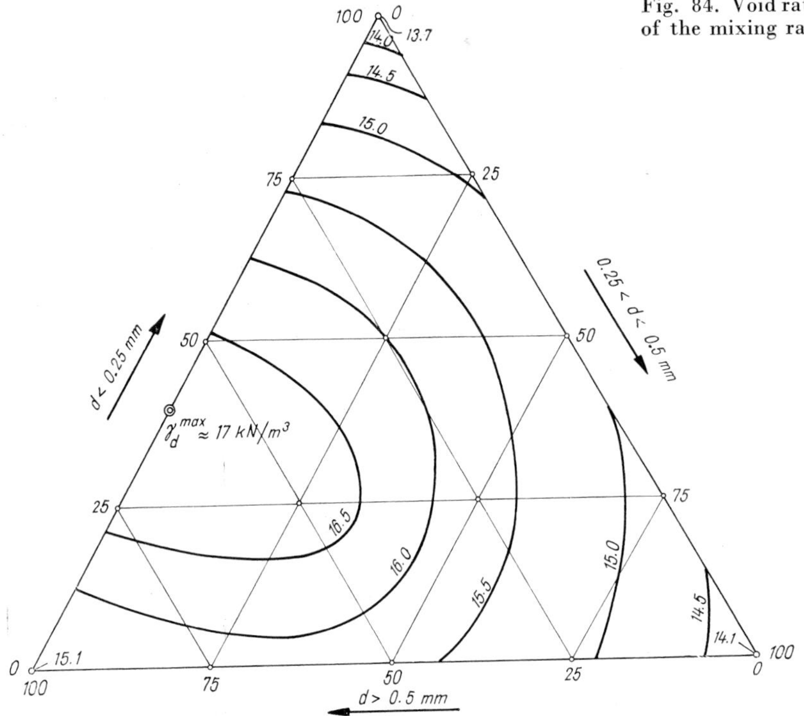

Fig. 84. Void ratio of a three-component system as a function of the mixing ratio

refers to the mixture in which the voids of the relatively coarse components (sand) are filled with small clay particles. The normal porosity for the separate components is $n = 50\%$; the greatest density occurs, for each value of d_1/d_2, at an intermediate mixing ratio. The porosity is always smaller than that of either of the components separately.

A three-component system, as stated by Findley (1951), possesses two degrees of freedom, and the state of the system is dependent on the relative concentration of the components and on the porosity (temperature). Taking three different sizes of particles and determining the void ratios of their mixtures at different compositions, but all subjected to the same compactive effort, gives a set of diagrams like those in *Fig. 84*. This diagram is the counterpart of the phase diagram for mixtures of three different liquids.

Interesting test results are presented in *Fig. 85* (V. Domján, 1965). The maximum and minimum void ratios for mixtures of sand grains of three different sizes are plotted in triangular diagrams as functions of the composition of the mixtures (Figs 85*a* and *b*). In diagram *c*, the possible range of variation in void ratio ($e_{max} - e_{min}$) is shown, again as a function of the relative proportions of the components.

If it is required to produce an artificial soil mixture with good compactibility for some practical purpose (e.g. for earth roads, subgrades of road pavements, earth dams) it is worthwhile to carry out a full set of compaction tests with various mixes of the available granular types of soils in order to find the optimum mixing ratio.

The void ratios belonging to the loosest and densest state of the soil do not, of course, provide the complete picture of the arrangement of the

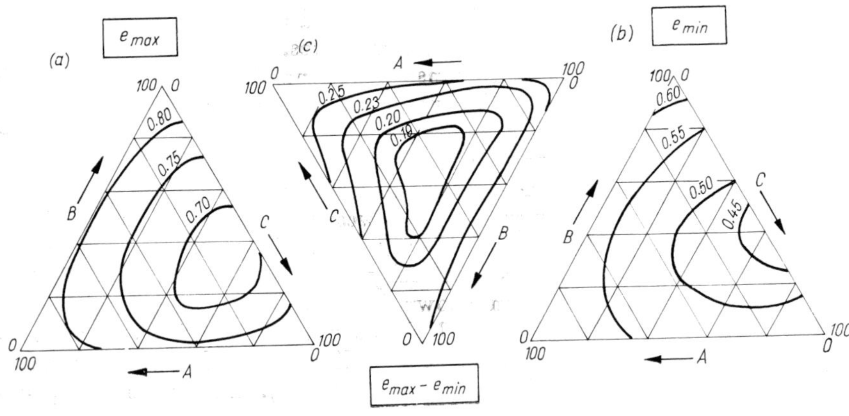

Fig. 85. Maximum (*a*) and minimum (*b*) void ratios of three-component mixtures as functions of the mixing ratio; (*c*) — the range of variation in void ratio of the mixtures

particles, i.e. the true structural pattern of the soil. In fact, besides the distribution, by size or by volume, of the grains it is rather more important to have a clear idea of the shape and size of the voids between the grains, and of the manner in which water and air are distributed within the voids. As the relative volume of the solids, *s*, reveals nothing about the grain-size distribution, likewise the values of *m* and *g*, by themselves, do not characterize the arrangement of the particles in relation to one another and the structural make-up of the constituent phases. Soils described by the same (*s*, *m*, *g*) values may possess very diverse structures and exhibit quite different behaviour.

Let us consider first the two-phase solids–air system. In the case of a coarse-grained material the shape characteristics and the distribution, by volume, of both the solid grains and the voids can be studied by the following test. The volume distribution of the solid grains is to be determined first, in the way described in Ch. 2. Then, in both the loosest and the closest packing of the assemblage, the voids are filled with molten paraffin wax. After the wax has solidified the saturated blocks are broken up, the angular particles of wax filling the voids separated, and their volumes measured. Typical void formations are shown in *Fig. 86*. The results of the test are plotted in volume distribution curves. Actual tests with coarse Danube-gravel and crushed limestone are presented in *Fig. 87*. From the difference between the curves for to the loose and the dense states of the soil, the voids most reduced in size can be determined.

In the three-phase state of granular soils, the particles of the constituent phases may be arranged into various structural patterns. Typical examples are shown in *Fig. 88*. In case *a*, the solid grains and the voids are uniformly dispersed. This type of structure is characteristic of granular soils in the two-phase state (*solids–air* or *solids–water*), when the small normal forces and frictional resistance having their seat at the points of contact are the only forces between the grains.

If the voids are filled mostly with air, only a small amount of water being present, the structure shown in Fig. 88*b* is likely to occur. Contact pressure due to the capillary action of the water ring around

Fig. 86. Characteristic void shapes in granular system:

a — packing of uniform spheres; *b* — fairly uniform particles; *c* — packing of mixed grains

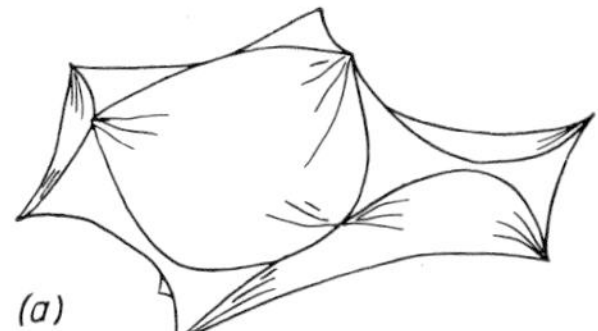

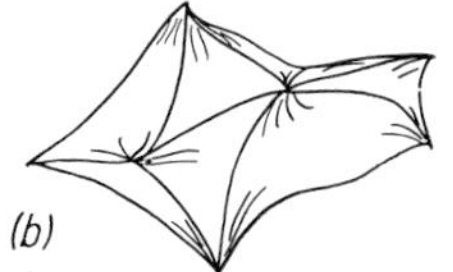

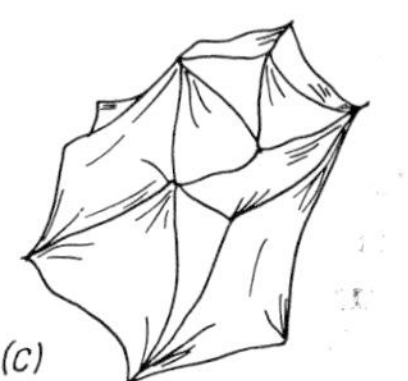

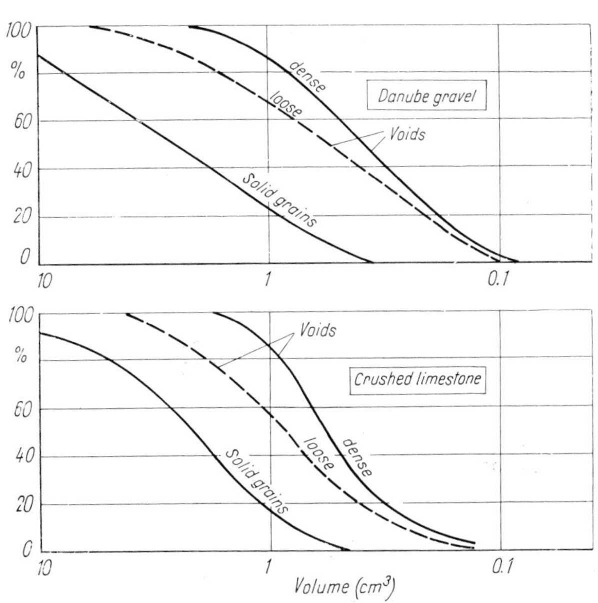

Fig. 87. Volume-size distribution curves for solid particles and voids

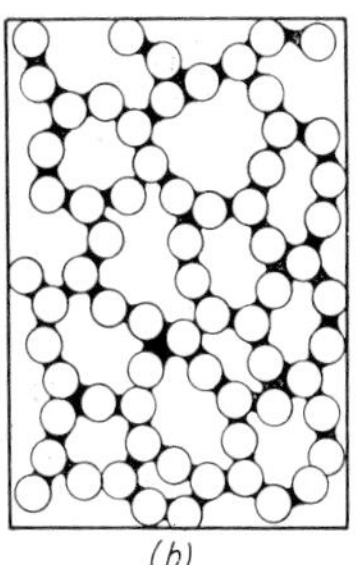

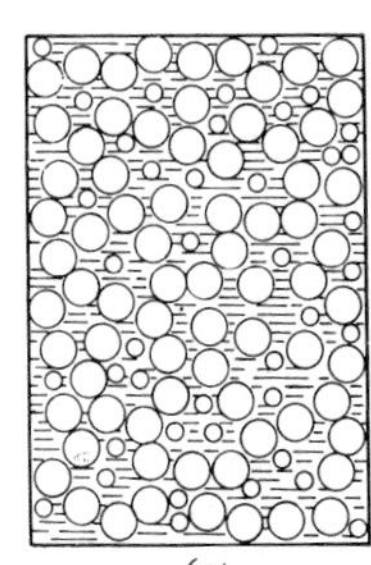

Fig. 88. Possible structures of granular systems:

a — fairly uniform dispersion of the solid particles; the voids are continuously filled with either water or air; *b* — the coarse particles are unevenly distributed, the volume of the voids is variable. Water is located in the corners between the particles; *c* — the voids are filled with water, the air occurs in fully enclosed tiny bubbles

the points of contact makes it possible for the grains to be stacked into a very loose yet stable arrangement enclosing large cavities several times the size of the grains. (Cf. bulking of damp sand, Fig. 81.) Such a soil is pervious to air since the pores filled with air form a continuous network of "channels".

As the moisture content increases, the surface tension acting at the air/water interface will contract the pore air into tiny globules with the result that water becomes the continuous "matrix" holding the particles of the other two phases embedded (Fig. 88*c*).

As far as three-phase soils are concerned, only the volume distribution curves for each of the separate phases would provide a true picture of the phase composition of the soil. This concept is illustrated by the curves in *Fig. 89*. By comparing the curves, certain critical values of saturation could also be determined. The actual performance of such tests, however, would be possible on very rough assumptions only, and this sets limitations to the wider use of the method.

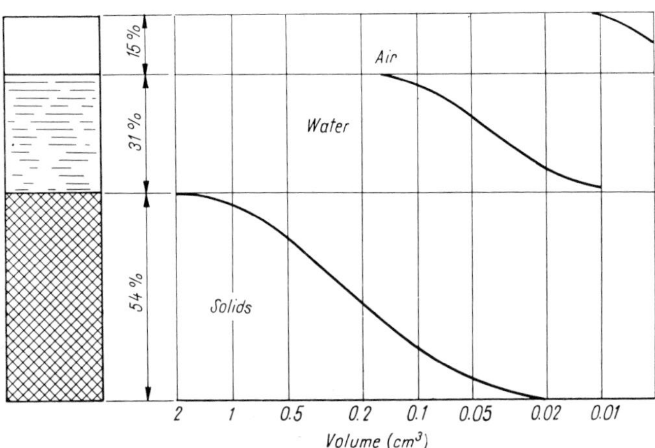

Fig. 89. Volume-size distribution curves for all phases (solid particles, water and air) in a dispersed system

The knowledge of soil structure is of primary importance in the study of movement of water and air through the soil (Ch. 6).

3.3 Properties of fine soil particles

The behaviour of assemblages composed of very fine particles is controlled by forces and physical effects associated with the surface of the particles, because of the high specific surface of this fraction. In order to understand the nature and the practical implications of these phenomena, it is necessary to discuss briefly the atomic structure of matter and to get acquainted with a few basic concepts and laws of physical chemistry and colloid chemistry.

3.3.1 Elementary particles and structure of atoms

The chemical elements of the rock-forming minerals are composed of atoms, the ultimate units with discrete chemical identity. An atom is made up of a positively charged nucleus around which rotate a definite number of negatively charged electrons. The nucleus in turn consists of positively charged protons and uncharged neutrons. The number of electrons in orbit is equal to that of the protons in the nucleus. The electrons rotate in orbits of different radii forming

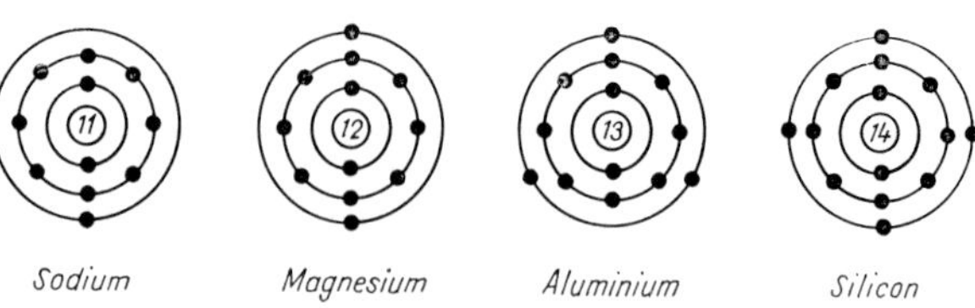

Fig. 90. Structure of the electron shell of a few elements. Full circles are the electrons; numbers in the kernel give the number of protons

so-called electron shells. Each shell accomodates a permitted number of electrons; there is "*space*" for two electrons in the innermost orbit, and for 8, 18 and 32 electrons successively in the outer orbits. The mutual electrical attraction between the protons and the electrons is balanced by the centrifugal force of the electrons. The chemical properties of an element are determined by the number of the protons in the nucleus and the number of the outermost or valency electrons. The electron shell models for some elements commonly occurring in soil-forming minerals are shown in *Fig. 90*. An electron shell comprising electrons in the highest allowed number is said to be *saturated;* and the shell is called *complete* if the number of electrons is *eight*.

The size of the nucleus is of the order of 10^{-11} mm, and that of the whole atom 10^{-7} mm = = 0.1 mμ = 1 Å (Ångström). Thus, atoms cannot be made discernible even under the electron-microscope, whose resolving power is approximately 4 Å.

Atoms may combine with other atoms, or may be transformed into ions by loosing or gaining electrons. Saturated electron shells may be formed in two ways: first, by the sharing of electrons by adjacent atoms, which thus form electrically neutral molecules; secondly, by the gain or loss of electrons, in which case the products, called *ions*, carry electric charges. The positively charged ions are called *cations*, and those carrying negative charges are *anions*.

The forces which bind individual atoms into molecules, or link one molecule with others to build up the structure of substances, are primarily of electrical nature. When the force is due to the electrostatic attraction of unlike charges, an *ionic*

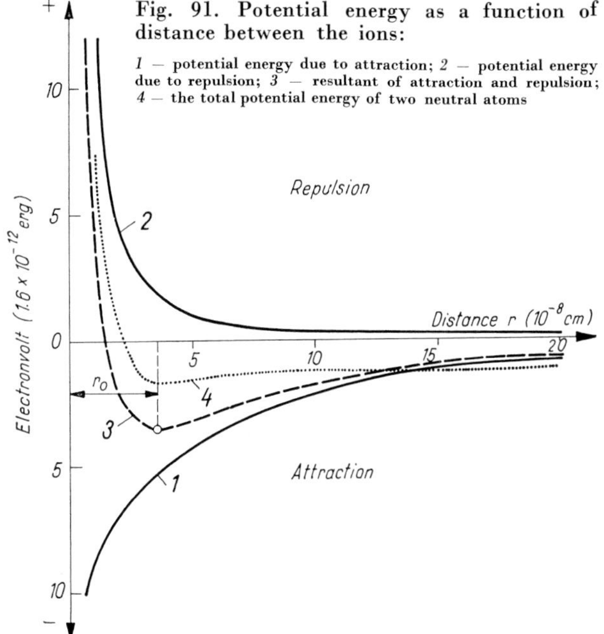

Fig. 91. Potential energy as a function of distance between the ions:

1 — potential energy due to attraction; *2* — potential energy due to repulsion; *3* — resultant of attraction and repulsion; *4* — the total potential energy of two neutral atoms

bond is formed. Another type of bond which involves the sharing of a pair of electrons between two atoms is called a ***covalent bond***. The attractive forces between a swarm of electrons dissociated from atoms and the positive charges of the atoms produce ***metallic bonds***. Finally, non-symmetrical arrangements of the electrical charges within a molecule give rise to the ***van der Waals forces***.

Let us now examine more closely those types of bond which are of importance for soil mechanics.

Ionic or heteropolar bonds are formed between oppositely charged ions. The attraction between cations and anions is directly proportional to the magnitude of their charges and inversely proportional to the square of the distance between them. The resulting value is negative, since it requires work to force the ions apart. The potential energy due to attraction is inversely proportional to the distance of the ions (curve *1* in *Fig. 91*). Besides attraction, repulsive forces are also developed because the outer electrons of two adjacent ions tend to repell each other. This repulsion prevents the interpenetration of the electron shells of adjacent atoms. The force of repulsion is inversely proportional to a higher power of the distance between the ions. The resulting potential energy is positive, and it diminishes with the distance according to curve *2* in Fig. 91. The net potential energy of the two ions is equivalent to the algebraic sum of the forces of attraction and repulsion: this is presented in curve *3*. The force of attraction is largest, i.e. potential energy is at a minimum, at a finite small distance r_0. For the sake of comparison, the variation of the total potential energy of two neutral atoms with their distance is given in curve *4*. It is seen that the potential energy is always greater than in the case of ionic bond, with the consequence that the ionic bond is more stable.

Atoms held together by ionic bonds form ***ionic compounds***. For example, common salt and the majority of inorganic crystalls fall into this group. In the ionic form of crystal structure, positively charged cations (mostly metallic in character) and negatively charged anions are arranged in a regular lattice pattern. The structural make-up of the crystals depends on the size of the ions. Under the action of electrostatic attractive forces, the ions tend to occupy an arrangement in which each ion is surrounded by the greatest possible number of neighbouring ions. Ionic compounds in the solid state do not conduct electricity and their melting point is high. In the molten state or in aqueous solutions they are good conductors.

If the elementary particles which come into contact are not electrically charged ions, but neutral atoms, they are not capable of bonding each others' electrons. In this case, the chemical bond is due to forces of different character; two adjacent atoms of a molecule both donate one electron to make both electron shells complete. In other words, these valency electrons belong

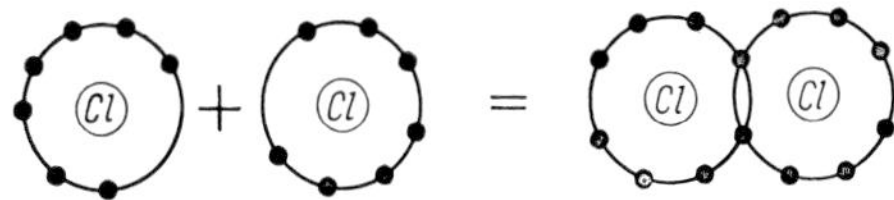

Fig. 92. Formation of covalent bond

simultaneously to both atoms. For example when two Cl atoms, each of them having seven electrons (i.e. one less than a complete shell) in its outermost shell, combine into a Cl_2 molecule, a pair of electrons, one from each atom, is involved in rising the number of the electrons in the outer shell of both atoms to eight (*Fig. 92*). This type of bond is called a ***covalent*** or ***homopolar*** bond. Compounds in covalent bond possess low melting point and do not conduct electricity either in the solid or in the liquid state.

The ionic bond and the covalent bond represent two extreme cases between which several tran-

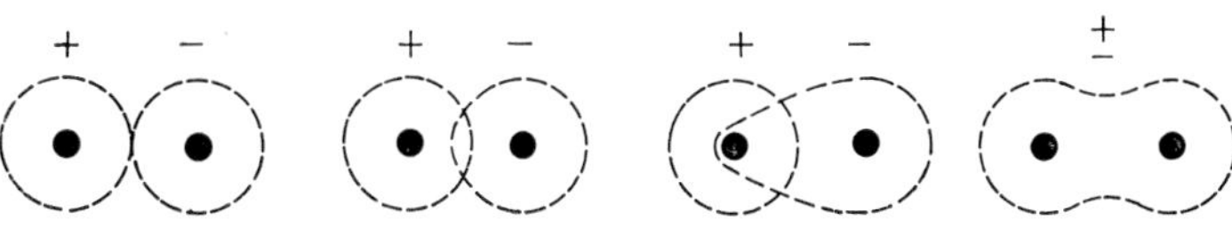

Fig. 93. Ionic, covalent and transitions bonds

sitional types of bond are possible. Examples of these are shown in *Fig. 93*.

The original true spherical shape of the electron shells of the atoms involved is, as a rule, deformed in the molecule. This phenomenon can be accounted for by the polarization effect, and is more pronounced the greater the charge of the cation, and the greater the ion radius of the anion. The outermost electrons of the anion may get near enough to the nucleus of the cation to be forced, by its greater attraction, into an orbit which is greatly elongated in the direction of the cation. In the end, the distortion may reach the extent when the path of the electrons of the anion includes the cation itself, and a covalent bond results.

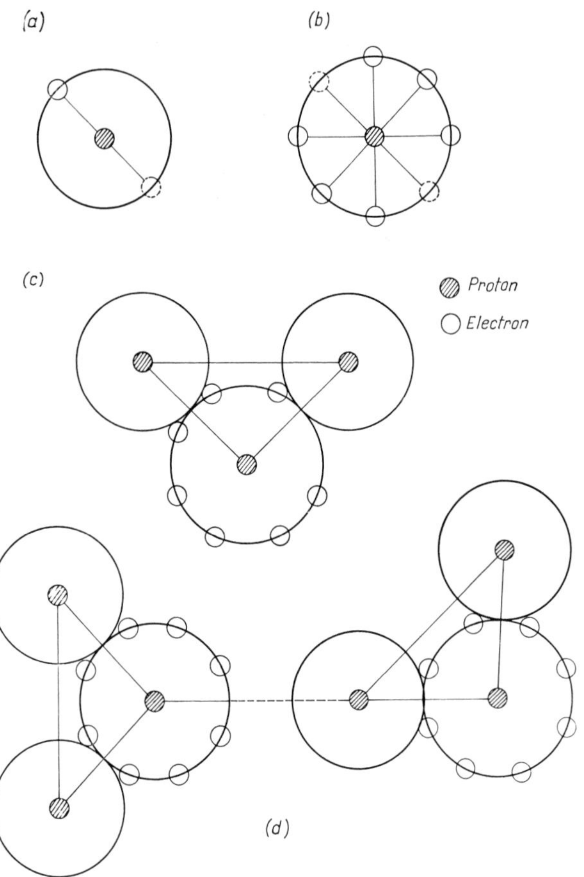

Fig. 94. Development of hydrogen bonds. *a* — hydrogen atom consisting of one positively charged proton and one negatively charged electron. The oxygen atom (*b*) has eight electrons, six of which are arranged in an outer shell. The shell of H has room for one more electron (dashed circle) and the outer shell of O has room for two more electrons; therefore, the atoms have an affinity for each other. In the water molecule (*c*; cf. Fig. 39a) the electrons of the hydrogen atoms are shared by the oxygen atom. The positively charged proton of the H atom now sticks out from the H_2O molecule, and it has an attraction for the negatively charged electrons of a neighbouring water molecule (*d*). This relatively weak force (dashed line) is called a hydrogen bond. Each H atom belongs, therefore, to an O atom, forming a H_2O molecule, and is linked to another O atom by the relatively weak hydrogen bond. Each water molecule is, therefore, linked to its neighbour by four hydrogen bonds. However, if the number of hydrogen bond per molecule were counted, there would be only two instead of four, since each H bond is shared between two molecules. Thus, the energy of one mole of hydrogen bonds is half of the energy required to vaporize one mole (18 g) of water, i.e. 6000 calories

The types of bond so far discussed are called *primary valence bonds*. They combine the atoms into molecules. The forces that link atoms in one molecule to atoms in another, produce secondary valence bonds. These are much weaker then the primary valence bonds. The secondary valence bonds are of two kinds: *van der Waals forces* and *hydrogen bonds*.

The van der Waals forces exist between molecules and are made up of three components: first, the force of attraction between the oppositely charged ends of permanent dipoles (orientation effect); secondly, the attractive force between permanent dipoles and dipoles induced by these in adjacent, originally non-polar, molecules (induction effect); thirdly, interaction between instantaneous, fluctuating dipoles due to the constant oscillation of the electrons (dispersion effect).

The contributions of these three effects to the van der Waals forces are as follows:

orientation	77%
dispersion	19%
induction	4%.

The hydrogen atom is capable of forming but one primary valence bond. When this is formed between hydrogen and oxygen, the oxygen atom appears to accumulate an excess of electronic charges and the hydrogen with a deficiency of electrons appears to have a positive charge. The positively charged hydrogen is capable of attracting an oxygen atom from another molecule thus forming a *hydrogen bond* which is a weak bond compared to the primary valence bond. The hydrogen atom in this bond lies between the two oxygen atoms, but not at equal distances from the two. It is much closer to the oxygen atom to which it may be considered to belong (*Fig. 94*). Water is a liquid with a structure such that the molecules are linked by hydrogen bonds so that they are not free to rotate. The relative strengths of the various kinds of bond are approximately the following

van der Waals forces	1 to 10
hydrogen bond	10 to 20
ionic and covalent bonds	40 to 400

The fields of force between electrically charged soil particles, the water surrounding them and the ions adsorbed at the surface have a decisive influence on the behaviour of fine-grained soils. Before we proceed to the study of these interactions, we have to deal with the individual mineral particles, those building elements, which fundamentally determine the characteristics of the fine-grained soils.

Fig. 95

a — tetrahedral unit of clay minerals; *b* — silica sheet; *c* — symbols; *d* — building-block symbol

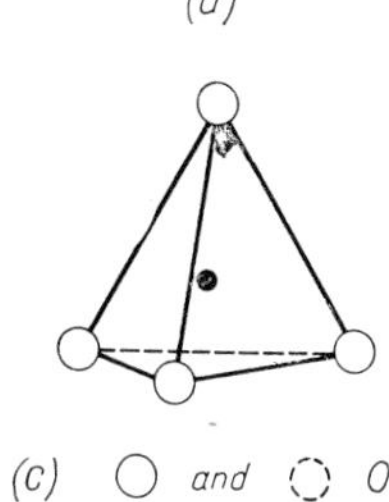

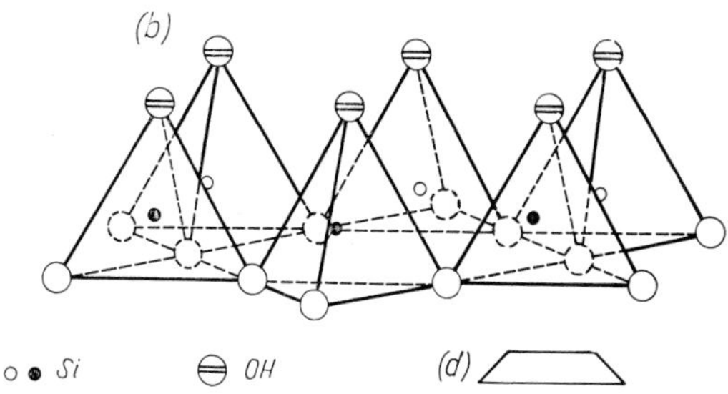

(c) ○ and ◌ O ○ ● Si ⊖ OH (d)

3.3.2 Structure of clay minerals

The crystals and rock fragments that form the grains of coarse-grained soils, sand and gravel, consist of hard materials whose atoms are held together by strong bonds so that the strength of the mineral particles themselves is usually of little concern. The surface activity of these grains is low owing to their strong inner structure and small specific surface. As a consequence, both the gross and the specific effects of the surface activity of the grains are slight. This accounts for the lack of plasticity. Minerals which fall into this group are quartz, calcite, muscovite, biotite and feldspars. They form the bulk of the grains of inorganic silts, and occur in some varying proportions in clays too. Their crystals exist in an atomic lattice form, with one neutral atom at each corner of the lattice, held together by covalent bonds, i.e. some electrons from the outer shells are shared by adjacent atoms. Such crystals make hard substances, melt only at high temperature, and do not conduct electricity.

Another group of rock-forming minerals includes the *clay minerals*. The characteristic properties of these minerals are high surface activity and high specific surface. The behaviour of assemblages composed of such minerals is governed largely by surface phenomena. It is therefore necessary to gain an insight into the atomic structure of these minerals.

The term clay is applied to the fraction of grains whose equivalent diameter is less than 0.002 mm. The individual grains are fragments of a single mineral, i.e. a solid compound with definite chemical composition and unique crystalline structure. The overwhelming majority of clays that occur in nature is made up of only a few minerals known by the group name *clay minerals*. Up to quite recently they were thought to be composed of amorphous gels and crystalline zeolites. It was only the application of X-ray analysis to the investigation of clays (Hendricks and Fry, 1930) that revealed the crystalline structure of the clay minerals.

A common character of all clay minerals is their poor crystallizing power. They, therefore, occur in the form of extremely small crystals, which exhibit a low degree of symmetry. A soil is often composed of a mixture of several compounds related in structure but differing in chemical composition. In the following, we shall discuss the structures of the most common clay minerals, together with some related facts which are of importance from a technical point of view.

The minerals composing clays were formed by the weathering of rocks. The igneous rocks consist, to the extent of 75 to 80%, of oxygen, silicon and aluminium. A further five elements, iron, calcium, sodium, potassium and magnesium, compose about 16%. These elements combine into minerals. More than 90% of all minerals fall into a few groups.

According to Vendl (1951) the constituent minerals of clays may be divided into four main groups. The most important, as regards physical behaviour, are those minerals which are newly formed by the chemical decomposition of the parent material; these are the genuine clay minerals. The second group includes those minerals which formed the constituent parts of the aggregate of the parent rock. In the course of weathering they underwent physical disintegration only, but resisted chemical decomposition. Quartz, feldspar, micas and some other rare minerals belong to this group. The minerals of the third group were formed during or after the process of the sedimentation of the clay as e.g. pyrites, dolomite, glauconite. Finally, the minerals in the last group consist of

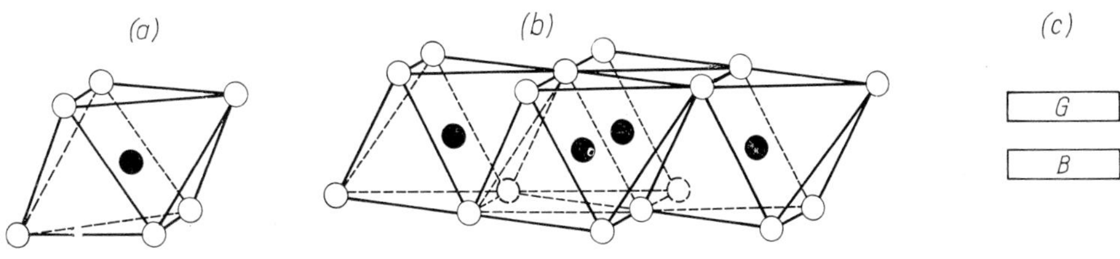

Fig. 96. Octahedral unit of clay minerals:

a — basic unit and symbols; *b* — octahedral sheet; *c* — building-block symbol (gibbsite and brucite)

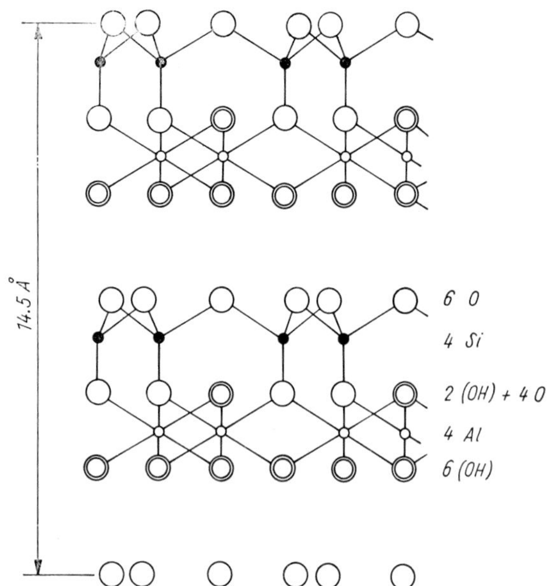

Fig. 97. Structure of kaolinite composed of the elements given in Figs 95 and 96

the solid fragments of the inorganic skeleton of plants and organisms.

The atomic structures of clay minerals are built up of two fundamental building blocks. One is a silica tetrahedral unit (*Fig. 95a*), in which four oxygen atoms arranged in the form of a tetrahedron enclose a silicon atom positioned at the centre. The units combine into a sheet so that the free tips of all the tetrahedra point in the same direction (Fig. 95*b*). Each oxygen atom is shared by two adjacent tetrahedra. The tip of each unit is occupied by a hydroxyl radical OH^-, so that a sheet consists of a layer of oxygen and a layer of hydroxyl with a layer of silicon between them. A simplified representation of the sheet structure is shown in Fig. 95*d*.

The second building block is an octahedral unit in which an Al, Fe, or Mg atom is enclosed by six OH^- radicals arranged in the form of an octahedron (*Fig. 96a*). Again the units are combined in such a manner as to form a sheet (Fig. 96*b*). The atom at the centre may be aluminium, and the resulting formation is then the *gibbsite* structure, or magnesium, forming the *brucite* structure.

The structure of *kaolinite* is shown in *Fig 97*. Its chemical composition is $Al_4Si_4O_{10}OH_6$. The atoms are held together by ionic bonds. Silicon and oxygen atoms form an octahedral sheet with attached $Al(OH)_6$ units. The charges of the silica tetrahedral sheet are almost completely balanced within the structure by the Al^{3+} cations and the OH^{-1} anions, thus kaolinite has no unbalanced charges at the surface and cannot absorb ions except along the edges of its flakes.

Kaolinite is a friable earthy material which becomes lightly plastic on wetting. When heated to a temperature of about 500 to 550 °C it gives off water which originates from the OH^- radicals. Its structure contains no lattice water.

Figure 98 shows a diagrammatic sketch of the structure of a typical kaolinite mineral. Successive unit sheets are held together by hydrogen bonding, which is, as shown in the previous section, a relatively strong bond. This is the reason why kaolinite, although composed of layers stacked one above the other, does not cleave readily along the planes between the layers. The kaolinite structure is stable with no interlayering of water molecules between the units. As a consequence, kaolite exhibits little swelling and shrinking.

Another important mineral in the kaolinite group is *halloysite* $[Al_4(Si_4O_{10})(OH)_8\,4H_2O]$. A unit layer in its structure is composed of one alumina octahedral sheet and one silica tetrahedral sheet in which the tips of the tetrahedra point alternately in opposite directions. The distance between the unit layers is thus slightly increased. At the tips of the inverted tetrahedra the O^{2+} ions are replaced by OH^- radicals which hydrate easily.

In the second group of clay minerals the most prominent member is *montmorillonite*. Its chemical composition, if the cations Ca and Mg which occasionally occur between the unit layers are also considered elements of the structure, is n(Ca, Mg)

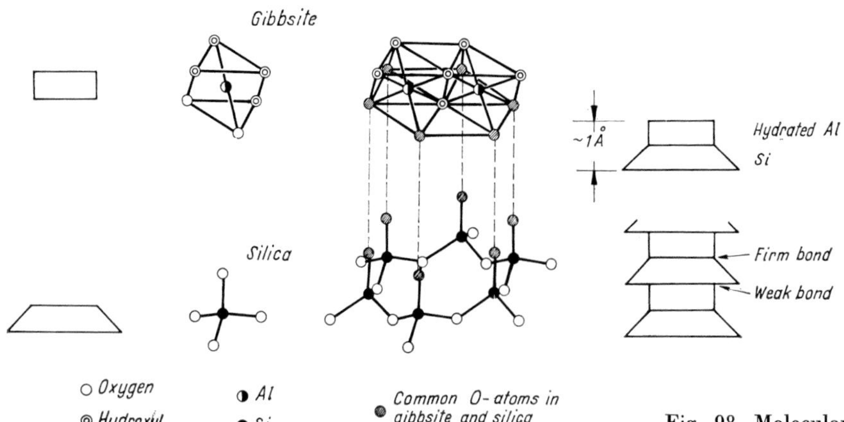

Fig. 98. Molecular structure of silica and gibbsite

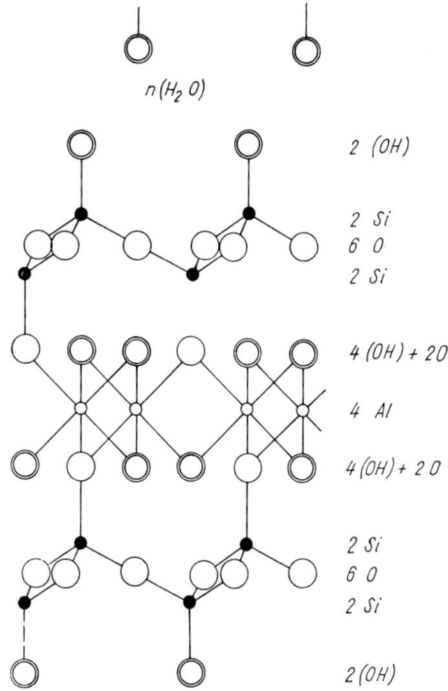

Fig. 99. Molecular structure of montmorillonite (HOFFMANN and ENDELL, 1934)

$O \cdot Al_2O_3 4SiO_2 \cdot H_2O + xH_2O$. Its lattice structure is made up of two silica tetrahedral sheets with an octahedral sheet of hydrargillite [aluminium hydroxide, $Al(OH)_3$] between them. In the hydrargillite sheet magnesium, iron, etc. may be substituted for aluminium, this giving various members of the montmorillonite group such as ***saponite*** (MgO approx. 25%), ***hectorite*** (MgO $\sim$ 20%, $Li_2O \sim 3\%$) and ***nontronite*** ($Fe_2O_3 \sim 30\%$).

The lattice structure of montmorillonite, as suggested by HOFFMANN and ENDELL (1934), is shown in *Fig. 99*. In both silica tetratedral sheets, only every second tetrahedron is linked to the central octahedral sheet, while the intermediate units are directed "outward". The tips of the latter are occupied by OH^- radicals which hydrate readily. The spacing of the unit layers varies according to the number of the molecular water layers between the unit layers (interlayer water). If increasing amounts of water are allowed to enter between the layers, the montmorillonite expands like the accordion. The spacing of the unit layers is influenced also by the chemical composition of the species of the montmorillonite and the occasional occurrence of exchangeable cations between the layers.

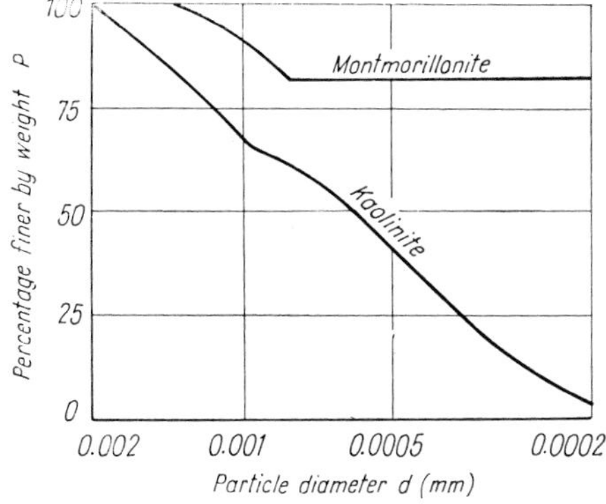

Fig. 100. Grain-size distribution curves of kaolinite and montmorillonite

The expansion of the lattice structure gives rise to substantial volume changes. In a moist state, montmorillonite is highly plastic and has slight internal friction. Its excessive swelling capacity may seriously endanger the stability of overlying structures and road pavements. Slopes cut in montmorillonite-containing clays have a tendency to flow in wet weather. *Bentonite*, which is a clay formed from volcanic ash, is mainly composed of montmorillonite.

Some other members of the montmorillonite group are beidellite, and nontronite (ferri-montmorillonite); these are all similar in their crystal structure to montmorillonite.

The grain-size distribution curves of pure kaolinite and montmorillonite are shown in *Fig. 100* (CAQUOT–KÉRISEL, 1956). In kaolinite, the grain size gradually decreases from 2 μ, and the percentage of grains smaller than 0.2 μ is negligible, whereas in montmorillonite some 80% of all grains is smaller than 0.2 μ.

Glimmerton, a mica-like mineral, has been recognized to be a distinct type of the clay minerals by HOFFMANN and MAEGDEFRAU (1937). It occurs in large quantities in the state of Illinois, US., hence the currently accepted name, *illite*. It is very similar in structure to the micas, and consists of exceptionally small particles. In comparison with the micas, however, illite has a lower content of alkalies and a higher water content.

Some information about the sizes and the proportions of the main dimensions for the commonest clay minerals is given in *Table 10*.

Table 10. Size of typical clay minerals

Mineral	Ratio of dimensions	Diomensions, Å*		Specific surface m² per gram
		length	thickness	
Kaolinite	10× 10×1	1000–2000	100–1000	10
Illite	20× 20×1	1000–5000	50– 500	80
Montmorillonite	100×100×1	1000–5000	10– 50	80

* 1 Å = 10^{-10} m = 10^{-7} mm

3.3.3 Identification of clay minerals

The composition of clays is of great importance, not only in mineralogy, but also from the point of view of industries using clay as a raw material

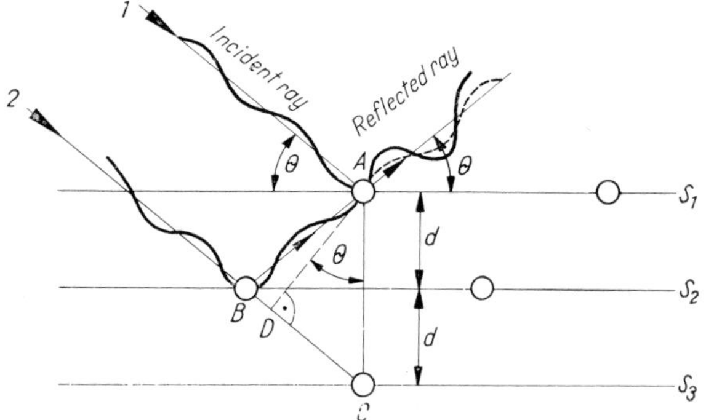

Fig. 101. Impact and interference of X-rays on the layers of crystals

(for example, the ceramics industry). A number of methods have, therefore, been developed for the investigation of clay structure. These methods have also become part of present-day soil mechanics and soil physics investigations, because of the growing recognition that many important engineering properties of clays are directly related to their mineral compositions.

Since 1923, *X-ray analysis* has become the most widely used method for the investigation of clay minerals. The method is based on the fact that X-rays be diffracted by interference effects. X-rays, though of considerably shorter wavelengths, may be diverted from their original direction by letting them pass through a diffracting grating in the same way that interference phenomena are observed when ordinary light waves, in the wavelength range of 400 to 800 mμ are projected onto a suitable optical grating. The spacing of the grating must be of the same order of magnitude as the wavelength of the X-rays. Besides the primary incident rays a secondary scattered radiation (diffracted rays) arises from the atoms exposed to the primary radiation. From these scattering centres, an infinite number of spherical waves enters the surrounding space. By interference these rays, for the most part, cancel each other, but some are enhanced in certain directions, which are determined by the distribution of the atoms. Now if a beam of parallel X-rays is passed through the parallel sheets of a crystal which acts as a diffraction grating, the electrons in the shells of the atoms exposed may be brought to a state, producing secondary radiation. For example, the ray *1* in *Fig. 101*, having interacted with an atom at point A in the plane S_1, has already produced a secondary radiation. The same thing happened to ray *2* except that it struck an obstacle only in the plane S_2. Reflected rays result only in those directions, where diffracted rays meet in phase and therefore intensify each other. The conditions for this depend on the wavelength and the angle of incidence of the primary ray, and on the lattice spacing of the mineral. Reflected rays are obtained, when the path difference of rays *1* and *2* is equal to a multiple of the wavelength, i.e.

$$\overline{BA} - \overline{BD} = \overline{DC} = n\lambda = 2d \sin \theta \, .$$

If the angle of incidence θ and the wavelength λ are known, the spacing d of the reflecting atomic planes can be determined from the above equation. The lattice spacing is characteristic of the mineral, thus, the X-ray diffraction method can be used for the identification of minerals and for the analysis of mixtures of minerals.

Another frequently used method of clay mineral investigation is the *differential thermal analysis* (DTA). It was first employed by Le Châtelier (1904). Essentially, the method consists in the gradual and continuous heating of the test sample to an elevated temperature, and examining the character of the endothermic and exothermic reactions which may take place during the process. Since the thermal effects to be recorded are very small, differential thermocouples are used. One junction of the differential thermocouple is fixed on the metallic specimen holder, and the other junction on the holder containing an inert material (i.e. one which shows no thermal reaction when heated). A galvanometer connected to the differential thermocouple circuit indicates electromotive force only when there is a temperature difference between the junctions. The small electromotive force is usually amplified before it is recorded by means of a sensitive galvanometer (*Fig. 102*). Apparatus of up-to-date design is usually provided with automatic recording devices which furnish continuous records. Platinum–platinum rhodium or platinum rhodium thermocouples are the most widely used types today. The inert material is calcined aluminium oxide. The specimen holder and the inert material, together with the thermocouples, are placed in an electric furnace in which the temperature is raised continuously at a constant rate of 10 to 15 °C per minute. As long as there is no difference between the temperature of the test sample and that of the inert

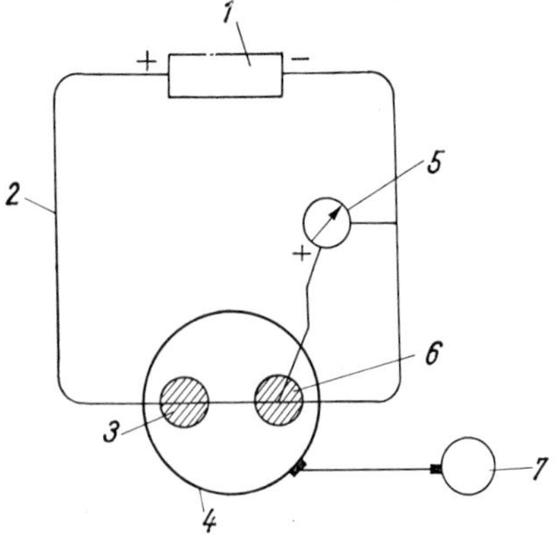

Fig. 102. Differential thermal analysis-device:

1 — galvanometer; *2* — platinum wire; *3* — melting pot with sample; *4* — electric furnace; *5* — millivoltmeter; *6* — melting crucible for inert material; *7* — temperature controller

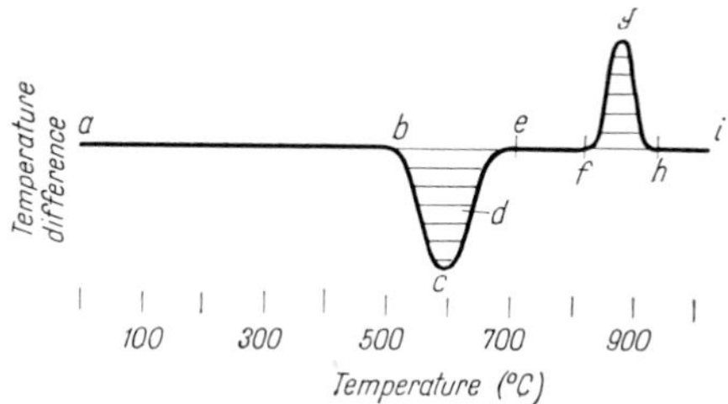

Fig. 103. Differential thermal analysis; endothermic and exothermic processes

material, no electromotive force is recorded by the galvanometer (section *a — b* in *Fig. 103*). When an endothermic reaction takes place, i.e. heat is absorbed in the test specimen, the thermocouple terminal attached to the sample becomes colder. The decrease in temperature continues up to the peak *c*, beyond which the absorbed heat is less than the conducted heat, and the curve rises again. The process ends at *d*, which is, however, not a sharply marked point, and only point *e*

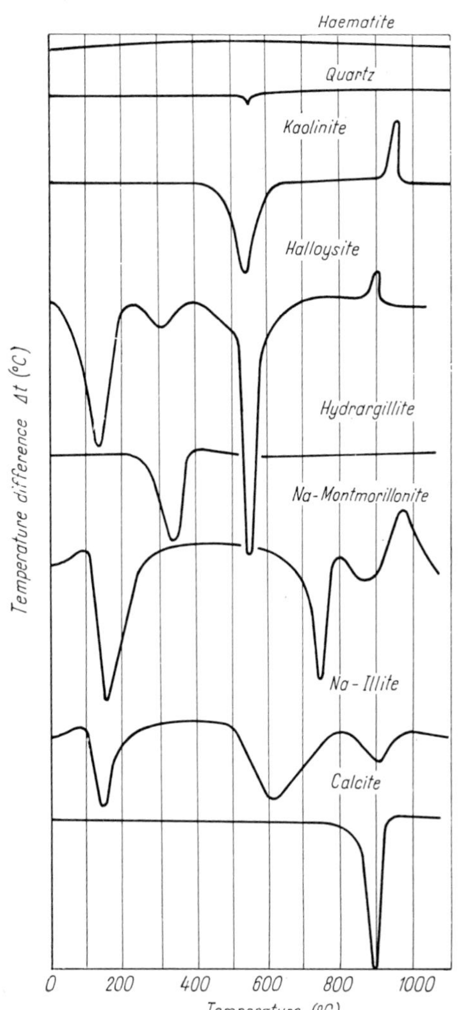

Fig. 104. DTA-curves (thermograms) for the most important clay-minerals

can be clearly observed. The area *b c e* is proportional to the active material content of the test specimen. The upward deflection *f g h* from the base line indicates an exothermic process i.e. one in which heat is released.

Experimental data obtained from differential thermal analysis of clay minerals have led to the conclusion that every clay mineral possesses a characteristic DTA curve. The thermal reactions take place at certain definite temperatures. The magnitude of the thermal reaction (measured by the area under the curve) is also significant. DTA curves for typical clay minerals and for some other minerals normally present in clays are shown in *Fig. 104.*

The accuracy of the method depends on a variety of factors such as particle size, degree of orientation of ions, shrinking and fusing phenomena, the presence of an amorphous hull round the particles, etc.; and it is also influenced by the method of the analysis (the fineness of the powdered test specimen, rate of heating, impurities, etc.). For that reason a reference curve, obtained from a test on a chemically pure material, must be determined for every DTA apparatus before it is used for routine analyses.

Besides X-ray diffraction studies and DTA analysis, ***microscopic*** examination of thin slices and ***electron micrographs*** have proved useful tools in the identification of clay minerals.

3.3.4 Surface activity of clay minerals

Research on the chemical composition and crystal structure of clays has revealed that clay minerals have the property of adsorbing certain ions at the crystal surfaces. This is an evidence that the crystal surfaces are polar, i.e. capable of producing electric fields around themselves without any contributory external cause. Crystalline compounds with ion-lattice structure and substances composed of polar molecules, e.g. water, possess this property. "Active" surfaces are those containing ions, polar molecules or polar radicals, which in turn can attract ions or polar molecules from the environment. This adsorption is due to electrostatic attraction between opposite charges. The process of adsorption of ions on a polar surface is illustrated in *Fig. 105*. A homogeneous distribution of the positively and negatively charged spots on the crystal surface is reflected in a similarly homogeneous distribution of the negative and positive ions in the adsorbed layer (Fig. 105*a*). If, however, the distribution of the unbalanced charges on the surface of the adsorbent is heterogeneous or the ions in the solution are of different valences, the available cations and anions will not be adsorbed to the same degree. As a consequence, a so-called *electric double layer* will be developed at the surface, manifested by a definite and easily measurable

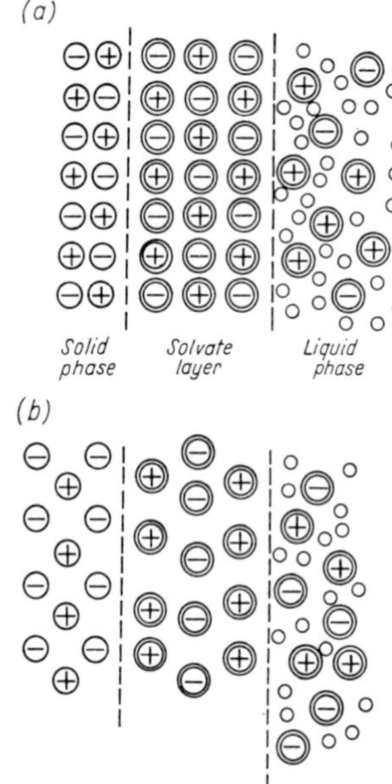

Fig. 105. Adsorption of ions on polarized surface:

a — homogeneous; *b* — heterogeneous distribution

difference in electric potential between the particle surface and the solution in the pores.

Of all ions adsorbed at clay mineral surfaces, H^+ and OH^- are the most abundant. Monovalent ions, e.g. Na are attached to the crystal surfaces by relatively weak bonds. Bivalent ions such as C^{++} and Mg^{++} are held more strongly, and finally the monovalent hydrogen, H^+, and certain trivalent ions, Al^{3+} and Fe^{3+}, are held by the strongest bonds. Cations held around the silica tetrahedral sheet of a clay mineral are shown in *Fig. 106*.

Besides the ions adsorbed, dipolar water molecules may also be tied to both the crystal surfaces and the cations present in the water. A portion of the water molecules can enter between the layers of a crystal (interlayer water), as was shown for montmorillonite. The water molecules around the polar surface of a clay mineral are arranged in a highly oriented pattern, as illustrated in *Fig. 107*. Proceeding outward, successive layers of water molecules are linked together to form a chain-like configuration, but whose degree of orientation diminishes farther away from the crystal surface.

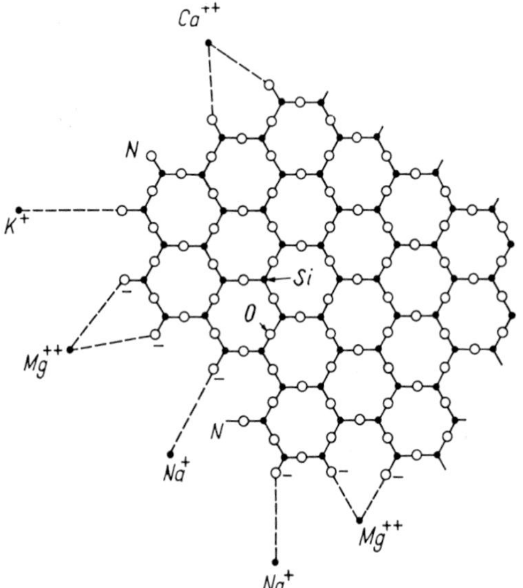

Fig. 106. Cations attached to the tetrahedral sheet

The magnitude of the electrostatic attractive forces may be very great close to the negatively charged surface of the mineral particle, but it decreases rapidly as the distance from the surface increases. The sphere of influence of the attraction depends also on the composition of the pore water and may extend to a layer several molecules thick. A schematic representation of how dipolar molecules are arranged in positions oriented toward the particle surface is given in *Fig. 108*. The attractive force per unit surface area is, according to theoretical calculations by WINTERKORN and BAVER (1934–35), of the order of magnitude of 20,000 kp/cm². Subjected to such immense pressure, the adsorbed water assumes the properties of solids, forming stable "*ice*" of various types. As

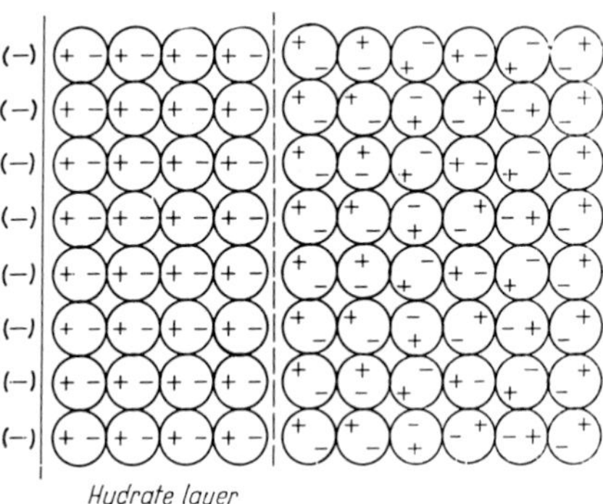

Fig. 107. Adsorption on polarized surface of a clay mineral with ionic lattice; molecular chain effects

the distance from the surface increases, the water shows the consistency of thick, viscous asphalt, and reverts to ordinary liquid water only beyond several molecular layers, at a distance of about 0.5 μ.

A water-saturated dispersed soil can thus be visualized as composed of solid mineral particles surrounded by a "liosphere", i.e. a thin hull of adsorbed water. Within this hull, the pressure, the ion concentration and other physical properties vary with the distance from the surface of the solid particle.

The process of adsorption of the water molecules is accompanied by heat production. This heat is termed the ***heat of wetting***.

The negatively charged particle surfaces, as discussed previously, can adsorb positive cations which in turn, when left with unbalanced charges, can again bond water molecules to the surface.

The number of the water molecules that can be bonded by a cation depends on the magnitude of the electric charge and on the ionic radius of the cation. Thus, for example, a Ca^{++} ion can bond more water molecules than a Na^{+} ion, their respective atomic radii being 1.06 Å and 0.93 Å. On the other hand, two Na^{+} ions can be substituted for every Ca^{++} ion at the lattice surface. The total volume of the two Na is 7.88 Å^3, against 4.99 Å^3, the volume of a single Ca^{++}. Therefore, the water-film adsorbed by the Na^{+} ions is thicker, giving rise to the formation of an exceptionally thick layer of hydrated cations around the clay particle.

The hydration shell, i.e. the water molecules together with the adsorbed cations around clay mineral particles together with the solid is referred to as an *adsorption complex*. Exchange reactions whereby certain cations are replaced by others take place within this adsorption complex. This process is known as *cation exchange*. Such reactions take place when the "foreign" cations, i.e. those in the outside solution, have a higher replacing power than that of the cations already held on the particle surface. For example, hydrogen clays, i.e. those having predominantly hydrogen as exchangeable ions in their adsorption complex, are transformed into Na clays when leached by a solution of Na^{+} percolating through the soil. Not all cations present are exchangeable. The

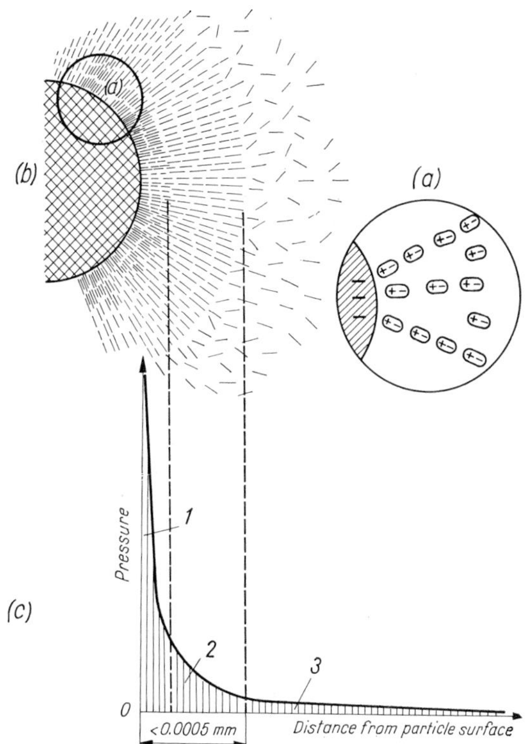

Fig. 108. Molecular effects in the solid particle-water system:

a — oriented water-dipole on the particle surface; *b* — water molecules in the lyosphere; *c* — molecular force as a function of the distance from the particle surface

quantity of all exchangeable cations is termed the cation exchange capacity. A clay with a predominance of a particular cation, H^{+}, Ca^{++} or Na^{+}, in its adsorption complex is usually indicated by the symbol of that cation, thus; H clay, Na clay, etc.

The cations adsorbed at clay mineral surfaces have a far-reaching influence on the properties of clays. Járay (1955) has shown the significant effect of the adsorbed cations on the liquid limit of the soil. Their influence on the shearing strength of the soil was investigated by Sullivan (1939). He transformed a natural clay into a H clay and then, having treated the latter with various cations, he measured the shearing strength at constant phase composition of the soil. With various cations adsorbed, the shearing strength showed a decreasing tendency in the following order:

$$NH_4^+ > H^+ > K^+ > Fe^{+++} > Al^{+++} > Mg^{++} > \\ > Ba^{++} > Ca^{++} > Na^+ > Li^+ .$$

Plasticity showed a tendency to increase in approximately the same order. The probable explanation for this phenomenon is that the exchangeable cations cause a thin, highly viscous film of water to be adsorbed on the surface of the clay particles. As mentioned before, the adsorbed water layer may be of considerable thickness in the case of Li^{+} or Na^{+} ions, while it is negligibly thin with H^{+} ions. As a consequence, Li and Na clays, their particles being forced apart by the viscous layer of adsorbed water, are highly plastic and have low shearing strength. Natural Na clays were deposited from the sea, or came into contact with salt water after sedimentation. Ca clays are fresh-water deposits. H clay is formed when a natural clay deposit is subsequently leached by pure or acid water. It follows from the foregoing that any such change in the character of the clay will result in substantial changes of the physical properties, too. The replacing cations can be conveyed to the active particle surfaces through the percolation of the pore water, through diffusion, i.e. the Brownian movement of the molecules, or through the effect of electric current. It is thought that certain exchange reactions are reversible but the general law controlling replaceability is not yet fully understood. At present, it cannot be predicted for certain how a given soil will be affected by a particular chemical treatment.

The electric fields of force between charged particles and the water surrounding them, and the associated ions, exert a great influence also on the strength properties of fine-grained soils. The alteration of any of these factors, e.g. the type or the concentration of the ions, the temperature, the pore liquid, etc., will lead to changes in the physical properties. The electric field around charged particles was described theoretically, with many simplifying assumptions, by Gouy (1910)

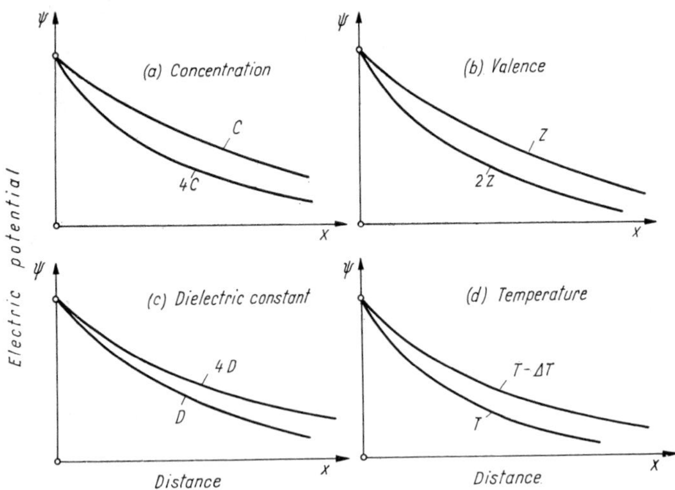

Fig. 109. Variation of electric potential with the distance from the particle surface:

a — concentration only variable; *b* — valence only variable; *c* — dielectric constant only variable: *d* — temperature only variable

and Chapman (1913). [For derivation see Lambe (1958).] The main conclusions derived from the theory are presented in *Fig. 109*. By varying the factors that determine the change of the electric potential with the distance from the particle surface, the curves shown in the figure were obtained. The existence of an electric potential in the interparticle space simply means that it requires work to be done to move the two adjacent charged plane surfaces closer to each other. This work is the product of the force acting against the repulsion and the decrease in the distance.

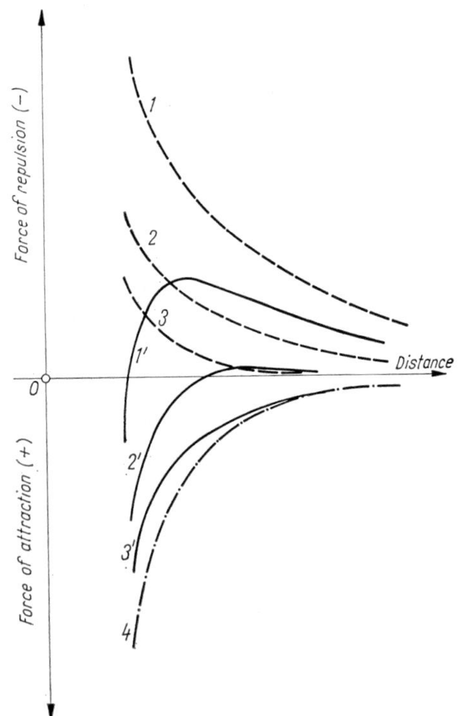

Fig. 110. Net force between two surfaces as a function of distance between surfaces and concentration of dissolved electrolyte

The repulsive force acting between two parallel planes in an electrolyte is a function of the distance. The resultant force between the two surfaces is the algebraic sum of the forces of repulsion and attraction. Attraction arises from the *van der Waals* effect already discussed. *Figure 110* shows the variation of the resultant force with the distance between the two surfaces at varying ion concentrations. If, in a solution, two particles are brought, either through Brownian movement or through pore water flow, sufficiently close together for interparticle forces to act, then the behaviour of the particles is conditioned by the magnitude and sense of the resultant force. If the net force is repulsion, the particles will be dispersed. If this occurs in the course of sedimentation, a so-called *dispersed clay* is formed (*Fig. 111a*). The conditions for this are satisfied when the concentration of the electrolyte is very low, giving rise to a net repulsive force as represented by curve *1* in Fig. 110. By gradually increasing the concentration of the electrolyte, successive stages being represented by curves *2* and *3*, such conditions can be created as will give a net attraction at any distance between the particles. If the particles in such a solution have a chance of getting closer together, they adhere to each other and form large clusters or flocs. These clumps of particles settle out more quickly, forming a flocculated structure (Fig. 111*b*) with very high porosity.

Many suspensions and colloid solutions possess the peculiar property of hardening into a gel-like substance when at rest, but liquifying again when agitated. This property is known as *thixotropy*. A probable explanation of the phenomenon can be given as follows.

The particles of a suspension are subjected simultaneously to gravitational attraction, on the one hand, and to repulsive forces due to the electric field of the particle surfaces, on the other. These forces are different functions of the distance between the particles, and it is therefore conceivable that there is a position for the particles in which

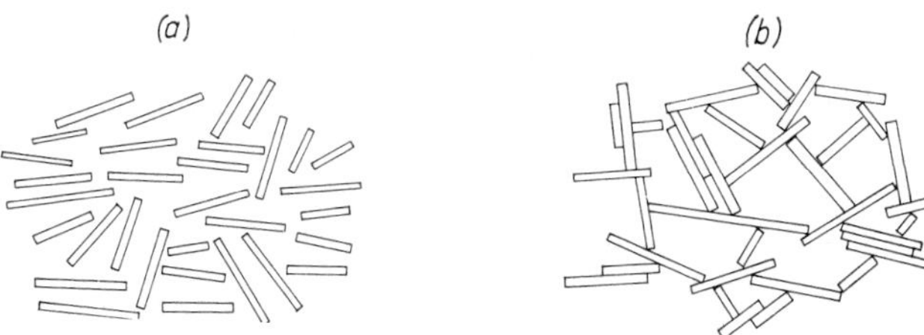

Fig. 111. Structure of clay consisting of flat particles: *a* — in dispersed state; *b* — in flocculated state

they are in the state of equilibrium, held in a stable situation without actually being in contact with each other. This balanced state can be achieved at varying interparticle distances depending on the degree of the chemical saturation of the surfaces. If the attractive force is greater, the suspension coagulates which means that the particles adhere to each other and are precipitated from the suspension. If, on the other hand, repul-

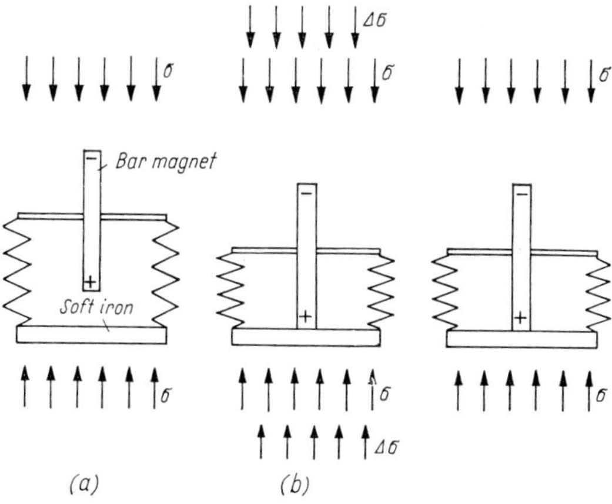

Fig. 112. Model to explain the behaviour of clay particle

sion is greater, peptization occurs. The particles surrounded by adsorption hulls are in a highly dispersed state with free water filling the interparticle spaces, and the suspension remains stable. Between these two limits there must be one intermediate position in which the adsorption hull of one particle just comes into contact with that of another. When such a suspension is vigorously shaken, the orientation of the water molecules within the adsorption hull may temporarily cease to exist. After a time, the oriented arrangement of the water molecules may be restored and the suspension reverts to its former state, having a higher strength (*"rest-hardening"*).

The physico-chemical relationships discussed here have many important practical consequences. The interparticle forces are non-linear, therefore the deformations that may have taken place between irregular particles are irreversible. In the same way, a normally loaded clay will not regain its initial void ratio after the removal of the pressure but experiences a permanent deformation. The final void ratio depends on the initial state of the soil, and on the magnitude and duration of the previous pressure. The normal pressure produces new points of contact between the particles, giving rise to strong bonds which remain when the pressure is removed.

This peculiarity of clay structure can be simulated by the model shown in *Fig. 112*. The bar magnet and the piece of soft iron represent two adjacent soil particles in *"edge-to-face"* arrangement, and the springs simulate the electrostatic repulsion between the particles. Provided that both the bar magnet and the soft iron are sufficiently long, the attractive force is inversely proportional to their distance (*Fig. 113*). This character is indicated by the hyperbolic type load-displacement diagram of the magnet, whereas the elastic behaviour of the springs is described by the straight line shown on the upper part of the figure. The resultant force is obtained as the algebraic sum of the forces of attraction and repulsion. As seen from the figure, the resultant is also a function of the distance between the parts of the model.

Let the system at rest be acted upon by an external pressure σ. The resulting compression will

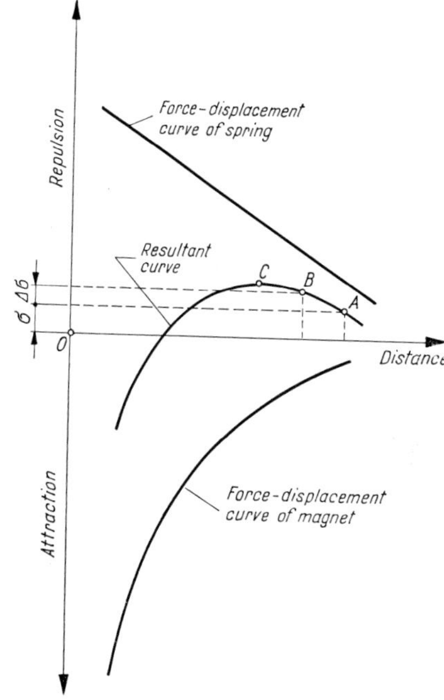

Fig. 113. Force-displacement characteristic of spring-magnet system in Fig. 112

be helped by the attraction of the magnet, and resisted by the forces in the springs. Let the equilibrium state, at given σ, be represented by point A on the resultant curve. If the pressure is further increased by $\Delta\sigma$, the distance also decreases, yet, at this stage, the deformation continues to be reversible. Should, however, the pressure be increased to the extent that the resulting conpression reaches point C, the equilibrium of the system can no longer be sustained, and even the smallest displacement past C results in the closing of the gap between the magnet and the soft iron (Fig. 112*c*). The pressure increment $\Delta\sigma$ may be subsequently removed, but the system can no longer recover its original state, because the magnetic attraction, increased tremendously by the direct contact, prevents the breakdown of the freshly formed bond, so the two bars become permanently fixed in their new position. In the new system of forces, tensile stresses are induced in the magnet bar, analogously to the stresses in real clay particles.

The concept of clay structure and the associated system of forces offers in part an explanation for the not perfectly elastic rebound of clays, subsequent to compression. Clearly not all particles in a clay mass can be forced by compression to form such tight bonds as are simulated by the magnet-soft iron model. Nevertheless, even a moderate increase in the number of interparticle contacts may cause irreversible deformations.

3.3.5 Forms of existence of soil water

This section deals with the various forms in which water is present in soils.

If an observation well is driven into the soil, it normally reaches the ground water at a certain depth. This means that the water draining from the surrounding soil enters the well and rises in it, coming to rest at a "free" water level, called the water table. Below that level all the pores in the soil are filled with water. The soil, however, may be in saturated state also above the water table, due to capillary action. Approaching the ground surface, the soil changes into the three-phase state, its pores being occupied by both air and water. The configuration of the two different substances, water and air, in the voids is determined by their respective attraction to the solid phase. Of the two, that having the smaller attraction to the rock material makes tighter contact with the solid grains; in soils this is always the water. Under stable equilibrium conditions, the water tends to assume a configuration such as to wet the grains over the largest possible surface area. The pore air, on the other hand, has a tendency to assume a position in which the contact surface area between the air and the liquid is a minimum. As a result, the liquid phase always occupies the smaller voids, whereas the air accumulates in the larger voids, forming bubbles as closely spherical in shape as possible.

On a pure phenomenological basis, soil water can be classified into the following kinds.

1. *Ground water*, i.e. the subsurface water that fills the voids continuously and is subjected to no force other than gravity. The free surface of this ground water is observed in wells and boreholes.

2. *Capillary water*, that is lifted by surface tension above the free ground water surface. To a certain distance above the water table, the capillary water fills all the pores in the soil (zone of capillary saturation). Above this level the zone of open capillaries is located. Within this zone the capillary water held in the finer pores is connected through a network of pores with the ground water, whereas in the larger pores water is displaced by air. The water-saturated voids are, in accordance with the capillary theory, bounded by menisci.

3. *Adsorbed water* held on the surface of the grains. Its properties are different from those of ordinary water.

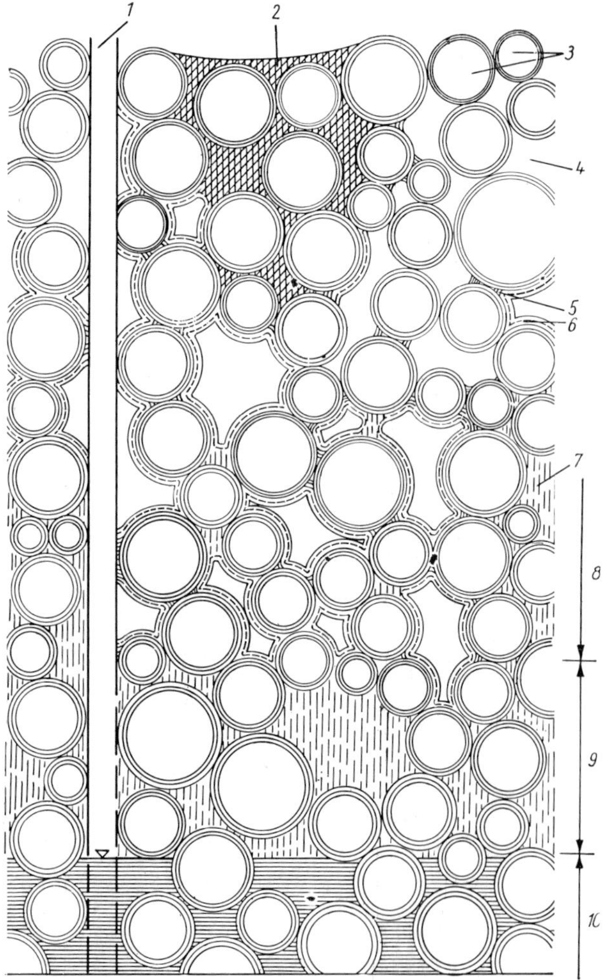

Fig. 114. Forms of water occurring in soils:
1 — observation tube; *2* — perched gravitational water; *3* — solid particle surrounded by hygroscopic water; *4* — air saturated with water vapour; *5* — contact water; *6* — water film; *7* — open capillary water; *8* — air + water; *9* — zone of capillary saturation; *10* — ground water

4. *Film water*—ordinary water surrounding the grains in the form of a very thin film due to surface tension.

Contact capillary water is the water held in the angles around the points of contact between the grains. Its surface forms concave menisci. Water that may completely fill groups of the pore-channels above the ground water table, without being connected with the latter, is called *suspended* water (perched water).

5. *Infiltered water*—this term applies to that portion of surface precipitation which soaks into the ground moving downwards through air-containing zones. It is subject to capillary forces.

From a *structural* aspect, the following kinds of soil water can be distinguished.

1. *Pore water*. It exhibits the physical and chemical properties of ordinary liquid water. This kind of water is capable of moving under hydrodynamic forces unless restricted in its free movement as e.g. when entrapped between air bubbles, or by retention due to capillary forces that in fine pores may overcome the hydrodynamic forces.

2. *Solvate water*. Water that is subject to polar, electrostatic and ionic binding forces forms a hydration shell presumably not more than 200 molecules thick around the soil grains. Its density and viscosity are greater than those of ordinary water, yet it remains mobile under hydrodynamic forces.

3. *Adsorbed water*—an extraordinarily thin layer of water attracted to the external surfaces of clay minerals, or held as interlayer water by minerals with an expanding-lattice structure. It varies from 1 to 10 molecules in thickness. Because the adsorptive forces are extremely great, this kind of water cannot be moved by normal hydrodynamic forces.

4. *Structural water*. In essence, this is not water, since the term refers to hydroxyl groups that constitute parts of the crystal lattice. This kind of "water" can only be driven off at such high temperatures as would cause the destruction of the crystal structure.

These four kinds of water are illustrated in *Fig. 115*. Every one of them has a particular role in the formation of soil structure and in determining soil properties. It is thought, however, that the physical phenomena associated with changes in the water content (i.e. swelling, shrinkage, strength changes etc.) are largely due to changes in the pore water and in the solvate water, for adsorbed water and structural water, being both held by immense bonding forces, are not likely to undergo appreciable changes in quantity under normal variations in temperature and pressure.

The understanding of the forces involved makes it possible to influence—by artificial means—the soil–water interaction. A number of soil stabilization processes are based on this principle. The latest development in this line involves, for example, methods by means of which the sensitivity of the soil to water can be completely reduced and the soil rendered practically unwettable. This can be done by ion-exchange, since by the bonding of non-hydrated positively charged ions the hydrated shell can largely be eliminated. Methods using polyacrylates work on the principle that ion clusters of great molecular weight hold the solid grains together by electrostatic and polar bonding forces with the result that the soil becomes resistant to the effect of water and preserves its stable structure. Polyvalent ions adsorbed on the particle surfaces can be transformed by treatment with certain chemicals, e.g. polyphosphates, into complex compounds with the consequence that the thickness of the hydration shell is greatly increased, the cohesion and the friction are reduced, so that the grains can readily change position even under slight external loads. The soil becomes easily compactible and tends to liquify even at low water contents.

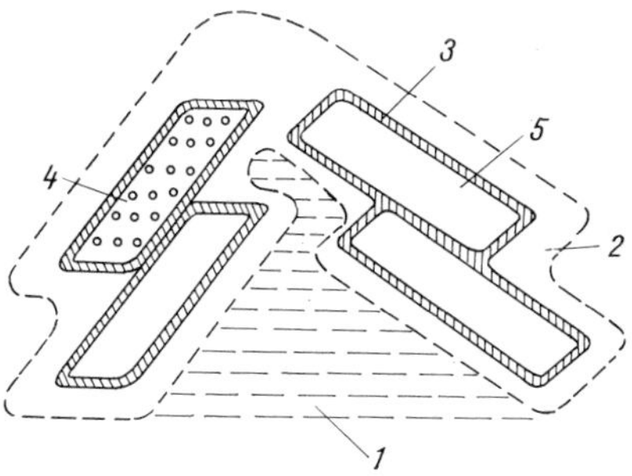

Fig. 115. Classification of water in soil:
1 — pore water; *2* — adsorbed water; *3* — solvatation water; *4* — structural water; *5* — solid particle

One advantage of these methods is that only very small quantities of chemicals are required to achieve the desired alteration in soil behaviour. Future research may develop new methods, by which soil properties can be altered almost at will, and thus soil materials for specific purposes (soil stabilization etc.) can be produced, or the behaviour of natural soil masses can be favourably influenced.

3.3.6 Gaseous phase in soil

In three-phase soils, a gaseous phase is also present beside the solid grains and the water.

Air, or other gas, in soils may originate from various sources. It is a natural concomitant of wind-laid deposits whose grains, having settled from air, are covered with only a thin molecular water hull. Such deposits show a characteristic low degree of saturation. In the upper layers, the moisture content is subjected to seasonal variations. In the lower depths, the pores are partly saturated with water drawn by capillarity from the ground water. In contrast to Aeolian deposits, water-laid sediments, particularly clays, are found under normal conditions in a state of near 100% saturation. Favourable conditions for air invasion exist only when the surface of the soil is exposed to drying. The voids in a soil vary greatly in size. In the larger voids, the water is subjected to molecular forces only to a limited degree. Therefore, evaporation from the large pores is greater than that from the smaller pores, which are filled with a kind of water more strongly bonded to the particle surfaces.

A major factor in the formation of the three-phase soil structure is the variation of the water table. The free water drains out of the pores as the water table is lowered, and after a subsequent rise of the ground water the voids refill again. The water movement caused by these variations does not take place uniformly, for the pore channels are not to be pictured as tubes of constant cross-section but rather as series of narrow spaces alternating with more bulbous ones, as they are randomly formed between adjacent soil particles. In such tubes, the velocity of the capillary rise varies from point to point. The energy of the rising water (*capillary potential*) decreases in the wider spaces, and increases greatly in the narrow ones. As a consequence, the upward movement of the menisci will be intermittent and non-uniform even in adjacent capillaries. It may occur that the moving water cannot pass a large void, while the rise of the water still goes on in neighbouring narrower channels.

The capillaries communicating with each other form a network of voids interconnected in every possible direction; vertical tubes are thus connected by transverse passages, through which the water rising to higher elevations can bypass the large voids and begin to fill them from above, thereby blocking the free communication of the pore air with the atmospheric air (*Fig. 116*).

In addition to entrapped air bubbles, the pore water always contains a certain amount of dissolved air or other gas. The maximum quantity of air which dissolves in water at given temperature and pressure is strictly determined by physical laws. If the pore water is, under given conditions, in a saturated state, an increase in temperature or decrease in pressure may cause the dissolved gas to escape from the water. The freed gas will then either disappear into the atmosphere or get trapped in the capillaries in the form of bubbles.

According to this, the gaseous substance contained in the voids of the soil may be of two kinds:

gas connected with the atmosphere—this can freely communicate between the pores and the atmosphere;

gas separated from the atmosphere—this is in continuous interaction with the surrounding pore water.

The pressure of the gas communicating with the atmosphere is equal to the atmospheric pressure; should the latter change, air will move in, or out of, the pores until the pressure is balanced.

The water and the entrapped air bubbles form dispersed matter whose general physical properties are governed by the interactions between the gaseous and liquid phases. From a physical point of view, this mixture is a gas emulsion, with the gas as disperse phase and the liquid as dispersion medium. In the case of flow of the mixture, the two phases may move differently. At constant temperature and pressure, both substances being in a stable physical condition, the presence of the air exerts no influence on the hydrostatic equilibrium. If, however, the physical conditions change, the entrapped air turns out to act as a separate factor which may greatly influence the behaviour of the soil. Because of the presence of a separate gaseous phase, free interfaces are formed between the solid grains, the water and the gas bubbles, giving rise to surface tension and capillary forces. Air bubbles contained in the pore water tend to reduce the permeability of the soil to water. An increasing gas content of the pore water generally causes the soil to behave elastically to an increasing degree, since gases exhibit perfect elastic deformations under increasing pressure.

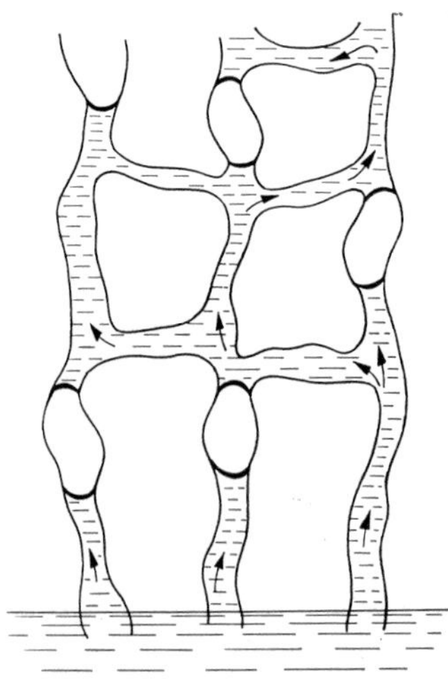

Fig. 116. Air bubbles enclosed in capillary tube system

3.4 Structure of cohesive soils

3.4.1 Structure of clays

The fine soil particles formed by clay minerals may, according to the characteristics of the adsorbed water layer, aggregate into clusters in a manner dictated by the surface forces. The formation of such particle aggregates may take place, as the final stage of sedimentation, at the bottom of a lake; the particles will then be arranged in a pattern known as *honeycomb structure*. If, however, the electrical forces acting at the particle surfaces change during the process of

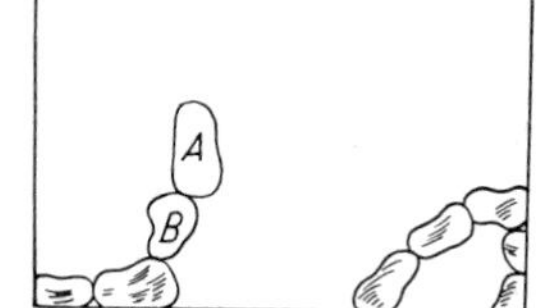

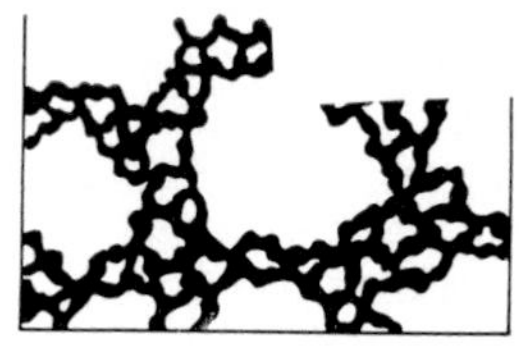

Fig. 117. Honeycomb and flocculent structure of soil

sedimentation, the aggregation (flocculation or clogging) due to interparticle attraction may start while the particles are still in suspension, so that afterward large flocs, instead of individual grains, will settle to form a *flocculent structure* (*Fig. 117*).

Our present knowledge in respect of the bonds between individual soil particles is rather scanty. As far as flat, plate-shaped particles of kaolinite and illite are concerned, it is generally assumed (Ford, Loomis, Fidiam, 1939; Van Olphen, 1951; Tan Tjong Kie, 1957) that the flat mineral surfaces are negatively charged and can thus adsorb cations, whereas the edges of the plates carry positive charges. Schofield and Samson (1953) attribute the flocculation of kaolinite to this circumstance. Experimental verification was given by Iler (1956). On the basis of the results of these investigations, *Tan Tjong Kie* suggested the following concept of clay structure.

A clay "*micelle*" carries negative charges on the flat surfaces and positive charges around the edges and at the corners. Therefore, interparticle forces of attraction are developed whenever edges or corners chance to come into contact with flat surfaces. The constants may be of end-to-face or edge-to-face types (*Fig. 118*, *A* and *B*). The bonds are due to *Coulomb* forces, *van der Waals* forces and binding forces of cations and adsorbed water molecules. When two particles come into contact in a face-to-face arrangement (Fig. 118, *C*), the Coulomb forces of repulsion between like charges become dominant, but this may be overcome by the combined effect of external loads, van der Waals forces, cation bonds and hydrogen bonds of the adsorbed water molecules. In this manner, the thin clay particles which can be likened to razor blades build up a loose, card-house structure. The skeleton thus formed is fairly stable and rigid owing to the firm interparticle bonds and to the stiffness of the particles themselves. A schematic picture of the clay skeleton as suggested by *Tan Tjong Kie* is shown in *Fig. 119*. The stiffness of the whole structure depends upon the strength of the bonds, which in turn is determined by the character of the bonds and the magnitude of the binding forces. The type of contact between the particles is largely determined by their geometrical shape. For example, montmorillonite plates invariably form contacts of type *A*, while with kaolinite and illite particles the much stronger contacts of type *B* are more common.

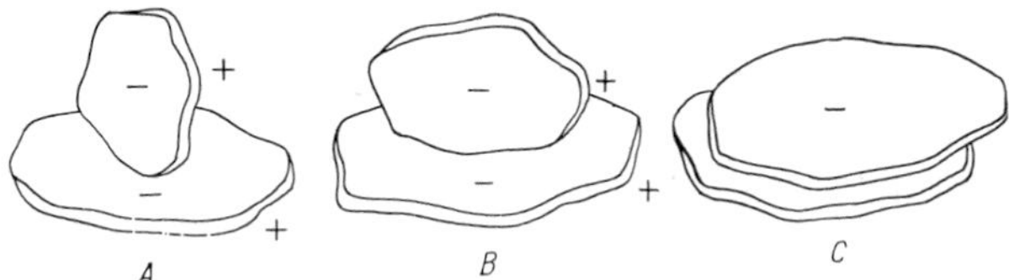

Fig. 118. Types of bond between plate-like clay particles: *A* — bond on a tip; *B* — bond along an edge; *C* — contact along a plane

Fig. 119. Cardhouse structure of flocculated clay particles, according to Tan Tjong Kie

If a natural clay contains coarser grains of sand and silt as well as fine clay particles, bonds may be established on the surfaces of the large grains, too. A skeleton structure is formed, made up of sand and silt grains with large voids between them, which are filled with a clay matrix of the kind described in the preceding paragraph. The micropores of the clay matrix, in turn, are occupied by water (*Fig. 120*). If such a deposit is subjected to compression, the pressures due to the load are transmitted through the skeleton, i.e. from one large silt grain to another, with the result that the clay particles in the narrowest gaps between adjacent silt particles are highly compressed, whereas the clay matrix in the interior of the large voids remains practically unstressed. The particles trapped between the coarse grains may develop fairly

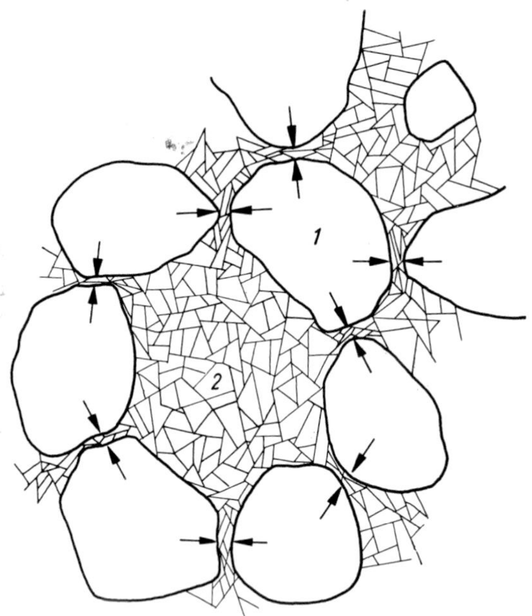

Fig. 120. Structure of clay when coarse particles are present: *1* — sand particle; *2* — clay particles

strong bonds of type *B* and *C* (Fig. 118), which may be augmented by the cementing effect of carbonates and iron oxides. Under certain loads the rigidity of the skeleton may cease to exist, deformations take place, and the loads are transmitted to the originally unstressed clay matrix. At that stage, the compression of the soil is greatly increased. Changes in the soil structure offer an explanation also for the loss of strength of clays on remoulding. In the course of time, as the particles are rearranged in a new structural pattern and new bonds are formed, the remoulded clay can regain a part of its original strength, partly due to the effect of the van der Waals forces. Clays with this type of structure experience an increase in strength under all-around compression, because water is squeezed out of the pores and the forms of bond are transformed; thus type *A* changes into *B*, and *B* into *C*.

The orientation of platy or lath-shaped clay particles may essentially influence the behaviour of clays. Particles deposited from still water are arranged in an irregular *card-house* pattern, and, although the role of the force of gravity is not entirely negligible, the manner of the formation of the structure is dictated primarily by surface forces. If an external load begins to act on a deposited sediment it will initiate a process of consolidation, i.e. the layer will undergo a gradual compression normally prolonged over a long time.

As a consequence, the flocculated structure of the clay will also be modified in that the degree of orientation of the grains gradually increases, since the particles tend to occupy positions perpendicular to the direction of the applied load. If now this soil with highly oriented particles is remoulded, e.g. by kneading, it will break down into small clods. The clay particles keep their parallel orientation within each lump, but the lumps are arranged randomly among themselves (*Fig. 121b*). Shearing stresses acting on the mass over a long period of time give rise to gradual reorientation of the structure (Fig. 121*c*).

The degree of orientation of the particles plays an important role in the study of such properties of clays as consolidation, compaction, shear strength, shrinkage and swelling.

The thin, platy particles of clay are deposited with their flat surface more or less horizontal, i.e. a tendency towards horizontal orientation arises from the manner of sedimentation. The degree of orientation is, as seen before, further increased by consolidation due to overburden pressure. That is why clays, deposited in horizontal layers, show as a rule much lower compressibility in the vertical direction, i.e. perpendicularly to the plane of orientation of the platy particles, than in the horizontal direction. The same holds for permeability. Most cohesive soils show similar signs of anisotropy.

The effect of remoulding on the strength of clay is characterized quantitatively by the term sensitivity, which was introduced by TERZAGHI (1948). The degree of sensitivity (S) is defined as the ratio of the unconfined compressive strength of an undisturbed soil specimen to the unconfined compressive strength of the same specimen in the remoulded state. I.e.

$$S = \frac{\sigma_u^{\mathrm{nat}}}{\sigma_u^{\mathrm{rem}}} . \tag{51}$$

For overconsolidated (preloaded) clays the degree of sensitivity is about 1.0. In extrasensitive clays with a tendency to flow (*quick clays*) sensitivities as high as 100 or even 1000 are encountered. For normally loaded clays, the values of S range between 2 and 4, for sensitive clays they range from 4 to 8. In Hungary, no values of S exceeding 3 to 4 have been found yet. High sensitivity values occur in glacial clays with flocculent structure, and in clays derived from volcanic ash.

Extrasensitive marine clays occur in Norway. As BJERRUM (1954) stated, if a marine clay is subsequently uplifted to dry ground and its salt content is removed by leaching, then it experiences a substantial decrease in shear strength accompanied with a significant increase in sensitivity. Sensitivity was found to have a logarithmic relationship with the consistency index (I_c) of the soil in its natural state.

Under field conditions remoulding of a clay may be caused by the sliding of large soil masses or by pile driving. As the sliding begins, the shearing strength of the soil in the vicinity of the sliding surface is radically decreased leading to the development of what is called *progressive failure.*

Likewise, in sensitive clays pile driving causes complete remoulding and loss of almost the entire strength in the soil adjacent to the surface of the pile. A portion of the loss in strength may be regained later. This accounts for the slow gradual increase with time of the bearing capacity of piles driven in clays.

The structure can also affect the behaviour of clays subjected to stresses. As has been shown in Fig. 111, clay structures can be classified into two kinds: dispersed and flocculated structures. In a dispersed clay, the particles repell each other and are arranged randomly without actually being in contact. When shearing stresses are applied to the soil, the particles will be oriented to some degree until with increasing stress they become, at least in the zone of shear, fully oriented, i.e. parallel to each other. Therefore, to maintain a constant rate of shear strain requires gradually increasing stresses until a maximum is reached. On the stress-strain curve there is no characteristic point that would indicate an essential change in soil behaviour (*Fig. 122a*). No frictional resistance is developed during the process, as the soil possesses no internal friction. If, on the other hand, the total stress due to external loads is increased, the distances between the particles are decreased, with the result that a greater shearing stress will be required to maintain the same constant shear deformation velocity ($\mathrm{d}s/\mathrm{d}t$).

If a clay with a flocculated structure is subjected to shear, some of the interparticle bonds break down in the course of shear deformations, while new ones are formed continuously. If the breakdown of the bonds becomes predominant, the clay suffers an essential change in its structure resulting in a substantial decrease in its structural strength, so that constant shear strain can be maintained even by greatly reduced stresses. Interparticle forces of adhesion cease to exist, since the bonds themselves are destroyed. Thus, the part of the shearing stress due to cohesion decreases whereas the frictional resistance increases.

The phenomenon illustrated by the stress-strain curve in Fig. 122*b* is even more accentuated if the electrolyte concentration of the pore water or the valence of the electrolyte have changed as compared with their values at the time of the deposition of the clay. In such cases, the structure of the clay may change completely in that the flocculated structure changes into a dispersed one. Because of the high void ratio of the original flocculated structure, the particles in their new arrangement will be at relatively great distances from each other. This results in a radical decrease in the shearing strength with increasing shear deformations, since the interparticle forces of repulsion, latently present in the undisturbed state of the system, have a chance to come into play through the destruction of the former bonds, and prevent the formation of new bonds. This is the case with the extrasensitive marine clays known

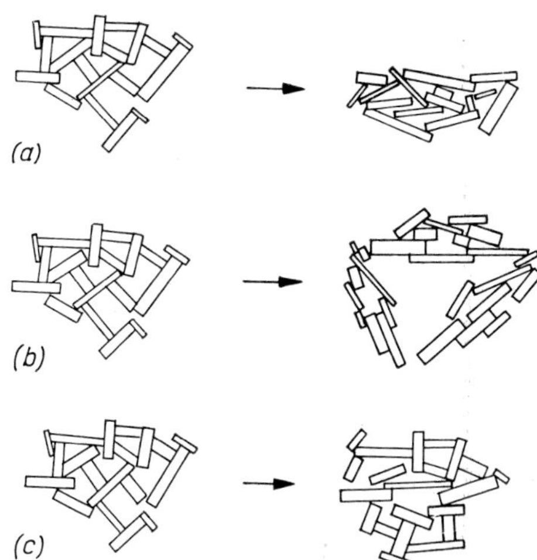

Fig. 121. Structures of clay particles with different degrees of orientation:

a — consolidated state, highly oriented particles; *b* — irregular pattern after remoulding; *c* — increase of the degree of orientation due to the action of shear stresses

as *quick clays*, which are found in Norway, Canada etc.

What has been stated in the foregoing about the structure of clays and other cohesive soils relates to the primary structure, that is to the connection between the elementary constituent parts. But cohesive soils may, under external influence, develop a different kind of structure known as a *secondary structure*. This is, in contrast to primary structure, a macro-phenomenon, visible to the naked eye. The secondary structure is formed by *post-sedimentation* effects. In an originally dense clay that would appear homogeneous under macroscopic inspection, exposure to drying causes the formation of shrinkage cracks. In a subsequent wet period swelling takes place and the cracks close again. But once the sides of the fissures have

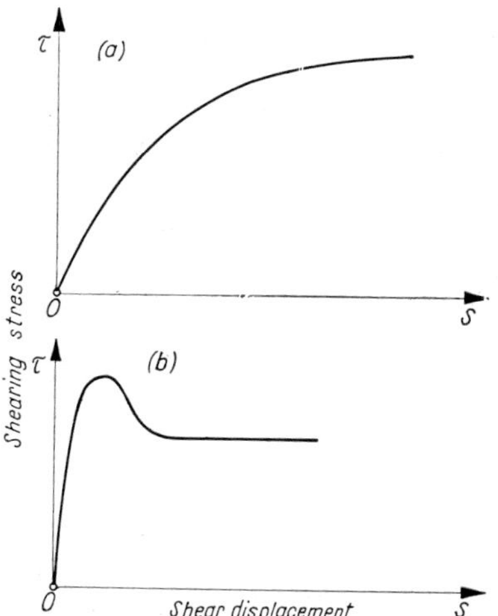

Fig. 122. Behaviour of clays in shear:

a — dispersed clay; *b* — flocculated clay

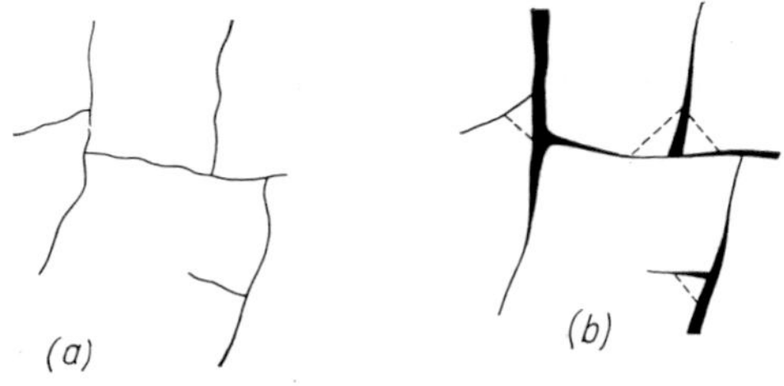

Fig. 123. Section through stiff fissured clay mass:

a — old fissures closed before relief of stress by excavation; *b* — relief of stress causes fissures to open whereupon circulating water softens clay adjoining the walls

parted, the original coherence between them can never be completely restored. Thus, repeated drying and wetting lead to a radical destruction of the strength of the soil. Fine hair cracks develop in all those flood-plain deposits which have been temporarily exposed to atmospheric effects. The clay mass readily breaks up into chunks along these hair cracks, widened fissures or joints with shiny wall surfaces are formed, and large blocks of seemingly sound soil disintegrate into small angular fragments. Fissures may be produced also by tectonic action or electrical effects. Lateral pressure, electroosmotic effect in the freshly deposited sediment as well as certain colloid-chemical phenomena can contribute to the formation of cracks. When subject to slow deformation or creep, the fissured clay may loose a substantial part of its original strength. A constant load causes the fine hair cracks to open resulting in the complete breakdown of the mosaic structure. The hazard of the gradual diminution of the shearing strength is increased even more when an excavation is made in a fissured clay. The relief of the overburden pressure by the excavation gives rise to the expansion of the soil adjoining the slope. The fine fissures open, and water may enter them, causing the stiff mosaic of pieces to swell and disintegrate completely. The average shearing strength of the soil will be radically reduced. This effect may account for a loss of as much as 70 to 80% in the strength of the soil.

Fig. 124. Fossil sliding surfaces (slickensides) caused by previous landslides in stiff fissured clay

The process of the deterioration of the mosaic structure is illustrated in *Fig. 123*. Water enters the cracks that opened wide because of the relief of stress, and softens the soil adjoining the walls. Uneven swelling produces fresh fissures, and causes

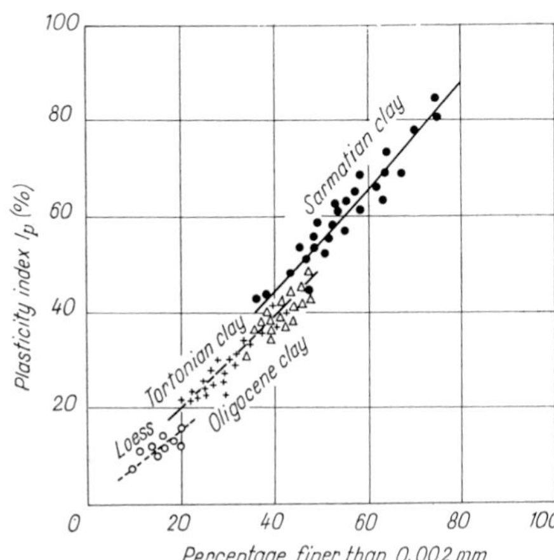

Fig. 125. Relationship between index of plasticity and clay content

the larger chunks to disintegrate until the clay is softened in its whole mass, and a soft matrix completely surrounds the harder pieces. In the course of the process the sloping mass moves with a slow, creeping movement. The velocity of the movement gradually increases up to the critical point when the value of the shearing strength is reduced below the acting shearing stresses and a slide occurs.

In Hungary, clays with mosaic structure occur in the Sajó valley, on the Eastern slopes of Mountain *Bükk* and near the town *Rudabánya*. These are high-swelling clays of high plasticity. They often contain concealed, so-called fossil sliding surfaces produced by earlier slides (*Fig. 124*).

In clays there are often found shiny, smooth, almost polished surfaces known as slickensides. These were formed by volume changes due to chemical processes or by deformations caused by tectonic or gravitational forces, or by faulting.

The plasticity of clays arises almost exclusively from the interaction between the clay fraction and the water. Provided the mineral composition of the clay-size fraction is constant, the plasticity index I_P is in direct proportion to the clay content of the soil. On the other hand, plasticity is greatly influenced by the mineral composition of the clay fraction. The ratio of the plasticity index to the clay content (i.e. the percentage of particles smaller than 2μ in size) is called the *activity* of the soil (Skempton, 1948). A sediment is said to be inactive when this number is smaller than 0.75; normal when it is between 0.75 and 1.25, and active when it is greater than 1.25. Activity values for the most common clay minerals are given in *Table 11*. Measured values for some typical soils in Hungary are shown, after Szilvágyi (1955), in *Table 12*. The relation between plasticity index and clay content for the same soils is presented in *Fig. 125*.

Table 11. Activity of clay minerals

Type of mineral	Activity I_P/clay content
Quartz	0.—
Calcite	0.18
Muskovite	0.23
Kaolinite	0.46
Illite	0.90
Ca-montmorillonite	1.5
Na-montmorillonite	7.5

Table 12. Activity of some clays according to Szilvágyi (1955)

Soil type	Origin	Activity I_P/clay content
Loessial silt	Dunapentele	0.72–0.90
Oligocene clay	Budapest	0.9 –1.02
Tortonian clay	Budapest	0.92–1.05
Sarmata clay	Budapest	1.01–1.23
Bentonite	Nagytétény	2.6

3.4.2 Structure of loess

Loess, the typical wind-laid deposit, is known for its peculiar structure. The characteristics of true loesses are determined primarily by the manner of formation rather than by the properties of the constituent particles. Their petrographical properties and grain-size distribution are remarkably uniform. About 50% of the grains fall between the sizes of 10 and 50 μ. The uniformity coefficient U is about 10. The grains of the loess are made up of quartz (50 to 75%), feldspars (10 to 25%) and calcite (10 to 30%). Clay minerals are represented in a few percent only. The liquid limit of loesses ranges from 25 to 30% and the plasticity index from 4 to 12%. As a consequence of the manner of the deposition (Section 1.1), so-called macropores i.e. large holes well visible even to the unaided eye can be observed in the structure of the loess (*Fig. 126*). At the time of deposition, moisture content produces weak interparticle bonds that later gradually gain in strength probably because a calcareous film and other cementing materials are precipitated around the points of contact as a portion of the water film evaporates. Thus, the extraordinarily loose, macroporous structure can remain stable even under increasing overburden pressure. Certain investigations attribute the structural bond of loess to the action of clay minerals.

If a loess is loaded, very fine hair cracks may develop in the calcareous films and in the cementing bonds, yet the stability of the structure is preserved mainly through cohesion due to contact moisture. If, however, the loess under load is completely soaked through with water, the beneficial effect of this cohesion ceases, and the loose structure, already weakened, collapses and a sudden settlement occurs. The size of the settlement depends on the initial porosity and water content, on the magnitude of the superimposed load and on the quantity of the water. If the initial void ratio e is smaller than 0.8, the settlement due to submergence is negligible. The value of the initial void ratio, on the other hand, is determined by the climatic conditions prevailing at the time of the deposition of the loess and by any modification due to events in its subsequent history. Because of the presence of vertical root holes in its structure, loess is highly anisotropic in both strength and permeability. It also readily splits along vertical planes.

Fig. 126. Structure of macroporous soil (loess)

3.4.3 Structure of silt

In respect of grain size, silt is a transitional soil type between sand and clay. Therefore, its properties and structure also occupy a mid-position between the two main soil types discussed in detail. Silt contains only a relatively small amount of submicroscopic particles, and the colloidal and electrical phenomena associated with clays are therefore of minor importance. Nevertheless, silt exhibits, to some degree, such properties typical of clays as plasticity, swelling, sensitivity to chemical effects etc. This is so because even a comparatively small percent of "fines" causes the specific surface area of the whole assemblage to increase substantially. Silt when dried possesses fairly high strength that is chiefly due to the capillary effect. Dry silt, therefore, quickly disintegrates when submerged in water.

3.5 Organic constituent parts in soils

The properties of soils are strongly influenced by any organic matter present. There are soils that consist almost entirely of organic matter, mainly of plant origin (lignite, peat, bog and swamp soils); and there are others, composed predominantly of mineral grains, which may contain varying amounts of organic matter as well. Peats, bogs and swamps consist of the decaying, or decayed, remains of ancient vegetation, transformed, in the absence of air, by anaerobic decomposition, and usually compacted to some degree. In peats the fragments and the fibrous structure of the decayed plants are well recognizable even by visual inspection. They are brown or black in colour, and made up of a loose, or sometimes compressed, mass of poorly humified vegetable matter. Isolated from the air, the vegetable debris undergoes a process similar to dry distillation. The formation of peat is the first stage of this process, accompanied by the release of various hydrocarbon products rich in carbon. Essentially the process is diagenesis brought about, in part, through the agency of microorganisms. *Renault* suggests the following simplified formula for the decomposition of cellulose:

$$4\,C_6H_{10}O_5 \rightarrow C_9H_6O + 7\,CH_4 + 8\,CO_2 + 3\,H_2O\,.$$

Methane (marsh gas) and carbon dioxide are produced during the formation of peat. *Peat-bogs* are formed in shallow, undrained inland depressions when, on the one hand, the depth of the water gradually diminishes because of the accumulation of *sapropel*, and on the other, the vegetation belt (sedge, rush, reed) growing along the shore encroaches upon the open water surface, and overgrows it completely in the end. These low-lying peat deposits are called *low moors*, or *muskeg*. In contrast to these, *high moors* are areas with a deep-lying water table and are overgrown mainly by mosses and heather. If a peat deposit is later overlain by fresh clayey sediments and compressed under the increased overburden pressure, *pitch-peat* and *liver-peat* are formed.

Peat, as already mentioned, has an extremely high water content, often amounting to as much as 80 to 90% of the total weight (i.e. $w = 400$ to 900%). Its unit weight ranges from 5.0 to 13.0 kN/m^3. The solid matter is made up of hydrocarbons, lignin and nitrogen compounds. On the average, peats contain 40 to 60% carbon, 5 to 7% hydrogen, 25 to 45% oxygen and 1 to 6% nitrogen. In the process of anaerob decomposition, the greater part of the carbon gets separated from the other constituents. Various organic acids are also produced in the course of rotting, and peat, therefore, usually shows an acidic reaction. In an acidic environment carbon is uncapable of forming water-soluble compounds and, thus, peat becomes richer in carbon as time passes. During the process of humification the loose mass of vegetable debris is gradually compacted, and is often reduced to one tenth or even one hundredth of its original volume. In the absence of air the process is extremely slow. Its degree can be assessed by macroscopic inspection and by calorific value measurements.

If decomposition takes place in the presence of air, the process is greatly accelerated, and it is also much more detrimental to the state of the soil. Aerobic decomposition (humification) produces water-soluble compounds. These will soon be washed away, and the soil quickly disintegrates into its elements. Various microorganisms play an important role in the process. These questions are dealt with in detail by soil science. During the process of aerobic rot carbon is oxidized, hydrogen combines with oxygen into water, and nitrogen and sulphur form various water-soluble salts (nitrites, nitrates, sulphates).

Aerobic and anaerobic decomposition may take place simultaneously under certain conditions. For example, inside a soil lump completely surrounded by water decomposition takes place under anaerobic conditions, while at the exposed outer surfaces the process is aerobic decomposition. The two processes may also supersede each other. Thus, when air is allowed, by the lowering of the water table, to penetrate the peat, the process of carbonization is halted and the rapid oxidation of the peat begins. In excavations, long exposure to the atmosphere can also cause the rapid decay of the peat.

If organic plant fragments and mineral grains are deposited simultaneously or are mixed in some way after sedimentation, soils with high organic contamination are formed. Organic silt and organic clay belong to this group of soils. The organic matter contained by them is subject to continuous changes. The process of transformation may again be either slow carbonization or quick decomposition depending on whether it takes place in the absence or in the presence of air. Because of the decomposition of the organic matter, these soils are highly compressible and possess very low strength.

A commonly used method of determining the quantity of organic matter present in the soil is the so-called weight loss on heating test. In this, a sample is heated to red heat and the loss in weight is considered as a measure of the organic content of the soil. The loss in weight is usually expressed as the percentage of the dry weight. The test is based on the assumption that the organic constituents of the soil are combustible, in contrast to the mineral constituents. The weight of the residual ash is negligible, and the difference between the weights before and after the test can thus be taken as the organic content of the soil.

The test procedure is as follows. First a sample of soil of about 15 to 20 g is dried in an oven. The drying temperature must be kept well below the usual 105 °C to prevent the slow oxidation of organic matter. The drying is therefore effected at a temperature of about 50 °C. The dried specimen is powdered, weighed (W_d), and then placed in a heat-resistant crucible over a gas burner and heated at about 600 °C for a period of 1 to 3 hours. After the burning is complete, the sample is allowed to cool and then weighed again W_b. The loss on heating is calculated as the percentage of the oven dry weight:

$$p\% = \frac{W_b - W_d}{W_d} \cdot 100 . \qquad (52)$$

According to the soil classification system adopted by the Hungarian Standards a soil is considered organic if the loss on heating is more than 10%.

Recent investigations (Biczók, 1955; Silfverberg, 1957) have pointed out that this method is unsuitable for the determination of the true organic content of the soil, since heating at 600 °C not only removes organic matter by burning, but also drives off the lattice water of clay minerals and other crystalline particles present, and the resulting reduction in weight also appears incorrectly as organic content. Preference should, therefore, be given to those methods by means of which the quantities of organic matter, adsorbed water and carbonic acid present in the soil can be determined separately. *Biczók* suggests the use of the *Dennstedt* apparatus for this purpose. A small porcelain boat containing the powdered sample is inserted into a long glass pipe. The sample is then heated while pure oxygen is allowed to flow through the pipe. The products of combustion are collected, and their weights determined separately. According to *Biczók*, a soil should be classified organic if its organic content as determined by the above method is higher than 5%.

There are other methods in which the organic content is determined by a wet oxidation process, e.g. by treatment of the soil with hot sulphuric acid, with chromic acid or with hydrogen peroxide (Clare, 1949).

The organic content may have a marked influence on the physical properties of the soil. The natural water content of organic soils is high. The liquid limit and the plastic limit tend to increase, whereas the plasticity index decreases, with increasing organic content. The changes in plasticity are, in part, due to the fact that clay minerals, particularly montmorillonite, are capable of adsorbing organic molecules (Grim, 1948; Hendricks, 1941). Another cause is that in an organic soil the mineral grains are coated with organic mineral colloids with the result that the soil becomes more hydrophilic. Organic soils are, as a rule, highly compressible. All the same, when these soils occur at considerable depths and are submerged in ground water, and are therefore permanently sealed from the air, the use of deep foundation to support overlying structures is not always justified, since even these soils are capable of carrying some load. For example liver peat, a black colloidal clay formed through humification, often has adequate bearing capacity. In lime-containing soils ulmic acid and humic acid are bonded by Ca ions and this has a beneficial effect on the properties of the soil. Many old buildings in *Budapest* which were actually built on poor ground with a layer of organic soil beneath the foundations show no sign of deterioration attributable to foundation failure. At any rate, it always needs a careful study of the soil conditions to make a realistic estimate of bearing capacity and the magnitude of settlement, and to decide whether, allowing for the type and function of the structure, shallow foundations on the organic soil strata are at all feasible. In a study of the compressibility of organic soils particular attention should be paid to the secondary consolidation, which may be significant because of the physico-chemical changes brought about by the pressure acting on the soil (Salas and Serratosa, 1953).

3.6 Harmful constituents in soils

Both the soil and the ground water often contain chemical substances which are injurious to the building materials of engineering structures and may cause their serious deterioration. This effect is known as *soil corrosion*. The design of adequate protective measures against corrosion often requires soil chemical tests. However, the harmful constituents can exert their influence only when they are dissolved in the ground water and are thus, brought into direct contact with the structure; i.e. they represent real danger only when they exist in an ionized state. The degree of the hazard of soil corrosion depends, on the one hand, on the quality of the ground water, and on the other, on the properties of the building material. The intensity of the corrosive effect is also influenced by a number of other factors such as temperature, the mode of contact, alternating exposure to water and air flow of ground water, water pressure etc. An effective prevention can only be designed if the role of the factors involved and their order of importance are correctly established.

One of the most important building materials is *concrete*. Understandably, the investigation of substances that are aggressive to Portland cement concrete has been widely dealt with in the literature (BICZÓK, 1956). The commonest cause of concrete corrosion is the high *sulphate* content of the soil. By sulphate is meant the ion SO_4. In soils, sulphate ion is present in aqueous solutions of sulphuric acid or of various salts of sulphuric acid. Of the latter the most dangerous are *gypsum* ($CaSO_4 \cdot 2\,H_2O$), Glauber salt ($Na_2SO_4 \cdot 10\,H_2O$), Epsom salt ($MgSO_4 \cdot 7\,H_2O$) and $(NH_4)_2SO_4$. Sulphate ion occurs in all natural waters, including ground water, particularly in clays. Sulphate ion in soils is derived from either chemical transformation of sulphur-containing minerals, or decomposition of sulphur-containing organic matter. The initial reaction is mostly the oxidation of pyrites (FeS_2), (VENDL, 1952):

$$2\,FeS_2 + 2\,H_2O + 7O_2 = 2\,FeSO_4 + 2\,H_2SO_4$$

or

$$4\,FeS_2 + 2\,H_2O + 15O_2 = 2\,Fe_2(SO_4)_3 + 2\,H_2SO_4\,.$$

The sulphuric acid produced reacts with the calcium carbonate contained in the soil to form gypsum:

$$H_2SO_4 + CaCO_3 = CaSO_4 + CO_2 + H_2O\,.$$

The sodium- and potassium-containing compounds of feldpars are converted into sulphates of Na and K. Since the latter readily dissolve in water, the sulphate ion concentration of the ground water and, thus, its reaction with Portland cement concrete will be substantially increased. Similarly, sulphates are formed by the decay of nitrogenous organic substances. Swamps, marshes and undrained shallow ponds are also rich in sulphate ion. In stagnant waters hydrogen sulphide H_2S is converted into sulphate by the sulphur bacteria. Ground water may be strongly polluted by industrial refuse, foul water, coal and slag. Sulphur dioxide present in the smoke produced by industrial plants can also contaminate the soil in the surrounding area.

Sulphate ion reacts with the free lime (slaked lime, $Ca(OH)_2$) content of Portland cement to form crystalline gypsum. The concomitant volumetric expansion of the gypsum exerts a splitting effect on the concrete. This effect is even more intense when the tricalcium aluminate component of Portland cement reacts with gypsum derived from the reaction of sulphate ion with calcium carbonate. The product is hydrated potassium sulphoaluminate. The resulting expansion of the crystals exert immense pressure and can eventually wedge apart the concrete. The reaction is as follows:

$$3\,CaSO_4 + 3\,CaO \cdot Al_2O_3 + 3\,H_2O =$$
$$= 3\,CaO \cdot Al_2O_3 \cdot 3\,CaSO_4 \cdot 3\,(H_2O)\,.$$

Dangerous soil expansion may occur when concentrated sulphuric acid penetrates the soil in an industrial plant.

Protective measures against soil corrosion due to high sulphate content are provided for, in almost every country, by building regulations and bylaws. The properties of commercially available cements, the type of concrete, the cement content of the mix, the density and strength required of the concrete, exposure to still or percolating ground water are, among others, the main factors to be considered. A comprehensive survey of relevant regulations now in force in various countries was given by BICZÓK (1956).

Various *magnesium* salts (e.g. Epsom salt, $MgSO_4 \cdot 7\,H_2O$) are also dangerous; in concretes affected by them magnesium is substituted for the calcium content of the binder and as a result the concrete disintegrates. Ammonium ions occur in industrial wastes. Its reaction with cement is accompanied by the liberation of ammonia gas, and the Ca salt produced is leached out of the concrete.

Industrial wastes may carry a number of other dangerous compounds into the soil. A detailed study of them would be beyond the scope of this book and the reader is advised to consult the numerous reference books available on the subject (BICZÓK, 1956; GÁSPÁR, 1952; KIND, 1955; KLEINLOGEL, 1950).

Ground water with high alkali content is also harmful since it dissolves free lime from the concrete.

Free carbonic acid present in the ground water causes the cations of the binder to pass into solution, whereby the bonding effect between the grains of the aggregate is lost.

Carbonic acid becomes harmful when its concentration is high enough to render the soil acidic ($pH < 7$).

The harmful effect of various aggressive chemical substances is very much intensified by percolating ground water that continuously supplies fresh active material to the concrete. Likewise, under the conditions mentioned, high temperatures and fluctuation of the ground water table promote the corrosion of concrete.

Recommended reading

BERNATZIK, W. (1947): Baugrund und Physik. Schweizer Druck- und Verlagshaus, Zürich.

BICZÓK, I. (1956): Betonkorrózió—betonvédelem (Concrete corrosion and concrete protection). Műszaki Könyvkiadó, Budapest.

BÖLLING, W. H. (1971): Bodenkennziffern und Klassifizierung von Böden. Anwendungsbeispiele und Aufgaben. Springer-Verlag, Wien, New York.

GLÉRIA–KLIMES–SZMIK–DVORACSEK (1957): Talajfizika és talajkollodika (Soil Physics and Soil Colloidics). Akadémiai Kiadó, Budapest.

GRIFFITHS, J. C. (1967): Scientific method in analysis of sediments. McGraw-Hill Book Co., New York.

GRIM, R. E. (1953): Clay Mineralogy. Mc Graw-Hill, London.

GROFCSIK, J. (1956): Kerámia elméleti alapjai (Theoretical fundamentals of Ceramics). Akadémiai Kiadó, Budapest.

KÉZDI, Á. (1966): Grundlagen einer allgemeinen Bodenphysik. VDJ-Zeitschrift, Düsseldorf, Bd. 108, Nr. 5.

KÉZDI, Á. (1967): Stabilizált földutak (Stabilized earth roads). Akadémiai Kiadó, Budapest.

LAMBE, T. W. (1953): The structure of inorganic soil. Proc. ASCE. Sep. No. 315, Oct.

ROSENQVIST, I. TH. (1955): Investigation in the Clay-Electrolyte-Water System. Norvegian Geotechnical Institute.

SCOTT, R. F. (1963): Principles of Soil Mechanics. Addison-Wesley Publishing Co. Inc.

Symposium on Exchange Phenomena in Soils (1953). American Society for Testing Materials, Spec. Techn. Publ. No. 142. Philadelphia.

SANKOV, A. A. (1950); Geokhimiya. Moscow. (In German: Geochemie, Berlin, 1953.)

WINTERKORN, H. F. (1948): Physico-chemical Properties of Soils. Proc. 2nd Int. Conf. on Soil Mechanics and Foundation Engineering. Rotterdam, Vol. I. pp. 23–30.

Chapter 4

Classification of soils

4.1 General considerations

It is the inherent endeavour of every science to classify the things that constitute its subject matter by arranging them into a system of certain mental categories. If this is done on a genetic basis which truly reflects the division by nature of the things under consideration into certain genera and species, a natural classification results. On the other hand, any arbitrary grouping of the subject for some utilitarian purpose can only be valued as an artificial classification. If, in the first instance, we succeed in finding a natural system of classification we may get nearer to the objective essence of the things i.e. to those attributes from which all other properties follow.

So far none of those branches of science concerned with soil have been successful in classifying soils in a natural system in spite of all the effort made towards this. Soil mechanics is no exception to this either, as is best evidenced by the very existence of the wide variety of current classification systems that all attempt, in different ways but always on more or less arbitrary bases, to find a universal qualifying principle, based on which the vast diversity of natural soils could be classified into a unified system. The value of most of these systems is highly questionable. It cannot be denied, of course, that a classification system allowing a given soil to be classed on the basis of a few quick and simple identification tests into a distinct and well-defined category would be of obvious advantage in every field of civil engineering activity. Moreover, the general acceptance of a particular system would greatly facilitate the correct interpretation of soil descriptions used in various technical documents such as boring logs, reports etc. But supposing a workable classification system is found, it would be entirely wrong to expect that all we need then to solve a particular soil mechanics problem is simply to perform a few identification tests based on which we can group the soils into certain classes and obtain the desired data from tables and graphs. Soil behaviour is much too complex to be described adequately unless a large number of physical characteristics is used. But any classification, of necessity, uses as few basic characteristics as possible. Therefore, as Taylor (1948) pointed out, "... any classification is worthless, even dangerous, unless the characteristics it is based on are the ones which are important in the problem under consideration". A further obstacle to developing a dependable soil classification is that one can hardly ever acquire a perfect ability to distinguish, beyond doubt, between different soils by means of mere visual inspection or feel; carrying out at least a few so-called identification tests, mechanical analysis and plasticity index being the most common ones, cannot be dispensed with. But the results of these tests, presented in numerical or graphical form, usually provide better information regarding the behaviour of a given soil than could be learned from simply stating the class to which that soil belongs. It should also be noted that whatever classification we have to deal with, there are always intermediate types between the main soil groups.

For these reasons, the question of classifying soil will be dealt with here only to such an extent as is necessary to meet the need for a uniform nomenclature of soils, for use in field reports, bore hole logs, soil profiles etc.

4.2 Main soil types; their identification and nomenclature

In Section 2.1.1, we have discussed the division, by size, of the soil-forming grains into fractions, and their respective names. The terms used to designate the individual grain-size fractions are, in general, identical to the names commonly adopted for naturally occurring soils. As a rule, a soil is named after the predominant fraction in it.

Before we set out to discuss, in detail, how a given soil could be classified into one group or another, let us deal with those simple criteria which allow soils to be described and identified by means of visual examination.

We have seven simple identification tests that are useful in estimating the role of the constituent fractions in a composite natural soil and thereby facilitate the classification of the soil.

1. *Visual examination of coarse-grained fractions* (sand and gravel). The grains may be described

as angular, rounded, flat etc. in shape. (Cf. Fig. 15). A note should also be made of the hardness of the grains, degree of weathering and any earthy surface coating. To assess the grading of a soil, a representative sample is spread on a flat surface, and the uniformity of the grain sizes is noted.

2. *Dilatancy or shaking test.* This test serves to distinguish between sand, fine sand (Mo) silt and clay. A small wet pat of soil is first shaken in the palm of the hand, then squeezed in the fist between the fingers. If the soil surface becomes shiny, "livery" on shaking, but the sample turns again stiff and crumbly when squeezed, the soil is a fine sand (Mo) or silt. Rapid reaction indicates low plasticity, the soil being a very fine sand or Mo. Slow reaction indicates sandy silt or silt with slight plasticity. Lack of reaction indicates fine silt or clay.

3. *Dry strength.* A small lump of dry soil is crushed by finger pressure. If the sample breaks apart easily under slight pressure, the soil is Mo, silty sand or coarse inorganic silt. High breaking strength indicates that the soil is a sandy or silty clay, or clay of low to medium plasticity. If the sample resists crushing by finger pressure altogether, the soil is an inorganic clay of high plasticity, or a coarse-grained soil aggregate cemented by some high-strength binder (e.g. calcium carbonate or iron oxide).

4. *Plasticity.* Coarse grains larger than 5 mm must be removed from the sample to be tested. A small amount of soil, with some water added, is kneaded between the fingers until thoroughly mixed. It is then rolled on a smooth surface into a thread, repeatedly if necessary, until the thread begins to crumble. With soils of high plasticity the thread is tough, and the crumbling pieces when remolded into a lump can still be deformed by strong finger pressure. A medium tough thread that crumbles when again lumped together indicates medium plasticity. Finally, a weak thread that can be rolled only with care, breaks easily, and cannot be formed into a lump below the plastic limit indicates low plasticity.

5. *Colour and odour.* Wet organic soils have a characteristic pungent odour due to decaying organic matter they contain. On gentle heating of the soil, the odour becomes more intense. Clays when moistened have a distinctive earthy smell.

The colour or any staining of the soil, as indications of its origin or of any contamination present, should always be noted.

6. *Treatment with acid.* A few drops of 20% hydrochloric acid are applied onto the surface of a soil sample; if effervescence results, the soil is calcareous. From the intensity and duration of the reaction the carbonate content can be estimated. A simple correlation suggested by the Swiss Society for Testing Materials is as follows:

Intensity	Duration	Approximate carbonate content
of effervescence		
nil	nil	less than 1%
weak	short	1 to 2%
fairly strong	not lasting	2 to 4%
very strong	lasting	more than 5%

A soil which has high dry strength but exhibits no effervescence in the acid test is probably a highly cohesive clay, whereas a strong reaction indicates that the strength is mainly due to the calcium carbonate content of the soil.

7. *Shine test.* The surface of a lump of dry or slightly moist soil is stroked, under strong pressure, with the flat of a knife blade or with a fingernail. If the surface becomes shiny, the soil is a clay of high plasticity. With silts the surface remains dull.

Coarse-grained soils are denoted by the names *gravel* (when more than half of the grains is larger than 2 mm), or *sand* (with a predominance of grains between the limits 2 mm $>$ d $>$ 0.1 mm). The individual grains can be well distinguished, and their sizes easily gauged with the unaided eye. These soils have high permeability, water can drain freely from their pores, permitting compaction and consolidation under pressure to take place in practically no time. The effect of adsorbed water is negligible. Coarse-grained soils possess no cohesion, but their internal friction is high.

Very fine sand (also referred to as *Mo*) consists predominantly of grains ranging from 0.1 to 0.02 mm. It does not show much resemblance to sands and would be mistaken by the inexperienced observer for a cohesive soil. The grains are just distinguishable with the naked eye. When rubbed in dry state it forms a fine dust that can be easily blown off the fingers, and when wet it can be rolled into a thread only with difficulty, or not at all. It can be easily distinguished from true cohesive soils by the shaking test. Surface phenomena have no appreciable influence on the behaviour of fine sand. Its internal friction is still fairly high, but its permeability is much lower than that of coarse sands. When dry it has a slight cohesion which, however, is highly susceptible to changes in the moisture content, and in wet saturated state the soil easily becomes liquid. Consolidation can take place in this soil practically unhindered, and loads applied to it are transmitted from grain to grain. Fine silty sands may gain appreciable strength due to desiccation in contrast to coarse sands which, on drying out, collapse into a fine powder.

Silt is a fine-grained soil composed predominantly of particles within the size range of 0.02 to 0.002 mm. Water films have an appreciable effect on this soil giving rise to cohesion between the grains. The individual grains can no longer be felt when rubbed between the fingers. The surface of a sample of silt becomes shiny in the shaking test, but the reaction is very slow. A freshly cut surface has a dull, velvety appearance. Unlike sands, silt possesses low internal friction. Settlements under load occur in it at a relatively slow rate of time. Owing to its slight permeability and good compaction properties, silt is a suitable constructional material in such cases where a fair degree of impermeability is required. The capillary rise in silt is quite significant.

The finest particles ($d < 0.002$ mm) constitute the dominant fraction in *clays*. The adsorbed water surrounding the particles has a decisive role in the behaviour of this soil. Increasing water content causes the adsorbed water layer to increase in thickness giving rise to swelling of the clay. Conversely, clay shrinks intensely on drying. When cut with a sharp blade it shows a smooth, shiny surface, but when sprinkled with water it has a soapy, slippery touch since only a very thin superficial layer sucks in water. Clay exposed to water and worked, at the same time, by some mechanical means becomes soft, plastic, and easily deformable. The manner in which applied loads are transmitted in a clay mass is not from grain to grain but from one adsorbed water hull to another. This accounts for the very low internal friction and high compressibility of clays. But because of its low permeability, deformations under load take place at an extraordinarily slow rate and the settlement of buildings founded on clay may often continue for years or even decades. Destroying the skeleton structure typical of some clays may cause a substantial decrease in strength leading to a potential danger of failure, since once a sliding movement has started remoulding inevitably occurs and the shearing strength of the clay may be reduced to a fraction of its initial value. Clays mostly occur as saturated two-phase systems because their formation was possible under water only, and the height of capillary rise in clays is so large that it can prevent the soil from subsequent desiccation except in the uppermost layer near the surface.

Besides the main soil types just discussed, countless intermediate varieties are found in nature. These are mostly mixed soils with variable grain sizes. The properties of such soil mixtures are extremely difficult to determine by mere visual inspection. Let us consider, for example, a composite soil made up of gravel, sand and fines, in a natural packing permitting the direct contact between the coarser grains whilst fine material is filling the interstices between them. As long as the original structure remains intact, this soil exhibits strength and deformation characteristics associated with coarse-grained soils, but behaves like clays as regards permeability. Should, however, the fines be present in excess so that coarser grains can no longer form a firm skeleton but "float" in a clayey matrix, the mixture assumes the properties of pure clays. In such cases reliable information as to the behaviour of the soil can only be obtained from tests on undisturbed samples. Once again, it must be emphasized that the behaviour of a given soil is influenced to a much greater degree by the fine fraction than it contains by the coarser ones.

Coarse gravel and sand are of common occurrence as talus deposits at the foot of mountains and as alluvial valley deposits in the higher reaches of rivers. Sand also occurs as a wind-blown sediment in the form of dunes. A characteristic feature of these soils is the complete lack of fine fractions. Gravel and sand occur also as coastal deposits that often are highly cemented.

Medium-grained soils, fine sand (Mo) and silt, are found mainly in the plains either as water-laid and, thus, water-logged sediments, or as predominantly dry, aeolian deposits known as loess. These sediments built up, in immense thickness, the subsoil of the Great Hungarian Plain, the central basin of the country. Also large areas in the western regions (Transdanubia) are covered with aeolian deposits.

Pure silts are chiefly of glacial origin and were formed during the Ice Age. They generally occur as water-logged deposits.

Clays of high plasticity (fat clays) are, in the main, lake or sea deposits. If the coarser fractions were completely absent, sedimentation took place through flocculation. Such deposits have a flocculent structure and when disturbed in their natural state may suffer a substantial loss of strength; i.e. they are highly sensitive.

It is perhaps little realized how mistaken even experienced engineers with a firm soil mechanics background can often be in estimating grain size and identifying soils by mere visual examination. In this connection an interesting test was made in Berlin in 1950. On a working committee of DEGEBO (German Society of Soil Mechanics) dealing with matters of soil classification, the members — all of them experienced engineers and geologists—were called upon to determine, solely on the basis of visual inspection, grain sizes and gradings for an assortment of soil samples. The test brought an unexpected result in that many of the participants came to conflicting and apparently erroneous conclusions. With fine sands and silts, they generally put grain sizes too high, misjudged the amount of fines and failed to distinguish correctly between fine sand and Mo, or Mo and silt. The test has led to the important conclusion that even the practising soils engineer, however experienced, cannot do without the results of appropriate soil tests if he is to avoid gross mistakes in identifying soils. Such tests should, of course, not be confined, particularly in the case of cohesive soils, to the determination of grain-size distribution. It has become also clear from the test described above that any further refinement in the division of grain-size fractions would be in vain, since too detailed a system would only prove

Table 13. Classification of soils by visual inspection and simple identification tests
[Abridged from Earth Manual, U. Bureau of Reclamation, Fig. 2]

Major divisions				Field observations: Total natural sample	Field observations: Fractions smaller than 3 mm	Typical names	Data required for describing soils
Coarse-grained soils	Gravel and gravelly soils	Gravels	Well-graded	All coarse sizes are present to even proportions (up to maximum in sample)	Enough clayey fines to bind material together very well	Gravel with sand-clay binder	Maximum size, shape, surface texture and hardness of grains; approximate percentage of gravel
					Clean material. Grains visible to the unaided eye. No clayey binder. Material not plastic or cohesive when wet	Well graded gravel and sandy gravel	
			Poorly graded	One or more coarse sizes predominate		Poorly graded gravel and sandy gravel	
		Gravelly soils		Dirty material. Coarse grains with an excess of fines. A considerable proportion of grains not visible to the eye	A large proportion of grains not visible to the eye. Material has little or no cohesion or plasticity when wet	Gravel with excess silt; poorly graded gravel-sand-silt mixtures	Approximate percentage and maximum grain size of gravel; fines very silty or moderately silty
					A large proportion of grains not visible to the eye. Material has distinct cohesion and plasticity when wet	Gravel with excess clay; poorly graded gravel-sand-clay mixtures	Approximate percentage and maximum grain size of gravel; approximate amount of binder
	Sand and sandy soils	Sands	Well-graded	All sizes are present to even proportions (up to maximum size in sample)	Enough clayey fines to bind material together very well	Sand with clay binder	Shape and hardness of sand grains; approximate percentage of gravel; colour of soil
					Clean material. A large proportion of the grains visible to the eye. No binder. Material not plastic or cohesive when wet	Well-graded sand and gravelly sand	Shape and hardness of sand grains; approximate percentage of gravel. Cleanness of sand sizes, colour of soil
			Poorly graded	One or more sand sizes predominate		Poorly graded sand	Sand: coarse, medium or fine; angular or rounded clean or slightly dirty. Colour of soil
		Sandy soils		Dirty material. Coarse grains with an excess of fines. A considerable proportion of the grains are not visible to the eye	A large proportion of particles not visible to the eye. Material has none to little cohesion or plasticity when wet	Sand with excess silt; sand-silt mixtures	Sand: coarse, medium or fine. Silt: much or little. Clay: trace or no appreciable amount. Approximate percentage of gravel. Colour of soil
					A large proportion of particles not visible to the eye. Material has distinct plasticity and cohesion when wet	Sand with excess clay; sand-clay mixtures	Sand: coarse, medium or fine. Clay: much or little. Very, moderately or slightly plastic. Approximate percentage of gravel. Colour of soil

Major divisions		Simple identification tests: Behaviour when wet	Dry strength	Plasticity	Dilatancy test-type of reaction	Typical names	Data required for describing soils
Fine-grained soils	Silt and clay with low compressibility	Non-plastic and non-cohesive	None powdered soil gritty between teeth	Very weak thread easily crumbled	Fast	Very fine sand, rock flour, silty or slightly clayey fine sand	Presence of trace of clay, organic matter should be noted
		Moderately plastic and cohesive	Significant powdered soil may be gritty between teeth	Medium to tough thread. Material adheres to the hand	None	Lean clay; sandy clay; silty clay	Degree of plasticity should be noted. Approximate sand content
		Slightly plastic and cohesive	Slight	Soft weak thread. Material adheres somewhat to the hand	None to very slow	Organic silt	Organic remains. Odour and colour
	Silt and clay with high compressibility	Slightly plastic and cohesive	Slight	Soft weak to medium thread	Slow to moderate	Silt, micaceous or diatomaceous; fine silty or sandy soils	Normally odourless; light pastel colour. Trace of clay if any should be noted
		Very plastic and cohesive	High to very high. Cannot be powdered by finger pressure	Very tough thread. Material very sticky to the hand	None	Fat clay	Strong earthy odour. Presence of organic matter or sand, if any, should be noted
		Moderately plastic and cohesive	Moderate to high	Moderately tough thread. Often soft and fibrous	None	Organic clay	Dark; pungent odour

an impractical and useless tool in the hand of the foremen in charge of the boring.

By making use of the previously described simple identification tests, soils can be classified with the aid of *Table 13*. It gives comprehensive and detailed information in every respect. Its intelligent use, however, presupposes a certain skill in manual soil tests and competence in fundamental soil mechanics investigations.

4.3 Soil classification based on grain size

The basic problem in soil classification by grain size is to decide by what criteria a given mixed-grained soil should be classed as gravel, sand, fine sand (Mo) or silt. The answer to this question, according to *Jáky*, is what he calls the predominant grain size. This is defined as the diameter of the grains that occur with the greatest frequency, i.e. the diameter pertaining to the peak of the grain-size frequency curve. The soil will then be designated by the name of that fraction to which the predominant grain size belongs. This criterion, however, proves workable only if the grain-size distribution curve shows a fairly close resemblance to the normal probability curve. Otherwise, with mixed, irregularly graded soils, the method does not furnish satisfactory results.

Soil classification standards often adopt the simple rule of designating a natural soil by the name of its principal constituent fraction, i.e. the one represented in the highest percentage by weight.

In the past, triangular classification charts have been rather widely used. This method is based on the division of soils into three principal fractions, e.g. "sand", "silt" and "clay". The sum of the percentage of the three components is 100%. The division, by the same principle, of coarse-grained soils into "coarse", "medium", and "fine" grades, together with the corresponding triangular diagram, has been previously discussed (Section 2.1.4 and Fig. 33). Two further examples, *Pietkowsky*'s suggestion and the soil classification chart adopted by the U.S. Public Road Administration, are given in *Fig. 127*.

Another method of classification takes into consideration the entire grain-size distribution curve by comparing it with standard curves. These are arranged in two sets, the individual curves in each set being drawn parallel to one another (i.e. the uniformity coefficient U within one set is constant). The regions between the curves with the lower U value are denoted by even numbers, and those between the curves with the higher value by odd numbers. By drawing the grain-size distribution curve of the soil to be tested on a sheet of tracing paper and placing it over the plot of standard curves, the region into which the curve of the given soil falls can be stated. The soil will then be given the symbol of that region. The method is illustrated in *Fig. 128*. The standard curves may be very useful especially in the classification of coarse-grained, non-cohesive soils of similar geological origin.

4.4 Soil classification based on consistency limits

Soils with almost identical grain-size distribution may widely differ in other physical properties as is strikingly demonstrated by the comparison of the grading curves shown, after A. Casagrande (1936), in *Fig. 129* with the corresponding physical characteristics given in *Table 14*. With soils of almost the same grading (see e.g. curves *1*, *3*, *4*, *5* or *7* and *8*), the *Atterberg* limits and the strength in dry state may show

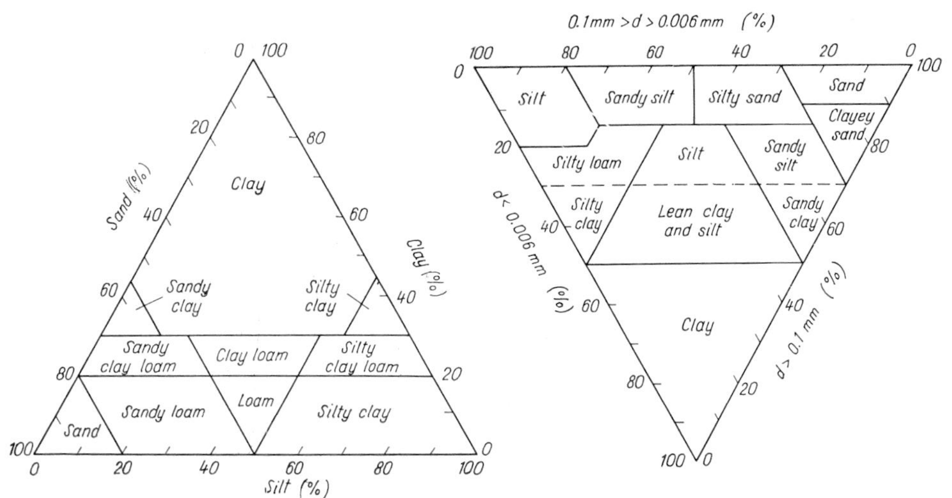

Fig. 127. Triangular soil classification charts:
a – proposal of Pietkowski *b* – Public Road Administration (U.S.A.)

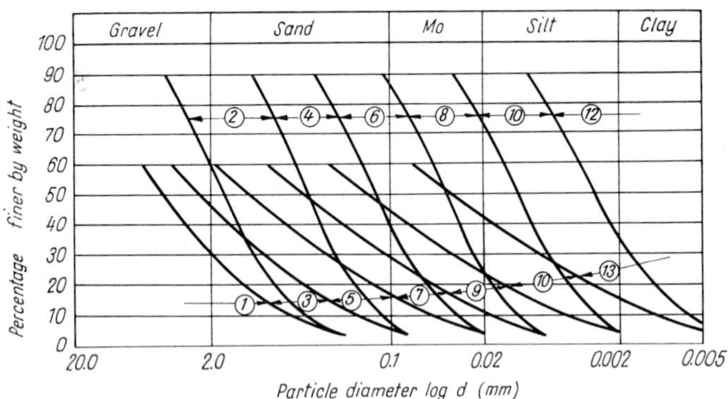

Fig. 128. Soil classification using standard grain-size distribution curves

extraordinary variations. Since these latter properties are much more significant in respect of the behaviour of soils under loads, the classification of cohesive soils should not be based on grain-size distribution. Instead, the consistency limits—plastic limit, liquid limit and plasticity index—can more appropriately be used as a basis of classification.

A useful means of correlating the principal physical characteristics of cohesive soils with the *Atterberg* limits is *Casagrande*'s plasticity chart (CASAGRANDE, 1932). In a rectangular coordinate system, values of the liquid limit are plotted as abscissae and the corresponding plasticity index values as ordinates. Thus every soil is represented by a point. The chart is divided into six regions,

Table 14. Physical properties of soils given by grading curves in Fig. 129

No.	Percent finer than the size given in mm				Liquid limit w_L%	Plastic limit w_P%	Plasticity index I_P%	Dry strength	Soil description
	0.05	0.005	0.002	0.001					
1	90	75	54	35	96	29	67	High	Tough clay of high plasticity
2	97	77	62	50	55	26	29	Medium	Soft clay of medium plasticity
3	92	70	50	37	37	24	13	Medium	Soft, slightly organic clayey silt
4	96	76	50	34	55	31	24	Low	Kaolin clay, typically non-tough
5	96	70	47	34	122	80	42	Very low	Diatomaceous earth
6	92	52	27	15		nonplastic		Almost none	Diatomaceous earth, rock flour like
7	90	26	16	10	35	22	13	Medium	Lean, clayey silt
8	88	20	10	6	24	21	3	Very low	Typical rock flour

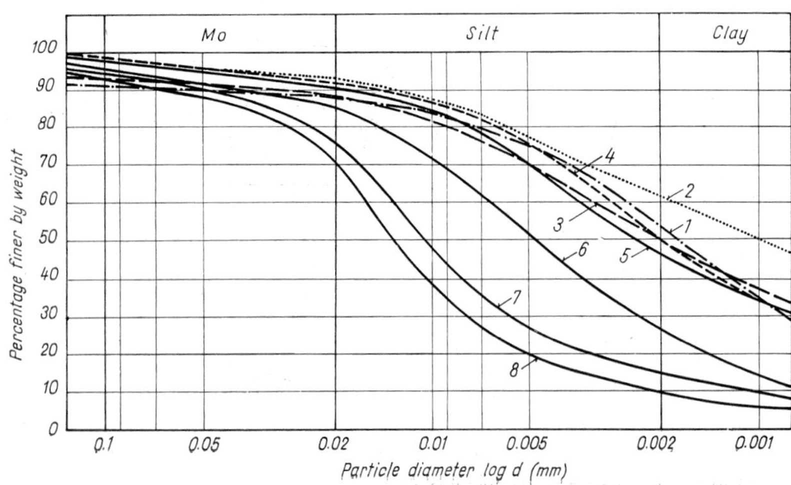

Fig. 129. Soils with different properties but with similar grain-size distribution curves

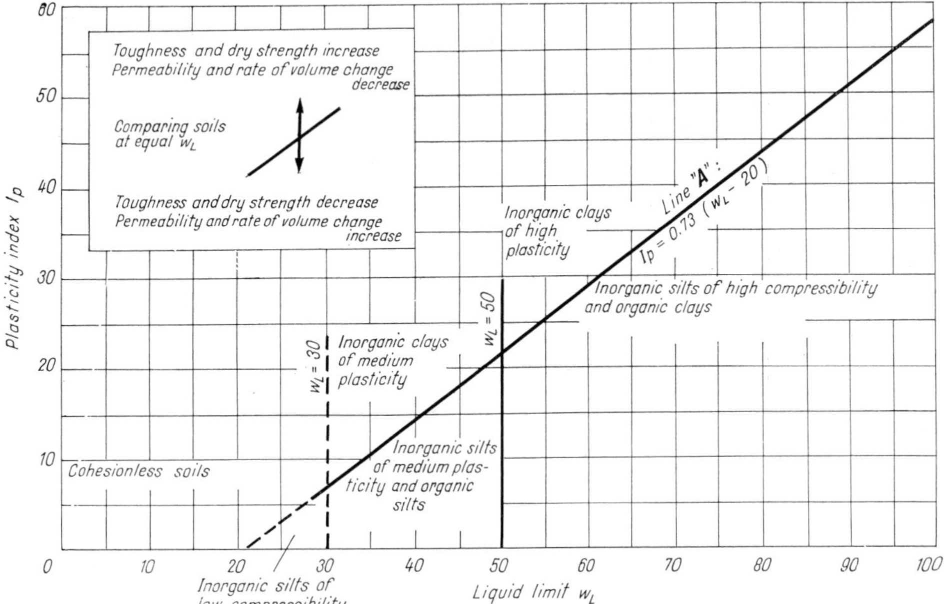

Fig. 130. Plasticity chart by CASAGRANDE

each representing certain soil groups or classes, and a given soil is named after the region into which the point representing the soil falls. Three regions lie below, and three above, a main division line known as line *A* which is defined by the equation

$$I_P = 0.73(w_L - 20)\,.$$

Inorganic clays give points above line *A*, and inorganic silts give points below it. For organic soils such a sharp division line cannot be drawn. Points representing organic clays usually fall within the region assigned to inorganic silts of high compressibility. The organic character of a soil can be easily detected by its distinctive odour and dark colour. In doubtful cases it is advisable to perform the liquid limit test on a natural soil sample as well as an oven-dried one. If the value of the liquid limit for the latter is at least 30% less than that obtained with the natural sample, the soil is classified as organic.

Experience has shown that test results obtained on soil samples which were taken from the same stratum usually give a straight line roughly parallel to line *A*. The position of such a line permits important conclusions to be drawn concerning the behaviour of the soil. These are summed up in *Table 15*.

This classification system furnishes, on the basis of the easily performed *Atterberg* limit tests, valuable information relating to the behaviour of the soil, which may be profitably used in many fields of civil engineering construction.

Plasticity characteristics for typical soils of different geological origin are given in *Fig. 131*.

In addition to the classification systems hitherto discussed, numerous other proposals are dealt with in the literature. Many of them have been devised to suit special requirements e.g. the design of roads and air fields. Based on entirely different principles, there are also geological and pedological classifications.

Table 15. Behaviour of soils

Soil characteristics	At constant liquid limit and increasing plasticity index	At constant plasticity index and increasing liquid limit
Compressibility	about the same	increases
Permeability	decreases	increases
Rate of volume change	decreases	—
Toughness near the plastic limit	increases	decreases
Dry strength	increases	decreases

4.5 Soil classification standards

Undoubtedly, many reasons could be put forward against the standardization of soil classification. On the other hand, the introduction of a compulsory system could be instrumental in bringing the activity of soil mechanics institutions within a country to a uniform basis. As long as classification is not looked upon as a solution for everything but only as an aid to the uniform presentation of technical reports and as a source of rough information on probable soil behaviour, it can serve a very useful purpose. Such considerations have prompted the authorities in a number of countries to adopt standard soil classification systems.

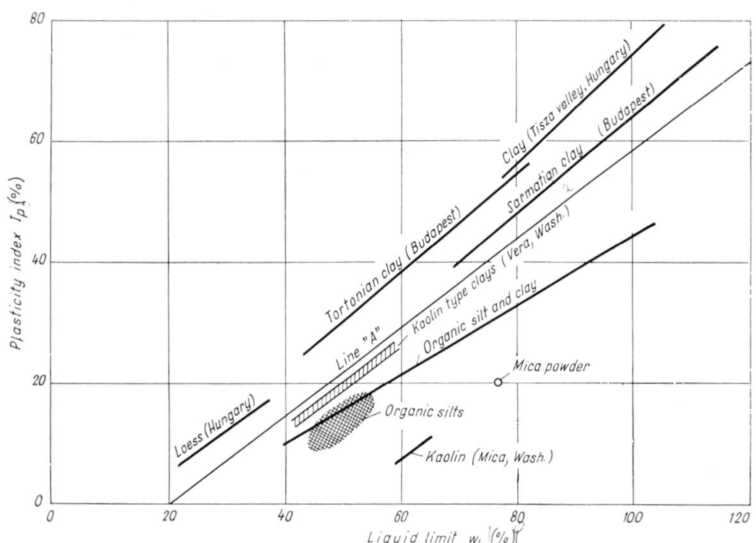

Fig. 131. Classification lines in the plasticity chart

The Hungarian Standards (MSZ 14045) classify all deposits, according to *Table 16*, into "rocks",

Table 16. Hungarian standard classification of geological deposits MSZ 14045

Number	Origin	Main groups	Subdivisions
1	Natural	Solid rocks	Sound Fissured or jointed Weathered
2		Detrital sediments (Soils)	Granular Cohesive Organic
3	Artificial	Fills	Earth fills
			Combustion wastes, slag and ashes Rubbish and debris Industrial wastes Agricultural topsoil

"soils" and "artificial fills". From the standpoint of our subject, the deposits under group 2 (circled in the table) are of particular importance. Further subdivisions of this group (soils), are given in *Tables 17* and *18*. According to the quoted Standard, the term "granular" is applied to soils whose grains form a "pourable" medium, i.e. one consisting of discrete particles that can be moved freely with respect to one another. The individual grains, or more rigorously grains within certain size ranges or fractions, are named according to Table 17. A given soil is designated by the name of that fraction which is represented in the highest proportion in the soil. If a granular soil contains more than 6% of grains finer than 0.02 mm but does not yet exhibit signs of plasticity, this should be indicated by an adjective in the name of the soil (e.g. silty sand).

Soils for which it is possible to determine the liquid limit and the plastic limit, are classified on the basis of the plasticity index according to Table 18.

In the United States, a new system known as Unified Soil Classification System (USC) was adopted in 1952 by the U.S. Bureau of Reclamation. The system is based on the *Casagrande* soil classification for airfield projects (Casagrande, 1947). Since this system has been subsequently adopted by several other states, it is appropriate to give herein a brief account of the essential features of it.

The limits between the grain-size fractions have been fixed in terms of U.S. Standard Sieve sizes. In respect of grain-size distribution, no distinction between silt and clay is recognized by the USC.

Table 17. Hungarian standard classification of grain-size fractions MSZ 14035

Grain-size limits mm	Denominations of grain-size fraction	
	for angular to subangular grains	for subangular to rounded grains
>200	Stones Rip-rap Rock debris	Boulders
200÷20	Coarse rock fragments	Coarsel gravel
20÷2	Fine rock fragments Grit	Fine gravel Cobbles and pebbles
2÷0.5	Coarse sand	
0.5÷0.25	Medium sand	
0.25÷0.10	Fine sand	
0,10÷0.02	Sand flour (very fine sand)	

Table 18. Hungarian standard classification of cohesive soils MSZ 14045

Soil type	Plasticity index I_p%	Standard name
Poorly cohesive	0÷5	Sand flour (very fine sand)
	5÷10	Silty fine sand
Moderately cohesive	10÷15	Silt
	15÷20	Lean clay
Highly cohesive	20÷30	Medium clay
	more than 30	Fat clay

The reasons for this are that the determination of the amount of fine particles by the hydrometer analysis is very inaccurate, and that the properties of cohesive soils are governed to a much greater extent by the surface characteristics of their fine fractions than by the size of the grains (see Ch. 3). The very fine sand (Mo) as a separate fraction has been omitted. The division of the soil fractions, which forms the basis of the classification system is as follows:

GRAVEL

passing 3 in. sieve retained on

		No. 4 sieve	7.62 to 4.76 mm
Coarse	3 in.	to 3/4 in. sieves	76.2 to 19.1 mm
Fine	3/4 in.	to No. 4 sieves	19.1 to 4.76 mm

SAND

passing No. 4 sieve, retained on

		No. 200 sieve	4.76 to 0.074 mm
Coarse	No. 4	to No. 10 sieves	4.76 to 2.00 mm
Medium	No. 10	to No. 40 sieves	2.00 to 0.42 mm
Fine	No. 40	to No. 200 sieves	0.42 to 0.074 mm

FINE-GRAINED COMPONENTS smaller than No. 200 sieve (0.074 mm)

Silt $w_L < 28\%$ $\quad I_P \leq 6\%$
Clay $\quad I_P > 6\%$

Natural soils that in general contain more than one fraction are divided, according to this system, into three main groups: coarse-grained, fine-grained and highly organic.

Coarse-grained soils are divided further into gravels or sands depending on whether half of the coarse fraction is larger or smaller than No. 7 sieve size (2.411 mm). They are classified as well-graded if both of the following requirements are satisfied

$$U = \frac{d_{60}}{d_{10}} \left\} \begin{array}{l} > 4 \text{ for gravels,} \\ > 6 \text{ for sands} \end{array} \right.$$

and

$$C = \frac{d_{30}^2}{d_{60}\, d_{10}} \text{ is between 1 and 3.}$$

With the introduction of this new factor, C, only a rather limited range of soils would qualify as "well-graded", as demonstrated by the examples in *Fig. 132.*

A fine-grained soil is called silt if its plasticity index, determined on a sample with grains larger than No. 10 sieve (0.42 mm) removed, is less than 4 or its *Atterberg* limits give a point below line *A* in the *Casagrande* plasticity chart.

The soil is classed as clay, if the plasticity index (with grains $d < 0.42$ mm) is greater than 7, and the point on the plasticity chart lies above line *A*.

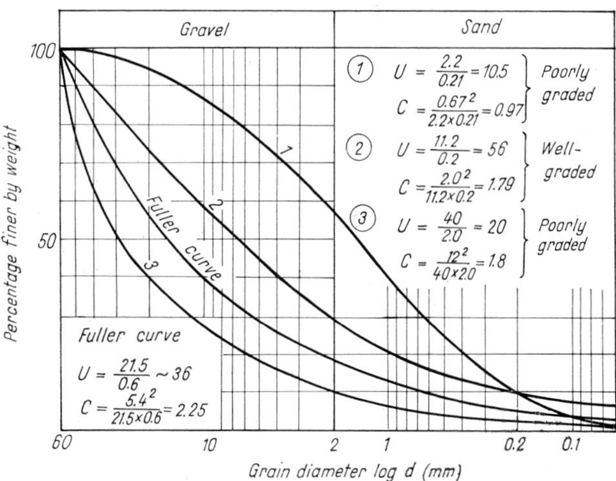

Fig. 132. Grading soils according to USC

On the basis of the above criterion, soils are divided into 15 groups. The six main groups are designated by prefix capital letters as follows:

Gravel	*G*
Sand	*S*
Silt	*M*
Clay	*C*
Organic silt and clay	*O*
Peat	*Pt*

Further suffix letters are used to indicate certain essential characteristics:

Well-graded	*W*
Poorly graded	*P*
Excess of fines	*F*
Low plasticity ($w_L < 50\%$)	*L*
High plasticity ($w_L > 50\%$)	*H*

Thus, each of the 15 groups is denoted by two letters (e.g. *GW*, *SM* or *CH*). Borderline cases are

Table 19. Unified Soil Classification

<table>
<tr><th colspan="8">Field Identification Procedures
(Excluding particles larger than 3 in. and basing fractions on estimated weights)</th><th>Group Symbols a</th><th>Typical Names</th></tr>
<tr><td rowspan="8">Coarse-grained soils
More than half of material is larger than No. 200 sieve size</td><td rowspan="16">(No. 200 sieve size is about the smallest particle to naked eye)</td><td rowspan="4">Gravels
More than half of coarse fraction is larger than No. 7 sieve size</td><td rowspan="8">(For visual classification, the 1/4 in size may be used as equivalent to the No. 7 sieve size)</td><td rowspan="2">Clean gravels (little or no fines)</td><td colspan="3">Wide range in grain size and substantial amounts of all intermediate particle sizes</td><td>GW</td><td>Well-graded gravels, gravel-sand mixtures, little or no fines</td></tr>
<tr><td colspan="3">Predominantly one size or a range of sizes with some intermediate sizes missing</td><td>GP</td><td>Poorly graded gravels, gravel-sand mixtures, little or no fines</td></tr>
<tr><td rowspan="2">Gravels with fines (appreciable amount of fines)</td><td colspan="3">Nonplastic fines (for identification procedures see ML below)</td><td>GM</td><td>Silty gravels, poorly graded sand-silt mixtures</td></tr>
<tr><td colspan="3">Plastic fines (for identification procedures, see CL below)</td><td>GC</td><td>Clayey gravels, poorly graded gravel-sand-clay mixtures</td></tr>
<tr><td rowspan="4">Sands
More than half of coarse fraction is smaller than No. 7 sieve size</td><td rowspan="2">Clean sands (little or no fines)</td><td colspan="3">Wide range in grain sizes and substantial amounts of all intermediate particle sizes</td><td>SW</td><td>Well-graded sands, gravelly sands, little or no fines</td></tr>
<tr><td colspan="3">Predominantly one size or a range of sizes with some intermediate sizes missing</td><td>SP</td><td>Poorly graded sands, gravelly sands, little or no fines</td></tr>
<tr><td rowspan="2">Sands with fines (appreciable amount of fines)</td><td colspan="3">Nonplastic fines (for identification procedures, see ML below)</td><td>SM</td><td>Silty sands, poorly graded sand-silt mixtures</td></tr>
<tr><td colspan="3">Plastic fines (for identification procedures, see CL below)</td><td>SC</td><td>Clayey sands, poorly graded sand-clay mixtures</td></tr>
<tr><td rowspan="8">Fine-grained soils
More than half of material is smaller than No. 200 sieve size</td><td colspan="6">Identification Procedures on Fraction Smaller than No. 40 Sieve Size</td><td></td><td></td></tr>
<tr><td colspan="3" rowspan="4">Silts and clays liquid limit less than 50</td><td>Dry Strength (crushing characteristics)</td><td>Dilatancy (reaction to shaking)</td><td>Toughness (consistency near plastic limit)</td><td></td><td></td></tr>
<tr><td>None to slight</td><td>Quick to slow</td><td>None</td><td>ML</td><td>Inorganic silts and very fine sands, rock flour, silty or clayey fine sands with slight plasticity</td></tr>
<tr><td>Medium to high</td><td>None to very slow</td><td>Medium</td><td>CL</td><td>Inorganic clays of low to medium plasticity, gravelly clays, sandy clays, silty clays, lean clays</td></tr>
<tr><td>Slight to medium</td><td>Slow</td><td>Slight</td><td>OL</td><td>Organic silts and organic silty clays of low plasticity</td></tr>
<tr><td colspan="3" rowspan="3">Silts and clays liquid limit greater than 50</td><td>Slight to medium</td><td>Slow to none</td><td>Slight to medium</td><td>MH</td><td>Inorganic silts, micaceous or diatomaceous fine sandy or silty soils, elastic silts</td></tr>
<tr><td>High to very high</td><td>None</td><td>High</td><td>CH</td><td>Inorganic clays of high plasticity, fat clays</td></tr>
<tr><td>Medium to high</td><td>None to very slow</td><td>Slight to medium</td><td>OH</td><td>Organic clays of medium to high plasticity</td></tr>
<tr><td colspan="5">Highly Organic Soils</td><td colspan="3">Readily identified by colour, odour, spongy feel and frequently by fibrous texture</td><td>Pt</td><td>Peat and other highly organic soils</td></tr>
</table>

Information Required for Describing Soils		Laboratory Classification Criteria		
Give typical name; indicate approximate percentages of sand and gravel; maximum size; angularity, surface condition, and hardness of the coarse grains; local or geologic name and other pertinent descriptive information; and symbols in parentheses For undisturbed soils add information on stratification, degree of compactness, cementation, moisture conditions and drainage characteristics Example: *Silty sand*, gravelly; about 20% hard, angular gravel particles 1/2-in. maximum size; rounded and subangular sand grains coarse to fine, about 15% nonplastic fines with low dry strength; well-compacted and moist in place; alluvial sand; (*SM*)	Use grain-size curve in identifying the fractions as given under field identification	Determine percentages of gravel and sand from grain-size curve Depending on percentage of fines (fraction smaller than No. 200 sieve size) coarse grained soils are classified as follows: Less than 5% — *GW, GP, SW, SP* More than 12% — *GM, GC, SM, SC* 5% to 12% — *Borderline* cases requiring use of dual symbols	$C_U = \frac{D_{60}}{D_{10}}$ Greater than 4 $C_C = \frac{(D_{30})^2}{D_{10} \times D_{60}}$ Between 1 and 3	
			Not meeting all gradation requirements for *GW*	
			Atterberg limits below "A" line, or I_P less than 4	Above "A" line with I_P between 4 and 7 are *borderline* cases requiring use of dual symbols
			Atterberg limits above "A" line, with I_P greater than 7	
			$C_U = \frac{D_{60}}{D_{10}}$ Greater than 6 $C_C = \frac{(D_{30})^2}{D_{10} \times D_{60}}$ Between 1 and 3	
			Not meeting all gradation requirements for *SW*	
			Atterberg limits below "A" line or I_P less than than 5	Above "A" line with I_P between 4 and 7 are *borderline* cases requiring use of dual symbols
			Atterberg limits below "A" line with I_P greater than 7	
Give typical name; indicate degree and character of plasticity, amount and maximum size of coarse grains, colour in wet condition, odour if any, local or geologic name, and other pertinent descriptive information, and symbol in parentheses For undisturbed soils add information on structure, stratification, consistency in undisturbed and remoulded states, moisture and drainage conditions Example: *Clayey silt*, brown: slightly plastic; small percentage of fine sand; numerous vertical root holes; firm and dry in place; loess; (*ML*)		Plasticity chart (see below)		

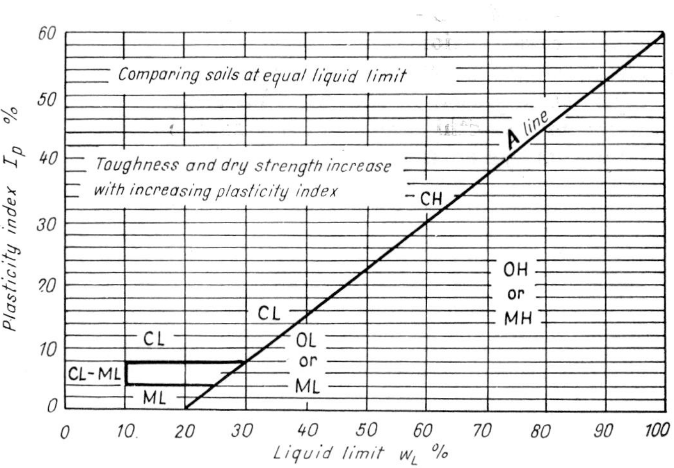

Plasticity chart for laboratory classification of fine grained soils

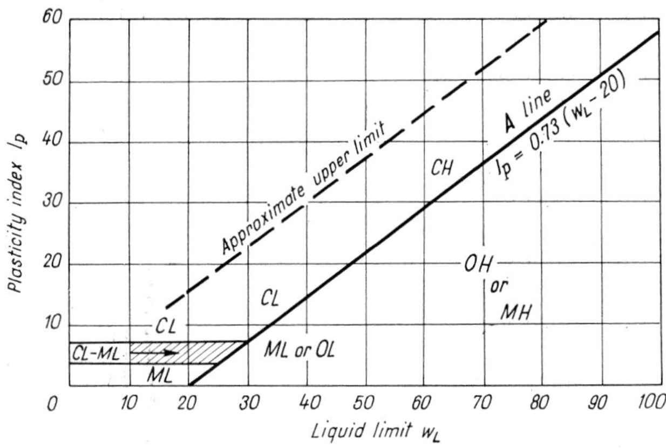

Fig. 133. Position of USC soil groups in the plasticity chart

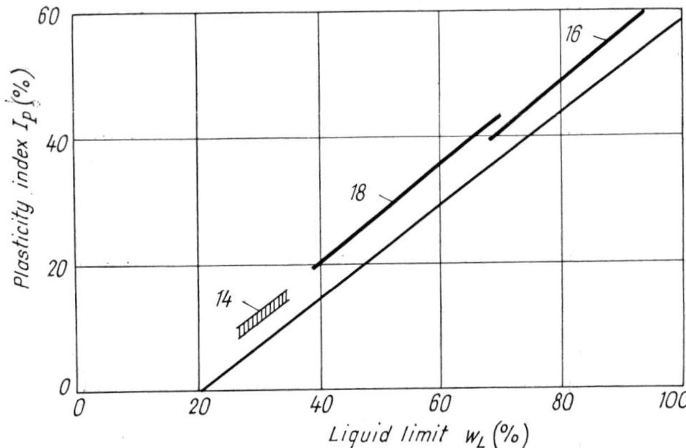

Fig. 134. Classification of the soils given with grain-size distribution curves in Figs 35 and 36 in the plasticity chart

represented by combinations of the group symbols such as *GW — GM, SC — CL, SM — SC* etc.

The complete classification system is shown in *Table 19.* The plasticity chart showing the position of the soil groups is presented in *Fig. 133.* Soils represented by points within the shaded area are designated by dual symbols.

Recommended reading

Bureau of Reclamation: Earth Manual, Denver, 1950.

Bureau of Reclamation: Earth Manual, Denver, Colorado, 1960.

Casagrande, A. (1947): Classification and Identification of Soils. Transactions, Am. Soc. Civil Engrs. Vol. 113.

Childs, E. C. (1969): An Introduction to the Physical Basis of Soil Water Phenomena. J. Wiley & Sons Ltd: London, New York.

DIN 18197 (Entwurf) 1967: Bodenklassifikation für bautechnische Zwecke und Methoden zum Erkennen von Bodengruppen.

Dücker (1957): Feldmässige Erkennungsverfahren zum Benennen der natürlichen, mineralischen Bodenarten nach DIN 4022. Bohrtechnik-Brunnenbau 8, p. 284 and p. 308.

Symposium on the Identification and Classification of Soils. American Society for Testing Materials, Spec. Techn. Publ. No. 113. Philadelphia, 1951.

Schultze, E.–Muhs, H. (1967): Bodenuntersuchungen für Ingenieurbauten, 2. Auflage: Springer-Verlag, Berlin/Heidelberg/New York.

Ungár, T. (1957): Üledék és talajosztályozások összehasonlítása (A comparison of sediment classification and soil classification systems.) Hidrológiai Közlöny, Vol. 37, No. 1, pp. 34–43.

van Olphen, H. (1963): Clay Colloid Chemistry. Interscience Publisher, New York—London.

Yong, N. R., Warkentin, B. P. (1966): Introduction to Soil Behavior. The Macmillan Company, New York.

Chapter 5

Stresses in soils

5.1 Effective and neutral stresses

Chapter 3 presented a general idea of soil structure. Soils were considered as disperse systems composed of solid grains, water and air. The solid grains consist either of fragments of composite rocks or of individual crystals of various minerals. In either case, the deformation of the grains themselves, when they are subjected to relatively low pressures normally encountered in engineering practice, is hardly appreciable, and therefore the grains can be assumed to be practically incompressible (see also Ch. 8). The same holds true for the water contained in the pores and the more so for the water adsorbed at the surface of the grains. All this is clearly demonstrated by the data of *Table 20*, in which values of modulus of

Table 20. Modulus of elasticity for various materials

Material	Modulus of elasticity E N/mm²
Air	—
Water	2,500
Quartz	20,000
Clay minerals	10,000
Steel	215,000

elasticity for various materials are listed. Consequently, the decrease in volume a soil mass will suffer when subjected to loads is in general not due to the compression of the constituent parts, but to excess water and air being squeezed out of the pores. In the course of the process, stresses are induced in all three phases giving rise to displacements of the separate phases in relation to each other.

The term "stress" is defined by the theory of strength of materials as follows:

Consider a rigid body acted upon by a system of external forces in equilibrium conditions (*Fig. 135*). Let us imagine the body divided into two parts by a plane section. Supposing that one of these parts with all the forces acting on it is removed, its action on the remaining part should be substituted by auxiliary forces in order to maintain equilibrium and restore the original state of deformation. It is assumed that these forces are continously distributed over the cross-section so that every elementary area ΔA is acted upon by an elementary force ΔP. The quotient $\Delta P/\Delta A$ approaches a limiting value as the elementary area approaches zero. This limiting value $\lim_{\Delta A \to \infty} \Delta P/\Delta A$ is referred to as stress.

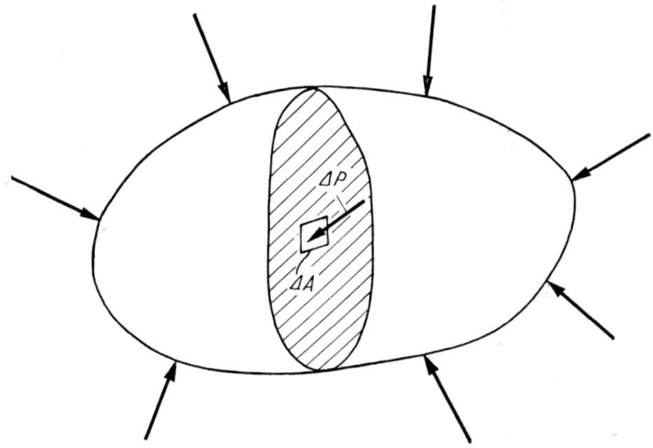

Fig. 135. Concept and definition of stress

The resultant stress p acting on an elementary area dA can be resolved, in the general case, into two components: a normal stress, σ, perpendicular to the area, and a shearing stress, τ, acting in the plane of the area.

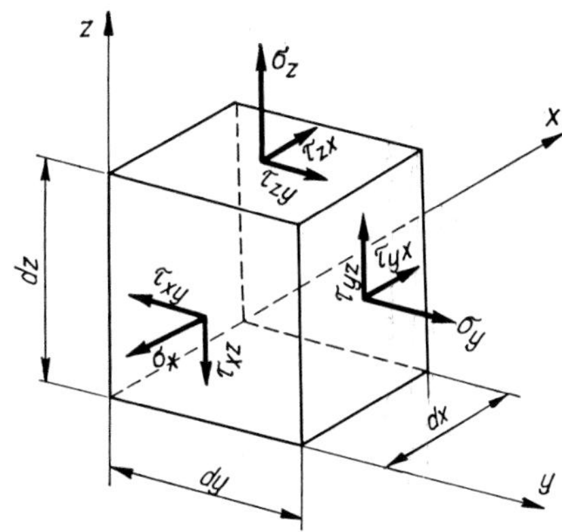

Fig. 136. Stresses acting on the sides of an elementary cube

Considering now plane sections passing through a given point at different inclinations, the stress acting on them will also be different. If we cut out an elementary prism from the body in the manner shown in *Fig. 136*, stresses will be acting on all six sides of it. On each side, the resultant stress can be resolved, parallel to the coordinate axes x, y, z, into three components. Of these, one will be normal and the other two tangential to the side in question. The stresses acting against opposite sides differ only by infinitesimally small amounts and can be, therefore, assumed to be equal. Thus, the number of stress components is reduced to $3 \times 3 = 9$. These are denoted by the symbols shown in Fig. 136.

In soil mechanics, contrary to the usual sign convention, compressive stresses are taken as positive, and tensile stresses as negative.

In strength of materials as well as in the theory of elasticity we deal with materials assumed to be homogeneous to which the concept of differential elements is applicable. Soils, on the contrary, are disperse systems in which loads are transmitted by intergranular pressure at the points of contact between the grains. Therefore, on a plane section through a loaded soil mass the actual distribution of stresses will never be continuous. Taking as the simplest case the stress distribution within a mass of completely dry sand or gravel, it will be seen from *Fig. 137* that loading induces high stresses at the points of contact whereas the stress increment in the pores will be zero. Now for a "continuous" stress distribution which would fit in with the theoretical study of granular media, "stress" can only be conceived as the average, with peaks levelled off, of the actual stresses within a sufficiently large portion of the cross-section. Consequently, when we talk of stress in soils, we are referring to a statistical mean value averaged over a sufficiently large area even if the actual contact stresses have been computed by infinitesimal calculus.

The pore space between the grains is, in the general case, filled partly with water and partly with gas. Any load acting upon the whole granular mass will, in general, be jointly carried by all three phases, stresses being induced in each one of them.

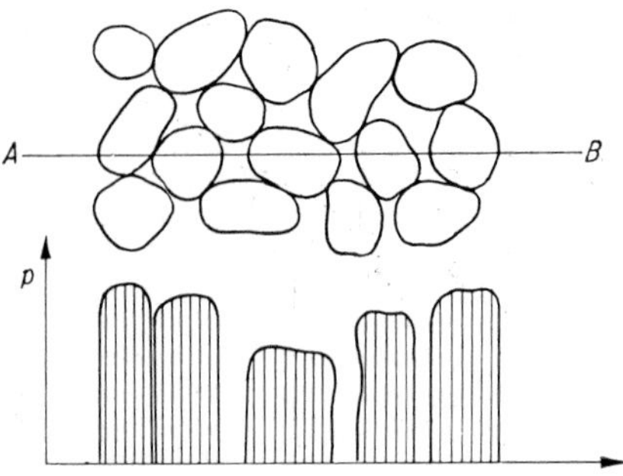

Fig. 137. Distribution of stresses on a plane section in soils

Let us now examine the state of stress of such disperse systems. First, it will be assumed that the grains are so large that surface forces of electrical nature can be neglected. This is indeed the case with sands and gravels.

Let us consider a plane section through a partially saturated granular mass (*Fig. 138a*). This plane will pass in part through mineral grains, in part through water, and in part through air. *Figure 139b* gives a simplified picture in which the minute contact surfaces are combined into a single contact area A_s and the assemblage of solid particles is fused into two grains. Likewise, the water present in the soil is depicted as contact moisture and the rest of the pore space is occupied by the air. The two grains in contact are pressed together by force P. We shall further assume that this force has no tangential component. Let the total area upon which the load P acts be denoted by A. This is composed of three parts: the contact area between the solid grains A_s and the surfaces A_w and A_g, passing through water and air, respectively. Let the respective portions of the stress carried by the separate phases be denoted by p_s, p_w and p_g. Equilibrium requires that

$$P = p_s A_s + p_w A_w + p_g A_g \,. \tag{53}$$

Introducing the following ratios

$$\varphi_s = \frac{A_s}{A}, \quad \varphi_w = A_w/A \quad \text{and} \quad \varphi_g = A_g/A \tag{54}$$

the total stress acting across the section can be written as

$$\sigma = \frac{P}{A} = p_s\varphi_s + p_w\varphi_w + p_g\varphi_g \,. \tag{55}$$

In saturated soils

$$\varphi_g = 0 \text{ and } p_g = 0,$$

hence

$$\varphi_w = 1 - \varphi_s$$

and

$$\sigma = p_s\varphi_s + (1 - \varphi_s)\, p_w \,. \tag{55a}$$

In practical cases the contact area ratio is usually very small and thus the factor $(1 - \varphi_s)$ differs only little from unity. Therefore, we may write

$$\varphi_w p_w = (1 - \varphi_s)\, p_w \cong p_w \,.$$

The contact stress p_s, on the other hand, is very high and may even reach the yield limit of the mineral grains: therefore, the product $p_s \varphi_s$ cannot be zero but has a finite value. This value is called in soil mechanics the effective stress. It is denoted by $\bar{\sigma}$. The second term in Eq. (55a) represents the pressure in the pore water and is known as pore-water pressure or briefly pore pressure. It is also called the neutral stress and is denoted by u.

Equation (55a) may now be written in the form

$$\sigma = \bar{\sigma} + p_w = \bar{\sigma} + (u_0 + u)\,. \tag{56}$$

Herein, u_0 denotes the hydrostatic pressure in the pore water in equilibrium conditions, and u is the unbalanced pore pressure (or excess pore pressure) due to external load or other causes.

In the other extreme case where the granular mass is completely dry, loading may again produce neutral stresses, this time however in the gaseous phase. If the excess air is allowed to move freely under the action of these neutral stresses out of the pores, the pore air pressure will rapidly diminish to the atmospheric pressure. In a closed system, however, the pore air pressure may be lasting, similar to the neutral stresses in the pore water in undrained conditions.

In the general case—partially saturated soil—stresses are developed in all three phases. The effective stress ($\bar{\sigma} = p_s \varphi_s$) i.e. the stress acting at the points of contact between the grains enables the system to resist shearing stresses. This general state of stress can be graphically represented in a triangular diagram (Fig. 139), namely in the form

$$\frac{\bar{\sigma}}{\sigma} + \frac{u}{\sigma} + \frac{p_g}{\sigma} = 1$$

wherein

$$\sigma = \bar{\sigma} + u + p_g$$

represents the total stress. This diagram is a very useful aid to illustrate changes in the state of stress when the total stress σ remains constant. For example, the initial pressures in the pore water u and in the pore air p_g due to a total pressure increment σ will dissipate according to the "vector" DB until the state $u = p_g = 0$ is reached. In two-phase systems (solids–air, solids–water), the process of pore pressure dissipation is represented by a point moving on one side of the triangle, from a to b, or from c to b, respectively.

In partially saturated soils then, we have pressures also in the air phase to be reckoned with. As discussed before, air may exist in an unsaturated soil either in the form of small entrapped bubbles that can move freely in the aqueous matrix filling the pore spaces, or else forming accumulations large enough to come into direct contact with the solid grains and to fill several adjoining pores. The volume of an air bubble depends on the pressure prevailing in the entrapped air which, in turn, is governed by the maximum curvature of the bubble surface, by the surface tension of the water and by the hydrostatic pressure in the water. The total gas pressure inside the bubble can be given as

$$p_g = u + u_0 + p_e + p_a \tag{57}$$

wherein u and u_0 are the components of the neutral stress in the pore water as defined by Eq. (56), $p_e = 2\,T_s/r$ is the capillary pressure derived from the surface tension T_s of the water [Eq. (19)], r is the radius of curvature of the bubble, and p_a is the atmospheric pressure. Variation of the pore water pressure will cause variation of the gas pressure inside the bubble and, hence, of the volume of the bubble as well. This is due partly to the isothermic volume change of the pore air according to the Boyle–Mariotte law, partly to the solubility of air in water governed by *Henry*'s law. By taking all the effects into consideration, a general equation of state expressing the variations with time of the stress components can be derived for the soil.

The simple mechanical picture presented in the foregoing applies only to those soils in which the influence of the surface forces can be neglected owing to their small specific surface. But for clays such an

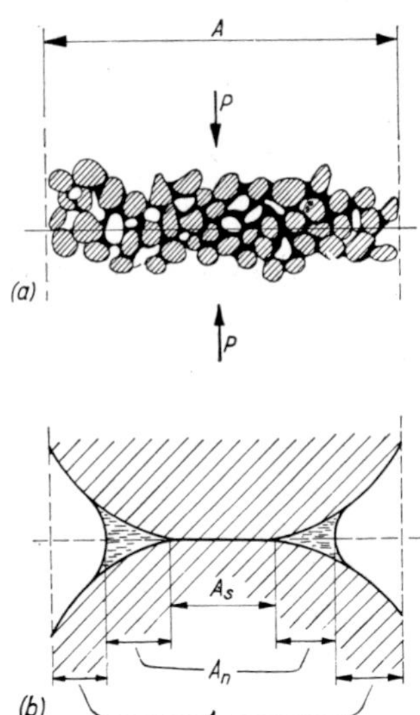

Fig. 138

a — section in granular medium; *b* — stresses in the different components of soil

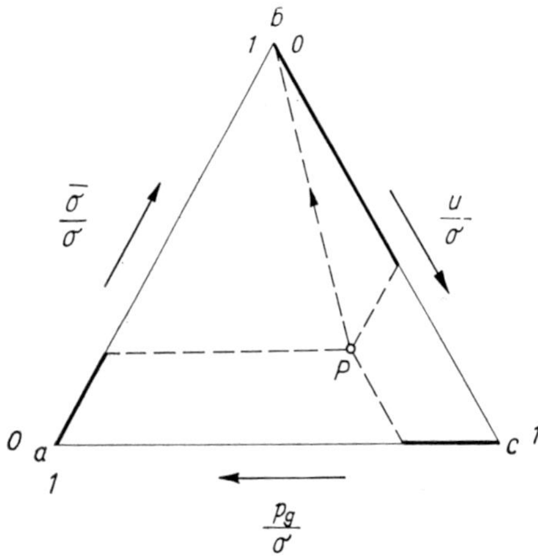

Fig. 139. Visualization of a general stressed state in a triangular diagram by plotting ratios of stresses acting in the different phases

approximation would not be justified and Eqs (55) and (57) must be rewritten to allow for the physicochemical effects between the fine particles. LAMBE (1950) suggested for saturated clays the use of the following modified formula:

$$\sigma = p_s\varphi_s + (u + u_0)(1 - \varphi_s) + \sigma_r - \sigma_a \, . \quad (58)$$

Herein σ denotes the total stress (per unit area of the cross-section) due to external load, p_s is the mean value of stresses acting at the points of direct contact between the grains, φ_s is the ratio of the sum of the contact areas to the total area of the cross-section, and σ_r and σ_a denote, respectively, the interparticle forces of repulsion or attraction per unit area of the total cross-section. Unfortunately, this relationship is much too intricate to permit a numerical evaluation of the terms involved, therefore, Eq. (58) should be primarily looked upon as one of theoretical significance. Nevertheless some practical conclusions can still be made.

Let the average intergranular stress be denoted by $\bar{\sigma}'$. If we further assume that the ratio φ_s is so small as to permit the approximation $1-\varphi_s \approx 1$, then for the total stress we have

$$\sigma = \bar{\sigma}' + \sigma_r - \sigma_a + (u + u_0) \, . \quad (59)$$

Comparing this with Eq. (56), it will be seen that the effective stress, in the general case, cannot be simply taken as equivalent to the stress acting at the contact surfaces between the grains but is obtained from the more composite expression $\bar{\sigma} = \bar{\sigma}' + \sigma_r - \sigma_a$. For a dispersed soil where there is no actual contact between the grains, $\bar{\sigma} = 0$ and, hence, the effective stress $\bar{\sigma} = \sigma_r - \sigma_a$.

For flocculated clays the equation of equilibrium takes the form

$$p_s\varphi_s = \sigma - (\sigma_r - \sigma_a) \, . \quad (60)$$

From the diagrams in Fig. 108 it is clear that when the grains are in actual contact or else the gap between them is very small, the repulsive force relative to the attractive force and the total stress is negligible (i.e. $\sigma_r \cong 0$ and $\sigma \cong 0$), and Eq. (60) simplifies to

$$p_s = \frac{\sigma_a}{\varphi_s} \quad (61)$$

i.e. we can obtain the intergranular stress p_s by dividing the stress due to the force of attraction by the contact area ratio φ_s. This stress value is equivalent to the yield limit of the mineral matter at the point of contact. Consequently in a flocculated clay the average contact stress between the grains is constant and remains unaffected by an increase of the external pressure.

The above discussion is of particular importance in the study of the shearing strength of clays.

In the following we present two examples intended to give another approach to the concept of the two kinds of stresses.

Neutral stresses in soils may arise from many different causes. First of all, neutral stress presents itself as the hydrostatic pressure of the water at rest. But compression, surface tension of the water, expansion or contraction during shear may also produce neutral stress. It can assume both positive and negative values. Since water in itself has practically no shearing strength, frictional resistance cannot arise either through the action of neutral stress. Therefore, in problems of stability and of deformation (Table 1) it is important to know the values of the neutral stress.

Now, let us turn to the examples.

Figure 140a shows the cross-section of a thin layer of soil placed onto the bottom of a container. The pores are saturated with water and the free water surface is level with the top of the layer. If a uniformly distributed load p is applied to the surface of the sample, e.g. by covering it with lead shot (Fig. 140*b*), then the sample will be compressed: the surplus water will be squeezed out of the pores and as a result the void ratio of the soil decreases from the initial value e_0 to e. Loading brings about marked changes also in other mechanical properties of the soil: water content and compressibility are decreased and density and shearing strength increased. The stress causing such effects is called the effective stress.

In a similar setup (Fig. 140*c*), we raise the water level to an elevation h_w above its original position so that the water load per unit of area is equivalent to the load applied in the first test, i.e. $h_w = p\gamma_w$. Then the pressure increment on the section $a - b$ is exactly the same as before. Yet, this time the increase in pressure does not produce any measurable compression and does not affect the other mechanical properties of the soil either, provided the water communicates freely with the atmosphere throughout the test. This kind of stress will not be carried by the skeleton; there will be merely an increase in the hydrostatic pressure in the pore water. This stress is known as neutral stress, u.

The pressure in the water is measured by the height h_w to which the water rises in a standpipe,

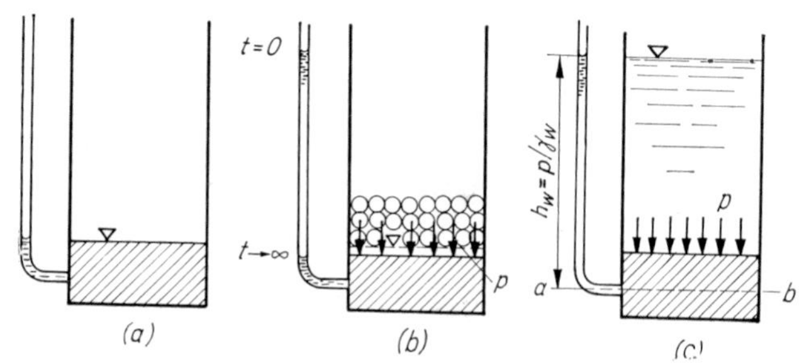

Fig. 140. Effective and neutral stresses in a soil layer

a — saturated sample with no surcharge; sample loaded with lead shots (*b*), and with the weight of a water column (*c*) called piezometer, installed at the point under consideration. The height h_w is termed the piezometric head. At the beginning of the test described above (case *a*), the water level in the piezometer is identical with the top surface of the sample. In case *b*, the water rises to the height $h_w = p/\gamma_w$, but as the excess water is being gradually squeezed out through the upper surface of the sample, the water level in the standpipe drops accordingly until it stands again in the original position, level with the top of the layer. In terms of stresses this means that in the first moment the total load is carried by the neutral stresses in the pore water; the value of the effective stress is zero. As time goes on, the neutral stresses gradually diminish to zero, while the effective stresses carry an ever increasing portion of the total load. At any stage during the process, the sum of the two kinds of stress is equivalent to the total applied load *p*. In the example, this sum is constant, the load itself being unaltered. In case *c*, the water in the standpipe rises again to height h_w, but this time remains permanently in that position indicating that the neutral stresses produced by the water load cannot be converted into effective stresses.

The neutral stress may assume positive and negative values alike. If *u* is positive, it is usually called the pore water pressure. Negative values indicate tension in the pore water, caused for example by capillary action.

According to the above considerations, the total normal stress acting on a section through a saturated soil mass consists of two parts: the effective stress $\bar{p}$ that is transmitted through the skeleton, and the neutral stress *u* that is carried by the water,

$$\boxed{p = \bar{p} + u\,.} \tag{62}$$

This is regarded as one of the fundamental equations of Soil Mechanics. From Eq. (62), the effective stress is obtained by subtracting the neutral stress from the value of the total stress: $\bar{p} = p - u$.

In the second example we shall consider the vertical stresses in a soil profile. *Figure 141* shows a vertical section through an extensive layer of homogeneous, saturated sand. The water table stands at height h_w above the horizontal ground surface. It is assumed that the water is at rest. Since there is no variation in the state of stress in the horizontal direction, the total stress acting upon a plane section at a depth *z* below the ground surface can be simply obtained as the sum of the weights, per unit area, of the soil and of the water:

$$p = h_w \gamma_w + z\gamma_{sat}\,, \tag{63}$$

γ_{sat} is the unit weight of the saturated soil.

The neutral stress is equal to the pressure in the pore water. Since the piezometric head above the

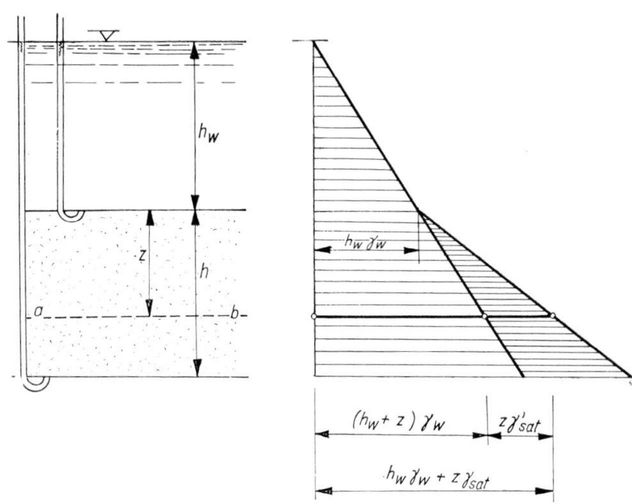

Fig. 141. Effective and neutral stresses in a saturated sand layer covered with water

plane *a* — *b* is $h_w + z$, the neutral stress

$$u = (h_w + z)\,\gamma_w\,. \tag{64}$$

In this case (static ground water conditions), the neutral stresses are solely due to the hydrostatic pressure in the water. Since the voids are completely saturated, this pressure is transmitted through the pore water. If the values of the total stress and neutral stress are known, the effective stress can be obtained from Eq. (62):

$$\bar{p} = p - u = h_w\gamma_w + z\gamma_{sat} - (h_w + z)\,\gamma_w = \\ = z(\gamma_{sat} - \gamma_w) = z\gamma'_{sat}\,, \tag{65}$$

where γ'_{sat} is the submerged unit weight of the soil. As seen from the above equation, the magnitude of the effective stress is independent of the piezometric head at the point under consideration.

The variation of the vertical stresses with depth in a soil profile is illustrated in Fig. 114.

5.2 Stresses in soil; steady flow of water

In Section 5.1 we considered the case where no movement of the pore water, or of the pore air, occurred, and we derived formulas for the stresses in the state of rest. In many practical cases, however, water does move in the pores. Its movement may take place without any volume change of the soil, or it is accompanied by a simultaneous expansion or compression. The stresses giving rise to, or caused by, these movements are of paramount importance to the understanding and quantitative analysis of those processes. Therefore, in the following we discuss the stresses arising from the flow of water through the soil. First we deal with the flow of water through a saturated sand mass.

Water flow occurs when the energy of the water differs at various points within the soil mass. The total energy is the sum of the potential and kinetic energy:

$$E = z + \frac{u}{\gamma_w} + \frac{v^2}{2g}. \tag{66}$$

In this formula, the first term is the elevation head, the second the pressure head and the third the velocity head. Because of the very low seepage velocities in soil, the last term can be neglected. The first two terms give the potential of the movement:

$$h = z + \frac{u}{\gamma_w}, \tag{67}$$

where u is the pore water pressure (*Fig. 142*). The lines drawn through points of equal potential are called equipotential lines. If there is a difference in the potential between two points in the soil, water flow results. The loss of potential per unit of length is the gradientin the given direction:

$$\bar{i} = -\frac{\Delta h}{\Delta s} \rightarrow -\frac{\mathrm{d}h}{\mathrm{d}s} = \mathrm{grad}\,(-h)\,. \tag{68}$$

$\bar{i}$ is a vector, which gives the direction and the magnitude of the steepest slope of the potential surface at the point under consideration.

To illustrate the meanings of the terms in the basic energy equation [Eq. (66)], two examples are given in *Fig. 143*. Here we are concerned with one-dimensional flow. The samples have the same length, and the elevations of head water and tail water are also the same in both cases. The values of the potential at the top and the bottom of the samples are indicated by piezometers. Pressure heads and elevation heads are also given for each case, in tabulated form, in Fig. 143. The gradients

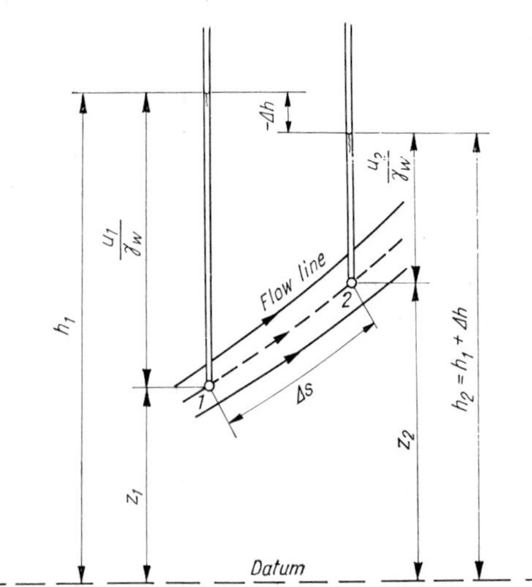

Fig. 142. Definition of terms in the energy equation

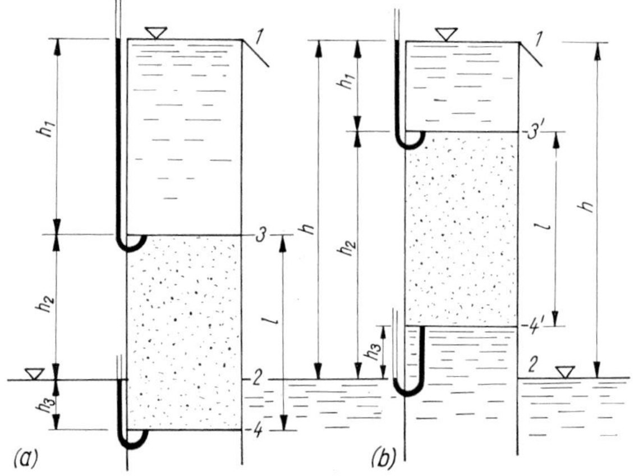

Case	Point	Pressure head	Elevation head	Total head	Head lost
a	3	h_1	h_2	$h_1+h_2=h$	h
	4	h_3	$-h_3$	$h_3-h_3=0$	
b	3'	h_1	h_2	$h_1+h_2=h$	h
	4'	$-h_3$	h_3	$-h_3+h_3=0$	

Fig. 143. Calculation of heads in the case of downward vertical seepage

are equal, $i = h/l$, in both cases. We can now compute the stresses in a soil mass through which flow of water occurs. We shall first consider the simple case of one-dimensional flow illustrated in Fig. 141. It will be further assumed that the sand layer is saturated, homogeneous and isotropic. As shown in Fig. 141, under a static ground water condition the water levels in the standpipes installed at the upper and lower limits of the sand layer are the same and, therefore, flow of water does not occur. If there is a difference of Δh in the total head (i.e. in the elevations of the water in the standpipes between two points in a soil mass), water flow takes place. If the total head is smaller at the lower boundary of the layer, the water flows vertically downwards (*Fig. 144a*). The neutral stress at this lower surface is then

$$u = (h_w + h - \Delta h)\,\gamma_w\,. \tag{69}$$

As compared with the state of rest, there is a decrease of Δh in the neutral stress. This decrease is not due to the velocity of flow, since the velocity term, $v^2/2g$, is negligible for the velocities occurring in soils. The total stress is determined exclusively by the weights of the soil and the water. The total stress remains therefore the same as it was in the case shown in Fig. 141, and a decrease in the neutral stress must, according to Eq. (62), result in an equal increase in the effective stress. This increment is called the seepage pressure; it is due to the frictional resistance to flow of water on the surface of the solid grains. The increment of the effective stress on a plane *a—b* at a depth *z* is, by proportionality, $z\gamma_w \Delta h/h$. Here $\Delta h/h$ is the hydraulic gradient, and the seepage pressure can thus be given as $zi\gamma_w$. The effective stress on the plane *a — b* can be written in the form

$$\bar{p} = z\gamma'_{sat} + zi\gamma_w . \tag{70}$$

If the total head at the bottom of a layer is greater than at the top, the water moves upward, and the neutral stress at the bottom increases by an amount of $\Delta h \gamma_w$ (Fig. 144*b*). The effective stress is reduced accordingly to

$$\bar{p} = z\gamma'_{sat} - zi\gamma_w .$$

Figure 144 gives the diagrams of the distribution of stresses for both cases discussed, with the stresses at rest being shown by dashed lines.

By increasing the total head difference Δh in the second case—upward flow—, a critical condition can be reached when the effective stresses in the soil are reduced to zero. Then

$$z\gamma'_{sat} - zi\gamma_w = 0 .$$

This condition arises when the upward gradient

$$i = i_{crit} = \frac{\gamma'_{sat}}{\gamma_w} . \tag{71}$$

i_{crit} is called the *critical hydraulic gradient.* If $i = i_{crit}$, the effective stress and, hence, the shearing strength and the bearing capacity of a non-cohesive soil become zero. The soil mass is said to be then in a *buoyant condition.*

In the normal case, when $i < i_{crit}$, the total head difference Δh is spent in driving the water through the pores of the soil. This energy is consumed in viscous friction, thereby exerting a force in the direction of motion. In flow of water through the soil, this frictional force, or "drag", acts on the surface of the grains which form the walls of the pores.

Consider the upward vertical flow illustrated in *Fig. 145.* The head lost in flow from a plane at depth *h* to the ground surface is Δh, and the upward seepage force acting on the bottom of a soil column with a height *h* and a unit cross-section area is, as shown before, $\Delta h \cdot \gamma_w \cdot 1$. In a uniform flow this force is uniformly distributed throughout the soil column, and the force per unit of volume of soil can be expressed as

$$\frac{\Delta h \cdot \gamma_w}{h \cdot 1} = i\gamma_w . \tag{72}$$

This represents a body force called *seepage force.* Like other body forces, it is usually given in units of N/cm^3 or kN/m^3. In an isotropic soil the seepage force always acts in the direction of flow, and its magnitude per unit soil volume is the hydraulic gradient times the unit weight of water. It is a vector and can thus be vertically added to the force of gravity. In one-dimensional vertical flow the resultant body force can be obtained in two

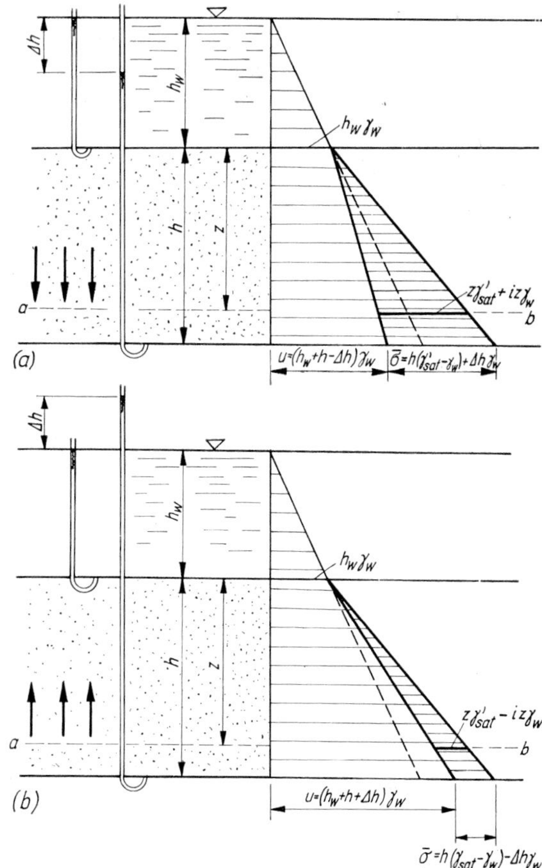

Fig. 144. Stresses in the case of one-dimensional seepage: *a* — seepage downwards; *b* — seepage upwards

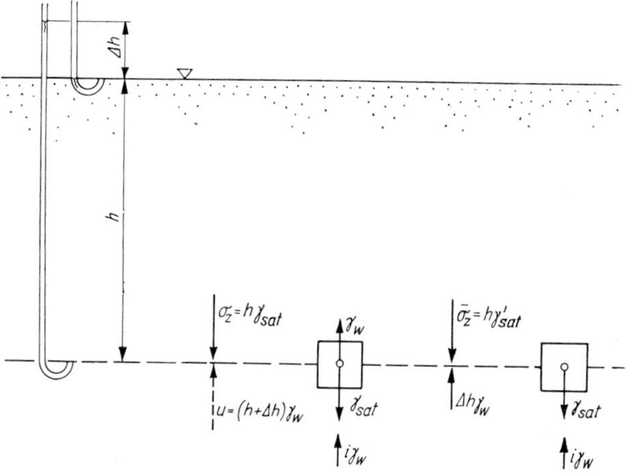

Fig. 145. Methods of calculation of the resulting mass force in one-dimensional seepage

ways: in the first method, we work with total (saturated) weight plus boundary neutral forces; in the second, with submerged weight (i.e. total weight less buoyancy force) plus seepage force. The two methods must give identical results.

This section has thus far been confined to the discussion of the simplest case: one-dimensional vertical flow. Let us now consider a more general case: the stresses in the soil under conditions of two-dimensional flow with no volume change. The two-dimensional potential flow will be dealt with in detail in Ch. 6.

Let the rectangular block *1 2 3 4* in *Fig. 146* represent an elementary volume of a saturated soil mass. Its volume is chosen as unity. The ground surface is assumed to be horizontal. Let us analyse first the forces in the soil under static equilibrium conditions. It is assumed, for convenience, that the water table is level with the ground surface. The water in every standpipe installed at the corners of the block stands at the same elevation indicating that there is no unbalanced head that would cause flow of water. A plot of the water pressures acting on each side of the soil block is shown in Fig. 146*a*, and the differences between the water pressures on opposite sides are given in the diagram *b*. The vectorial summation of all body forces is illustrated in the force polygon *c*. The resultant of the boundary water pressure is the buoyancy force, as indeed it must be according to the laws of hydraulics. As shown in diagram *c*, the total body force γ_{sat} (i.e. the weight of solid grains plus water) is reduced by a buoyancy force γ_w, giving a net body force

$$\gamma_{sat} - \gamma_w = \gamma'_{sat} = (\gamma_s - \gamma_w)(1 - n)$$

which is the submerged weight of the soil per unit of volume. This can be further resolved into two components, q_1 and q_2, which are transmitted as effective pressure through the soil skeleton.

In the second case we assume that flow of water is taking place in the soil (*Fig. 147*). This is possible only if there is an unbalanced head in the soil and the hydraulic gradient between any two points is, in general, different from zero. There will be, however, certain lines along which the potential is constant. These are called equipotential lines. Our volume element was chosen in such a manner as to have equipotential lines for two sides (*1 2* and *3 4* in Fig. 147*a*). In a soil assumed to be homogeneous and isotropic, water flows in the direction of the steepest gradient, i.e. at right angles to the equipotential lines. Consequently, the remaining two sides of the block, *2 3* and *1 4*, must be flow lines (see Ch. 6).

There is more than one possible choice relative to the system of forces considered to act on this soil element.

In the first method we consider the material within the element as a two-phase system (solid particles plus water). We then have the following forces:

the combined weight of solid particles and water contained in the volume element *1 2 3 4*;

the resultant of the water pressures acting against the boundaries of the element;

the resultant of the effective or intergranular pressures acting on the boundaries.

The second method considers the soil mass as a skeleton of soil particles. The forces to be taken into account are

the total weight of the solid grains in the volume element;

the resultant of the water pressures acting on the surfaces of the individual solid particles;

the resultant of intergranular pressures acting on the boundaries of the element.

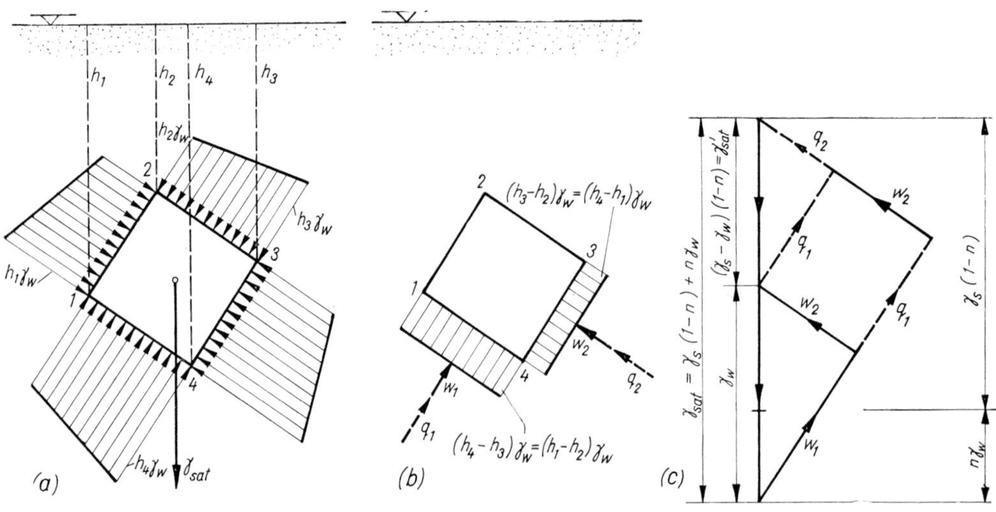

Fig. 146. Forces acting on the volume-element in soil with no water movement:
a — water pressures acting on the boundaries; *b* — difference in water pressures acting on the opposite sides; *c* — vector diagram of mass forces

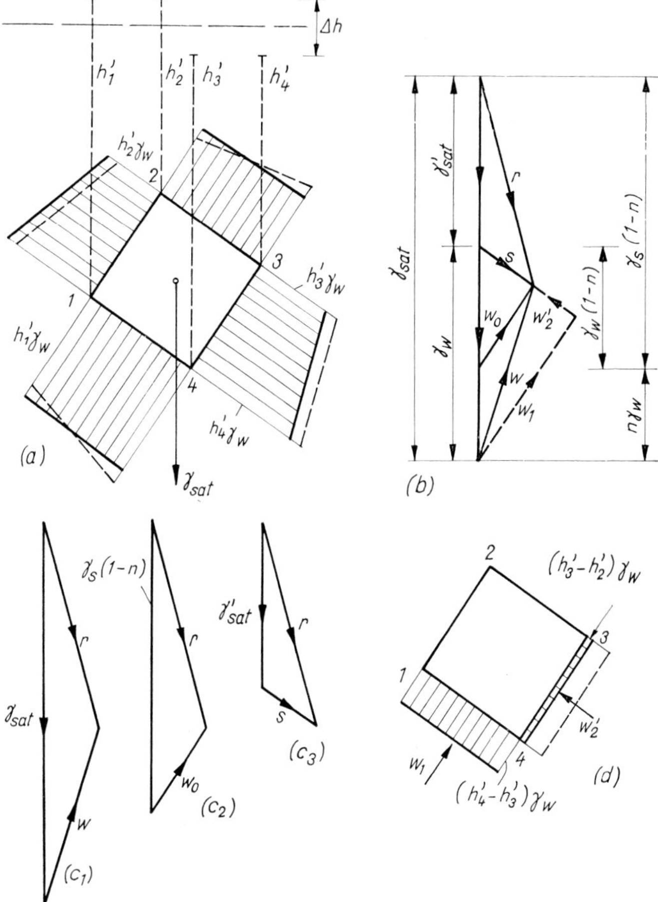

Fig. 147. Forces acting on the volume-element with water movement:

a — water pressures acting on the boundaries; *b* — combined vector diagram
c — three ways to find the resulting mass force; *d* — difference in water pressures acting on the opposite sides

As will be shown later, there is a third possible combination of the above-mentioned forces.

Whichever combination we choose, the vector sum of the boundary effective forces (which is just another term for the pressures transmitted from grain to grain at points of contact across the sides of the soil block) must, for equilibrium, always be balanced by the resultant of the remaining two forces. The latter is the resultant body force due to the combined effect of gravity and the presence of water in the soil.

As a first step we shall compute the vertical forces. Let us now analyse the equilibrium conditions for these forces in detail. The total weight of solid particles and water contained in the soil block is

$$\gamma_{\text{sat}} A ,$$

where

$$\gamma_{\text{sat}} = \gamma_s \frac{1 + w}{1 + w\gamma_s} = \gamma_s(1 - n) + n\gamma_w$$

and A is the cross-sectional area of rectangle *1 2 3 4*. (The dimension perpendicular to the plane of the drawing is taken as unity.) The weight of the solid particles is

$$\gamma_s(1 - n) A ,$$

wherein γ_s is the unit weight of the solids and n is porosity. The submerged (buoyant) weight of the solid particles is then

$$\gamma'_{\text{sat}} = (\gamma_s - \gamma_w)(1 - n) A .$$

The weight of water is

$$n\gamma_w A .$$

These vertical forces are represented in the vector diagram *b* in Fig. 147. (In the figures, we have $A = 1$.) The difference between the total weight γ_{sat} and the weight of the submerged solid particles equals the weight of a mass of water which has the same volume as that of the element. This is, as we have seen, the force of buoyancy, and the algebraic sum of the total weight and the buoyancy force gives the submerged weight of the soil:

$$(\gamma_s - \gamma_w)(1 - n) = \gamma_{\text{sat}} - \gamma_w = \gamma'_{\text{sat}} .$$

Note that

$$\gamma_{\text{sat}} = \gamma_s(1 - n) + n\gamma_w .$$

The second step of the analysis is the computation of the forces due to seepage of water. The water pressures acting upon the sides of the soil block are plotted in Fig. 147*a*. If seepage takes place through the soil, we have different levels in the piezometers on the sides *1 2* and *3 4*. For comparison, the pressure head diagrams for the static condition are also shown, by dashed lines, in the figure. Diagram *d* gives the differences between pressure heads on opposite faces. Comparing this with diagram *b* in Fig. 146, it is seen that the difference in pressure head shown on side *1 4* is constant and is the same for both the seepage case and the static case. But because of head lost in seepage from *1 2* to *3 4*, the pressure head difference that appears on side *3 4* is less for the seepage case than for the static case. It is also clear from the figures that, in relation to the seepage force acting on a soil element, the actual pressure heads are not important; what matters is the differences in total head. Since the difference in the pressure diagrams between sides *1 2* and *3 4* represents the pressure lost in seepage due to the viscous friction of water, it follows that an equal pressure, but one acting in the opposite direction, has been transferred by a drag on the particle surfaces to the skeleton formed by the particles. This force is called the seepage force. Its direction is always perpendicular to the equipotential lines and thus in isotropic soils it is parallel to the adjacent flow lines.

On the basis of the above considerations the force diagram can now be plotted. The resultant of the water pressures, acting across the surfaces *1 4* and *2 3* is the same for both the seepage case and the static case. This force is denoted by w_1. The resultant of the water pressures on the faces *1 2* and *3 4* in a static condition w_2 is, however, different from that under conditions of seepage w_2'. Their difference is the seepage force $s = w_2 - w_2'$. The overall resultant of the boundary water pressures is, for the static case, the buoyancy force $-\gamma_w$ (Fig. 146*c*); and for the case with flow of water, this resultant w is the vector sum of forces w_1 and w_2' (Fig. 147*b*). As can also be easily read from the figure, the resultant of all boundary water pressures in the seepage case is equal to the vector sum of the buoyancy force in the static state $-\gamma_w$ and the seepage pressure s.

Let us now plot the force diagram of all forces acting on the soil element in the seepage case (Fig. 147*b*). The resultant body force is r, so the resultant intergranular pressure must equal the reaction $-r$. This body force r can be determined, as was discussed in connexion with the static state, by various combinations of weight and water forces. The possible combinations are illustrated in Fig. 147*c*.

In the first method (1) we combine the total weight γ_{sat} and the resultant boundary neutral force w_0, the latter being the vector sum of w_1 and w_2'. This is the method most often used in practical problems.

The second possibility (2) consists in calculating the "effective" weight of the solid grains, $\gamma_s(1 - n)$, and adding vectorially the resultant of the water pressures on the surfaces of the individual grains, w. This force w in turn is obtained as the vector sum of the weight of water filling the pore spaces $n\gamma_w$ and the resultant of the boundary water pressures w_0.

The third choice (3) is a combination of the submerged (buoyant) weight γ_{sat}' and the seepage force s. This approach is perhaps the easiest to visualize and is the one which furnishes the clearest understanding of seepage effects.

It is important to realize that whichever combination we choose they all give the same result. In practical cases we may select the most suitable method.

The preceding consideration applied to an elementary block within a soil mass, but the analysis can be readily extended to any large mass of soil by the summation of forces for all the elements of the volume. To demonstrate the method, a simple example is given in the following. More complex cases connected with stability problems of earthworks will be dealt with in Vol. 2.

Let us consider an infinite slope in a homogeneous and isotropic cohesionless soil with a steady linear flow of water throughout (*Fig. 148*). The free water table is assumed to coincide with the sloping ground surface, and the flow lines to be parallel to the surface. The equipotential lines are therefore straight lines intersecting the surface at right angles. With the direction of flow stipulated, the value of the hydraulic gradient is given as $i = \Delta h/\Delta s = \sin \varepsilon$, where ε denotes the slope angle. The sides of the elementary soil block *1 2 3 4* are again, similar to the case shown in Fig. 147, either flow lines or equipotential lines. Employing the previously discussed method (3), the resultant body force r is obtained as the vector sum of the submerged (buoyant) weight of the soil block (for unit volume this is γ_{sat}') and the seepage force s (Fig. 148*b*). Or, if we choose method (1), the resultant r will be obtained by adding all boundary water pressures vectorially to the total (saturated) weight of the block γ_{sat}. In the latter case, the water pressures acting across faces *1 4* and *2 3* cancel out. (Note that the pressure heads in piezometers *1* and *2*, or *3* and *4*, are equal.) If the side lengths of the block taken as unity, the water pressure normal to side *1 2* is $h_1\gamma_w$. Similarly, the water pressure across side *3 4* is $h_3\gamma_w$. The algebraic sum of these two forces is thus the resultant boundary water pressure:

$$w_0 = (h_3 - h_1)\,\gamma_w = \gamma_w \cos \varepsilon\,.$$

This force acts upward in a direction perpendicular to the sloping surface. The vector sum of this force w_0 and the total weight γ_{sat} gives the same resultant r as the one obtained by method (3).

Finally, this example may also serve to demonstrate a third concept which, striving for exactness, works with a combination of the resultant neutral

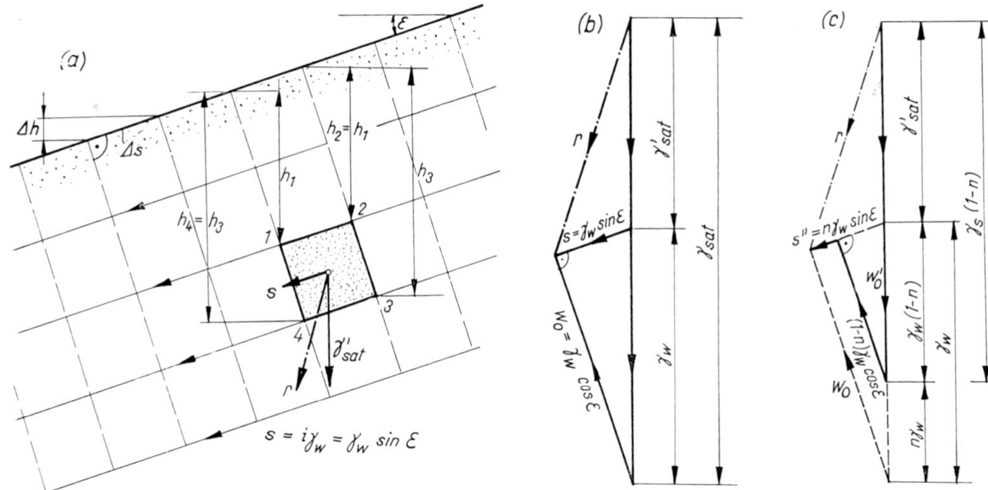

Fig. 148. Infinite slope in a cohesionless soil with steady linear flow of water, seepage parallel to the surface; investigation of mass forces

and effective forces acting on the individual grains plus the seepage force. According to this, the weight of the solid particles in the block *1 2 3 4* is $(1-n)\gamma_s$. The resultant of the water pressures on the individual particle surfaces is, by proportionality, $w_{0s}=\gamma_w(1-n)\cos\varepsilon$. (This could in fact be further resolved into two components not shown in Fig. 148*c*: an uplift, $\gamma_w(1-n)$, and a thrust acting on the individual grains parallel to direction of flow, $s'=(1-n)i\gamma_w=(1-n)\gamma_w\sin\varepsilon$.) The remaining force is the one acting on the pore water in the direction of flow, $s''=n\gamma_w\sin\varepsilon$. The seepage force transmitted by viscous drag to the soil skeleton is the sum of forces s' and s'':

$$s=s'+s''=i\gamma_w=\gamma_w\sin\varepsilon\,.$$

The vectorial summation of the forces is shown in Fig. 148*c*. The resultant *r* is clearly the same as that obtained by other methods.

Further numerical examples concerning the calculation of neutral and effective stresses under seepage conditions will be given at the end of Ch. 6.

5.3 Stresses in soils due to volume changes

In a three-phase granular medium that consists of solid grains, water and air, a load induces stresses in each phase separately. Under constant load and in undrained condition, when the free escape of both pore water and pore air is prevented, the state of stress does not vary with time. In the following we shall analyze this case of loading and determine the pressure in both pore water and pore air.

Let us consider first the case when the three-phase soil mass is subjected, in an undrained condition, to a constant all-around (hydrostatic) pressure as illustrated by the soil model in *Fig. 149*. Air is not permitted to escape the pores. But since air is highly compressible, the soil model will be reduced in volume. Assuming that the temperature remains constant throughout the process of compression (isothermic conditions), the change in volume of pore air can be described by the Boyle–Mariotte law for ideal gases according to which the product of volume and pressure is always constant. It must also be taken into consideration that air, under pressure, dissolves in water the amount of dissolved air being at constant temperature directly proportional to the pressure.

Let the initial volume of the air within the soil model shown in Fig. 149 be denoted by V_g and that of the water by V_w. The excess neutral stress that builds up in the pore water when the volume of the sample is decreased by ΔV is given, as was derived in Section 2.3, by Eq. (22):

$$u=\frac{p_1V_g}{V_g+jV_w-\Delta V}-p_1,$$

where p_1 differs only by a negligible amount from the atmospheric pressure, *j* is *Henry*'s solubility coefficient. Its variation with temperature is shown in *Fig. 150*.

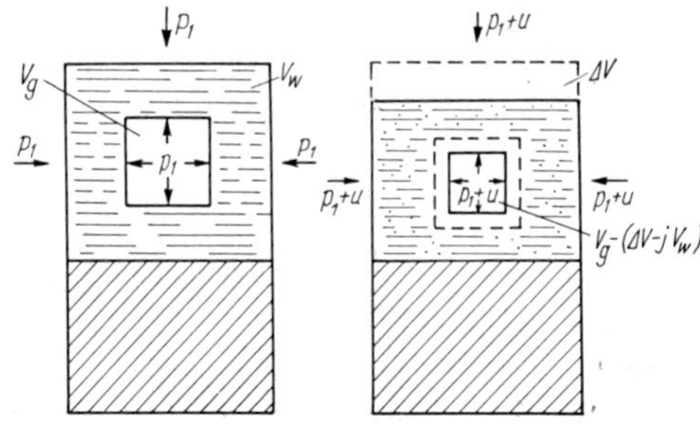

Fig. 149. Soil sample acted upon by hydrostatic stresses

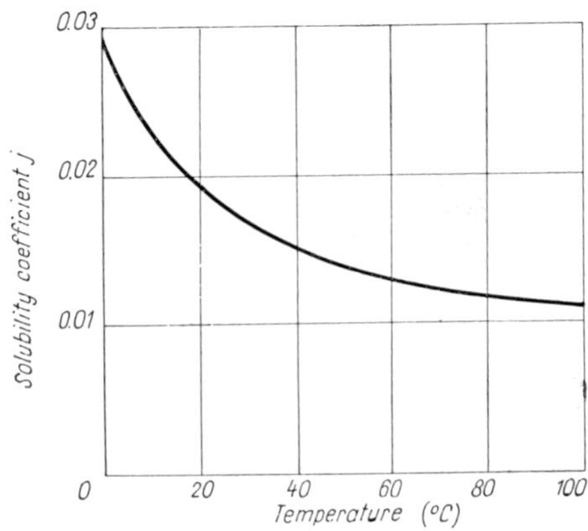

Fig. 150. Coefficient of solubility of water versus temperature

By making use of Eq. (22), we can calculate the total stress which would be necessary to produce, under undrained conditions, the same volume change in the soil as that caused by a given effective pressure in the usual drained consolidation test (see Ch. 7). The compression curve drawn by a solid line in *Fig. 151a* was obtained from a "drained" oedometer test, i.e. one continued long enough to permit the complete dissipation of pore water pressures generated by the applied load. The decrease in volume ΔV is thus given for any effective stress $\bar{p}$. The compression curve for the undrained case can now be determined as follows. Knowing the initial values of V_g and V_w we can calculate from Eq. (22) the excess neutral stress for any volume decrease ΔV. To obtain equal volume changes in both the drained and the undrained tests, the applied effective stresses $\bar{p}$ must also be equal. Hence, the total stress to be applied in the undrained test is $p = \bar{p} + u$. By plotting volume changes ΔV against total pressures p, the curve shown by the dotted line is obtained (Fig. 151*a*). The diagram in the lower part of Fig. 151 shows the effective stress $\bar{p}$ as a function of the total stress p. For a drained test the relationship is a straight line inclined at 45°. For an undrained test, the diagram is a slightly curved line which, however, can be replaced with fair approximation by a linear relationship ($\bar{p} = B\,p$).

The above theoretical consideration has also a practical application in the design of the impervious cores of earth dams since the conditions there are such as to hinder the free drainage of water, and the pore water pressures caused partly by the weight of the soil itself and partly by the compaction of the fill must also be taken into consideration in the stability analysis of the dam.

According to Eq. (22), there is an excess pore pressure $u = p_1 V_g / j V_w$ at which $\Delta V = V_w$. This means that at a certain load all air originally contained in the pores is completely dissolved in the water and the soil becomes saturated.

Thus far we have considered the case when a soil mass is subjected to a uniform all-around pressure ($\sigma_1 = \sigma_2 = \sigma_3 = p$). The general case when the stress increments are different in the principal directions was investigated by SKEMPTON (1954). For the two-dimensional case he derived the following formula for the computation of the excess pore water pressure Δu due to principal stress increments $\Delta\sigma_1$ and $\Delta\sigma_3$:

$$\Delta u = B[\Delta\sigma_3 + A(\Delta\sigma_1 - \Delta\sigma_3)] \qquad (73)$$

where A and B are called the *pore pressure parameters*.

Consider an element of soil (*Fig. 152*) that is completely consolidated under an initial hydrostatic pressure of p. Any change in the state of stress that would involve an increase of both the major and the minor principal stresses, σ_1 and σ_3, can be assumed to be applied in two stages. First, we increase the hydrostatic pressure by $\Delta\sigma_3$, then add a uniaxial pressure increment of $\Delta\sigma_1 - \Delta\sigma_3$, where $\Delta\sigma_1$ is the intended final increment of major principal stress. At each stage excess neutral stresses are built up in the pore water, and the total increment of pore pressure due to the principal stress increments of $\Delta\sigma_1$, and $\Delta\sigma_3$ can be obtained by superposition, i.e.

$$\Delta u = \Delta u_a + \Delta u_d\,. \qquad (74)$$

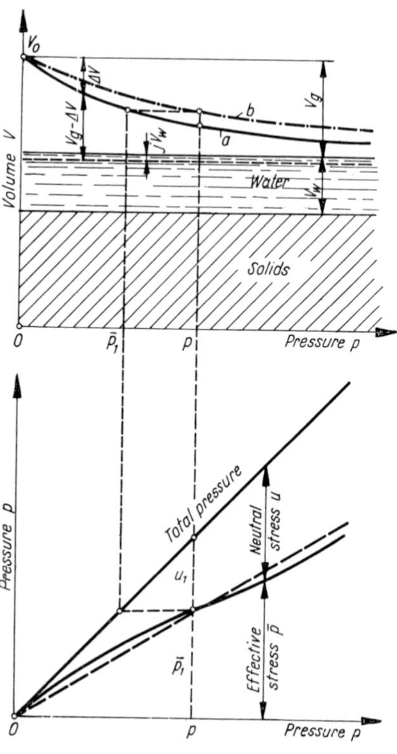

Fig. 151. Compression curves with and without drainage; neutral stresses acting during test

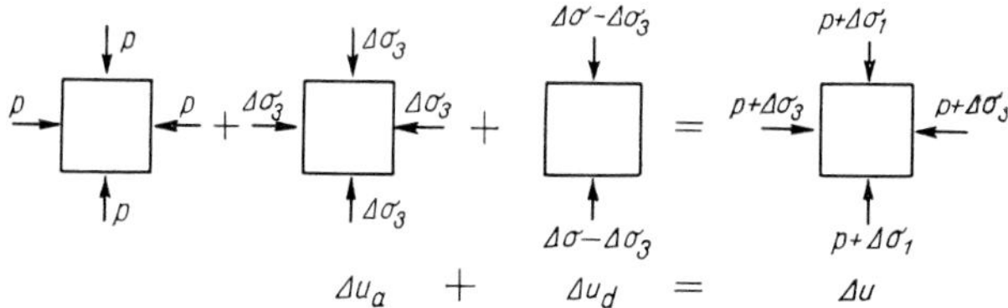

Fig. 152. Changes in pore water pressure due to changes in stressed state

The above-described changes in state of stress can be produced experimentally in a setup illustrated in *Fig. 153*. A cylindrical soil specimen encased by a thin flexible rubber membrane is placed into a sealed pressure cell, which can be filled with a suitable liquid. If a pressure is produced in the confining liquid, it will be completely transmitted also to the side of the specimen. The same confining pressure can be applied by a piston on top of the specimen. The application of the confining pressures causes an initial increase of the pore pressure, but with the drainage line kept open any excess water and air can freely flow out of the specimen permitting compression to take place; as a result the excess pore pressures dissipate completely after a certain period of time. The state of stress of the specimen consolidated under an all-around pressure corresponds to the initial stage of the loading test illustrated in Fig. 152. Now, with the valve closed, let us increase the pressure acting on the specimen in two steps. The pore pressures developed during loading are indicated by a manometer connected via a separate pressure line to the inside of the specimen.

The increment of effective stress $\Delta\bar{\sigma}$ due to an increase of $\Delta\sigma_3$ in the all-around pressure is

$$\Delta\bar{\sigma} = \Delta\sigma_3 - \Delta u_a .$$

Let

$$m_v = \frac{\Delta V}{V} \cdot \frac{1}{\Delta\bar{\sigma}}$$

denote the coefficient of volume change of the soil skeleton, then the change in volume of the sample is

$$\Delta V_v = -m_v V(\Delta\sigma_3 - \Delta u_a), \qquad (75a)$$

where V is the initial total volume of the specimen.

Similarly, if m_f denotes the coefficient of volume change of the pore fluid (water + air), and n is the initial porosity of the sample, then the decrease in volume of the pores due to an increment of pore pressure Δu_a is given by

$$\Delta V_f = m_f n V \Delta u_a . \qquad (75b)$$

Since the solid grains can be assumed to be incompressible, and free escape of water and air from the pores is prevented, the change in volume of the soil skeleton ΔV_v must be equal to the change in volume of the pore fluid ΔV_f.

Considering that at the first stage of loading $\Delta\sigma_1 - \Delta\sigma_3 = 0$, from Eq. (73)

$$\frac{\Delta u_a}{\Delta\sigma_3} = B .$$

Using Eqs (75a) and (75b) we have

$$\frac{\Delta u_a}{\Delta\sigma_3} = B = \frac{1}{1 + \dfrac{n\, m_f}{m_v}} \qquad (76)$$

and

$$\Delta u_a = B\,\Delta\sigma_3 . \qquad (77)$$

For a saturated soil $m_f/m_v = 0$, since the compressibility of the solid mineral particles is negligible relative to that of the air-water mixture filling the pores. In saturated soils, therefore, $B = 1$ which means that an increase in the hydrostatic pressure $\Delta\sigma_3$ causes an equal increment of pore pressure. TAYLOR (1944) and BISHOP and GAMAL ELDIN (1950) presented an experimental verification of this. If the soil is completely dry, the ratio m_f/m_v may have very high values, air being much more compressible than the soil skeleton. In dry soils $S = 0$, hence $B = 0$. For partially saturated soils $0 < B < 1$, and e.g. for soil specimens compacted at optimum moisture content, B varies from 0.1 to 0.5. The relationship between the pore pressure coefficient B and the degree of saturation S is shown for a fine sand in *Fig. 154*.

At the second stage of loading a deviator stress (i.e. a stress increment in only one of the three principal directions) of $\Delta\sigma_1 - \Delta\sigma_3$ is superimposed upon the hydrostatic state of stress. The resulting pore pressure increment is Δu_d. Hence, increments of the major and minor principal effective stresses, respectively, are

$$\Delta\bar{\sigma}_1 = (\Delta\sigma_1 - \Delta\sigma_3) - \Delta u_d$$

and

$$\Delta\bar{\sigma}_3 = -\Delta u_d .$$

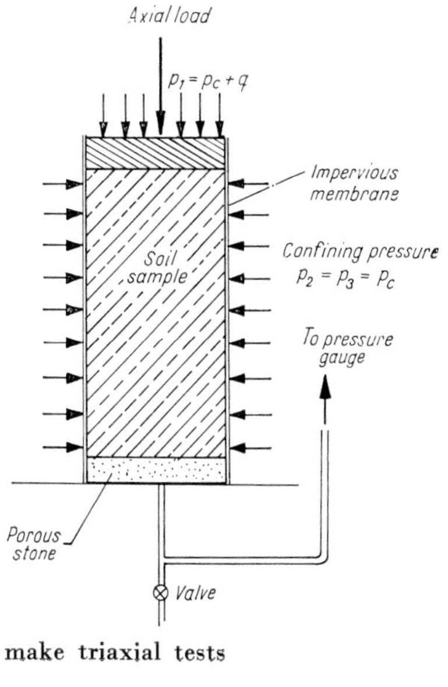

Fig. 153. Device to make triaxial tests

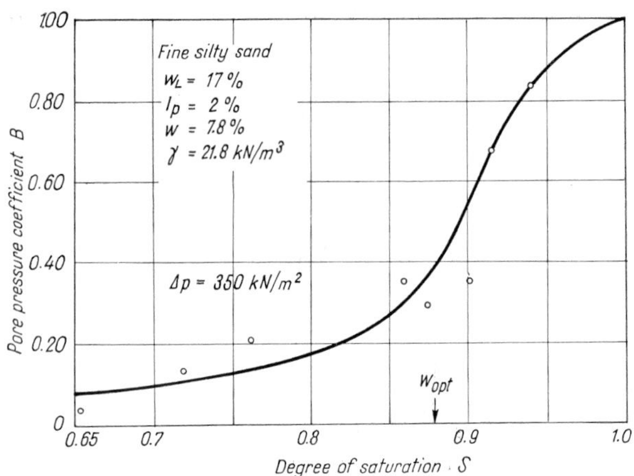

Fig. 154. Pore pressure coefficient B versus degree of saturation

Assuming, for a moment, that soil is an elastic medium, the change of volume of the soil model can be obtained according to the theory of elasticity as

$$\Delta V_v = -m_v V \frac{1}{3}(\Delta\bar{\sigma}_1 + 2\Delta\bar{\sigma}_3) =$$

$$= -m_v V \frac{1}{3}[(\Delta\sigma_1 - \Delta\sigma_3) - 3\Delta u_v].$$

The decrease in volume of the voids is

$$\Delta V_f = -m_f n V \Delta u_d.$$

Again, by reasons previously discussed, the two volume changes must be equal: $\Delta V_v = \Delta V_f$. Hence the pore pressure increment due to a deviator stress of $\Delta\sigma_1 - \Delta\sigma_3$ is

$$\Delta u_d = \frac{1}{1 + \frac{n\,m_f}{m_v}} \frac{1}{3}(\Delta\sigma_1 - \Delta\sigma_3) = B\frac{1}{3}(\Delta\sigma_1 - \Delta\sigma_3).$$

Soil, however, does not behave as a perfect elastic material, and the numerical factor 1/3 in the formula must, for real soils, be substituted by a general factor A.

Adding Eqs (77) and (78) gives the general formula for total pore pressure increment due to principal stress increments $\Delta\sigma_1$ and $\Delta\sigma_3$ as

$$\Delta u = B[\Delta\sigma_3 + A(\Delta\sigma_1 - \Delta\sigma_3)]. \qquad (73)$$

This is identical to *Skempton*'s formula presented in the preceding.

In saturated soils where $B = 1$,

$$\Delta u = \Delta\sigma_3 + A(\Delta\sigma_1 - \Delta\sigma_3). \qquad (73a)$$

(SKEMPTON, 1948.)

A practical application of Eqs (73) and (73a) is in the stability analysis of slopes. Their use, however, presupposes a knowledge of the state of stress in the problem under consideration, since the pore pressure parameters themselves are dependent on the stress and strain conditions in the soil. In problems of stability, A and B values pertaining to failure conditions in the soil are of particular importance; with their aid we can calculate effective stresses from total stresses and perform the stability analysis using drained shear strength values (i.e. those expressed in terms of effective normal stresses).

Typical values of the parameter A for saturated clayey soils are given, after *Skempton*, in *Table 21*;

Table 21. Values of pore pressure parameter A

Soil type	Volume change during shear	A
Very sensitive clay	Significant decrease	0.75–1.50
Normally consolidated clay	Decrease	0.50–1.00
Compacted sandy clay	Slight decrease	0.25–0.75
Lightly overconsolidated clay	None	0 –0.50
Compacted clayey gravel	Increase	– 0.25–0.25
Heavily overconsolidated clay	Increase	–0.50–0

which apply to failure conditions. *Skempton*'s investigations also showed the parameter A at failure to be a function of the stress history of a soil. Introducing, after *Skempton*, the ***overconsolidation ratio*** defined as the maximum past-consolidation stress divided by the effective consoli-

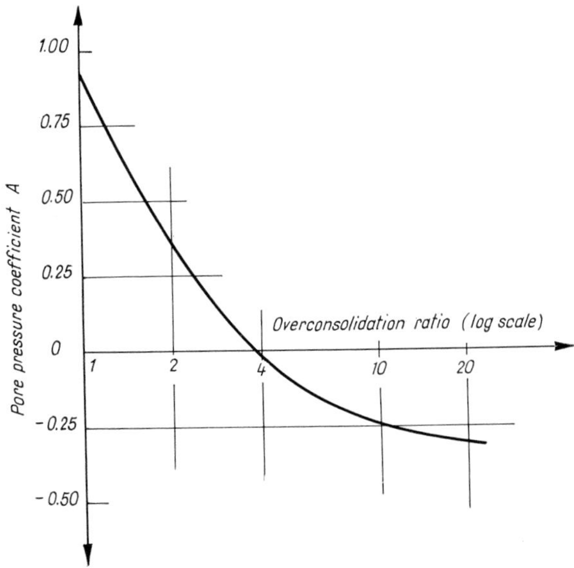

Fig. 155. Pore pressure coefficient A versus overconsolidation ratio

dation stress at which the value A was measured, we obtain the plot of pore pressure parameter A versus overcon solidation ratio shown in *Fig. 155.*

Another case of the state of stress in saturated soil was presented by KÉZDI (1970), omitting the assumption that led to the Eq. (55) and taking the real plane cross-section of the material, i.e. the section cuts solid grains and water-saturated voids as well. In this case, neutral pressures may arise in one part of the cross-section only, and the deformations and stresses depend also on the ratio of the compressibility of the porous skeleton to that of the material filling the voids. In the one-dimensional case ($\sigma_x = \sigma_y = 0$) we have:

$$E\varepsilon_z = \sigma_z - (1 - 2\mu)(1 - \alpha - n)\, u'$$

where μ is *Poisson*'s number, n is porosity, u is the neutral stress in the pore water and α is the above-mentioned ratio:

$$\alpha = \frac{\dfrac{E}{1 - 2\mu}}{\dfrac{E_w}{1 - 2\mu_w}} .$$

Recommended reading

BISHOP, A. W., HENKEL, D. J. (1957): The measurement of soil properties in the triaxial test. London: Edward Arnold, Ltd.

BJERRUM, L.–CASAGRANDE, A. (1953): From Theory to Practice in Soil Mechanics. J. Wiley & Sons, New York.

HARR, M. E. (1966): Foundations of Theoretical Soil Mechanics. McGraw–Hill Book Co., New York.

KÉZDI, Á. (1962): Semleges feszültség és áramlási nyomás (Neutral stress and seepage force). Vízügyi Közlemények. No. 4, pp. 570–571.

LAMBE, T. W.–WHITMAN, R. V. (1969): Soil Mechanics. J. Wiley & Sons, Inc., New York.

LEE, I. K. (editor): Soil Mechanics. Selected topics. American Elsevier Publ. Co. Inc., New York, 1968.

LEONARDS, G. A. (1962): Engineering properties of soils. Ch. 2. in Foundation Engineering. (Leonards, G. A., editor) McGraw-Hill Book Co., Inc., New York.

Proceedings of the Conference on Pore Pressure and Suction in Soils. London, 1960.

SCOTT, R. F. (1963): Principles of Soil Mechanics. Addison Wesley Publishing Company, Inc.

TAYLOR, D. W. (1948): Fundamentals in Soil Mechanics. J. Wiley & Sons, New York.

TERZAGHI, K. (1925): Erdbaumechanik. F. Deuticke, Wien.

TSYTOVICH, N. A. (1963): Mechanika gruntov. Moscow.

Chapter 6

Flow of water in soil

6.1 Introduction

In the preceding chapters we have already seen examples of the manifold effect of water on the properties of the soil. Effective unit weight, void ratio, behaviour of soil subjected to deformations, and state of stress of a soil mass under load are all greatly influenced by the ground water, under both static and flow conditions. The natural water content of a soil or the velocity of percolating water is often the decisive factor to be considered in a technical problem. Seepage through and beneath dams, stability of embankments holding water, drainage of open excavations, desiccation of soils and consolidation under load etc. are such problems where the water present in the soil has a prime role (*Fig. 156*).

In the coarse-grained soils water exists mainly as "free" pore water, while in the fine-grained soils the water adsorbed on particle surfaces becomes predominant. These two kinds of water, however, cannot be separated distinctly and this makes it difficult to determine accurately how they are apportioned within the pores. It is generally accepted that there is a continuous transition from the oriented pattern of water molecules around the solid particles to a dispersed arrangement as the distance from the particle surfaces increases, and that the thickness of the adsorbed layer is governed partly by the water of the mineral surfaces and partly by the environment. Because of the continuous transition in the physical state of the pore water, it would be extremely difficult to define what exactly is meant by "free" water. In the following discussion that portion of the pore water which can be moved by pressure acting either on the soil skeleton or on the pore water itself will be considered as free water. In cohesive soils, the amount of this free water depends on physicochemical conditions, on the magnitude of pressure and on temperature.

The effects that cause flow of water through the soil may be very diverse in nature. Gravity and capillarity are the most important, but besides them thermal effects, electric current and certain chemical processes may also initiate a migration of water molecules. Flow of water may take place with or without volume changes in the soil. In this chapter, various phenomena of water flow through soil will be studied to the extent dictated by their importance from a civil engineering point of view.

Soil as described earlier is a disperse system of solid particles, water and air. A general treatment

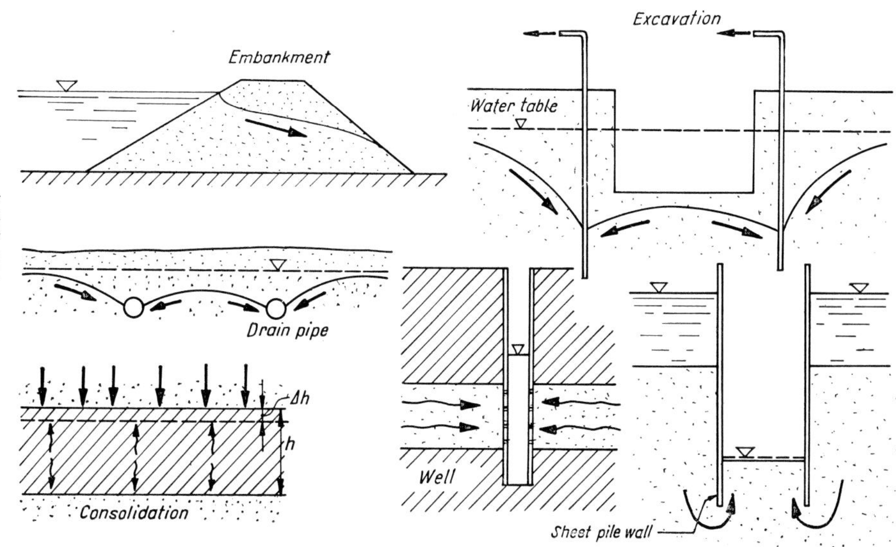

Fig. 156. Civil engineering problems associated with water percolation in soils

of problems of "motion" in such a multiphase system would involve a study of interphasal displacements and of their relative velocities. Only such an approach, with due consideration given to the physical laws that govern interactions between phases, could lead to the fundamentals upon which special cases of water flow could be dealt with. This new concept of studying soil action has, however, been advanced only very recently and only the first step has been made towards a universal theory of phase movements. It would seem therefore premature to treat, in this book, the problems of fluid flow in soil on the basis of these new principles, and the following notes are intended only to direct attention to the wide scope of this concept and to the many advantages inherent in it.

Let us consider first the case when the soil is saturated with water. There are two kinds of velocities that can be distinguished, namely velocity of water v_w and velocity of the solid particles v_s. By velocity is, of course, always meant a statistical average. If the two velocity vectors are equal and of the same direction, there will be no internal resistance generated in the course of the motion. This applies, for example, to the translation of a mass of soil without any deformation taking place within the mass, or to a steady flow of slurry in a pipeline. The special case when $v_s = 0$ and $v_w \neq 0$ represents seepage of water through a solid, porous and incompressible medium. When $v_w = 0$ and $v_s \neq 0$, we have solid particles sinking in still water (*Stokes*' law, Ch. 2). The motions just described are represented by straight lines in a plot of velocity of solids v_s versus velocity of water v_w (*Fig. 157*). Similarly, the motion of solid particles in a vertical current of water is represented by a straight line intersecting the coordinate axes at 45°. When a laterally confined and saturated mass of soil is subjected to a vertical load, water is squeezed out of the pores at a certain rate and, at the same time, the solid particles are forced to move with the same velocity in the opposite direction. Then $v_s = - v_w$ (consolidation). Analogous cases arise when the soil swells, or shrinks, except that in the latter case the movement of the solid particles ceases at the shrinkage limit while the movement of water may still continue on further drying. In a river bed, the load of transported sediment starts moving only at a critical flow velocity, but henceforth water and sediment move with the same velocity.

To Fig. 157 two more could be added; for it is conceivable to represent in a similar way the kinds of motion that occur in the other two possible two-phase systems: solid particles–air, and air–water. Examples of these are filtration of air through porous media, movement of dust clouds, rainfall, ascent of air bubbles in water, upward movement of water droplets in air as when spouted from a fountain etc. All these, seemingly unrelated, phenomena together with those occurring in the solids–water system should obey a common universal law which, however, still remains to be formulated.

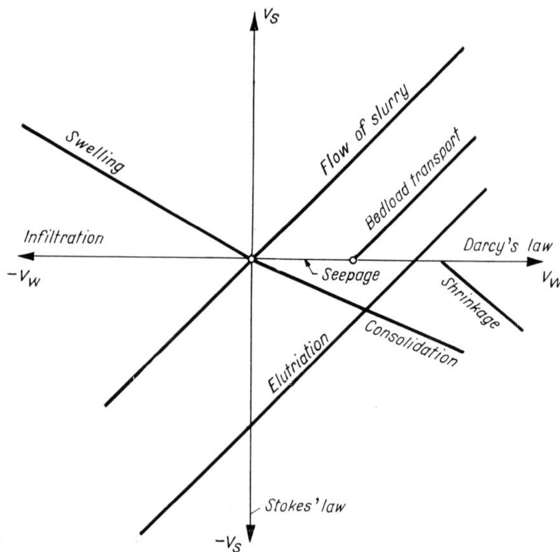

Fig. 157. Possible transfer phenomena in two-phase media

In the general case, a three-phase system, a motion is characterized by three velocities v_s, v_w and v_g which, respectively, describe the rates of displacement of the solid, liquid and gaseous phases separately. The possible combinations of these represent an almost infinite variety of relative displacements among the three phases. If we introduce the notation

$$v = v_s + v_w + v_g \tag{78}$$

and use the velocity ratios v_s/v, v_w/v and v_g/v, any motion in the three-phase system can be represented graphically by a plot in the triangular diagram (*Fig. 158*). Examples given in the figure show

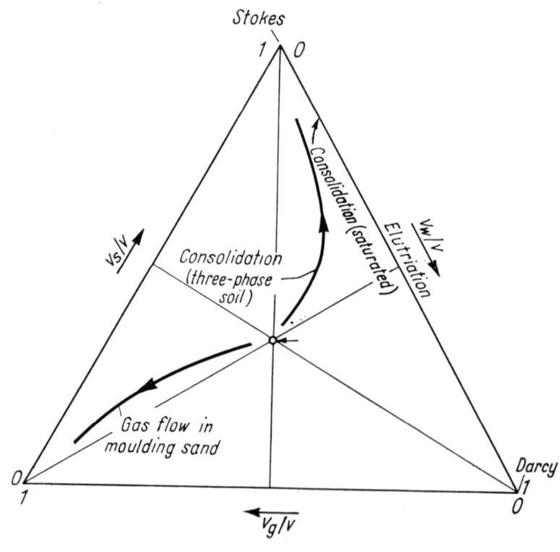

Fig. 158. Movement phenomena in three-phase systems, plotted in the triangular diagram

translation of a soil mass with no volume changes occurring ($v_s = v_w = v_g$), consolidation due to loading of a saturated mass of soil, and percolation of gases through moist moulding sand. Ore dressing by flotation process, internal erosion of fine soil particles, suffusion phenomena and the like can also be dealt with by this method. The author's intention is to present in a future book a comprehensive treatise on this subject.

6.2 Flow of water through soil due to gravity

6.2.1 Fundamentals

Between soil particles of various sizes and shapes, the pores are also irregular, forming an extremely intricate network of tortous channels with cross-sections and wall friction changing from point to point. In the study of flow of water through such a network we must always consider a large mass of soil as a whole, for which such terms as tube diameter and flow velocity should be understood as statistical averages.

In the following we deal, at first, with the simple case, when the soil is saturated and its particles are not moved by the flowing water but form a solid skeleton.

First of all, let us discuss the laws of fluid flow in a circular tube of constant but small diameter. As known from Hydraulics, the flow of water is, in general, of two kinds. In a *laminar* flow each particle of water moves along its own definite path without interfering with the paths of other particles. In a *turbulent* flow the paths are irregular and they also intersect each other.

The fundamental laws that determine whether in a given case the flow is laminar or turbulent were formulated by *Reynolds*. The basic results of his experiments with fluid flow in a tube are shown in *Fig. 159*, where the hydraulic gradient i, which is the head lost in flow due to friction, per unit of length of tube, is plotted against flow velocity v.

Apart from very low velocities at which the effect of gravity is negligibly small relative to the surface effects due to secondary physicochemical forces, the flow velocity is proportional to the hydraulic gradient (Section I). Such state is defined as laminar flow.

As the velocity is increased, a critical state is reached at which eddies begin to form near the wall of the tube and owing to these irregularities the relationship between flow velocity and hydraulic gradient becomes indefinite (Section II). A further increase in velocity leads again to a definite relationship, this time, however, represented by a continuous curved line (Section III). If, in a reverse process, the velocity is decreased, the relationship characterized by the curved section of the graph remains valid up to a velocity value

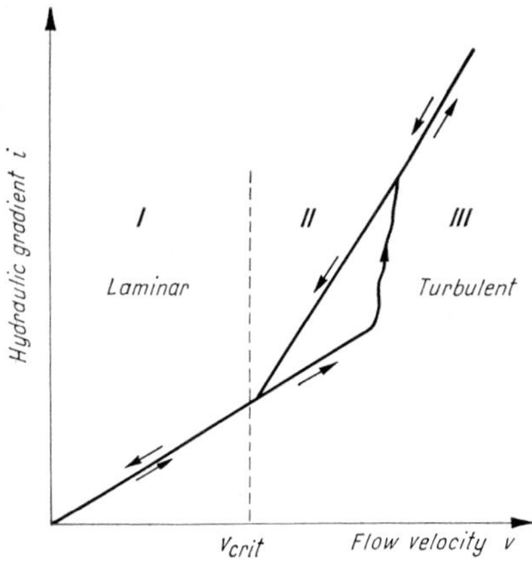

Fig. 159. Laminar and turbulent movement of liquids

much lower than that at which the eddy formation was previously observed. At a further decrease in velocity, the motion reverts to laminar flow. The value at which this occurs is called the *lower critical velocity* v_{cr}. Below this limit the flow in a tube is always laminar. Reynolds found that this critical velocity is inversely proportional to the diameter of the tube, and he derived the following general formula which is independent of the kind of fluid and system of unit

$$\frac{v_{cr}\, d\gamma_w}{\eta g} = 2000. \qquad (79)$$

Here d is the tube diameter, γ_w is the unit weight of water, η is viscosity and g is the acceleration of gravity. Assuming a constant temperature of 10 °C and substituting numerical values expressed in the metric system, we have $v_{cr}(\mathrm{cm/s}) \times d(\mathrm{cm}) = 28$. For a grain size of $d = 1.0$ mm $= 0.1$ cm, $v_{cr} = 2.80$ cm/s.

In most soils the pores are so small that the flow of water through them is always laminar. This case will therefore be discussed first in the

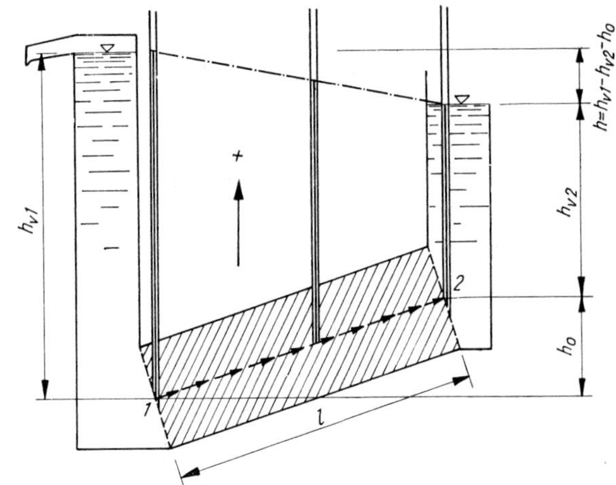

Fig. 160. Basic definitions in the theory of seepage

following section. Capillary effects will, for the time being, be disregarded.

Figure 160 shows a section through a soil sample. Connected to each end of the sample are vertical tubes, in which the water levels can be kept constant by means of overflows. The difference between the water levels in the two tubes is denoted by h. The path taken by a drop of water in passing through the soil is called a *flowline*. If such paths form a bundle of straight and parallel lines, the flow is described as linear. The flow demonstrated by Fig. 160 is laminar and also one-dimensional, which means that the flow velocity is the same at any point of a cross-section taken perpendicular to the flow lines. Let 1 and 2 mark the two ends of a flow line. If we insert standpipes at these points, the heights to which water rises in the pipes indicate the pressures in the water. The vertical distance h_0 represent the elevation head at point 2 with reference to the elevation of point 1 selected as datum. The piezometric head in pipe 1 is h_{v1}, and that in pipe 2 is h_{v2}. If the water stands at the same elevation in both pipes, no flow of water occurs. Flow takes place only if a difference in total head between two points is produced e.g. by increasing the hydrostatic pressure in only one of the piezometers shown in Fig. 60. In the special case when the elevation head h_0 is zero, the difference in pressure head that causes water to flow through the soil is equivalent to the difference in piezometric head between points 1 and 2. Otherwise, as in the figure, it is equivalent to $h = h_{v1} - h_{v2} - h_0$. The unbalanced hydrostatic pressure $h\gamma_w$ per unit length of the flow lines is called the *pressure gradient*:

$$i_p = \frac{h\gamma_w}{l}.$$

Dividing this by γ_w, a dimensionless ratio is obtained which is the hydraulic gradient:

$$i = \frac{i_p}{\gamma_w} = \frac{h}{l}.$$

(Note that Eq. (68) defined the gradient as a vectorial quantity.)

The *velocity of flow* is defined as the flow in unit time per unit of cross-sectional area perpendicular to the flow lines (v cm^3/s/cm^2 $= v$ cm/s). It should be noted that v does not mean the true velocity of the water particles passing through the pores, since water flows only through a part of the total cross-sectional area in a soil mass, the rest of the area being occupied by the solid grains. The actual velocity can be determined as follows. Consider an elementary rectangular prism of soil with side lengths a, b and c. Assuming that in any cross-section perpendicular to side c the ratio of the void area to the total cross-sectional area is constant, i.e. $n' = A_{\text{voids}}/A_{\text{total}} =$ constant, the total volume of voids within the prism is $n'\,abc$. On the other hand, the volume of the voids can be calculated by using the value of porosity n as $n\,abc$. By equating the two, it follows that $n = n'$. Let v_s denote the average actual flow velocity, termed the *seepage velocity*, of the water particles and v denote the *discharge velocity* which is calculated on the assumption that water flows through the total cross-section. The two velocities can then be related as follows

$$v = nv_s\,.$$

For fluid flow in fine-grained soils (fine sand, silt), the velocity v can be expressed fairly well by the following empirical formula

$$v = \frac{K}{\eta}\,i_p$$

where K is an empirical constant dependent on the type of soil, porosity etc., but independent of the physical properties of the seeping fluid. K is expressed in units of area (cm^2). Substituting i_p/γ_w for i, we have

$$v = \frac{\gamma_w}{\eta} Ki\,. \tag{80}$$

In normal civil engineering practice we are mainly concerned with seepage of ground water at constant temperature. The physical quantities η and γ_w can therefore be regarded as constants, making it possible to introduce the following substitution:

$$\frac{K}{\eta}\gamma_w = k\,.$$

Equation (80) can thus be written

$$\boxed{v = ki\,.} \tag{81}$$

In words: *the flow velocity is directly proportional to the hydraulic gradient*. The factor k is called the *coefficient of permeability*. It has the dimension of velocity and is usually expressed in cm/s. Physically it can be interpreted as the flow velocity under unit hydraulic gradient.

The relationship expressed by Eq. (81) was found experimentally by Darcy (1856), and is known after him as *Darcy's law*. It is the fundamental law of linear, one-dimensional flow of water in soil due to gravity.

The coefficient of permeability is the soil property that varies between the widest limits, and may assume values ranging from 10^2 cm/s for a coarse gravel to 10^{-12} cm/s for a colloidal clay. The following example will help to visualize how extremely wide the range of variation of k values is. Consider a layer of soil, 10 cm thick, through which water flows under an unbalanced pressure head of 1 m. Let $k = 10$ cm/s, then the rate of discharge in 1 second through a cross-sectional area of 1 m^2 is 1 m^3. Now, if $k = 10^{-8}$ cm/s, it would take 5 years for the same quantity of water to flow through the layer, under the same pressure head. It is these low k values that have lead to the wide-spread misconception that dense clay and concrete are perfectly impervious to water. Strictly, this is not true, but because of the extremely low permeability of these materials any water that may seep through them evaporates immediately from the free surface so that it has a dry appearance. It must also be remembered that the gradient and not the velocity of flow is the decisive factor that determines seepage forces (cf. 5).

6.2.2 Measurement of permeability

The coefficient of permeability can be determined in three ways: by field tests, by laboratory tests on soil samples and by empirical formulas.

The laboratory determination is based on the measurement of the quantity of water that flows under a given gradient through a soil sample of known length and cross-sectional area in a given time. According to the order of magnitude of the coefficient k one of the following two methods can be used: *constant head* and *falling head permeability tests.*

Constant head permeameters are particularly suited to the testing of highly pervious coarse-grained soils. For soils with medium to low permeability, the falling head permeameters are used.

The principle of the constant head apparatus, together with sketches of some commonly used types are shown in *Fig. 161.* Designs with either downward-flow (a) or upward-flow (b) are available. Water levels are kept constant by means of overflows. The hydraulic gradient i can be computed as the difference between the constant water levels h divided by the length of the sample l. With less pervious fine sands and silty sands more reliable results can be obtained by measuring the actual water pressure in the sample by means of piezometer tubes (Fig. 161b). The discharge Q during a given time t is collected in a graduated cylinder. By *Darcy*'s law

$$Q = A\,vt = A\,kt\frac{h}{l}.$$

Hence

$$k = \frac{Ql}{A\,th}, \qquad (82)$$

A denotes the cross-sectional area of the sample.

For soils of low permeability measurable flow through the sample could not easily be obtained by the constant head test. Moreover, test results would be unreliable due to evaporation. In such cases the *falling head permeameter* is more suitable. A commonly used arrangement is shown in *Fig. 162.* A cylinder containing the soil sample is placed onto a base (wire screen or perforated disc) fitted with a fine gauze. On top, the cylinder is plugged with a rubber stopper inserted into which is a graduated standpipe. During the test, the water level will continuously drop in the pipe. The quantity of water that flowed through the sample need not be collected and measured in this test, since it can be calculated from the fall of the water level.

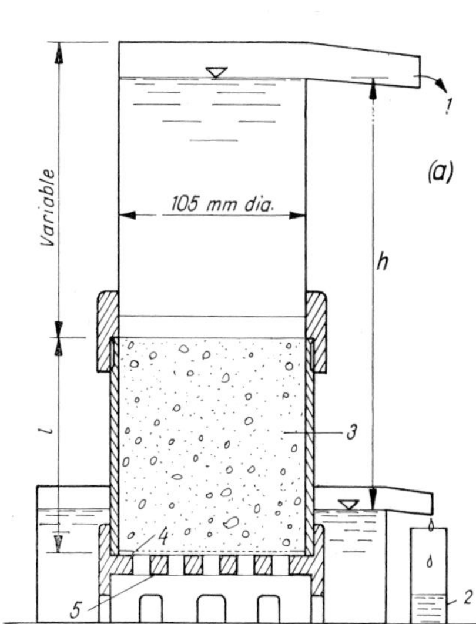

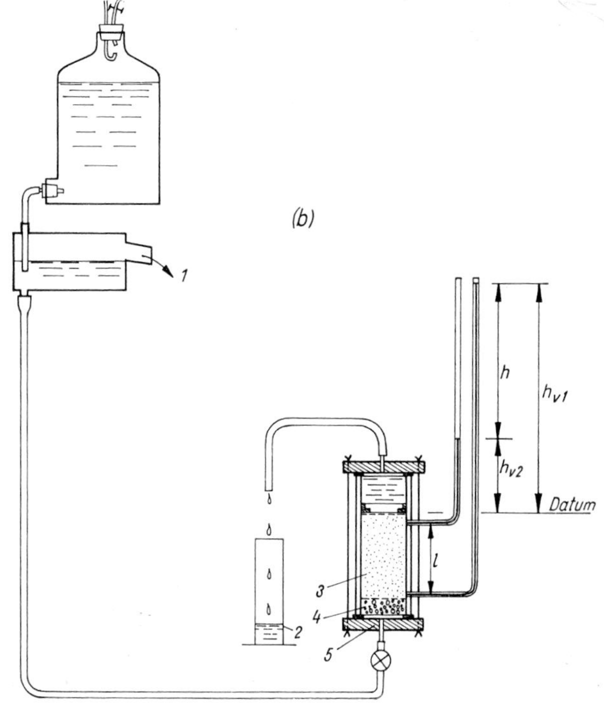

Fig. 161. Constant-head permeameters:
1 — overflow; 2 — measuring cylinder; 3 — soil sample; 4 — screen and filter layer; 5 — base plate

Let h_1 be the height of the water in the standpipe above the lower constant level at an initial time t_1 and h_2 that at the time of a second reading t_2. Let h be the height at an intermediate time t. The flow in unit time through the sample can be obtained as the cross-sectional area of the standpipe a multiplied by the velocity of the fall. This velocity is $-dh/dt$. The negative sign is to indicate that the water level decreases as time increases. By *Darcy's* law the discharge in unit time can also be expressed as the cross-sectional area of the sample A multiplied by the velocity of flow of water through the sample kh/l, where l is the length of the sample. Equating the two expressions, we get

$$-a\frac{dh}{dt} = k\frac{h}{l}A.$$

Separating the variables and integrating both sides,

$$-a\int_{h_1}^{h_2}\frac{dh}{h} = k\frac{A}{l}\int_{t_1}^{t_2}dt.$$

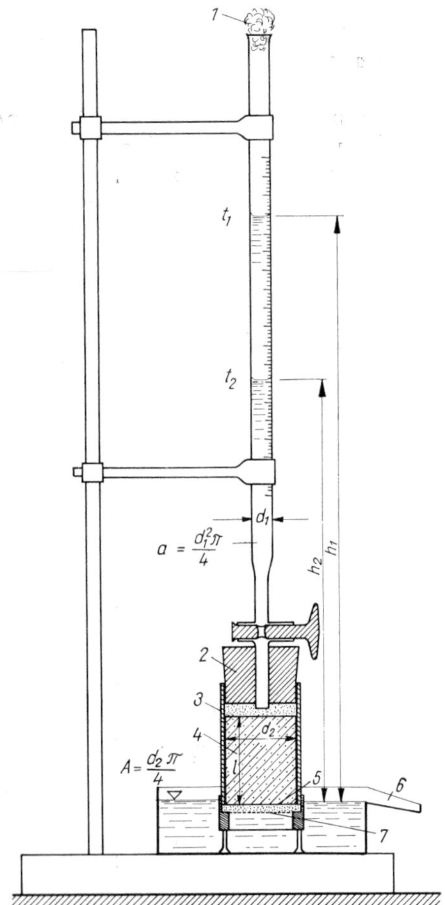

Fig. 162. Falling-head permeameter:
1 — wet cotton ball; *2* — rubber stopper; *3* — copper cylinder; *4* — soil sample; *5* — base plate; *6* — overflow; *7* — screen and filter

Hence

$$k = \frac{al}{A}\cdot\frac{1}{t_2 - t_1}\log_e\frac{h_1}{h_2}. \tag{83}$$

In order to increase the accuracy of the test, usually more than two readings are taken and the coefficient k is obtained by the statistical evaluation of the test data.

k values obtained from laboratory tests often differ considerably from the true coefficient of permeability of the soil (see e.g. MANSUR, 1957). There are many causes for this discrepancy, the most important being that the permeability of natural soil masses is greatly influenced by the heterogeneity and delicate, or even undetectable, stratification of the soil and by the uneven distribution of the fine fractions within it. Such effects, however, can hardly be simulated in normal laboratory testing.

For fine-grained soils with very low permeability the coefficient k is determined by an indirect method from the results of the consolidation test (see Section 6.6).

The mere fact that the coefficient k may have numerical values over an extremely wide range is an indication that soil permeability is very sensitive to any effect that influences the state of the soil. Even a clay stratum may be highly pervious, e.g. when it is interspersed with seams of sand, or has a fissured structure, or its original compact texture is broken up by root channels. Then water does not flow through the voids between the grains, but finds its way through the faults in the soil structure. With sands, even a slight change in the void ratio or the washing-out of a portion of the fine fractions may substantially impair the results of the permeability test. Therefore, the tests must be performed on "undisturbed" soil samples which maintained their natural *in situ* state. But this is where the main difficulty and source of error lies in permeability tests, since in those soils for which the determination of permeability is important (sand, fine sand, and silty sand), sampling, especially under water, is often met with almost insurmountable difficulties. To run tests, instead, on disturbed samples would not furnish dependable results. In cases of major importance, the coefficient k should, therefore, be determined by in situ test. There are various ways of measuring soil permeability in the field. The most suitable method, in a given case, is the one that simulates best the conditions of the particular problem for which the permeability values will be used.

One of the most important applications of k values is in solving problems of seepage (drainage of open excavations, storage losses due to seepage through dams etc.). An appropriate field test for such investigations consists in pumping water from a well and evaluating soil permeability from observations on the resulting drawdown surface. This procedure is called the *pumping test*. This method particularly lends itself to the design of ground water lowering where the drainage of a foundation pit is accomplished by pumping water from a series of filter wells.

A great advantage of the pumping test is that it eliminates a source of error inherent in all laboratory permeability tests, namely, that owing to the relatively small dimensions of the testing apparatus a soil sample is never truly representative of large soil masses whose properties often show strong local variations.

The range of applicability of the pumping test can be set, in terms of k values, to approx. 10^{-3} to 10^{-4} cm/s. If the permeability of the soil is too low, the flow towards the well is not sufficient to permit continuous operation of the pump, one of the fundamental assumptions of the test. Fortunately, drainage problems in clay seldom require knowledge of numerical values of permeability, since the little water that may enter an excavation can easily be collected in ditches and sumps and removed from them by pumping.

Let us now examine what changes occur in the ground water around a test well when water is being pumped out at a steady rate. *Figure 163* shows a section through a well. The theoretical investigation is based on the following assumptions (DUPUIT, 1863).

1. The water table is of infinite extent in horizontal direction.
2. The water-bearing stratum is homogeneous, horizontal, and it is of constant thickness.
3. The test well is a "perfect" well. This means that it extends to the bottom of the permeable stratum and is perforated over the section which is below the water table.
4. In its original state the ground water is at rest, and there is no flow into or out of the system during the test.

The pumping generates a radial flow of water toward the filter well and as a result the water table assumes a curved surface called *drawdown surface*. If the pumping is continued for a sufficiently long time, a state of steady flow is eventually attained. This means that the rate of radial flow through any cylindrical surface at radius x is constant, and equal to the amount of water pumped out of the well in unit time. Thus the rate of flow

$$q = vA_r \tag{84}$$

where A_r is the area of a cylindrical surface of radius x and height z and v is the flow velocity at that surface. By *Darcy*'s law $v = ki$. As indicated by the curved drawdown surface, the hydraulic gradient i varies with the distance from the centre of the well. To simplify the theoretical treatment, let us assume that the flow lines are horizontal and that the flow velocity at every point of a vertical flow section is the same. We further assume as a first approximation that the flow is two-dimensional. (Note that these assumptions mean only a rough approximation of the actual conditions.) The hydraulic gradient can now be expressed as the slope of the water surface:

$$i = \frac{dz}{dx}.$$

Substituting this in Eq. (84),

$$q = 2\pi x\, zk \frac{dz}{dx}.$$

Separating the variables and integrating both sides,

$$z^2 = \frac{q}{\pi k} \log_e x + C.$$

The constant of integration C can be determined from the boundary conditions as follows. When $x = r$, $z = h$. Hence

$$C = h^2 - \frac{q}{\pi k} \log_e r.$$

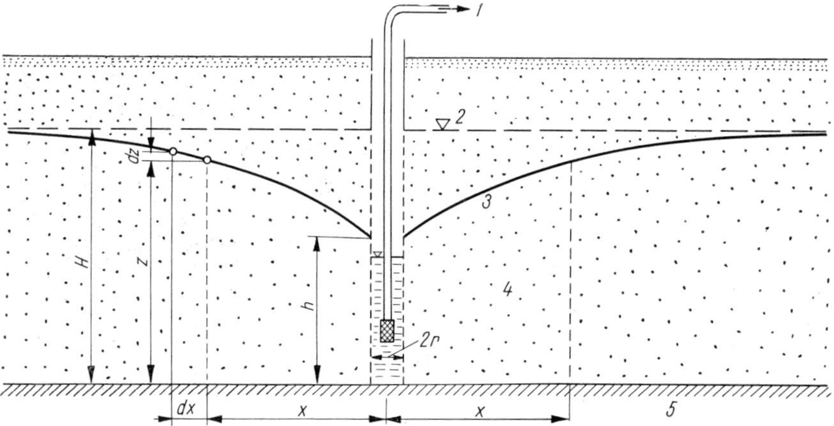

Fig. 163. Effect of pumping on the ground water level:
1 — to pump; *2* — original ground water table; *3* — lowered ground water surface; *4* — permeable layer; *5* — impervious layer

Substituting this for C,

$$z^2 - h^2 = \frac{q}{\pi k}(\log_e x - \log_e r)\,. \qquad (85)$$

Solving this for k,

$$k = \frac{q}{\pi}\,\frac{\log_e \dfrac{x}{r}}{z^2 - h^2}\,. \qquad (86)$$

As pointed out before, this derivation is based on very rough simplifying assumptions. The permeability coefficient k so obtained can therefore be used only in such cases where the conditions correspond as closely as possible to those of the pumping test. The highest deviations from the theoretical assumptions exist in the immediate vicinity of the test well. Therefore, instead of using in Eq. (86) the height h measured adjacent to the well (at $x = r$), it is advisable to make two observation wells located at some larger distances x_1 and x_2, and use the heights z_1 and z_2 to determine the coefficient k (*Fig. 164*). The formula of calculation then becomes

$$k = \frac{q}{\pi}\,\frac{\log_e \dfrac{x_1}{x_2}}{z_1^2 - z_2^2}\,. \qquad (87)$$

Even higher accuracy can be attained by observing the water table in more than two bore holes. There is usually a scatter in the observation data and the correct value of k can be obtained by the method of least squares.

Before observations are taken in the test, pumping at a steady rate must be continued for a considerably long time until a steady flow is reached. This may take days or sometimes weeks. Daily variations of the water table may also influence the accuracy of the measurement, and a relatively quiet period should be chosen for observations when the water levels in the bore holes are practically constant.

Further details and evaluation of test data will be dealt with in Vol. 3: Soil testing practice.

To obtain reliable data, forced pumping should be avoided and the observations should be made at relatively small values of drawdown in the well. Furthermore, it should not be overlooked that the water level in the well always breaks away some distance from the exit point of the lowered water table (BREITNÖDER, 1942; NAHRGANG, 1954; ÖLLŐS, 1957). This difference between the two levels varies with the drawdown in the well. ÖLLŐS (1957) found, on the basis of model tests, an empirical relationship (*Fig. 165*) between the height of water in the well h_i and that at the point of exit of the drawdown line h_e. The relationship fits the following equation

$$h_e - h_i = C\,\frac{(H - h_i)^2}{H}\,. \qquad (88)$$

The empirical constant C has a value of approx. 0.5.

A correct solution for the flow net around the well could be developed by the three-dimensional potential theory of flow (BOULTON, 1951).

If a well does not penetrate to the lower impervious boundary of the water-bearing stratum, water may enter the well also through the bottom. As a result the yield of the well increases, or else the drawdown required to produce the same yield as that in the case of a fully penetrating "perfect" well decreases. A number of approximation formulae are also available for the computation of yield of such "non-perfect" wells (NAHRGANG, 1954, SZÉCHY, 1955).

6.2.3 Factors affecting permeability

The coefficient of permeability of a soil depends on many factors. To understand the composite effect of these, let us study first the general laws governing fluid flow in a small-bore capillary tube.

If laminar flow takes place in such a tube, flow velocity varies from point to point in a cross-section of the pipe. This is due to the frictional resistance between the elements of the fluid. Consider a circular tube of radius r and length l. Let one end of the tube be subjected to an unbalanced hydrostatic pressure of $h\gamma_w$. Let ϱ be the radius of any concentric cylinder of liquid within the tube (*Fig. 166*). The thrust acting on one end

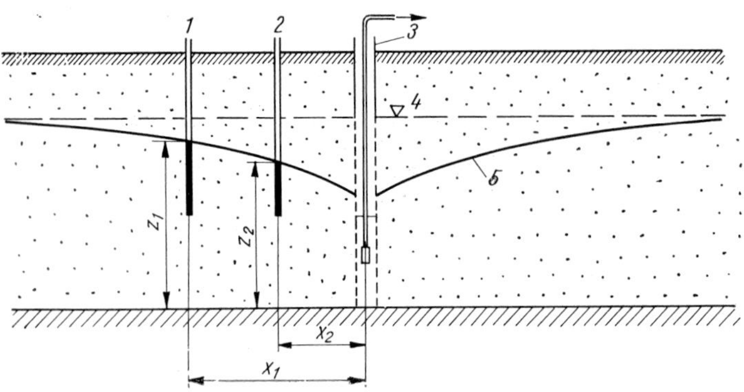

Fig. 164. Use of piezometer tubes to determine the draw-down curve:
1,2 — observation tubes; *3* — well; *4* — original ground water table; *5* — lowered surface

Fig. 165. Step between the water level in the tube and on the surface of the filter mantle (ÖLLŐS, 1957)

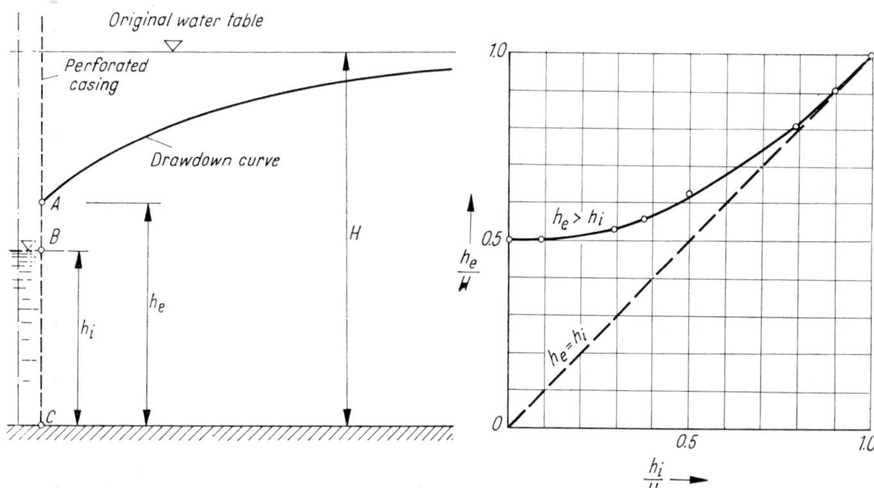

of this cylinder is thus equal to $\pi\varrho^2 h\gamma_w$. This force is balanced by the frictional resistance on the side of the tube. This shearing resistance is due to the viscous friction of the fluid, and is equal to $\eta(-\,dv/d\varrho)$. The negative sign is used to indicate that the velocity decreases as the radius ϱ increases. The total shearing force is obtained as the shearing stress multiplied by the surface area $2\varrho\pi\, l$. For equilibrium of the liquid cylinder

$$\pi\varrho^2 h\gamma_w - 2\pi\varrho\, l\eta(-\,dv/d\varrho) = 0\,.$$

Separating the variables,

$$dv = -\frac{h\gamma_w}{2\eta l}\varrho\, d\varrho\,.$$

By integration

$$v = \frac{h\gamma_w}{4\eta l}\varrho^2 + C\,.$$

In laminar flow the velocity on the inside surface of the tube is zero, i.e. $v = 0$, when $\varrho = r$. Using this boundary condition, the constant of integration C can be eliminated. Also substituting i for h/l, the expression of velocity becomes

$$v = -\frac{i\gamma_w}{4\eta}(r^2 - \varrho^2)\,. \tag{89}$$

If we plot the velocity vectors determined by Eq. (89) at all points of a cross-section, the resulting surface which represents the velocity distribution is a paraboloid. The rate of flow through any section can be obtained by integration:

$$q = \int_0^r v 2\pi\varrho d\varrho = \frac{r^2\gamma_w}{8\eta} iA$$

where A is the total cross-sectional area of the pipe. The average velocity can be expressed as

$$v_a = \frac{q}{A} = \frac{r^2\gamma_w}{8\eta} i = \frac{V_{\max}}{2}\,. \tag{90}$$

Let us now employ the expression of hydraulic radius R which is defined in Hydraulics as the ratio of cross-sectional area A to wetted perimeter U. For a circular tube with fluid flow through the total cross-section $R = r^2\pi/2r\pi = r/2$. After substituting in Eq. (90), the discharge in unit time through the tube becomes

$$q = \frac{1}{2}\frac{R^2\gamma_w}{\eta} iA\,. \tag{91}$$

A similar derivation gives the expression for the rate of flow between two flat parallel plates:

$$q = \frac{1}{3}\frac{R^2\gamma_w}{\eta} iA\,. \tag{92}$$

Equations (91) and (92) differ only in the values of the numerical constant. From this it may be inferred that the rate of flow through a tube of any geometrical cross-section can be given by a similar general formula:

$$q = C_a \frac{R^2\gamma_w}{\eta} iA \tag{93}$$

where C_a is a dimensionless shape factor.

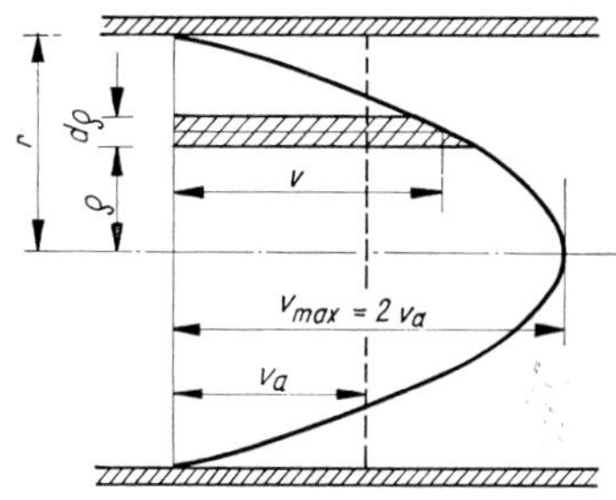

Fig. 166. Velocity distribution in a cylindrical capillary tube

Figure 167 shows a tube with constant but extremely irregular cross-section, similar to the flow passages formed between soil grains. The ratio of the void area to the total cross-sectional area A is, as was shown before, equal to the porosity n. Hence the rate of flow through an irregular flow passage is

$$q = C_a \frac{\gamma_w}{\eta} R^2 n\, iA\,. \tag{94}$$

The concept of hydraulic radius may also be applied to soils in view of the following considerations. The expression of R can be rewritten

$$R = \frac{A}{U} = \frac{Al}{Ul}$$

wherein l is the length of the tube. The numerator represents the total volume of void spaces in which flow of water takes place. This again may be expressed as the volume of solids V_s multiplied by the void ratio, i.e. $Al = nV = eV_s$. The denominator represents the surface area of the flow channel which, for soils, is equal to the total surface area of the solid grains A_s. Hence the hydraulic radius may be written

$$R = e\frac{V_s}{A_s}\,. \tag{95}$$

The surface area of the grains can be expressed in terms of the harmonic mean of grain sizes d_h. Using Eqs (5) and (6)

$$\frac{V_s}{A_s} = \frac{\alpha_v}{\alpha_f} d_h\,. \tag{96}$$

This quantity can be used as a grain shape characteristic.

A general formula for the permeability coefficient can now be derived as follows. Substituting Eq. (96) in (94) gives

$$q = C_a \left(\frac{\alpha_v}{\alpha_f}\right)^2 e^2 n \frac{\gamma_w}{\eta} d_e^2\, iA\,.$$

Considering that $n = e/(1+e)$ and replacing the expression $C_a(\alpha_v/\alpha_f)^2$, in which all terms depend on grain shape, by a composite shape factor C, the rate of flow may be written

$$q = d_e^2 \frac{\gamma_w}{\eta} \cdot \frac{e^3}{1+e} C\, iA\,.$$

Comparing this with *Darcy*'s law gives

$$k = Cd_e^2 \frac{\gamma_w}{\eta} \cdot \frac{e^3}{1+e}\,. \tag{97}$$

This equation is very useful in that it reflects the effect of the factors that influence soil permeability. Among these the most important is grain size. d_e in the formula is the *effective grain size*, i.e. the diameter of the sphere for which the ratio of its volume to its surface area is the same as the similar ratio for a given assemblage of soil particles. The factor γ_w/η depends on the kind and physical state of the pore fluid. Since both γ_w and η vary with temperature, k also is dependent on temperature. Other factors affecting permeability are the void ratio, the manner of packing of the grains and the shape characteristics of the voids. The collective effect of these is represented by the factor $Ce^3/(1+e)$.

For the calculation of the permeability coefficient a number of empirical formulas has been suggested in the literature. The value of these formulas should, however, not be overestimated, they serve only to give a rough estimate of permeability. All these formulae may be expressed in the following general form

$$k = cf_1(e)\, f_2(d) \tag{98}$$

where functions $f_1(e)$ and $f_2(d)$ indicate the effect of the void ratio and of the grain size, respectively, on soil permeability. Since these formulas must work with a single grain size, conveniently called the "average" or "effective" diameter and regarded representative of the whole mass of a soil, they cannot truly reflect the often considerable influence on permeability of such soil characteristics as the amount of fines or grading. Also, the composite effect of the shape and surface roughness of the grains is usually represented by a single shape factor c. In this respect Eq. (97), originally proposed by KOŽENY (1927), seems to provide the most exact approach. But even this formula is of limited value for practical calculations and serves mainly to demonstrate clearly the effect of the various factors on permeability.

Some of the most widely used empirical formulae are presented in the following. When choosing any one of them to calculate the k value of any given soil, we must always make sure that the conditions of flow and the type of soil are the same as those from which that particular formula has been derived experimentally. Hints on the range of validity will be given for each formula.

One of the earliest and simplest formulas was proposed by ALLEN HAZEN (1893). According to

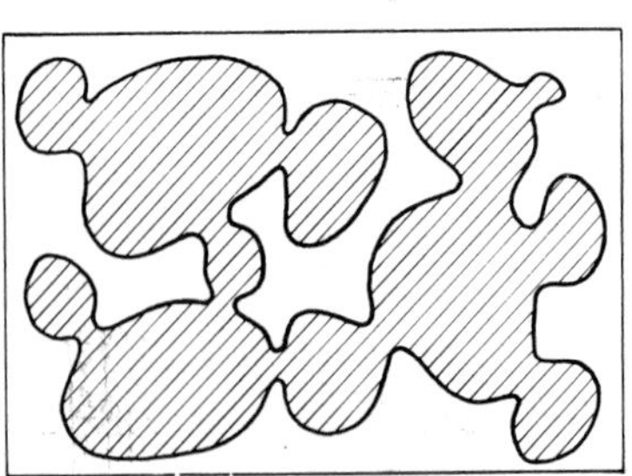

Fig. 167. Tube with irregular cross-section

him

$$k = 116\, d_e^2,$$

d_e is the "effective" grain size that corresponding to the 10% ordinate on the grain-size distribution curve. The formula applies to clean filter sands that occur in loose packing. It does not take account of the effect of density in spite of the fact that porosity, which may vary according to *Slichter*'s investigations between limits as wide as 26 to 47%, must exert a substantial influence on soil permeability. Also, the use of the effective size is entirely arbitrary.

Jáky (1944) found that a fair estimate of the order of magnitude of k can be obtained for all soils form the formula

$$k = 100\, d_m^2$$

where d_m denotes the grain size that occurs with the greatest frequency. Again, *Jáky*'s formula may be used only for a first rough estimate of soil permeability.

Terzaghi (1955) developed a formula for fairly uniform sands, which reflects the effect of grain size and void ratio:

$$k = 200\, d_e^2\, e^2.$$

He also suggested that for soils not falling within the range of validity of the above formula, the coefficient of permeability should be determined by means of direct measurement.

Numerous other relationships have been put forward by various investigators. But these do not offer greater accuracy either, and therefore they are not dealt with here.

The marked influence of porosity on the permeability of soils is clearly demonstrated by the *Koženy* formula [Eq. (97)] given above. The relationship may also be used to predict, for a given soil, the value of the coefficient of permeability at any void ratio, if the k value at another void ratio is already known. The conversion factor may be written

$$k_1 : k_2 = \frac{e_1^3}{1+e_1} : \frac{e_2^3}{1+e_2}. \qquad (99)$$

Equation (97) was obtained by the application of the concept of hydraulic radius to the pore channels. Zunker (1930) followed a different approach and found the following relationship:

$$k_1 : k_2 = C_1' e_1^2 : C_2' e_2^2.$$

In sands, the factor C' changes slightly with the void ratio. Assuming that C' is constant, the relationship reduces to

$$k_1 : k_2 = e_1^2 : e_2^2.$$

From experiments on clean sands with bulky grains, Casagrande (1948) developed the following formula

$$k = 1.4\, k_{0.85}\, e^2.$$

Here $k_{0.85}$ is the coefficient of permeability at a void ratio of 0.85.

Permeability also depends greatly on whether a soil is composed mainly of bulky grains or of flaky ones. The significant influence of the grain shape on soil permeability is illustrated in *Fig. 168* were the ratio $k/k_{0.85}$ is plotted against the void ratio e.

Chardabellas (1940) found that the function $k = f(e)$ is best fitted by the equation

$$k = Ae^B.$$

It follows that the plot of k against e on a log–log scale is a straight line. The constant A represents the coefficient of permeability at a void ratio of 1.0 and shows the order of magnitude of k. The constant B is indicative of the rate of change of k with the change of e. B has values ranging from 2 to 5.

It is mainly sands for which the permeability formulae give satisfactory results. With clays, the deviations of the calculated k values from the true permeability of the soil are generally more significant. A probable explanation for this lies in the behaviour of the bound water present in clays. As discussed previously, the fine particles of clay are surrounded by a film of adsorbed water whose properties differ essentially from those of normal water. A portion of the pore water also exists in the form of bound water. As a result the pore space open to seepage is greatly reduced.

In clays, the influence of structure on permeability is also important. Two clays, one with a dispersed structure and the other with a flocculated structure, may exhibit strongly differing permeabilities at the same void ratio.

There is some experimental information available on the permeability of clay minerals (Cornell University, Final Report, Soil Solidification Research, Vol. 2, 1951). k values for clay minerals with different adsorption complexes are listed in *Table 22*. Further tests on montmorillonite at a

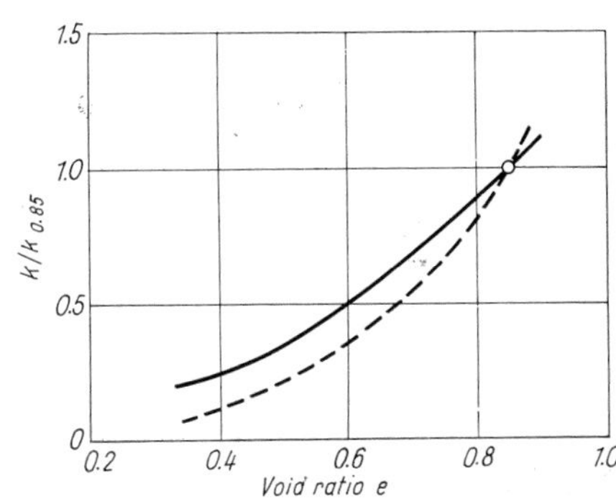

Fig. 168. Relationship between permeability coefficient and void ratio for sands with bulky grains (solid line) and flaky particles (dotted line)

Table 22. Permeability of clay minerals

Clay mineral	Exchange-able cation	Void ratio e	Coefficient of permeability k cm/s
Montmorillonite	Na	15	8×10^{-8}
		3	1.5×10^{-8}
	K	11	5×10^{-8}
		7	0.8×10^{-8}
	Ca	8	1000×10^{-8}
		4	10×10^{-8}
	H	9	200×10^{-8}
		3	10×10^{-8}
Kaolinite	Na	1.6–0.5	$150–80 \times 10^{-8}$
	K	1.6–1.1	$300–90 \times 10^{-8}$
	Ca	1.6–1.3	$1000–150 \times 10^{-8}$
	H	1.4–1.0	$1000–150 \times 10^{-8}$

constant void ratio of $e = 8$ showed permeability to increase, depending on the type of cation adsorbed, in the following order

$$\mathrm{K} < \mathrm{Na} < \mathrm{H} < \mathrm{Ca}\,.$$

For kaolinite at $e = 1.5$

$$\mathrm{Na} < \mathrm{K} < \mathrm{Ca} < \mathrm{H}\,.$$

Dispersion effect causes the permeability of a clay to decrease. The influence of the adsorption complex on permeability may have a bearing on the engineering uses of clays also. The fact for example that Na clays have relatively low permeabilities suggests that clays to be used as core material in earth dams can be rendered less pervious through cation exchange, e.g. by treating them with salt water.

Summing up the main conclusions, the coefficient of permeability was found to depend

(i) on the square of the harmonic mean of grain diameters,

(ii) on properties of the pore fluid, namely viscosity and unit weight, both in turn being dependent on temperature,

(iii) on the void ratio,

(iv) on the shapes and structural arrangement of the grains,

(v) on the amount of undissolved gas in the pore fluid,

(vi) on the chemical composition of the soil and nature of the adsorption complex.

6.2.4 Numerical values of the coefficient of permeability

As already mentioned, the range of magnitude of permeability is extremely wide. The value of $k = 10^{-6}$ cm/s indicates the approximate lower limit below which a soil may be considered "im-

Table 23. Numerical values of k; methods of determination

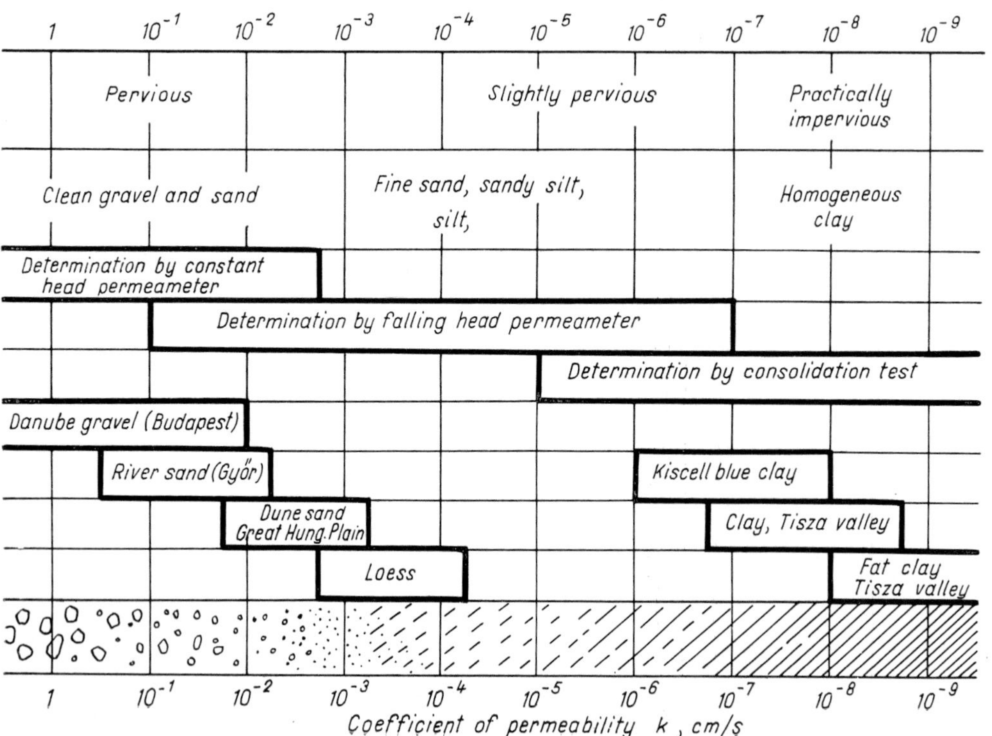

pervious" for all practical purposes. *Table 23* gives a survey of the permeability of natural soils and the suitable methods of permeability measurements.

In view of the wide range of variation of k values it would seem worth considering to express soil permeability in terms of the negative logarithm of k in the same manner as the hydrogen ion concentration of solutions is indicated by the number pH, instead of using complete numerical values. Test data furnish, in most cases, reliable information only on the order of magnitude of pH anyway.

Another point to be mentioned here is that the physical unit for the coefficient of permeability cm/s is often replaced in practical calculations by the unit m/year. To convert from cm/s to m/year, multiply by 3.15×10^5. (Conversely, 1 m/year = $= 3.17 \times 10^{-6}$ cm/s.)

6.2.5 Permeability of partially saturated soils

Permeability in the sense used in the preceding discussion cannot be spoken of in three-phase soils. The three phases may be arranged as was shown in Fig. 88, into various structural patterns and may move separately and at different velocities in relation to each other (see Section 6.1). Combined tests on the permeability of soil to air and to water can lead to an understanding of this problem. Let us introduce the concept of relative permeability. This term means the ratio of the coefficient of permeability measured in a three-phase state of the soil to that measured in the saturated state of it. This ratio is a function of the degree of saturation. A brief account of laboratory tests aimed in part at the determination of the composite air/water permeability of three-phase sands was already given in Section 6.1. Test results typical of the behaviour of sands are presented in *Fig. 169*, where relative permeabilities are plotted against the degree of saturation S. It is seen from the figure that below a critical value of saturation (approx. $S = 0.2$ in the case examined) the relative permeability to water $\varkappa_w$ is zero. In other words water cannot permeate through the soil unless the degree of saturation is increased above this critical value. A similar curve is obtained for the relative permeability to air $\varkappa_g$. If an assemblage of grains is sufficiently dry, the pore spaces filled with air form continuous passages permitting an almost unhindered flow of air and a high relative air permeability results. With increasing degree of saturation, the relative permeability to air reduces rapidly and may, at an upper critical value of S, diminish to zero. Beyond that point the pore air can exist only in the form of entrapped bubbles and is thus forced to move along with the pore water. In this state, a separate movement of the gaseous phase in relation to the liquid phase is possible only if the gas presses

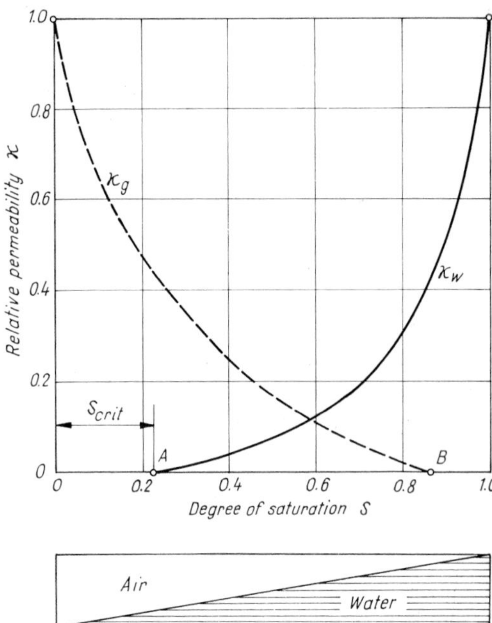

Fig. 169. Relative permeability to air and water of three-phase sands versus degree of saturation

water out of the pores, but then the degree of saturation must also change. The critical states represented by points A and B in Fig. 169 are of great importance from a practical viewpoint. If $S < 0.2$, the soil only sucks water in but does not let it through. In this state the apparent cohesion of sands is the greatest, and the bulking of dumped soils is also a maximum here. Beyond the upper critical state (point B, $S > 0.8$), the soil behaves as if it were saturated. The process of consolidation becomes prolonged. The peak of the *Proctor compaction curve* is here, and also the compaction of sands by vibration is the most effective in this state. In *Fig. 170*, the three characteristic states in air–water flow through partially saturated soils are represented in a triangular diagram.

Polubarinova–Kochina (1952) found the following relationship between the relative perme-

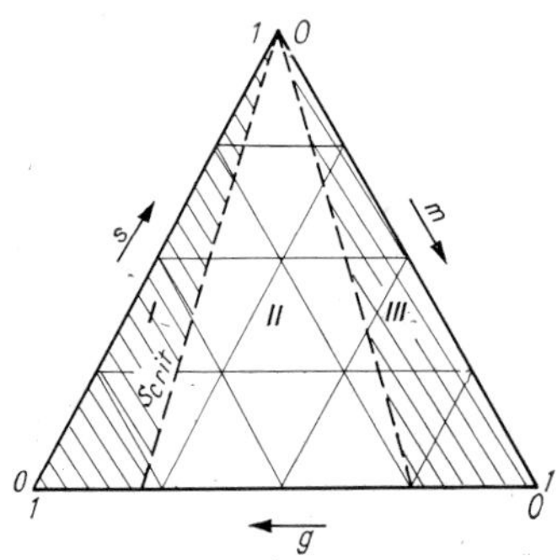

Fig. 170. Zones representing different types of water movement in the triangular diagram

ability to water and the phase composition of the soil, at a constant porosity

$$\varkappa_w \approx \left(\frac{m}{m+g}\right)^{3.5}.$$

In the range of $S = 0.8$ to 1.0, the expression becomes

$$\varkappa_w \approx 1 - \frac{Cg}{m+g}$$

where C is a numerical constant, with values ranging from 2 to 4. The lower value applies to uniform sands, and the upper value applies to well-graded sands (ORLOB and RADHAKRISHNA, 1958).

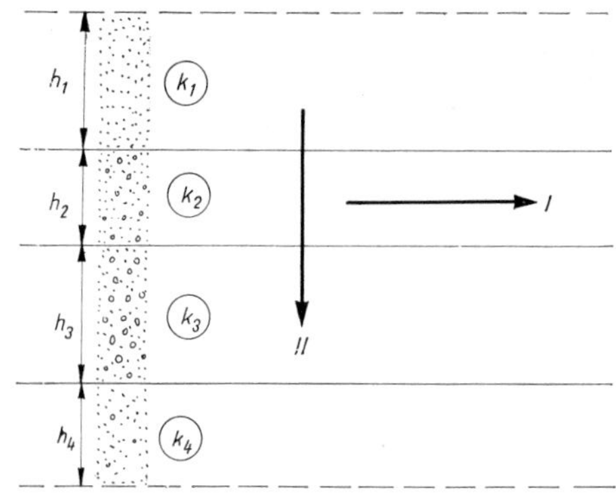

Fig. 171. Average coefficient of permeability in layered soils

6.2.6 Permeability of stratified and anisotropic soils

The subsoil is usually made up of several layers which have different permeabilities. For the purpose of seepage computations in such cases, undisturbed soil samples are taken from each of the layers and tested for permeability in the laboratory. Average values of k, representative of the whole of the stratified mass of soil, can then be computed after TERZAGHI (1943) by the following consideration.

Figure 171 shows the section of a stratified soil. Let $h_1, h_2, \ldots h_n$ denote the thicknesses of the individual layers and let $k_1, k_2, \ldots k_n$ be their respective coefficients of permeability. Let k_{I} denote the average coefficient of permeability parallel to the bedding planes and k_{II} the average coefficient of permeability perpendicular to the bedding planes.

If the flow is parallel to the bedding planes, the hydraulic gradient is the same at every point and, by *Darcy*'s law, the average discharge velocity may be written

$$v = k_{\mathrm{I}}\, i = \frac{1}{h}(v_1 h_1 + v_2 h_2 + \ldots\ldots + v_n h_n) =$$

$$= \frac{1}{h}(k_1 i h_1 + k_2 i h_2 + \ldots\ldots + k_n i h_n). \quad (100)$$

Herein

$$h = h_1 + h_2 + \ldots\ldots + h_n$$

is the total thickness.

Simplifying by i in Eq. (100),

$$k_{\mathrm{I}} = \frac{1}{h}(k_1 h_1 + k_2 h_2 + \ldots\ldots + k_n h_n). \quad (101)$$

If a steady flow occurs at right angles to the bedding planes, the continuity of flow requires that the discharge velocity be the same in each layer. It follows that the hydraulic gradient must change from layer to layer. Let the hydraulic gradients across successive layers be denoted by $i_1, i_2, \ldots i_n$, and let the total head lost in flow over the total thickness h be denoted by Δh.

The constant discharge velocity is

$$v = \frac{\Delta h}{h} k_{\mathrm{II}} = k_1 i_1 = k_2 i_2 = \ldots\ldots = k_n i_n.$$

Hence, for the losses in head across successive layers we have

$$\Delta h_1 = \frac{v h_1}{k_1},\ \Delta h_2 = \frac{v h_2}{k_2},\ \ldots\ldots,\ \Delta h_n = \frac{v h_n}{k_n}.$$

The total head lost is

$$\Delta h = \Delta h_1 + \Delta h_2 + \ldots\ldots + \Delta h_n = h_1 i_1 + h_2 i_2 + \ldots\ldots + h_n i_n.$$

Combining the first and the last equations

$$k_{\mathrm{II}} = \frac{h}{\dfrac{h_1}{k_1} + \dfrac{h_2}{k_2} + \ldots\ldots + \dfrac{h_n}{k_n}}. \quad (102)$$

A comparison of Eqs (101) and (102) shows that for every stratified system $k_{\mathrm{I}} > k_{\mathrm{II}}$.

In many problems, especially those concerned with seepage forces, subsurface erosion and piping, it is important to know the head lost in seepage in each layer of a stratified subsoil. As was shown in the preceding, the loss in head depends on the thicknesses and permeabilities of the individual layers. A general formula was developed by VARGA (personal communication):

$$\Delta h = \frac{h}{\dfrac{k_i}{h_i}\displaystyle\sum_1^n \frac{h_i}{k_i}}.$$

(For the symbols see Fig. 171). The formula is applicable to both horizontal and vertical flow cases.

From the considerations leading to Eqs (101) and (102) it follows that a stratified mass of soil may be replaced by an anisotropic system that exhibits different permeabilities in different directions. For this system *Darcy*'s law may be expressed in vectorial form as

$$\bar{v} = k\bar{i} \tag{103}$$

where the vectors $\bar{v}$ and $\bar{i}$ stand for the column matrices

$$\bar{v} = \begin{pmatrix} v_x \\ v_y \end{pmatrix} \quad \text{and} \quad \bar{i} = \begin{pmatrix} i_x \\ i_y \end{pmatrix}$$

and k is defined by the matrix

$$k = \begin{pmatrix} k_x & k_{xy} \\ k_{xy} & k_y \end{pmatrix}$$

x and y indicate any two directions at right angles to each other (*Fig. 172*). The horizontal and the

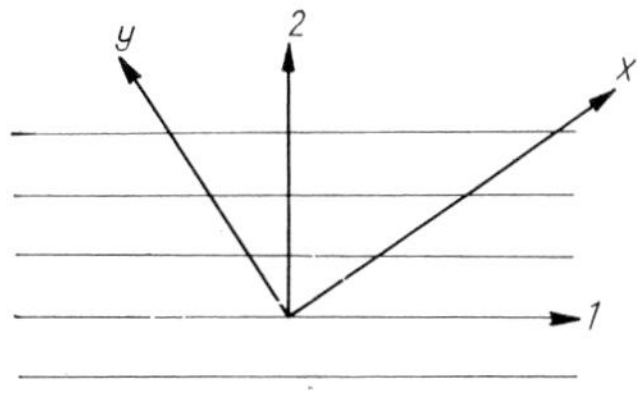

Fig. 172. Extension of the validity of Darcy's law to layered-anisotropic systems

vertical directions are principal directions, and the average coefficients of permeability k_{I} and k_{II} are the principal values of the k matrix. The vectors $\bar{i}$ and $\bar{v}$ are in general no longer collinear except in the special cases when the flow is either parallel or perpendicular to the bedding planes.

6.2.7 The validity of Darcy's law

According to *Darcy*'s law, the seepage velocity is proportional to the forces due to gravity, and inversely proportional to the frictional resistances acting against the flow. Resistances due to inertia forces that vary, according to the general *Navier–Stokes equations* of hydrodynamics, in direct proportion to the square or to a higher power of flow velocity are neglected for the low velocities normally encountered in seepage of water through soils. On this ground, many investigators disputed the validity of *Darcy*'s law and suggested, on the basis of experiments or theoretical considerations, more complex formulae that take all forces involved into account. A general formula based on analogy mechanics may be written as

$$i = av + bv^2 .$$

In this expression a varies in direct proportion to viscosity and inversely with the square of the grain size, and b is inversely proportional to grain size. For fine-grained soils b is negligible in relation to a, and the expression reduces to a linear form.

Forchheimer found analytically the following relationship which holds for seepage under medium gradients:

$$v^m = ci .$$

The exponent m was found to vary from 1 to 2.

Some investigators base their arguments against Darcy's law upon the observation that the yield of well tends to decrease during a prolonged pumping test. There can be, however, other causes that account for this tendency. One is that the ground water reservoir round the well is not of infinite extent as assumed for the theoretical analysis, and therefore it is being gradually exhausted as steady pumping goes on. In addition, dissolved air may be liberated from the water in the course of flow toward the well and plug up a portion of the pore passages. Similar effects may be caused by the migration of the finest particles toward the well and also by corrosion.

Returning now to the validity of *Darcy*'s law, it can be concluded that under the conditions of seepage through soil there is a slow transition from purely laminar flow to a slightly turbulent flow, and the range of validity cannot, therefore, be definitely established. FAUCHER, LEWIS and BARNES (1933) suggested the following approximating criterion for the limit of laminar flow

$$\frac{v d_a \gamma_w}{\eta g} \leq 1 \tag{104}$$

where v is the discharge velocity and d_a is the average grain diameter. Substituting numerical values for the physical constants, inserting $v = ki$, calculating k from *Jáky*'s approximate formula and assuming a hydraulic gradient of $i = 1$ gives the following expression for the limiting maximum grain size below which *Darcy*'s law holds:

$$\frac{100\, d^2 d \gamma_w}{\eta g} = \frac{100\, d^3}{10^{-5} \times 981} = 1 .$$

Hence $d = 0.5$ mm.

There has been much experimental evidence to verify the validity of the criterion given in Eq. (104). LEUSSINK (1957) found, from permeability tests on coarse gravels with grain sizes ranging from 20 to 350 mm and from 50 to 350 mm, the curvilinear relationships shown in *Fig. 173* between v and i. Also shown in the lower portion of the figure are the results of tests on a sandy gravel conducted at the Soil Mechanics Laboratory of the Technical University of Budapest. These relationships are of the parabolic type, and *Darcy*'s

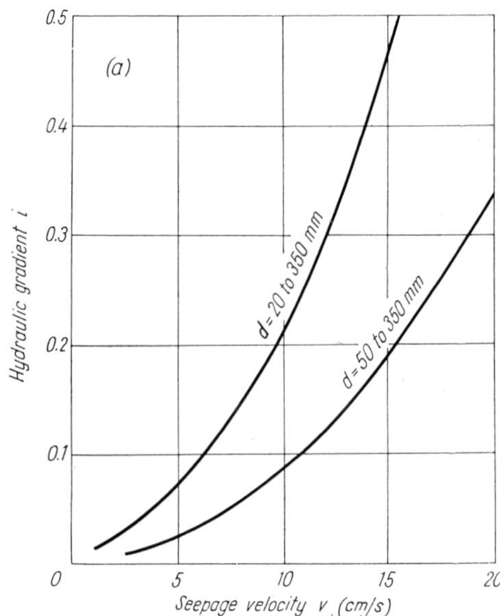

Fig. 173. Seepage velocities in coarse granular materials: *a* — Leussink's tests; *b* — author's tests

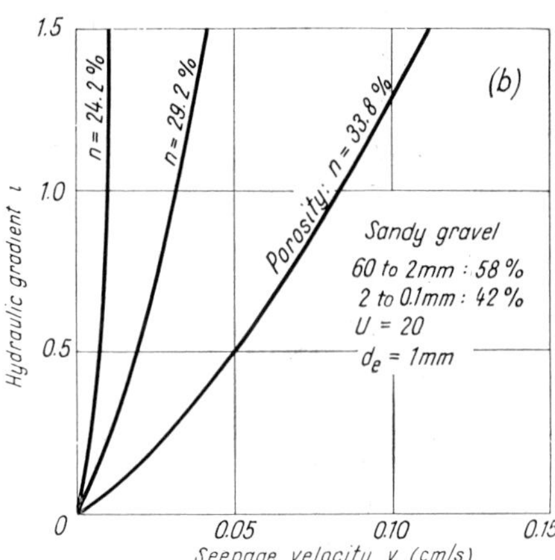

linear approximation can be seen to be valid, for the range of soils tested, only between the hydraulic gradients of 0.01 to 0.1. OHDE (1951) expressed the limit of applicability of *Darcy*'s law in terms of gradient and effective diameter d_e as follows.

Thus far we discussed the upper limit of the validity of *Darcy*'s law for coarse-grained soils. In sands and gravels the thickness of the water film adsorbed on the particle surfaces is negligible as compared with the size of the soil particles, and the relative amount of the bound water is also very small. In clays, as generally in cohesive soils, the greater part of the pore water exists in the bound state, and while forming an inert material itself from the point of view of seepage it also reduces the sizes of the pore channels available for the flow of the "free" water. Because of the intermolecular attraction due to gravity and to electrical effects on the solid–liquid interface, motion in the adsorbed water layer can be initiated only by the action of shearing stresses. Assuming, after KOVÁCS (1957), that the shearing stresses required to generate flow diminish linearly with the distance from the tube well (*Fig. 174*), fluid flow in a capillary tube comes to a standstill when $r_0 = 0$. In that state the molecular forces keep equilibrium with the force of gravity. The equation of equilibrium for a pipe of length l may be written

$$i\gamma_w \, l \, \pi r^2 = 2\pi r \, l \, \tau_0 \, .$$

Hence

$$i = i_0 = \frac{2\tau_0}{\gamma_w r} \, . \tag{105}$$

Considering that for unit length

$$i\gamma_w = p \, 1 \, .$$

Equation (105) is essentially identical to the Laplace equation, i.e. with the above assumptions, the shearing strength τ_0 is equal to the surface tension of water. The problem can therefore be treated as a fluid flow that is essentially resisted by capillary forces. These forces are responsible for the fact that fluid flow is not possible below a threshold gradient i_0. This critical gradient, however, cannot be determined from Eq. (105), since the numerical value of r is usually not known. We must therefore follow the experimental way. ROZA (1953) recommends for this purpose the use of the apparatus diagrammatically shown in *Fig. 175*.

Because of the threshold gradient, Darcy's law should for dense clays be rewritten in the following form

$$v = k(i - i_0) \, . \tag{106}$$

An experimentally obtained relationship between hydraulic gradient and flow velocity is shown in *Fig. 176*. The actual graph $0\acute{A}B$ can be replaced by the broken line $0AB$, which is the graphical representation of Eq. (106). The threshold gradient i_0 is read off as the intersect on the axis i.

The value of the critical gradient depends on the density of the soil, or, in saturated soils, on

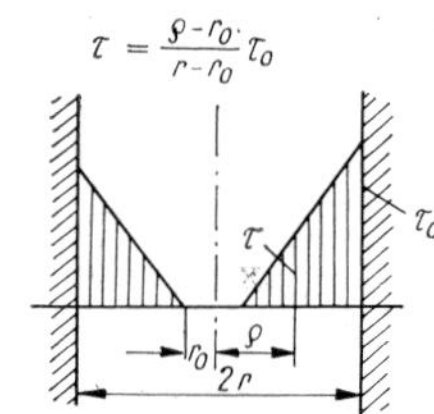

Fig. 174. Assumption for the distribution of shearing stresses required to bring adsorbed water on the surface of capillary tubes into movement

Fig. 175. Laboratory determination of the threshold gradient

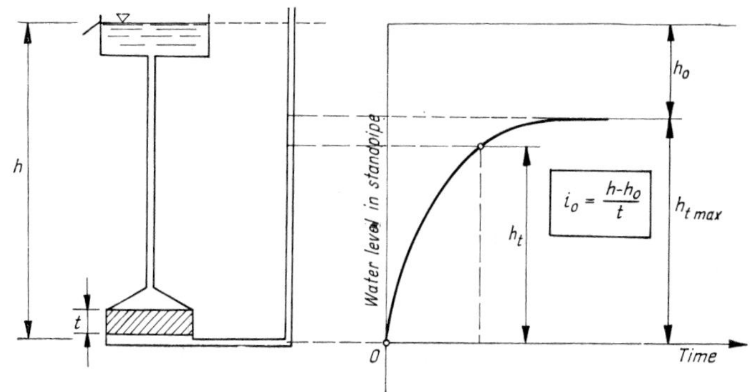

the moisture content (Roza, 1953). i_0 also depends on such factors as the mineral composition and adsorption complex of the soil, temperature etc. Measured values ranged from 0.2 to 0.5 for silts, and 12 to 18 to clays.

The existence of a threshold gradient greatly influences the process of consolidation in saturated clays under load, since the drainage of the water out of the pores and the compression of the soil are possible only if the actual gradient surpasses the threshold value i_0.

6.2.8 Examples

Example 1. A constant-head permeability test was run on a sand sample 15.4 cm in length and 58.2 cm² in cross-sectional area. Porosity was $n_1 = 44.0\%$. Under a head of 25 cm the discharge was found to be 35 cm³ in 15 s. Calculate the coefficient of permeability. Also determine the discharge velocity and the seepage velocity during the test. Estimate the permeability of the sand for a porosity of $n_2 = 38.9\%$.

Solution:

Using Eq. (82)

$$k = \frac{Ql}{Ath} = \frac{35\times15.4}{58.2\times15\times25} = 2.5\times10^{-2}\ \text{cm/s}\,.$$

The discharge velocity

$$v = k\,i = k\frac{h}{l} = 2.5\times10^{-2}\times\frac{25}{15.4} = 4.0\times10^{-2}\ \text{cm/s}$$

and the seepage velocity

$$v_s = \frac{v}{n} = \frac{4.0\times10^{-2}}{0.44} = 9.1\times10^{-2}\ \text{cm/s}\,.$$

For $n = 0.389$

$$k_2 = k_1\frac{\dfrac{n_2^3}{(1-n_2)^2}}{\dfrac{n_1^3}{(1-n_1)^2}} = 2.5\times10^{-2}\frac{\dfrac{0.389^3}{(1-0.389)^2}}{\dfrac{0.440^3}{(1-0.440)^2}} =$$

$$= 1.4\times10^{-2}\ \text{cm/s}\,.$$

Example 2. In a falling head permeability test on a sample 15.4 cm high and 58.2 cm² in cross-sectional area, the water level in the standpipe 2.7 cm² in area dropped from a height of 405 cm to 18.5 cm in 87 s. Find the coefficient of permeability.

Solution:

$$k = \frac{al}{A}\frac{1}{\Delta t}\log_e\frac{h_1}{h_2} = \frac{2.7\times15.4}{58.2\times87}\log_e\frac{40.5}{18.5} = 6.4\times10^{-3}\ \text{cm/s}.$$

Example 3. Permeability tests on a soil sample gave the following data:

Test No.	Void ratio e	Temperature °C	k
1	0.70	25	0.32×10^{-4} cm/s
2	1.10	40	1.80×10^{-4} cm/s

Estimate the coefficient of permeability at a temperature of 20 °C for a void ratio of 0.85.

Solution:

First, convert both test results to a temperature of 20 °C. Considering that

$$k_1 : k_2 = \frac{\gamma_{w1}}{\eta_1} : \frac{\gamma_{w2}}{\eta_2}$$

and substituting numerical values, we have

$$k_1^{20°} = 0.32\times10^{-4}\times\frac{8.95}{10.09}\times\frac{0.998}{0.997} = 0.284\times10^{-4}\ \text{cm/s}$$

$$k_2^{20°} = 1.80\times10^{-4}\times\frac{6.54}{10.09}\times\frac{0.998}{0.902} = 1.172\times10^{-4}\ \text{cm/s}.$$

Now, convert these values for a void ratio of $e = 0.85$. Using Eq. (95) gives

$$\frac{e_1^3}{1+e_1} = 0.202 \qquad \frac{e_2^3}{1+e_2} = 0.633 \qquad \frac{e_3^3}{1+e_3} = 0.331\,.$$

Hence

$$k_1^{0.85} = 0.284\times10^{-4}\frac{0.331}{0.202} = 0.465\times10^{-4}\ \text{cm/s}$$

$$k_2^{0.85} = 1.172\times10^{-4}\frac{0.331}{0.633} = 0.614\times10^{-4}\ \text{cm/s}\,.$$

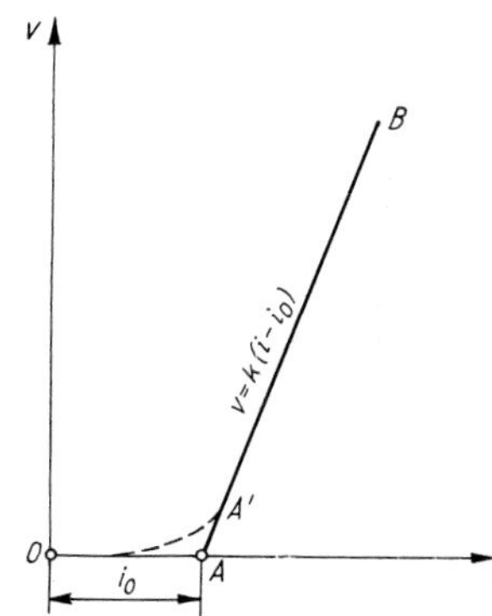

Fig. 176. Relationship between seepage velocity and hydraulic gradient

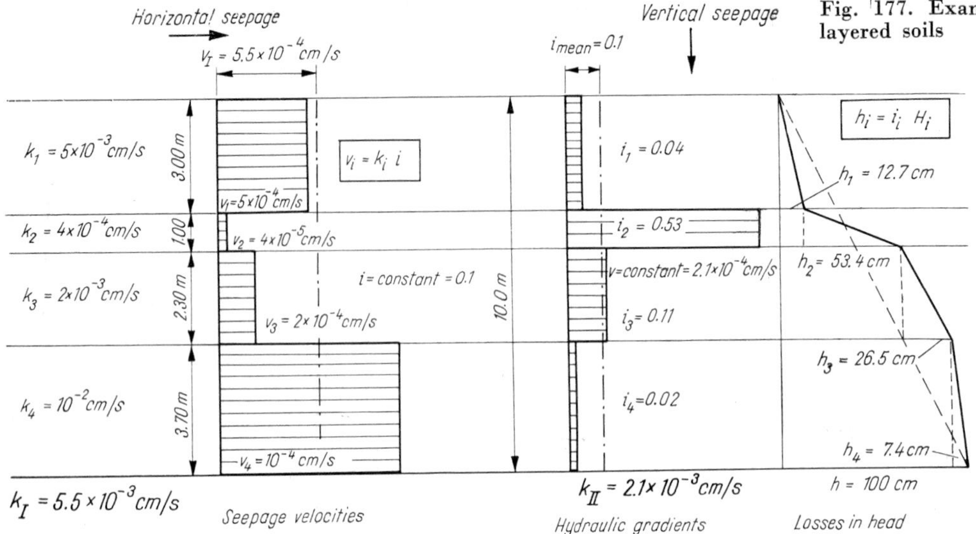

Fig. 177. Example 5: permeability in layered soils

Taking the average, the probable value is

$$k = 5\times10^{-5} \text{ cm/s}.$$

Example 4. A bank of sand is made up of three horizontal layers of equal thickness. The coefficient of permeability in the top and bottom layers in 1×10^{-4} cm/s, and that in the intermediate layer is 1×10^{-2} cm/s. Find the ratio of the average coefficient of permeability in the horizontal direction to that in the vertical direction.

Solution:

From Eqs (101) and (102)

$$k_{\mathrm{I}} : k_{\mathrm{II}} = \frac{k_1 + k_2 + k_3}{3} : \frac{3}{\dfrac{1}{k_1} + \dfrac{1}{k_2} + \dfrac{1}{k_3}} =$$

$$= \frac{\dfrac{1}{3}(2\times10^{-4} + 10^{-2})}{3(2\times10^{4} + 10^{2})^{-1}} = 22.8\,.$$

Example 5. Figure 177 shows the layers of a stratified subsoil. Also shown are the coefficients of permeability of the individual strata as obtained from laboratory tests. Determine the average permeability values for both horizontal and vertical seepage cases. Assuming an average hydraulic gradient of 0.1 in both cases, find the discharge value and the loss in head in each individual layer. Results are plotted in a soil profile in Fig. 177.

Example 6. Figure 178 is a vertical section through a test well. The discharge from the well was 1 liter/s. The drawdown at a distance of $x_1 = 15.0$ m from the centre of the well was found to be 1.2 m. Estimate the drawdown at a distance of $x_2 = 2.0$ m. The coefficient of permeability was 4×10^{-3} cm/s.

Solution:

From Eq. (86)

$$z_1^2 - z_2^2 = \frac{q}{\pi k}\, 2.3 \log \frac{x_1}{x_2}$$

and

$$z_2^2 = z_1^2 - \frac{q}{\pi k}\, 2.3 \log \frac{x_1}{x_2} = 6.8^2 - \frac{1\times10^{-3}}{\pi\times4\times10^{-5}} \times$$

$$\times 2.3 \log \frac{15}{2} = 46.2 - 16 = 30.2\,.$$

Hence $z_2 = 5.5$ m and $S_2 = H - z_2 = 2.5$ m.

Example 7. Figure 179 shows the section of a water-bearing sand stratum located between impervious clay

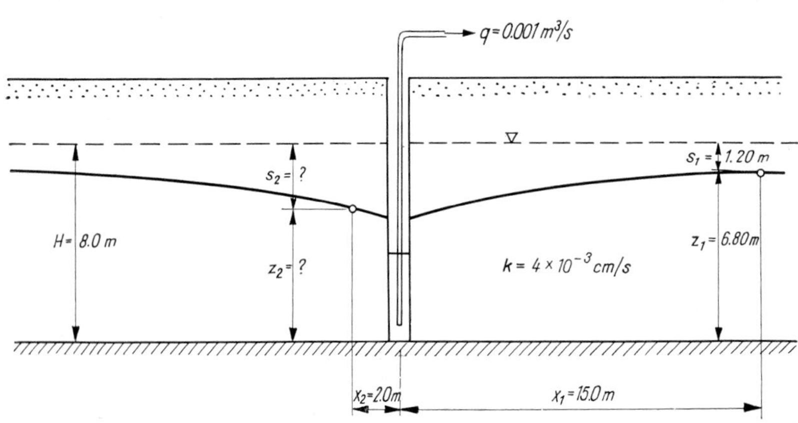

Fig. 178. Example 6: ground water lowering due to pumping

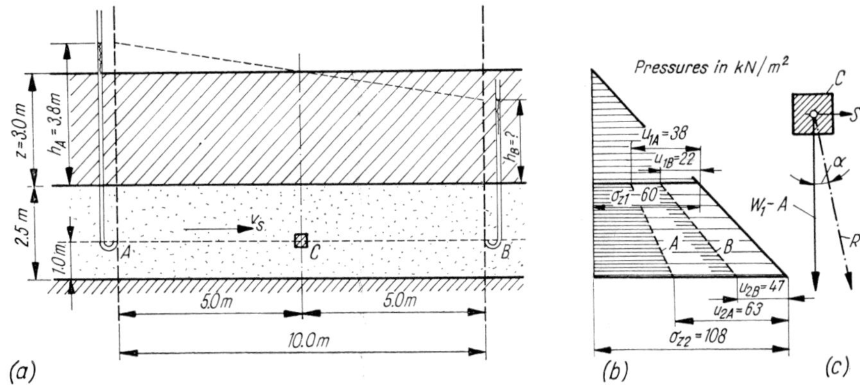

Fig. 179. Example 7: determination of the resulting mass force in a horizontal soil layer

strata. There is a hydraulic gradient in the pervious layer causing a strata horizontal flow of water. The seepage velocity is $v_s = 4 \times 10^{-6}$ cm/s. In a piezometer located at point A the water rises to a height of $h_A = 3.8$ m. To which level does the water rise in piezometer B?

Given the coefficient of permeability $k = 10^{-3}$ cm/s, the porosity $n = 40\%$ and the unit weight of the solids $\gamma_s = 26$ kN/m³, determine the resultant body force acting on a soil element with unit volume in the vertical section C located midway between sections A and B. Make a scaled plot of the distribution of total and effective vertical stresses along the vertical sections A and B.

Solution:

The discharge velocity:

$$v = nv_s = 0.4 \times 4 \times 10^{-4} \text{ cm/s} = 1.6 \times 10^{-4} \text{ cm/s}.$$

The saturated unit weight:

$$\gamma_{sat} = \gamma_s(1 - n) + \gamma_w n = 19.6 \text{ kN/m}^3.$$

From $v = ki$, the hydraulic gradient is

$$i = \frac{v}{k} = \frac{0.16 \times 10^{-3}}{10^{-3}} = 0.16 .$$

The pressure head at point B, with the velocity head neglected, is

$$h_B = 3.8 - 10 \times 0.16 = 2.2 \text{ m}.$$

The total vertical pressure at the top of the pervious layer ($z = 3.0$ m) is equal to the weight of the overlying clay stratum. Given $\gamma = 20$ kN/m³ for the clay,

$$\sigma_{z1} = 3.0 \times 20 = 60 \text{ kN/m}^2.$$

At a depth of 5.5 m

$$\sigma_{z2} = 60 + 2.5 \times 19.6 = 108 \text{ kN/m}^2.$$

The neutral stresses at the top and the bottom, respectively, of the pervious layer in the vertical section A are

$$u_1 = 3.8 \times 10 = 38 \text{ kN/m}^2$$

and

$$u_2 = (3.8 + 2.5) \times 10 = 63 \text{ kN/m}^3.$$

In the vertical section B

$$u_1 = 2.2 \times 10 = 22 \text{ kN/m}^2$$

$$u_2 = (2.2 + 2.5)\ 10 = 47 \text{ kN/m}^2.$$

The effective stresses are obtained from the formula $\bar{\sigma}_z = \sigma_z - u$. The distribution of the stresses with depth is plotted to scale in the lower portion of Fig. 179.

To obtain the resultant body force acting on the soil element in section C, calculate first the weight of saturated soil contained in the element

$$W_1 = 1 \text{ m}^3 \times 19.6 \text{ kN/m}^3 = 19.6 \text{ kN}.$$

The buoyancy force

$$F = 10 \text{ kN}.$$

The horizontal seepage force

$$S = 1 \text{ m}^3 \times 0.16 \times 10 \text{ kN/m}^3 = 1.6 \text{ kN}.$$

The resultant body force

$$R = \sqrt{(W_1 - F)^2 + S^2} = \sqrt{9.6^2 - 1.6^2} = 9.7 \text{ kN}.$$

The inclination of R to the vertical is

$$\alpha = \tan^{-1}\frac{1.6}{9.7} = 9° \ 30' .$$

Example 8. Determine the amount of filtered water that is produced in one day by the setup shown in *Fig. 180.* $k = 4 \times 10^{-3}$ cm/s. Also find the discharge velocity and seepage velocity.

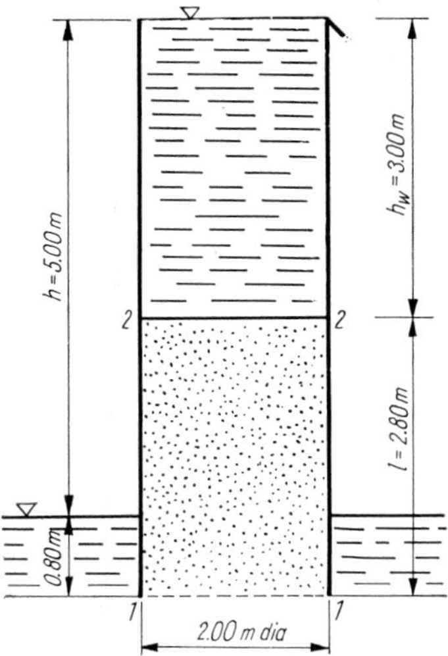

Fig. 180. Example 8: determination of the amount of filtered water:
1–1 screen; 2–2 top surface of soil sample

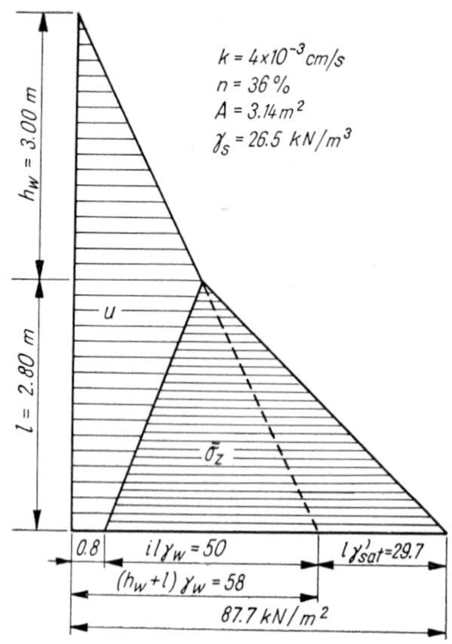

Fig. 181. Distribution of effective and neutral stresses in the filter layer of Example 8

Solution:

The discharge in one day

$$q = kA\,\Delta t\,\frac{h}{l} = 4\times10^{-3}\times\pi\times10^{2}\times8.64\times10^{4}\times\frac{500}{280} =$$

$$= 1.94\times10^{7}\ \text{cm}^3 \approx 20\ \text{m}^3 .$$

The discharge velocity

$$v = ki = k\,\frac{h}{l} = 4\times10^{-3}\times\frac{500}{280} = 7.2\times10^{-3}\ \text{cm/s}.$$

The seepage velocity

$$v_s = \frac{v}{n} = \frac{7.2\times10^{-3}}{0.36} = 2\times10^{-2}\ \text{cm/s}.$$

A scaled plot of the distribution of the effective and neutral stresses is given in *Fig. 181.* The saturated unit weight of the sand ($n = 36\%$, $\gamma_s = 26.5\ \text{kN/m}^3$) is

$$\gamma_{sat} = \gamma_s(1-n) + n\gamma_w = 26.5\times0.64 + 0.36\times10 = 20.6\ \text{kN/m}^3$$

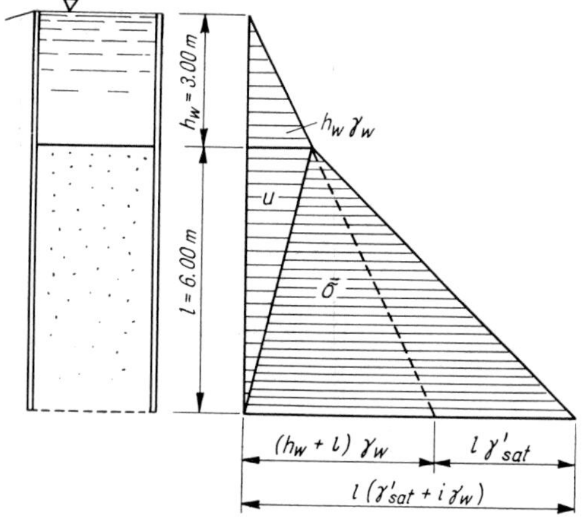

Fig. 182. Example 10: distribution of effective and neutral stresses

Example 9. How long would it take for a 30 cm-layer of water flooding flat ground to completely infiltrate into the soil? The ground water table is 2.5 m below the ground surface. $k = 3.6\times10^{-3}$ cm/s. The rising of the water table and the loss due to evaporation are assumed to be negligible.

Solution:

The seepage distance is $l = 2.5\times10^{2}$ cm. The initial pressure head with reference to the elevation of the water table is

$$h_0 = (2.5 + 0.3)\times10^{2} = 2.8\times10^{2}\ \text{cm}.$$

Using Eq. (83) and substituting $A = a = 1\ \text{m}^2$

$$k = 2.3\,l\,\frac{1}{\Delta t}\log\frac{h_0}{h_1}.$$

At complete infiltration the pressure head has dropped to $h_1 = 2.5\times10^{2}$ cm.

Thus

$$\Delta t = 2.3\,l\,\frac{1}{k}\log\frac{h_0}{h_1} = 2.3\times2.5\times10^{2}\,\frac{1}{3.6\times10^{-3}}\log\frac{2.8}{2.5} =$$

$$= 7.8\times10^{-4}\ \text{s} \cong 2.2\ \text{hours}.$$

Example 10. Determine the effective and neutral stresses for the filter shown in *Fig. 182,* and make a scaled plot of the results ($\gamma_{sat} = 21.0\ \text{kN/m}^3$).

Hints to solution:

Since at the lower screen the water discharges at normal atmospheric pressure, the excess neutral stress (i.e. that above atmospheric) at this level is zero. Assuming a homogeneous material in the filter, the neutral stress across the filter drops linearly. The effective stress is calculated from $\bar{\sigma} = \sigma - u$.

6.3 Two-dimensional potential flow

6.3.1 Derivation of fundamental equation

The seepage case discussed in the preceding section was described as one-dimensional and linear. The flow lines were parallel (see e.g. Figs 160 and 161), and the flow velocity at every point of a cross section perpendicular to the flow lines was the same. The study of one-dimensional fluid flow in soils served to demonstrate permeability as a physical phenomenon and to present methods by which the coefficient of permeability can be measured. In the following we shall deal with the theoretical study of the two-dimensional seepage case, in order to present the necessary fundamentals to the solution of practical seepage problems. The analysis is based on the following fundamental assumptions:

(a) *Darcy*'s law is valid.

(b) The soil is homogeneous and isotropic. The pores are completely filled with water. Capillary effects are negligible.

(c) Both the soil skeleton formed by the solid grains and the pore fluid are incompressible. No compression or expansion takes place during the flow.

The basic terms used in seepage theory are illustrated in *Fig. 183.* A horizontal sand stratum of even thickness is underlain by an impervious clay stratum. A sheet pile well is driven to a depth

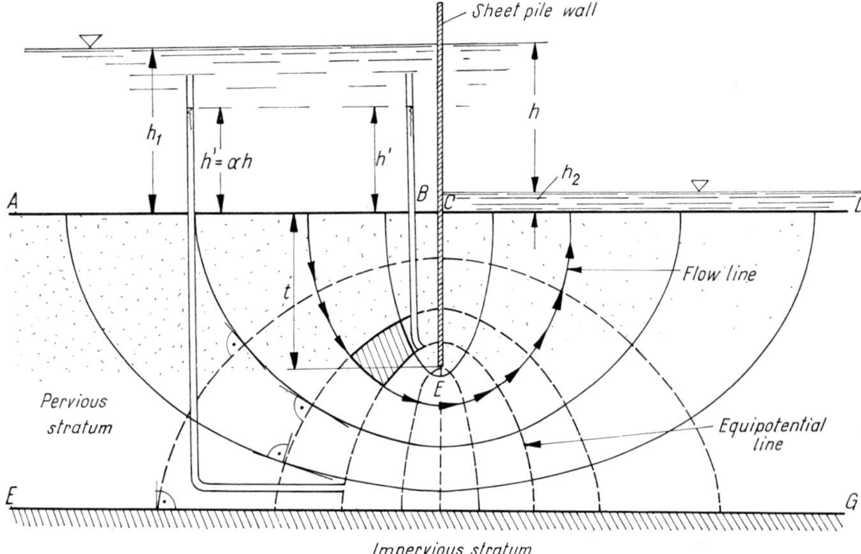

Fig. 183. Two-dimensional seepage around a sheet-pile wall driven into a horizontal sand layer

t into the sand. Water stands at height h_1 on one side of the wall and at h_2 on the other side. The water levels are assumed to remain constant. There is then a difference in head equal to $h = h_1 - h_2$ between the two sides of the wall. Under this unbalanced head the water enters the sand at the upstream surface, then passing round the lower edge of the wall it turns upwards and escapes at the downstream surface. The path taken by an elementary particle of water is called a *flow line.* In homogeneous soils, these lines must be, macroscopically, continuous smooth curves. Each of the flow lines begins at surface AB where a pressure of $h_1\gamma_w$ exists. This pressure is gradually dissipated in viscous friction as water passes through the soil. The flow lines terminate at surface CD where the pressure is $h_2\gamma_w$.

Since there is a continuous loss of head along the flow lines, points where the head dissipated in flow is equal to any given portion of the head $h = h_1 - h_2$ can be traced. Joining all such points, a continuous line called an *equipotential line,* is obtained. If piezometers are inserted into the soil at two points along an equipotential line, the water rises in both tubes to the same elevation. A graph showing flow lines together with equipotential lines is called a *flow net.*

Since the gradient in passing from one equipotential line to the next is equal to the potential drop Δh divided by the distance travelled, it is evident that the gradients are maximum when the flow lines are normal to the equipotential lines. If the soil is, as assumed, isotropic, flow must occur in the direction of the steepest gradient just as stones rolling down hill follow the steepest path. Therefore, the flow line must intersect the equipotential lines at right angles. If the conditions are such that in sections parallel to a given plane the flow nets are identical, the flow is called *two-dimensional.* Further, the flow is termed a steady state flow if the quantity of water passing through a surface element in unit time does not vary with the time.

Consider a prismatic element cut out of the permeable layer shown in Fig. 183. The lengths of the sides of this element are dx, dy and dz (*Fig. 184*). Let v_x be the component of the flow velocity in the horizontal direction, and let i_x be the hydraulic gradient in the horizontal direction. Let v_z and i_z be the corresponding values for the vertical direction. (Assuming that the flow is two-dimensional, the velocity in the y direction perpendicular to the plane of the diagram is zero.) By definition, the gradients in the directions x and z, respectively, may be written

$$i_x = -\frac{\partial h}{\partial x} \quad \text{and} \quad i_z = -\frac{\partial h}{\partial z}.$$

The total quantity of water that enters the element in unit time is

$$v_x \, dz \, dy + v_z \, dx \, dy \,,$$

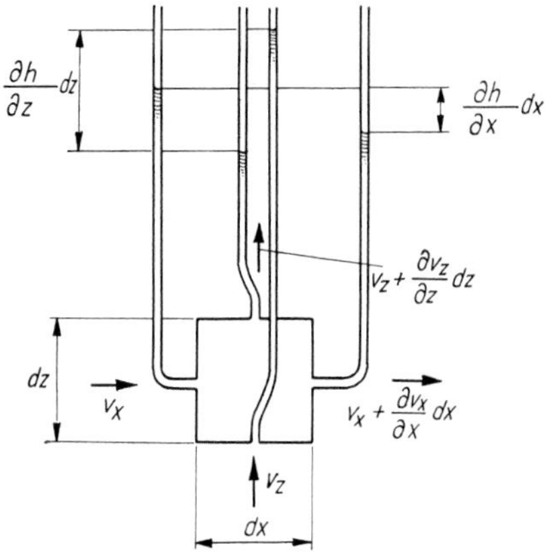

Fig. 184. Water entering and leaving the elementary tube

and that leaving the element is

$$v_x\, dz\, dy + \frac{\partial v_x}{\partial x}\, dx\, dz\, dy + v_z\, dx\, dy + \frac{\partial v_z}{\partial z}\, dx\, dy\, dz\,.$$

If we assume that the water is incompressible and that the total volume of voids also remains constant during the flow (assumption *c*, *p* . . .), the quantity of water entering the element must be equal to the quantity leaving. Equating the above two expressions we get

$$\frac{\partial v_x}{\partial x}\, dx\, dy\, dz + \frac{\partial v_z}{\partial z}\, dx\, dy\, dz = 0$$

whence

$$\frac{\partial v_x}{\partial x} + \frac{\partial v_z}{\partial z} = 0\,. \qquad (107)$$

Equation (107) represents the *continuity condition* for two-dimensional fluid flow.

By using *Darcy*'s law (assumption *a*), the two components of the flow velocity may be expressed as

$$v_x = -k\frac{\partial h}{\partial x} \quad \text{and} \quad v_z = -k\frac{\partial h}{\partial z}\,. \qquad (108)$$

Substituting in Eq. (107) gives

$$\frac{\partial^2 h}{\partial x^2} + \frac{\partial^2 h}{\partial z^2} = 0\,. \qquad (109)$$

This is known as the *Laplace* equation, for two-dimensional flow. In words, it states that for flow through an incompressible porous medium, gradient changes in the x direction must be balanced by gradient changes of opposite sign in the z direction.

Introducing the concept of velocity potential defined as $\Phi = kh$, Eq. (109) becomes

$$\boxed{\frac{\partial^2 \Phi}{\partial x^2} + \frac{\partial^2 \Phi}{\partial z^2} = \nabla^2 \Phi = 0\,.} \qquad (110)$$

The above fundamental equation can be much more easily derived by means of vector analysis. *Darcy*'s law expressed in vectorial form reads

$$\bar{v} = -k \operatorname{grad} \bar{h}\,.$$

Continuity [Eq. (107)] requires that

$$\operatorname{div} \bar{v} = 0$$

i.e.

$$\operatorname{div} \bar{v} = \operatorname{div}(-\bar{k} \operatorname{grad} h) = -k \operatorname{div} \operatorname{grad} \bar{h} = -k\nabla^2 h\,.$$

Hence

$$\boxed{\nabla^2 h = 0\,.} \qquad (110a)$$

Equation (110) represents the fundamental equation of potential theory of flow. Let us now examine how this equation can be used to describe the flow of water through soil.

From Eq. (108), using the expression of velocity potential, we have

$$\left.\begin{aligned} v_x &= -k\frac{\partial h}{\partial x} = \frac{\partial \Phi}{\partial x} \\ v_z &= -k\frac{\partial h}{\partial z} = \frac{\partial \Phi}{\partial z} \end{aligned}\right\}. \qquad (108a)$$

Integrating

$$\Phi(x, z) = -kh(x, z) + F_1(z)$$

$$\Phi(x, z) = -kh(x, z) + F_2(x)$$

x and z are independent variables, therefore,

$$F_1(z) = F_2(x) = \text{constant.}$$

The function Φ represents the variation of total head through the soil. The equation $\Phi(x, z) =$ = constant described a set of curves. Along any one of these curves the total head in the pore water (i.e. the potential) is constant; hence the name *equipotential lines*. In piezometric tubes installed at different points along an equipotential line the water would rise to the same elevation. At all the points the resultant hydraulic gradient is normal to the curve $\Phi(x, z) =$ constant.

Let us introduce a function $\Psi(x, z)$ such that the relationships between Φ and the velocity components in the directions x and z be

$$\frac{\partial \Psi}{\partial z} = v_x$$

$$-\frac{\partial \Psi}{\partial x} = v_z\,.$$

From Eq. (108a) the relationship between the functions Φ and Ψ are given by

$$\left.\begin{aligned} \frac{\partial \Phi}{\partial x} &= -\frac{\partial \Psi}{\partial z} \\ -\frac{\partial \Phi}{\partial z} &= \frac{\partial \Psi}{\partial x} \end{aligned}\right\}. \qquad (111)$$

Differentiating the first equation with respect to z and the second with respect to x, and adding the resulting two equations gives

$$\frac{\partial^2 \Phi}{\partial x\, \partial z} - \frac{\partial^2 \Phi}{\partial x\, \partial z} = \frac{\partial^2 \Psi}{\partial x^2} + \frac{\partial^2 \Psi}{\partial z^2} = \Delta \Psi = 0\,. \qquad (111a)$$

As is seen, the function $\Psi(x, z)$ also satisfies the *Laplace* equation. Let us determine the derivative dz/dx for the function Ψ

$$\left(\frac{dz}{dx}\right)_\psi = -\frac{\partial \Psi/\partial x}{\partial \Psi/\partial z} = \frac{v_z}{v_x} = \tan \alpha\,. \qquad (112a)$$

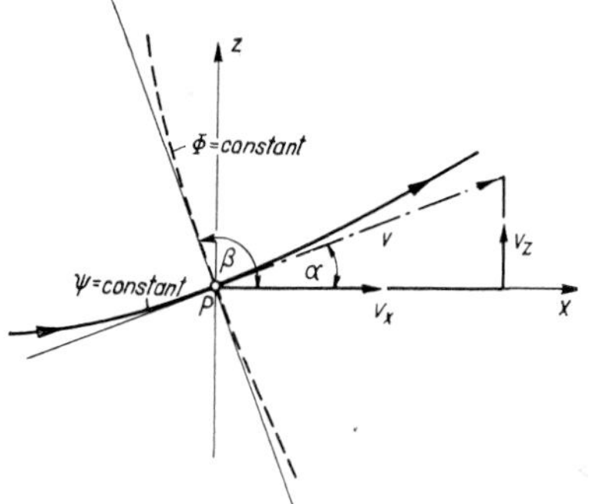

Fig. 185. Intersection of flow line and equipotential line; velocity components

The equations $\Psi(x, z) =$ constant again yield a set of curves. The slope of the tangent to a curve is given by Eq. (112a), which indicates the direction of the resultant velocity vector at any point P. The curves $\Phi =$ constant thus represent flow lines.

On the other hand, the slope of the tangent to the curves $\Phi =$ constant is given by

$$\left(\frac{dz}{dx}\right)_\varphi = -\frac{\partial\varphi/\partial x}{\partial\varphi/\partial z} = -\frac{v_x}{v_z} = -\frac{1}{\tan\alpha} = \tan(\alpha + \pi/2)\,. \tag{112b}$$

It follows that the equations $\Psi(x, z) =$ constant and $\Phi(x, z) =$ constant represent two sets of curves that intersect at right angles. The curves of one set are flow lines and those of the other set are equipotential lines.

If in a given flow case the functions Φ and Ψ that satisfy the boundary conditions are known, then all data required to solve a technical problem can be obtained either by graphical methods or by computation. Thus it is possible to determine the neutral stresses in the pore water, quantity of seepage, buoyancy force, flow velocity, seepage force, etc.

We may choose from various approaches to solve a problem. If the boundary conditions are relatively simple, mathematical methods may prove workable. With the aid of the theory of functions the equations for the two sets of curves —flow lines and equipotential lines—can be obtained. In more involved cases, which constitute the majority of practical problems, we often have to resort to graphic methods, or to carry out model tests. The latter include direct seepage tests, in which the pressure head at various points, the quantity of flow, etc. are actually measured. Other methods are based on the analogy between steady state fluid flow through porous media and other physical processes (heat conduction, current conduction, stress distribution in elastic bodies) that are all governed by the *Laplace* equation. It would be beyond the scope of this book to give a comprehensive and detailed discussion of all these methods, and only fundamental information required to solve flow problems connected with the stability analysis of earth masses will be presented in this section. After a brief introduction to the analytical method, the graphical determination of flow nets will be dealt with in some detail, since this is the method used almost exclusively in practical soil investigations.

The analytical solution is based on the correspondence between Eq. (111) and the *Cauchy–Riemann* equations, from which it follows that the functions Φ and Ψ may be regarded as the real part and the imaginary part of an analytical function. This function is defined as

$$w = \Phi(x,z) + i\Psi(x,z) \tag{113}$$

and is called the complex potential. Therefore, if an analytic function is resolved into a real part and an imaginary part, the real part represents the velocity potential function Φ, and the imaginary part represents the stream function Ψ for a two-dimensional potential flow. The solution to Eq. (111) consists in finding an analytic function w that satisfies the boundary conditions in a given problem.

For the case of seepage around the lower edge of a sheet-pile wall driven into an infinite permeable soil mass (*Fig. 186*), the complex velocity potential may be written in the

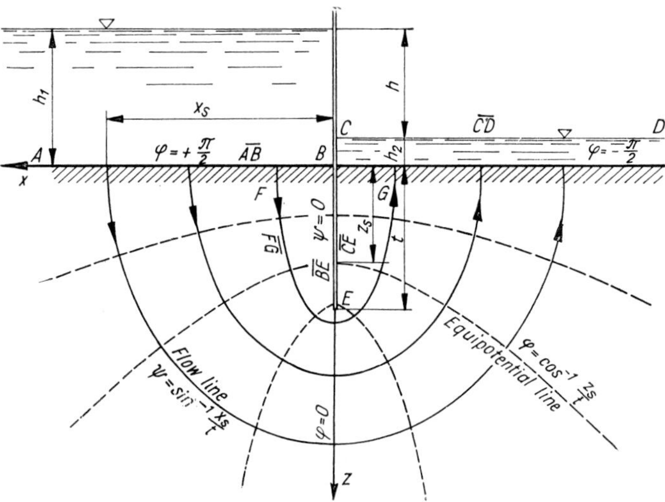

Fig. 186. Seepage around a sheet pile driven into a sand layer of infinite thickness

form

$$w = \cos^{-1}\frac{u}{t}.$$

Hence

$$\cos w = \frac{u}{t}$$

and since

$$w = \varphi + i\varphi \quad \text{and} \quad u = x + iz$$

therefore

$$\cos(\varphi + i\psi) = \frac{x}{t} + i\frac{z}{t}.$$

Expanding,

$$\cos\varphi \cosh\psi - i\sin\varphi \sinh\psi = \frac{x}{t} + i\frac{y}{t}.$$

Considering that the real parts on the left-hand side and on the right-hand side are equal as are also the imaginary parts

$$\left.\begin{aligned} x &= t\cos\varphi\cosh\Psi \\ z &= -t\sin\varphi\sinh\Psi \end{aligned}\right\}.$$

After squaring each of the two equations, by addition and substraction we have

$$\left.\begin{aligned} \frac{x^2}{t^2\cos^2\varphi} - \frac{z^2}{t^2\sin^2\varphi} &= \cosh^2\psi - \sinh^2\psi = 1 \\ \frac{x^2}{t^2\cosh^2\psi} + \frac{z^2}{t^2\sinh^2\psi} &= \cos^2\varphi + \sin^2\varphi = 1 \end{aligned}\right\}. \quad (114)$$

The equations for equipotential lines and flow lines are thus obtained. In Fig. 186 the equipotential lines are seen to be confocal hyperbolas and the flow lines to be confocal ellipses. The distance between the foci is $2t$ and the values of $\varphi =$ constant and $\Psi =$ constant are the parameters to the individual curves.

The conjugate case when the flow lines are ellipses and the equipotential lines are hyperbolas represents the two-dimensional flow beneath an impervious slab resting on an infinite permeable soil mass.

For further details of the rigorous analytical treatment of seepage problems see, inter alia, Polubaroniva-Kochina (1952), Pavlovski (1933), Harr and Deen (1961), Khosla (1954). Practical problems encountered in earth work and foundation engineering can normally be solved with sufficient accuracy by the trial-and-error method of sketching flow nets, which is the subject of the following section.

6.3.2 Construction of flow nets

One of the commonest and most effective methods of finding a solution to the *Laplace* equations is the graphical construction through successive approximation. By guess a few trial flow lines are drawn that satisfy the boundary conditions. Then several equipotential lines are drawn in such a manner that they always intersect the flow lines at right angles and the fields thus formed resemble squares in that their four sides are roughly equal. In this way a first very rough approximation to the flow net is obtained. This trial flow net has to be checked carefully, and it usually needs considerable adjustment. On a sheet of tracing paper placed over the first trial net a new flow net is sketched, so that gross errors of the first trial are possibly eliminated. After the third or fourth approximation the resulting flow net, which may be subdivided if necessary into smaller squares by interpolating further flow lines and equipotential lines, is usually accurate enough to permit the determination of the data required for the solution to the seepage problem concerned.

The free-hand sketching of flow nets is a tedious job which requires a great deal of skill and practice. Some hints to facilitate this work are given below.

The flow lines as discussed earlier represent the paths of individual water particles. It follows that these paths are smooth curves and that they do not cross each other. Along each of the flow lines the total head gradually decreases in the direction of flow, the loss in head being dissipated in work done against resistance to flow due to viscous friction (*Fig. 187*).

An equipotential line runs through all those points at which the total head h with reference to a datum—plane a—b is constant. The piezometric pressure at any point A is obtained as the product of the piezometric head and the unit weight of water:

$$u_A = h_A\gamma_w. \quad (115)$$

The piezometric head h_A equals the difference in height between the piezometrics surface (an imaginary surface indicating heights to which the water would rise in piezometers installed at the points concerned) and the elevation of point A.

It is clear that the piezometric head may change from point to point along an equipotential line; it is the total head that remains constant.

The two sets of curves intersecting at right angles form a complete flow net which thus represents the solution to the differential equation of flow that also satisfies the given boundary conditions. In a correct flow net the drop in head between any two successive equipotential lines should be the same. With this condition satisfied, the resulting fields bounded by two adjacent flow lines and two adjacent equipotential lines are curvilinear "squares". An essential feature of these squares is that inscribed circles that touch each other on all sides can be constructed into them (*Fig. 188a*). Any square can be subdivided into four smaller squares. By successive subdivisions the squares reduce in size to a point.

Before starting the construction of a flow net, the hydraulic boundary conditions should be fixed. This means finding surfaces that can be defined in advance as either flow lines or equipotential lines.

The tracing of the boundary conditions will be demonstrated by means of the following examples (*Fig. 189.*)

The flow case already shown in Fig. 187 is chosen as the first example. The top part of Fig. 189 is a simplified linear diagram of a concrete dam with an impervious base and a cutoff wall projecting from it to some depth into the pervious substratum. The water-bearing sand stratum,

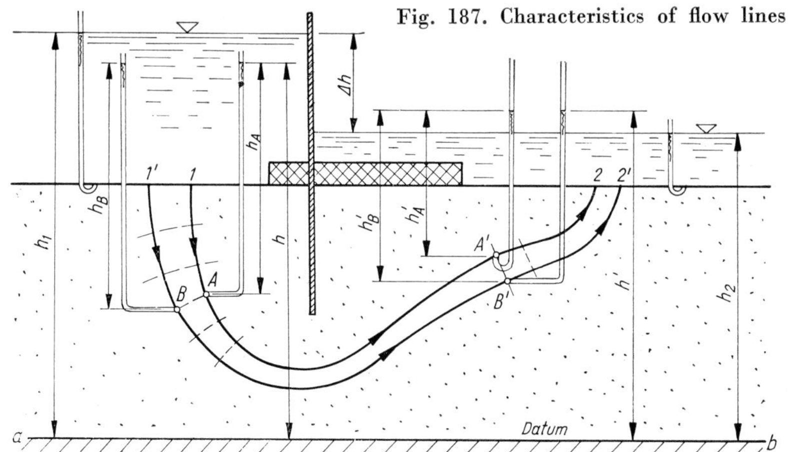

Fig. 187. Characteristics of flow lines

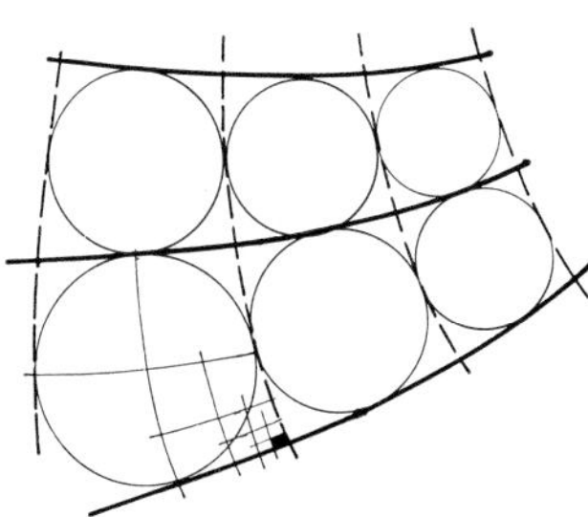

Fig. 188. Characteristics of curvilinear squares in the net of flow lines and equipotential lines

which is assumed to be isotropic and uniformly permeable throughout, is underlain by an impervious stratum. The boundary conditions are as follows.

(i) Line *1, 2* is an equipotential line along which the pressure equals $h_1\gamma_w =$ constant.

(ii) The base of the dam with the projecting cutoff wall (line *2, 3, 4, 5, 6*) represents the uppermost flow line.

(iii) Line *6, 7* is again an equipotential line along which the pressure is $h_7\gamma_w =$ constant.

(iv) Line *8, 9* which depicts the path of a water particle travelling from a very large distance is a flow line.

In general, there are four boundary conditions, some of which, however, are not always as readily defined as in the preceding example. If for instance the boundaries of the body of earth through which seepage is taking place are not definitely known on all sides, additional conditions should be present. This is the case e.g. when seepage occurs through an earth dam, and it is difficult to locate the top flow line (phreatic line) in advance. Such flow cases will be dealt with in Vol. 2.

A second example is illustrated in Fig. 189*b*. In order to reduce the flow toward a shallow trench excavated under the water table, single sheet-pile walls are driven into the permeable soil on both sides of the trench. The sheet piles do not quite extend to the impervious base so that the water can pass through around their lower edge. The boundary conditions for this flow case are graphically presented in Fig. 189*b*.

Figure 189*c* shows the cross-section of a retaining wall which stands on an impervious base. The backfill consists of a permeable material. A coarse-grained filter is placed against the back of the wall with outlets to the free surface. During lasting rainstorms the water infiltrating the backfill finds its way toward the backdrain and if the rain intensity is sufficiently high, steady-state flow conditions may be developed. The boundary conditions are as follows. The ground surface (*2, 3*) is an equipotential line and line *1, 4* is a flow line. The vertical face *1, 2* is, however, neither equipotential nor flow line. The prefixed condition here is that the neutral stress is zero everywhere along this surface.

Singular points in a flow net need some further examination. In the vicinity of such points the flow net does not satisfy the *Laplace* equation. There are several kinds of singularities; some typical examples are represented by points *3, 4, 5*

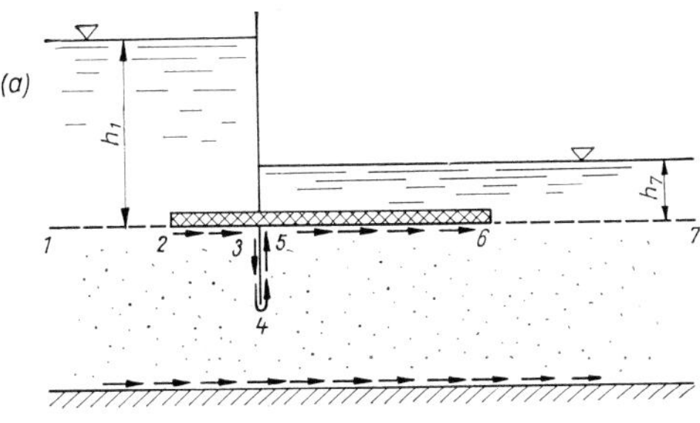

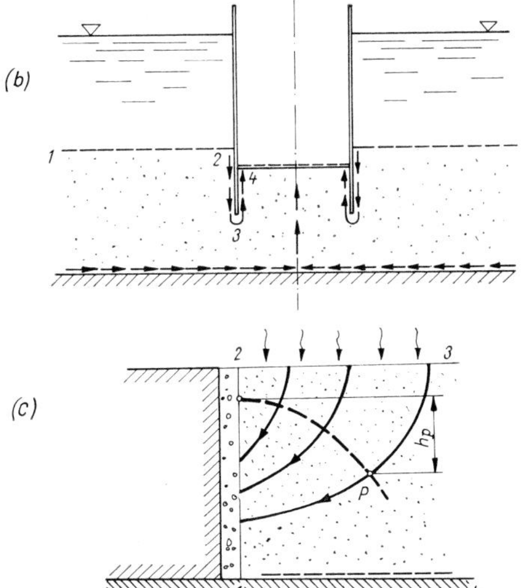

Fig. 189. Examples of boundary conditions of flow nets: (a) — weir with sheet pile and impermeable base; (b) — trench; (c) — retaining wall with backfill; rainstorm effect

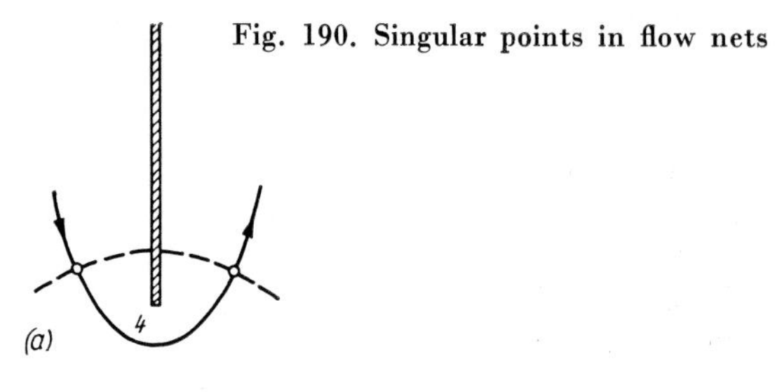

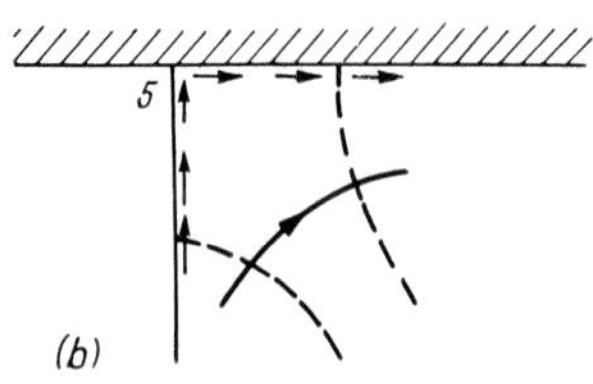

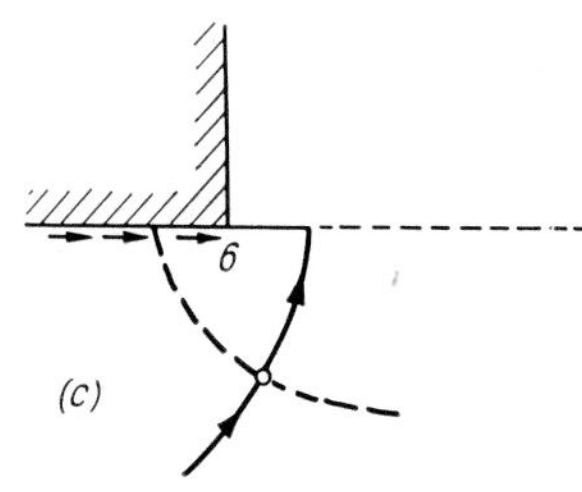

Fig. 190. Singular points in flow nets

and *6* in Fig. 189*a*. The portions of the flow net in the neighbourhood of these points are shown on a larger scale in *Fig. 190*. It is seen that none of the figures are curvilinear "squares". Around point *4*, for example, the small figure is bounded by only two curves. The figure around point *6* is triangular, and that around point *5* is pentagonal. If the flow net is correct, these figures should, after successive subdivisions, converge at the singular points.

Figure 191 shows two complete flow-nets. (For boundary conditions see Figs 189*a* and *b*.) Further flow cases will be presented in Vol. 2, in connection with the stability of slopes and dams and with drainage problems.

Once a flow net has been constructed, the quantity of seepage, the pore-water pressures, and the uplift pressures against the base of the dam can readily be determined, and also the hazard of failure by piping can be examined.

As we pointed out earlier, a flow net must fulfil the condition that the potential drops between successive equipotential lines are equal.

Consider a flow channel formed between two flow lines (*11'—22'* in Fig. 187). If the total head causing flow is Δh and this is dissipated in n_1 steps (n_1 is the number of "squares"), then the drop in head across one square is $\Delta = \Delta h/n_1$ (*Fig. 192*).

The hydraulic gradient is

$$i = \frac{\Delta}{a} = \frac{\Delta h}{n_1 a}.$$

According to *Darcy*'s law the flow velocity is $v = ki$ and the quantity of flow in unit time through a channel of width a is

$$\Delta q = k\frac{\Delta h}{n_1 a}a = k\frac{\Delta h}{n_1}.$$

The flow channels are n_2 in number. The total quantity of seepage per unit of width perpendicular to the section is, therefore,

$$q = k\,\Delta h\frac{n_2}{n_1}. \tag{116}$$

The pore-water pressures are easily determined by the flow net. The flow energy is lost continuously and in equal drops along a flow line.

Fig. 191. Flow nets for the cases *a* and *b* of Fig. 189

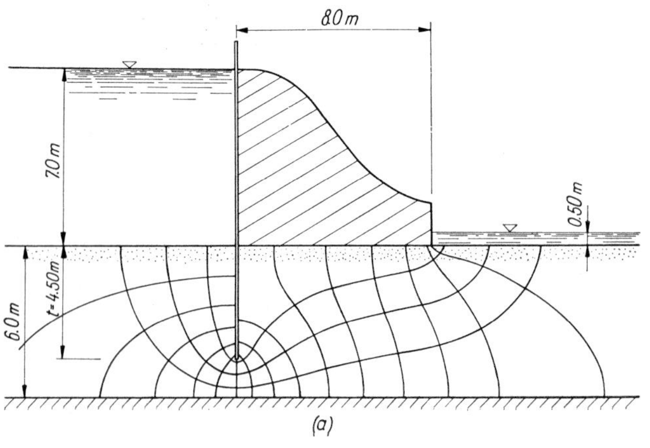

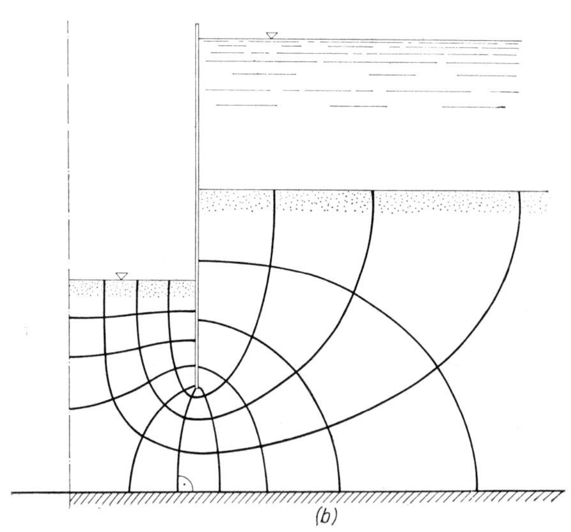

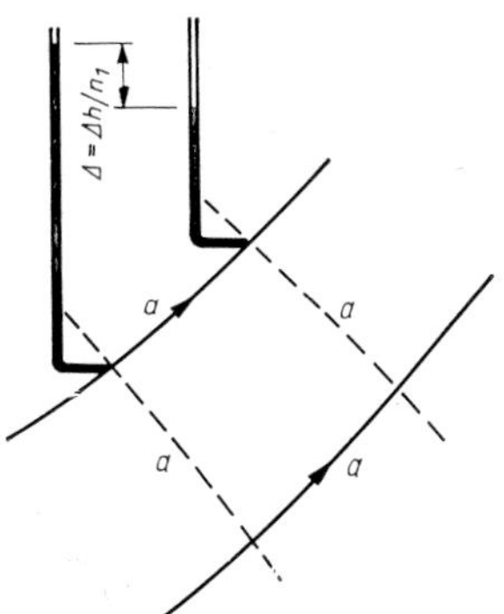

Fig. 192. Determination of quantity of seepage

Therefore, we only have to count the number of squares passed, up to the point concerned. Let this number be n. The ratio of n to the total number of squares along one of the flow channels equals the ratio of the excess pore-water pressure at the point concerned to the total excess hydrostatic pressure causing flow:

$$u = \Delta h \gamma_w \frac{n}{n_1}. \qquad (117)$$

If the point considered lies on an equipotential line which intersects the free surface at a point where the hydrostatic pressure is known (this pressure may of course be zero), the piezometric level existing there will be valid everywhere along that equipotential line, and the piezometric head h_P at any point P is easily read off on the flow net. (See e.g. Fig. 189*c*.) The pore-water pressure is then given by

$$u_P = h_P \gamma_w.$$

The uplift pressures acting against the base of a dam are determined in a similar manner. By counting the potential drops to the points where the equipotential lines cut the base, we can calculate the piezometric heads and plot the distribution diagram of uplift pressure across the base.

Besides the graphical process presented in this section, several numerical methods of successive approximation are also available to the solution of the potential flow equation. Such methods are particularly useful when the sketching of flow nets would prove too elaborate and cumbersome. This may be the case e.g. when seepage through an anisotropic soil is studied.

6.3.3 Two-dimensional flow through anisotropic soils

If a soil mass in which seepage takes place is made up of materials with different permeabilities, the solution of the flow equation becomes very complicated, though the fundamental laws that govern potential flow still hold. In the following some basic considerations for this case will be presented.

As was shown in Section 6.2.6, a horizontally stratified soil can be replaced, for the purpose of seepage computations, by a homogeneous but anisotropic system in which the permeabilities in the horizontal and vertical directions are different. Let us now examine what a flow net is like if a boundary, at which the permeability changes abruptly, exists across a soil mass. In the simplest case the flow is horizontal and the boundary is vertical (*Fig. 193*). Flow takes place from left to right. The flow lines are horizontal and the equipotential lines are vertical. On the left hand side they form true squares, but on the right hand side the equipotential lines are spaced at some distance b. Continuity requires that when the water crosses the boundary its velocity remain the same, and the quantity of flow through a channel be constant:

$$q = k_1 i_1 A_1 = k_2 i_2 A_2.$$

Let the loss in head between any two adjacent equipotential lines be Δh. The gradients in the first and second materials, respectively, are

$$i_1 = \frac{\Delta h}{a} \quad \text{and} \quad i_2 = \frac{\Delta h}{b}.$$

Substituting in the above equation and considering that $A_1 = A_2 = a \times 1$, we have

$$k_1 \frac{\Delta h}{a} a = k_2 \frac{\Delta h}{b} a.$$

Hence

$$\frac{b}{a} = \frac{k_2}{k_1} \qquad (118)$$

i.e. the squares are transformed to rectangles. If $k_2 < k_1$ then $b < a$. If the boundary is not perpendicular to the direction of flow, the flow lines deflect at the boundary in the way shown in *Fig. 194*.

By geometry

$$\text{MN} = \frac{a}{\sin \alpha} = \frac{c}{\sin \beta}$$

and

$$\text{NO} = \frac{b}{\cos \beta} = \frac{a}{\cos \alpha}$$

whence

$$a = c \frac{\sin \alpha}{\sin \beta} = b \frac{\cos \alpha}{\cos \beta}$$

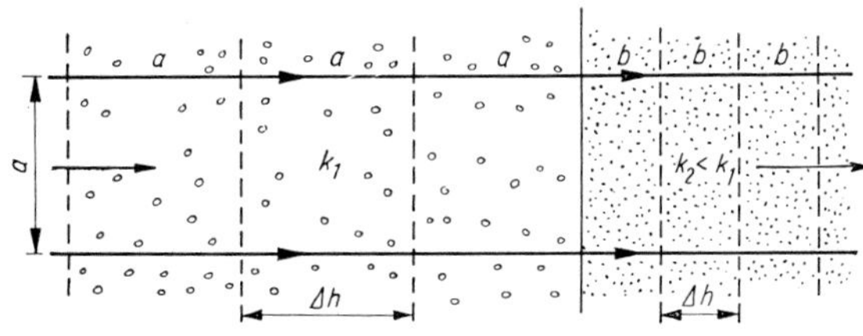

Fig. 193. Horizontal flow; flow net if there is a change in the permeability coefficient on a vertical face

and

$$\frac{b}{c} = \frac{\tan \alpha}{\tan \beta}. \qquad (119)$$

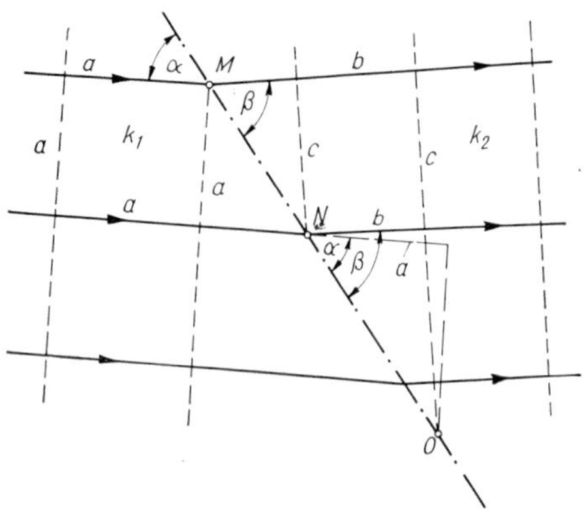

Fig. 194. The boundary is not normal to the direction of flow

To satisfy the continuity condition, the quantity of seepage must be the same in both materials, i.e.

$$q_1 = k_1 i_1 A_1 = k_2 i_2 A_2 .$$

Hence

$$k_1 \frac{\Delta h}{a} a = k_2 \frac{\Delta h}{b} c$$

and

$$\frac{b}{c} = \frac{k_2}{k_1}. \qquad (120)$$

By comparing Eqs (119) and (120), we have

$$\frac{\tan \alpha}{\tan \beta} = \frac{k_2}{k_1}. \qquad (121)$$

After deflection, the flow lines are not necessarily perpendicular to the equipotential lines.

Let us now return to the derivation of *Laplace*'s equation and assume that the permeability has different values k_x and k_z in the horizontal and vertical directions, respectively. Equation (107) expressing the continuity condition remains valid also for this case. But the velocity components become

$$v_x = -k_x i_x = -k_x \frac{\partial h}{\partial x}$$

$$v_z = -k_z i_z = -k_z \frac{\partial h}{\partial z}.$$

Substituting in Eq. (107) gives

$$k_x \frac{\partial^2 h}{\partial x^2} + k_z \frac{\partial^2 h}{\partial z^2} = 0$$

or $\qquad (109a)$

$$\frac{k_x}{k_z} \frac{\partial^2 h}{\partial x^2} + \frac{\partial^2 h}{\partial z^2} = 0 .$$

Introducing the following notations

$$\varkappa = k_z k_x \text{ and } X = \sqrt{\varkappa x}$$

the partial derivatives with respect to x become

$$\frac{\partial h}{\partial x} = \frac{\partial h}{\partial X} \cdot \frac{\partial X}{\partial x} = \sqrt{\varkappa} \frac{\partial h}{\partial X}$$

$$\frac{\partial^2 h}{\partial x^2} = \varkappa \frac{\partial^2 h}{\partial X^2}.$$

Hence

$$\frac{\partial^2 h}{\partial X^2} + \frac{\partial^2 h}{\partial z^2} = 0 . \qquad (122)$$

The derivation again results in a *Laplace* equation in which however x is to be replaced by the new variable

$$X = \sqrt{\frac{k_z}{k_x}}\, x . \qquad (123)$$

It follows that seepage problems in soils with permeabilities differing in horizontal and vertical directions can be solved by Laplace's equation; only the cross section through the soil parallel to the direction of flow has to be transformed. All horizontal dimensions on the section should be multiplied by the factor $\sqrt{k_z/k_x}$. On this transformed section the flow net is constructed in the ordinary manner. Finally, the flow net is redrawn to the natural scale by multiplying all horizontal dimensions by the reciprocal factor $\sqrt{k_x/k_z}$. In this new flow net the flow lines and equipotential lines no longer intersect at right angles. The expressions for the quantity of seepage [see Eq. (116)] becomes

$$q = \Delta h \frac{n_2}{n_1} \sqrt{k_x k_z} . \qquad (124)$$

The procedure will be demonstrated by means of practical examples in Section 6.3.5. Before that, we shall discuss the solution of the *Laplace* equation by numerical approximation methods.

6.3.4 Numerical solution of the Laplace equation

On the basis of the continuity equation, the general state equation of water, and *Darcy*'s law we derived Eq. (109a).

Figure 195 shows a part of a soil mass through which seepage occurs. Also shown in the figure are the variations of hydraulic head in two perpendicular directions x and z. Let the point considered be 0 and let 1, 3 and 2, 4 be two pairs of adjacent points at distances Δx and Δz, respectively, from 0. It is necessary to find, for chosen points in the soil mass, the values of a function $h(x, z)$ that satisfy the *Laplace* equation. To this end, the cross-section of the soil stratum is covered with a rectangular grid of size Δx. Δz.

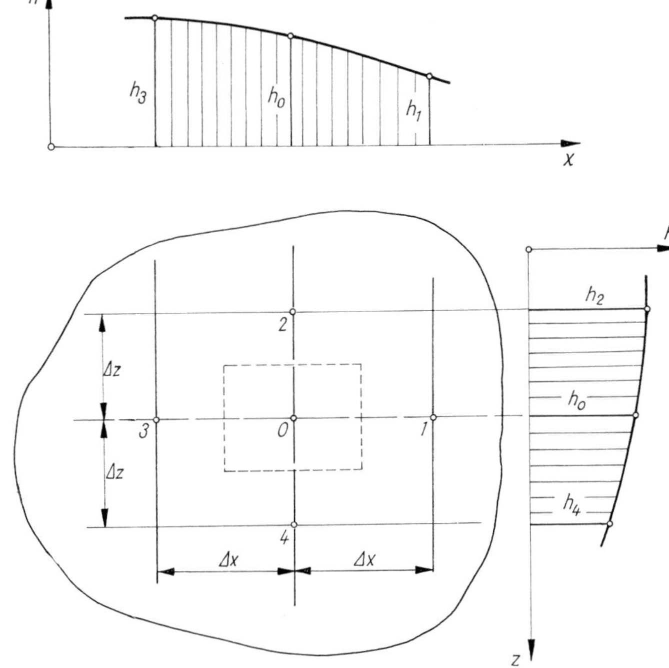

Fig. 195. Detail of two-dimensional flow; distribution of piezometric pressures

Points 0, 1, 2, 3, 5 in Fig. 195 are the lattice points of such a grid. We have to determine the values of h for each node in the grid. Using the *Taylor* expansion for the function h, the variation of hydraulic head over the width of a bay in the x direction may be written as

$$h_1 = h_0 + \Delta x \left(\frac{\partial h}{\partial x}\right)_0 + \frac{(\Delta x)^2}{2\,!}\left(\frac{\partial^2 h}{\partial x^2}\right)_0 + \\ + \frac{(\Delta x)^2}{3\,!}\left(\frac{\partial^3 h}{\partial x^3}\right)_0 + \ldots \qquad (125)$$

Similarly

$$h_3 = h_0 - \Delta x \left(\frac{\partial h}{\partial x}\right)_0 + \frac{(\Delta x)^2}{2\,!}\left(\frac{\partial^2 h}{\partial x^2}\right)_0 - \\ - \frac{(\Delta x)^3}{3\,!}\left(\frac{\partial^3 h}{\partial x^3}\right)_0 + \ldots - \ldots \qquad (126)$$

The subscripts 0 indicate partial derivative values at point 0.

Adding Eqs (125) and (126) gives

$$\left(\frac{\partial^2 h}{\partial x^2}\right)_0 = \frac{h_1 + h_3 - 2h_0}{(\Delta x)^2} - \frac{2(\Delta x)^2}{4\,!}\left(\frac{\partial^4 h}{\partial x^4}\right) - \ldots \qquad (127)$$

By neglecting the terms of higher powers, the second derivative of the hydraulic head with respect to x is given approximately by the expression

$$\frac{\partial^2 h}{\partial x^2} = \frac{h_1 + h_3 - 2h_0}{(\Delta x)^2}\,. \qquad (128)$$

The error is

$$\frac{(\Delta x)^2}{12}\left(\frac{\partial^4 h}{\partial x^4}\right)_0 + \ldots$$

The leading term indicates that the error is approximately directly proportional to the square of the grid spacing Δx.

By similar reasoning the second derivative of h with respect to z is obtained approximately as

$$\frac{\partial^2 h}{\partial z^2} = \frac{h_2 + h_4 - 2h_0}{(\Delta z)^2} \qquad (129)$$

in which the error is proportional to $(\Delta z)^2$. The *Laplace* equation can now be written in the following form

$$k_x \frac{\partial^2 h}{\partial x^2} + k_z \frac{\partial^2 h}{\partial z^2} = \frac{k_x}{(\Delta x)^2}(h_1 + h_3 - 2h_0) + \\ + \frac{k_z}{(\Delta z)^2}(h_2 + h_4 - 2h_0) = 0\,. \qquad (130)$$

A form more amenable to numerical work can be obtained for this equation by chosing a grid so as to satisfy the condition

$$\frac{k_x}{(\Delta x)^2} = \frac{k_z}{(\Delta z)^2}$$

or (131)

$$\frac{\Delta x}{\Delta z} = \sqrt{\frac{k_x}{k_z}} = \sqrt{\varkappa}\,.$$

We may proceed in two ways: either we chose a value for Δx, calculate Δz from Eq. (131) and lay out a grid of Δx. Δz over the cross-section of the layer; or, as by the graphical method, we first construct the cross-section to a distorted scale reducing the distances parallel to x by $\sqrt{k_z/k_x}$ and work with a square grid laid out on the transformed section.

For a square grid (which is used when $k_x = k_z$, or when we work with a transformed section) Eq. (130) becomes

$$h_0 = \frac{1}{4}(h_1 + h_2 + h_3 + h_4) \qquad (132a)$$

or

$$h_1 + h_2 + h_3 + h_4 - 4h_0 = 0\,. \qquad (132b)$$

This relationship holds for every node in the grid.

The first step in solving the *Laplace* equation by this method is to construct a square grid in the true, or in the transformed, section and to assign to each node in the grid some value of h. The numerical work is considerably reduced if the trial values are close to the final correct values. It is thus advisable to sketch a rough trial flow net first. The boundary conditions are also determined and known values of h at various boundary points are fixed in advance. The trial values at interior nodes are not likely to satisfy Eqs (132a) or (132b), and usually a difference called *residual*, R_0, results, i.e.

$$h_1 + h_2 + h_3 + h_4 - 4h_0 = R_0 .$$

The numerical procedure of approximation known as the *relaxation method* essentially consists of reducing the residual at each node to a sufficiently small value. This is accomplished by successively altering the h values assigned to the lattice points. The final values of h give a true picture of the variation in head through the soil. An important point to be noted is that R_0 changes, as is seen from Eq. (132), by four times any change made in h_0 and in the opposite sense. In the computation, R_0 is first altered in such a manner that the difference is a multiple of four and h_0 is then changed in the opposite sense by one-quarter of the change in R_0. The procedure is repeated until the residuals at all nodes are reduced to a desired small value.

It is advisable to start computation at the points with highest h values and to proceed then in a systematic order from point to point. One step in the relaxation procedure is illustrated in *Fig. 196.*

Having obtained final values of h for all nodes, the equipotential lines representing some round values of h can be drawn by interpolation. To complete the flow net, flow lines are drawn at right angles to the equipotential lines. The convergence of the approximation may, in some cases, be rather slow, and it may be an advantage to alternately employ the methods of relaxation and sketching of flow net, always refining the results obtained in the previous step.

The relaxation method shows a close similarity to the moment distribution method used in the analysis of statistically indeterminate structures, and it may be useful to adapt the terms and procedures used there to the determination of flow nets. Thus the ratio of the alteration in residual at one node to the variations induced at adjacent nodes corresponds to the distribution factor used in calculating "carry-over" moments. For the

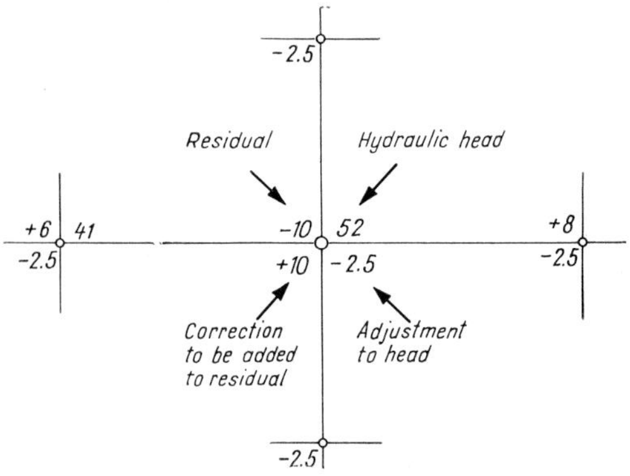

Fig. 196. Results of intermediate steps in the relaxation calculation

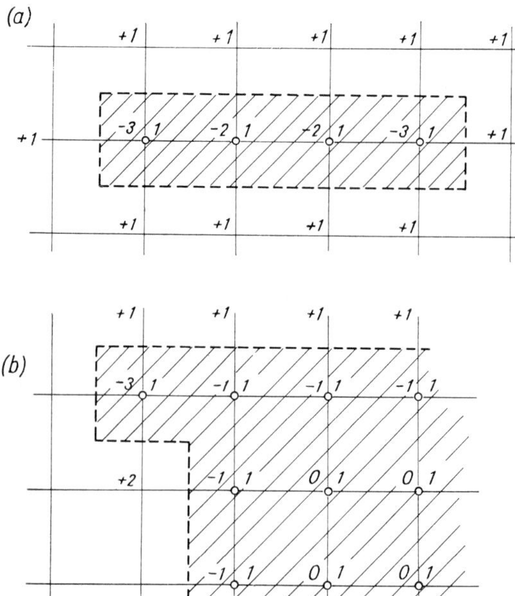

Fig. 197. Relaxation of a line (a) or a block (b)

relaxation method the distribution factor basically is, as defined by Eq. (132b), equal to 1/4. This means that if the residual at one node is adjusted by some value, the head at that point is correspondingly changed, in the opposite sense, by one quarter of the adjustment to the residual. The same adjustment is then applied to the residual at each of the four adjacent nodes. In the example shown in Fig. 196. the residual of -10 at the central lattice point is balanced by adding $+10$. Consequently, the head at that point should be altered by the adjustment times the distribution factor, $(-1/4) \times 10 = -2.5$. The distribution factor being constant, the residual at each of the four adjacent nodes is also adjusted by -2.5. The heads at adjacent nodes are, in this step, left unaltered; they will be adjusted when their own residuals are eliminated.

The approximation converges more rapidly, the more even the distribution of residuals across the grid, the lesser the variations, in absolute value, relative to the true values, and if residuals carrying positive and negative signs occur alternately and their algebraic sum within the examined domain is zero.

A more rapid convergency can be attained by "relaxing" a row or block of nodes, instead of individual nodes, at a time. An example of this is given in *Fig. 197.* The distribution factor is again constant throughout the grid. Values of head are simultaneously altered along conveniently chosen lines or over large blocks. Shown in the figure are the variations in the residuals and in the total heads.

For every interior node around which the head is altered in each of the four adjacent nodes, the residual remains unaltered. A boundary point is connected to three points inside the block in which the heads are altered plus one point outside the block in which the head remains unaltered after relaxation. Therefore, the residual at a boundary point within the block changes by (−1) times the variation in head. For an internal corner the corresponding factor is −2, and for an internal lattice point connected with three external plus one internal point the factor is −3. (These figures are given in Fig. 197.) A similar rule, except that the signs change to positive, can be developed for external points adjacent to the perimeter of the block: the factor is +1 if an external point has one neighbouring lattice point inside the block. If it has two inside the factor is +2, and if it has three the factor is +3. By skillfully choosing the blocks, numerical work can be substantially simplified as compared with the normal procedure of relaxing the nodes one by one.

It is a very common case in practical seepage problems, that the boundary of a permeable layer crosses the square grid obliquely (*Fig. 198*). A suitable approximation for this case is to replace the actual inclined boundary by a best fitting polygon. The heads at boundary points are defined from known boundary conditions.

It is worth noting that this numerical approximate method can be applied to heterogeneous systems as well. This usually involves a mass of numerical work which, however, may be reduced by using transformations. Further details are contained in the ample literature on relaxation methods.

To complete the discussion of approximate numerical methods, let us see how an experimental method developed from probability theory can be used to the solution of two-dimensional seepage problems.

Given is a domain in which seepage is occurring. Let us consider the path of an individual water particle. Let the initial position of the particle be the point (x, z). It is assumed that the particle can move freely but only with small finite displacements, Δx and Δz, in the directions parallel to the axes x and z, respectively. It has a finite velocity which is dependent on the coefficient of permeability. If the particle moves, in a time interval Δt, from the point (x, z) to either the point $(x + \Delta x, z)$ or the point $(x, z + \Delta z)$, then its velocity is, respectively,

$$v_x = \frac{\Delta x}{\Delta t} \approx k_x$$

or

$$v_z = \frac{\Delta z}{\Delta t} \approx k_z\,.$$

If we choose values of Δx and Δz such that $\Delta x/\Delta z = k_x/k_z$, then the values of Δt will be equal.

The probability that the particle moves in either of the directions $\pm x$ and $\pm z$ is 0.25. Suppose the particle has moved after a certain time to a point $(p - \Delta x, q)$. Let the probability of its arriving there be $P(p - \Delta x, q)$. The probability that the particle in the next step moves to the point (p, q) is, as was shown above, 0.25. The probability that the particle gets to point (p, q), given that it has already arrived at point $(p - \Delta x, q)$, is the product of the two probabilities, i.e. $0.25\, P(p - \Delta x, q)$. Similarly, $P(p + \Delta x, q)$, $P(p, q + \Delta z)$ and $P(p, q - \Delta z)$ are the respective probabilities that the particle gets to any one of the points $(p + \Delta x, q)$, $(p, q + \Delta z)$

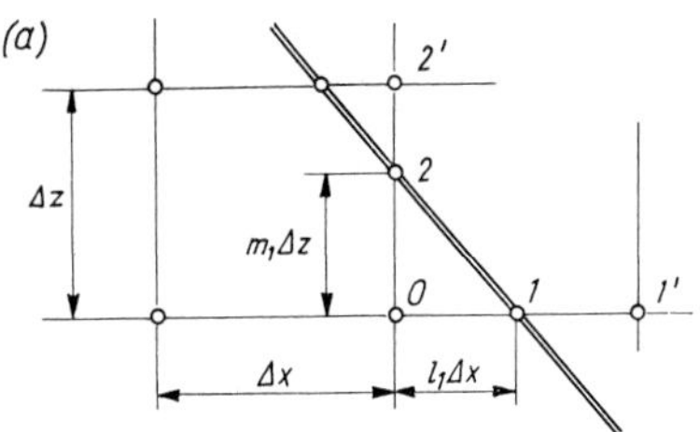

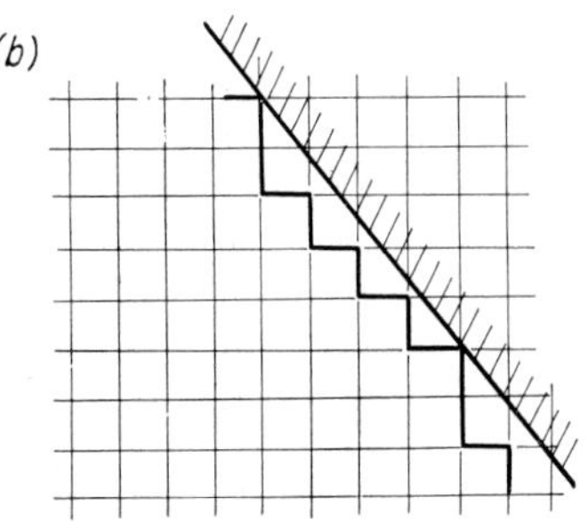

Fig. 198. The soil mass under consideration is bounded by an oblique surface

and $(p, q - \Delta x)$. Finally, the overall probability that the particle will arrive at point (p, q) may be written as

$$P(p, q) = 0.25\,[P(p - \Delta x, q) + P(p + \Delta x, q) + P(p, q - \Delta z) + P(p, q + \Delta z)]\,. \tag{133}$$

This equation is analogous to Eq. (132a) and, as can easily be verified, it is the expression of *Laplace*'s equation in terms of finite differences.

The above-described reasoning was first applied to the theory of absorption to find the probability that a particle starting from a given point arrives at a given absorptive surface. The procedure known as "random walk", "drunkard's walk" etc. is a special case of the general theory of *Markow* chains, notably the one in which the probability of the direction of the next movement of a particle is independent of the previous step, or steps, taken.

To give an idea of the application of probability theory, a simple case in which the method of random walk is used to find a concrete numerical solution to a given seepage problem will be presented in the following.

The method essentially consists of finding by trial-and-error what the probability is that a water particle starting from the point under consideration will finally arrive at a given boundary of known potential. Since Eq. (133) is analogous to the differential equation of flow, the probability value obtained by this method may be regarded as a relative measure of the potential at that point.

The "experiment" goes like a game of dice except that a special prism with a length several times the cross-sectional dimension is used so that only four casts, namely *1*, *2*, *3* and *4*, are possible.

Example: To determine the excess hydrostatic pressure at point A beneath the dam illustrated in *Fig. 199*.

The domain of seepage is confined in the vertical direction but is infinite in the horizontal direction. The permeable stratum is to be covered with a square grid so that the upper and lower boundaries are lattice lines, and also the point A coincides with a lattice point. By convention, the numbering of the directions of movement is as shown in Fig. 199. Instead of throwing the prism, we may use the last digits of telephone numbers in a directory. In this case only the numbers ending with the numerals 1, 2, 3, 4, or also with 5, 6, 7, 8, will be considered. We assign then the directions shown in Fig. 199 to these numerals. Now we put a marker at point A, and proceed as follows. We toss the prism, and move the marker in the direction corresponding to the number obtained. The tosses are then repeated and the marker is moved each time according to the numbering convention. When in this manner we arrive at a boundary of entry, or exit, (marked by 100 and 0 in the figure), the random walk is completed. We now repeat the walks several times, indicating on each occasion the mark of the particular free boundary at which the walk ended. If the marker arrived at an impervious boundary, (e.g. at point B on the base of the dam) and the number turning up in the next toss would require the crossing of that boundary, the marker is moved in the opposite direction (having obtained 2 in the example, the marker should, instead, be moved in the direction 4). For anisotropic or stratified soils a transformed rectangular grid is used according to Eq. (118).

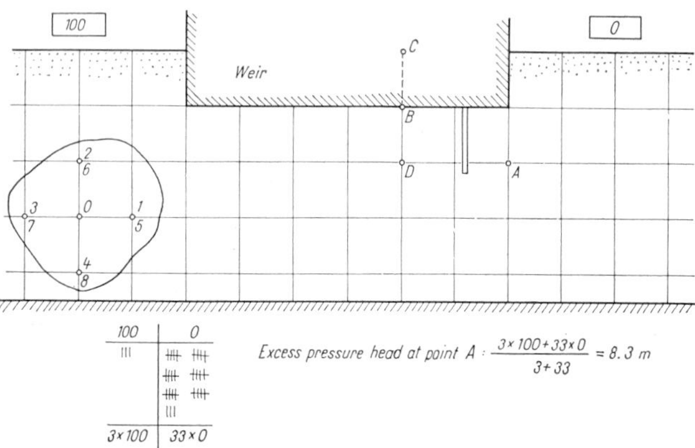

Fig. 199. Monte Carlo method to determine the water pressure at a given point

Having completed a "sufficient" number of random walks, we can statistically evaluate what the probability is of getting from the given point A to either of the free boundaries. The probability obtained is, as was shown earlier in this section, a measure of the potential at point A. Scores obtained in the example together with the calculation of the excess pressure head at point A are shown in Fig. 199. The greater the number of walks, the higher is the accuracy of the results. The probable error is inversely proportional to the square root of the number of walks. However, the convergency of the series of the results is slow, and the procedure is suitable mainly to find the excess pressure at single points in a flow domain rather than to furnish a complete flow net.

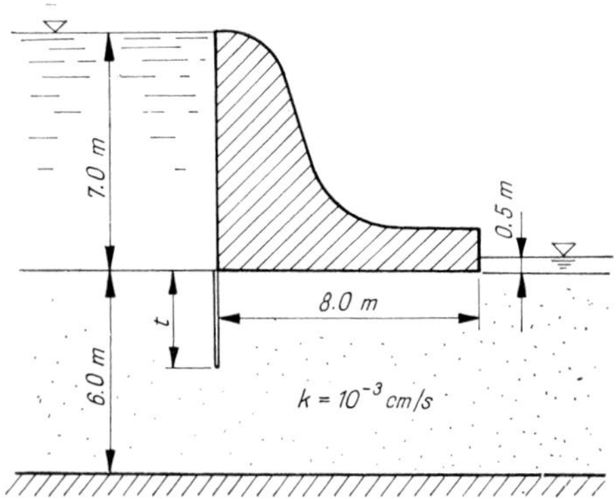

Fig. 200. Example 1: weir

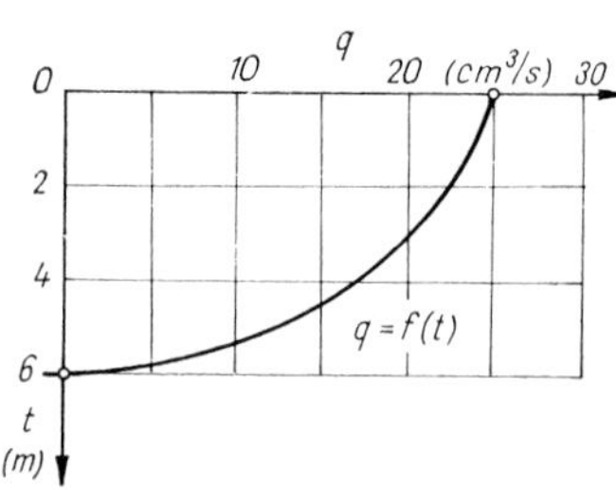

Fig. 201. Quantity of seepage for Example 1 as a function of the sheet pile length

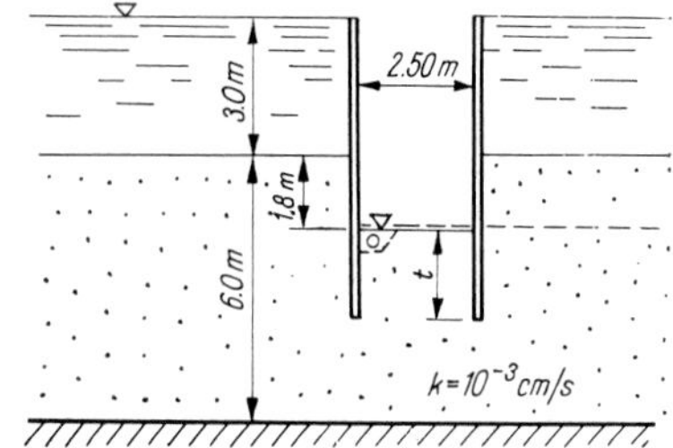

Fig. 202. Example 2: dewatered trench

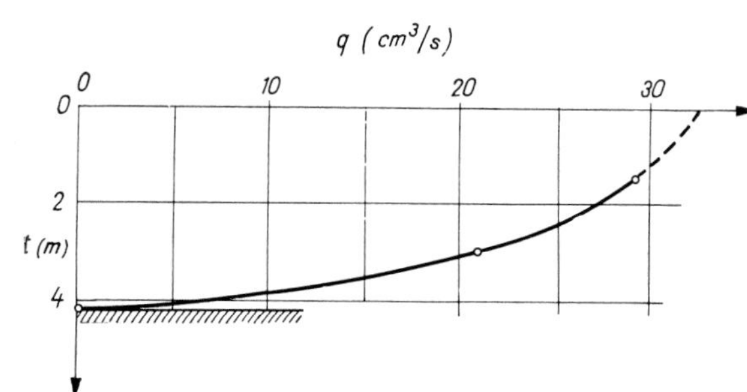

Fig. 203. Quantity of seepage water (Example 2) as a function of the sheet pile length

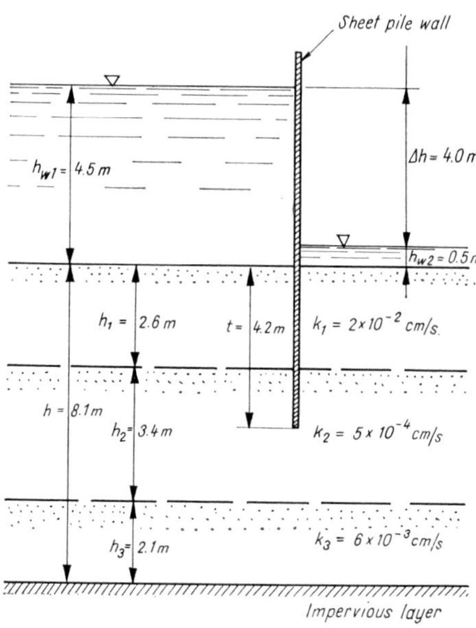

Fig. 204. Example 3: layered subsoil; calculation of the average values of the permeability coefficient

6.3.5 Numerical examples

Example 1. The foundation of a dam consists of a pervious stratum of sand of constant thickness, underlain by an impervious clay. To reduce seepage losses, a sheet-pile cutoff wall is driven, at the upstream end of the dam, to some depth into the sand. A sketch of the dam is shown in *Fig. 200*. A flow net worked out for this case was already presented in Fig. 191*a*. Assuming various lengths for the cutoff wall, similar flow nets can be constructed and seepage losses per unit of width of sheet piles and per unit of time can be computed by Eq. (116). The results of such computations are presented in *Fig. 201*, where seepage quantities q are plotted against the length of the cutoff wall. The graph gives an idea of the efficiency of cutoff walls.

Example 2. Figure 202 shows a cross-section through a double-wall sheet-pile cofferdam. The purpose of the sheet-piles is twofold: to support the sides of the excavation and to reduce, as a cutoff wall, the discharge entering the trench. The boundary conditions for this case were already given in Fig. 189 and a complete flow net was constructed in Fig. 191. Altering again the penetration depth of the sheet-piles, we can study the efficiency of the cutoff wall. Numerical results are plotted in *Fig. 203*.

Example 3. A similar case is shown in *Fig. 204* except that the subsoil is horizontally stratified. As a first step, we determined, using Eqs (101) and (102), the average coefficients of permeability for horizontal and vertical flow, then we transformed the cross-section by multiplying all horizontal dimensions by the factor $\sqrt{k_{II}/k_I}$ (where $k_{II} = k_x$ and $k_I = k_z$). A flow net with "square" figures was constructed on this transformed section (*Fig. 205*). Finally, we retransformed the figure to natural scale. The true flow net is shown in *Fig. 206*. The quantity of seepage was computed by Eq. (124).

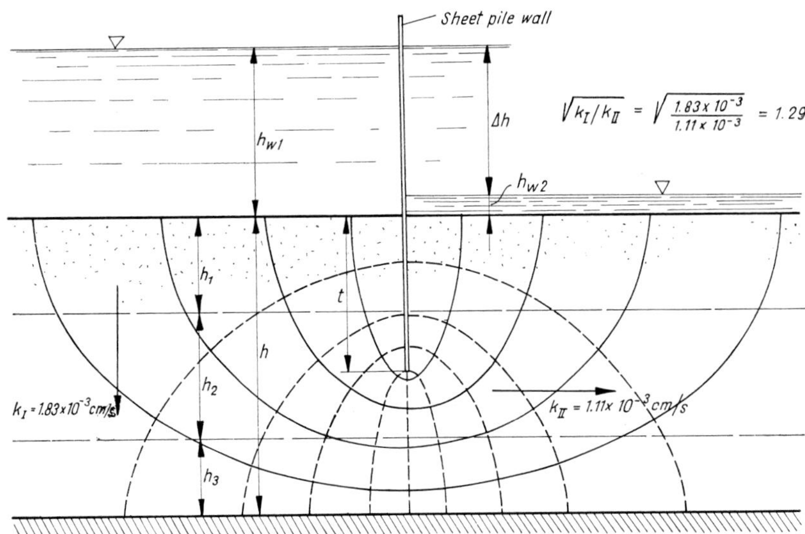

Fig. 205. Flow net construction in the transformed section

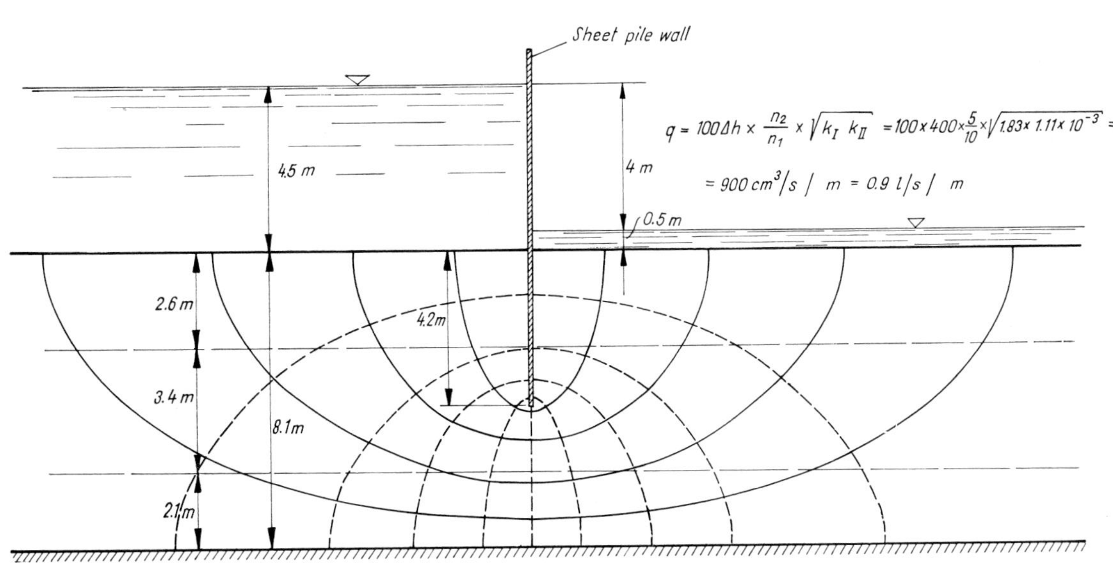

Fig. 206. Flow net in the aeolotropic medium

6.4 Capillary movement of water

6.4.1 Basic concepts

In Section 2.2 we have already discussed the phenomenon that water in a small-diameter capillary tube rises to a certain height above the free water surface and remains there lastingly. The weight of the water column held in the capillary tube is carried by forces acting on the surface of contact between the tube wall and the liquid. We have also dealt with the stress conditions of the curved liquid surface at the air–water interface and determined the maximum height of capillary rise [Eq. (120)]. In the following, we first examine the state of stress of the water in a capillary tube, then we treat, on a theoretical basis, certain capillary phenomena occurring in soils. Finally we deal briefly with the experimental determination of the height of capillary rise and derive a theoretical relationship for the capillary rise with time.

Figure 207 shows the section of a capillary tube whose lower end is immersed into water. The state of stress in the water in the capillary tube depends on the pressure acting in the air above the water. If the test is made in a perfect vacuum, the entire column of water located above the free water level is under tension. The film pressure of water is acting on the surface of the meniscus; this is in equilibrium with the weight of the water column of height h_c. This pressure was given by Eq. (19). Since the water is in a state of equilibrium, the pressure in the water is the same at every point in a horizontal section. The force acting downward is equal to $Az\gamma_w$. (A is the cross-sectional area of the capillary tube.) Since the sum of all the forces which act on the column is equal to zero, $u_z + z\gamma_w = 0$, thus

$$u_z = -z\gamma_w$$

i.e. we have tension in the entire water column.

If in the space above the meniscus atmospheric pressure p_a exists, the hydrostatic pressure in the water column at any height z above the free water surface is obtained as the algebraic sum of p_a and u_z:

$$u = p_a - z\gamma_w \,. \tag{134}$$

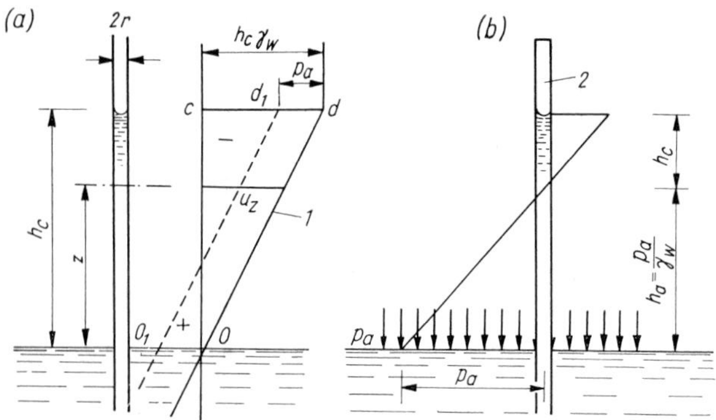

Fig. 207. Pressures in a capillary tube

It follows that u is negative, when $h_c > h_a = p_a/\gamma_w$ (about 10 m). If the height of capillary rise is less than h_a, there is no tension in the water but the pressure is less than atmospheric by $z\gamma_w$.

If the experiment illustrated in Fig. 207 is performed in a vacuum, the external pressure acting on the free water surface is zero and, according to Eq. (134), the hydrostatic pressure u is negative for all values of z, i.e. the water column above the free water surface is in a state of tension over its full length. The pressure diagram for this case is represented by the triangle *Ocd* in Fig. 207*a*. If the external pressure is atmospheric, the pressure diagram for an open-end capillary tube is transferred by an amount of p_a to the position O_1d_1 shown by the dashed line in the figure. The height of the capillary rise h_c, however, remains unaltered. If, finally, the upper end of the capillary tube is stopped up and we create a perfect vacuum inside the tube, while the outside pressure acting on the free water surface remains atmospheric, the water will rise in the tube to a height of

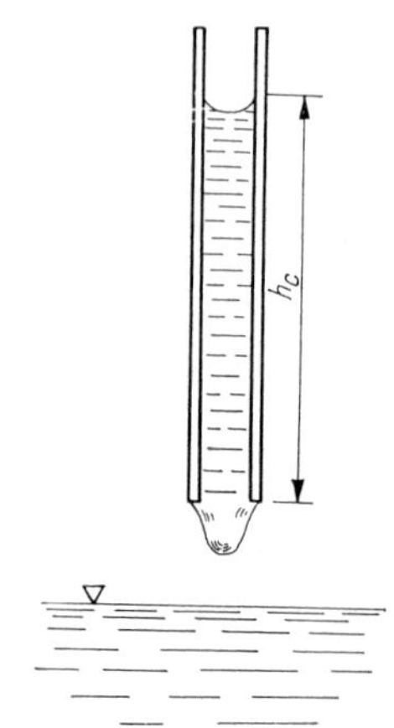

Fig. 208. Water droplet at the end of a capillary tube holding water

$$h_a + h_c = \frac{p_a}{\gamma_w} + \frac{0.15}{r} \cos\alpha \,.$$

(See Fig. 207*b*.)

Suppose the water in a capillary tube is forced to rise to a height greater than h_c, e.g. by pumping air out of the upper part of the tube. If the tube is lifted from the water pond and, at the same time, the vacuum is, the water begins to issue from the tube and comes to rest again only when its height above the end of the pipe has reduced to h_c. A droplet will be formed at the outlet (*Fig. 208*). The weight of the droplet is transferred by the surface film of the water, which acts like a stretched rubber membrane, to the tube wall.

Capillary rise of water is often observed in narrow flumes and *V*-shaped grooves. This can be demonstrated e.g. by putting two sheets of glass together so as to form an acute *V*-shaped groove and placing their lower ends under water. The relationship between the size of the gap and the height of capillary rise will strikingly show up (*Fig. 209*).

In natural soils the network of pore channels is very different from the simplified model made up of smooth capillary tubes of uniform cross-section. Between the soil particles, which themselves are very irregular in shape and which vary widely in size and are often surrounded by a film of adsorbed water, the voids form continuous but extremely

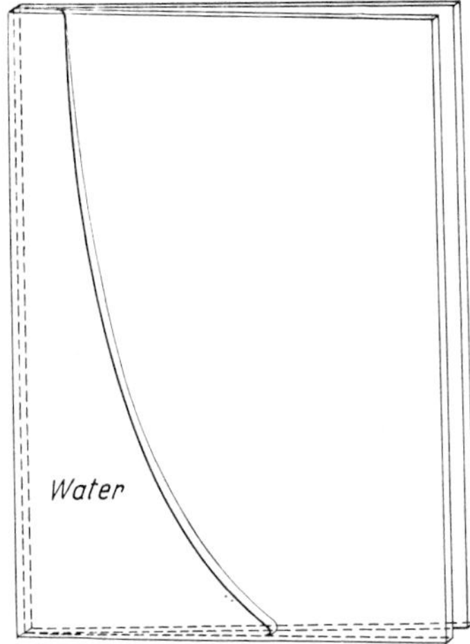

Fig. 209. Capillary rise between vertical glass plates forming a groove

tortuous "tubes" with narrow and bulbous sections alternating erratically. Nevertheless, the capillary flow of water in these voids is governed by the same laws as in uniform tubes. The basic fact that capillary rise is due to the surface tension of water remains valid, but the complexity of soil structure may give rise to intricate conditions. In the first place, the capillary rise will not be the same everywhere in the soil mass, at one place being greater than at others owing to the varying cross-sections of the voids. In the upper zone wetted by capillary action some of the pores become filled with water whereas other pores will still contain air. Thus a zone of partial capillary saturation is formed in which the soil is in a three-phase state. The distribution of the moisture within this zone after equilibrium is attained depends on the properties of the soil as well as on the elevation above the free water surface. It is also important whether equilibrium conditions were brought about in such a manner that the rising capillary water intruded into an originally dry soil from below following a rise of the water table, or else a portion of the pore water drained out of a formerly saturated soil by gravity upon a lowering of the water table. In either case the capillary water is subjected to the combined action of the force of gravity and of the water film pressure. But since the force of gravity is a mass force while the water film pressure is a surface force, the equilibrium conditions for rising or draining capillary water will not be the same.

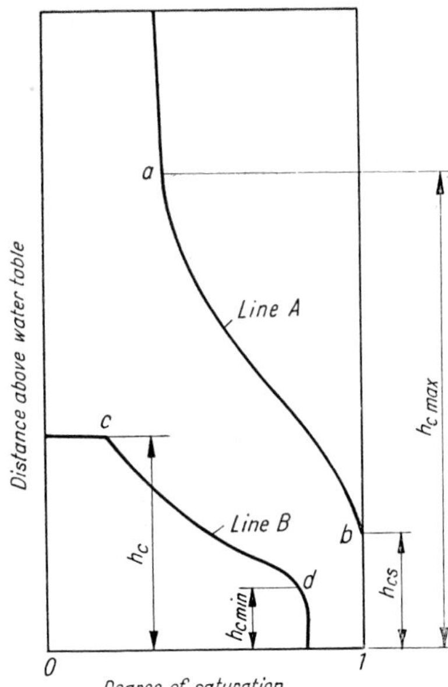

Fig. 210. Degree of saturation above the free water level in granular soil

In relatively coarse-grained soils (including coarse silt), the degree of saturation varies markedly with the elevation above the free water surface. Typical configurations are shown in *Fig. 210* (LAMBE, 1950). Here, the free water surface

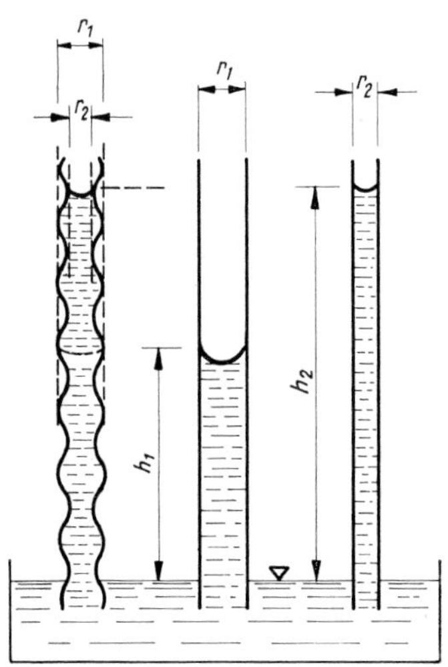

Fig. 211. Height of capillary rise in the Jamin-tube

is defined as the continuous surface at which the hydrostatic pressure in the water is equal to the atmospheric pressure p_a. Above that surface the pressure in the water is less than p_a (see Fig. 207).

As was pointed out earlier, the voids in a natural soil are infinitely variable in size and it is therefore impracticable to characterize capillary rise for a given soil by a single figure. The height to which water can rise in a capillary passage always depends on the radius of the meniscus at the top of the water column. This effect is clearly demonstrated by the bulbous tube model shown in *Fig. 211*. If the bottom of such a tube is immersed in water, the rising meniscus will stop at a bulb of radius r_1, the height of which above the free water surface is just equivalent to the height of capillary rise in a tube of constant radius r_1. If, on the other hand, the tube is filled by being placed below the water surface and is then gradually raised, or else the free water surface is lowered, the dropping meniscus will be caught at a narrow gap of radius r_2 at an elevation corresponding to the height of capillary rise in a tube of constant radius r_2.

Returning to Fig. 210, the case when the tube is raised resulting in free drainage of the water by gravity, is represented by line *A*, whereas line *B* corresponds to the case when water is drawn in from below into a dry soil. Point *a* on the drainage curve *A* represents the maximum height to which any continuous thread of water can exist. This distance therefore is taken as the maximum capillary head ($h_{c\,max}$). Point *b* marks the elevation to which the soil is completely saturated. This gives the height of capillary saturation (h_{cs}). On the curve of capillary rise (line *B*) there are two characteristic points. Point *c* indicates the highest elevation to which capillary

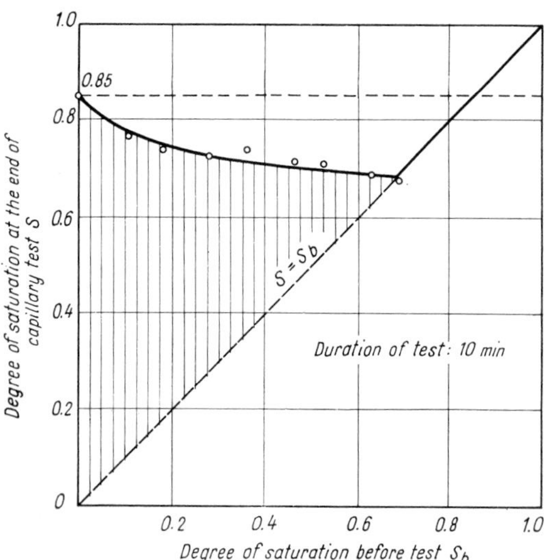

Fig. 212. Degree of saturation in a soil column after capillary rise, as function of the initial degree of saturation (RÉTHÁTI, 1956)

water would rise in a dry soil. This height is called the capillary rise (h_c). Maximum degree of saturation exists to a height above the water table, called the minimum capillary head, $h_{c\,min}$ (point *d*, Fig. 210).

These four heights represent the limits of the range of capillary heads for drainage and for capillary rise into a dry soil, respectively. It can easily be demonstrated by the bulbous tube model (Fig. 211) that $h_{c\,max}$ should be greater than h_c, and h_{cs} greater than $h_{c\,min}$.

We can conclude that the height of capillary rise depends, first, on the size and shape of the voids (consequently, on factors influencing the size of the voids such as grading of the soil, shape of particles, density, thickness of adsorbed water); second, on the way in which equilibrium conditions were reached; third, on the properties of the liquid (unit weight, viscosity, which in turn depend on temperature).

It is worth noting that line *A* in Fig. 210 characterizes, in a way, the distribution by volume of the voids within the soil. This means that for every height on the curve a void size (or at least some "equivalent" diameter) can be calculated, and hence the percentage distribution can be obtained. Such computations are, of course, based on the assumption that a real soil can be replaced by a system of capillaries.

Under natural conditions, also the initial water content and degree of saturation of the soil influence the height to which water eventually rises by capillary action. The wetter the soil and the higher the degree of saturation, the less is the amount of water that can be drawn into the soil by capillarity. RÉTHÁTI (1957) found e.g. that in a fine sand the capillary rise within a specified time, say 10 minutes, from the beginning varied with the initial degree of saturations S_b. He obtained the relationship shown in *Fig. 212*, where the degree of saturation after capillary rise, *S*, is plotted against the initial value S_b. It can be seen that an initially dry soil does not become completely saturated. The maximum degree of saturation, $S = 0.85$, was attained at $S_b = 0$. The higher the initial value S_b, the lower is the *S* value after 10 minutes of rise, and in the test shown there was no capillary intake of water above $S_b = 0.65$.

The foregoing statements apply mainly to coarse and medium-grained soils: sand, fine sand and cooarse silt. In fine silts and clays the movement of water has an entirely different character, partly because the voids are very small, and partly because a large portion of the pore water is bound to particle surfaces by much stronger electrical forces than the forces that would induce capillary flow. The pore space open to flow is therefore greatly reduced and the capillary movement of water can take place only at an extremely slow rate. In clays, movement of water against gravity that is not accompanied by either swelling or shrinkage of the soil is not likely to occur unless the soil had a secondary structure (as e.g. fissured clays). Such movement of water, however, can no longer be described as capillary flow. In the literature one often encounters the view, suggested by Eq. (20), that capillary rises of up to several hundreds of meters do actually occur in clays. But such figures prove unrealistic since the presence of bound water in the voids leaves practically no space open for the movement of water. Besides, the zone of capillary saturation cannot extend to heights greater than 10 m for the simple reason that at that height, the negative pore pressure being equal to the atmospheric pressure ($p = - u_z$), the air dissolved in the pore water is freed in the form of bubbles, and the soil becomes partially saturated.

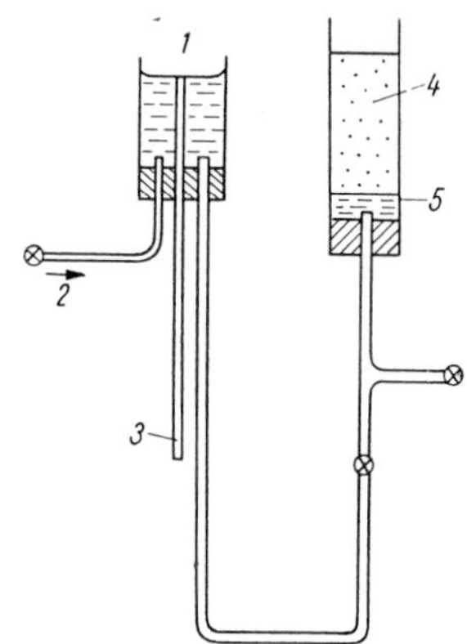

Fig. 213. Testing arrangement to determine the height of the zone of capillary saturation:
1 — pressure tank; *2* — water inflow; *3* — overflow; *4* — soil sample; *5* — screen

6.4.2 Determination of the height of capillary rise

According to the foregoing, the height of capillary rise for a soil cannot be given by a single figure. To get a complete picture it is necessary to determine all four capillary heads defined by Fig. 210. There are various capillary tests, each one suited to the determination of a particular capillary head.

The limit of capillary rise h_c (point *c*, Fig. 210) is determined by the following simple test. A soil sample, dried and powdered before the test, is poured and shaken vigorously into a glass tube 8 to 10 mm in diameter, care being taken that the void ratio is approximately equal to that in the natural packing of the soil. The base of the tube is then placed in a water-filled reservoir whereupon the water begins to rise into the sand. At various time intervals the distance from the free water surface to the top of the soil column wetted by capillarity is measured and taken as the height of capillary rise. By continuing the test for a sufficiently long time and plotting capillary rises against time elapsed, the limit of the capillary rise can be estimated. This method lends itself to the testing of coarse-grained soils (sand, fine sand, silty sand). But for highly cohesive soils it would be wrong to perform the above-described test on a sample prepared by grinding dry, hard lumps into a powder, and regard the capillary rise so obtained as a true soil characteristic.

The height of capillary saturation h_{cs} (see point *b* in Fig. 210) serves as a basis of comparison of the capillary behaviour of various soils and is especially useful in the study of soil drainage. This value represents the height of the zone that would not drain but remain saturated in the case of a drop in the water table. The height h_{cs} is determined by the apparatus shown diagrammatically in *Fig. 213*. Before the test the sample is completely saturated with water by applying a vacuum to the top of it. The initial atmospheric pressure in the pore water is then gradually reduced by lowering container on the left until the first air bubble appears beneath the sample. The saturation capillary head h_{cs} is measured as the vertical distance from the bottom of the sample to the water level in the container. For testing fine-grained soils mercury is used instead of water to reduce the pressure, and the drop in mercury level expressed as an equivalent head of water is taken as the value h_{cs} (Beskow's capillarimeter).

In addition to the two methods just described several other methods devised to determine the capillary head under specific conditions have been suggested in the literature. In the majority of practical cases, however, a knowledge of the values h_c and h_{cs} is sufficient. Only for a better understanding of the capillary properties of a soil it may be necessary to determine the full course of the curves *A* and *B* shown in Fig. 210. A simple experimental procedure of determining the capil-

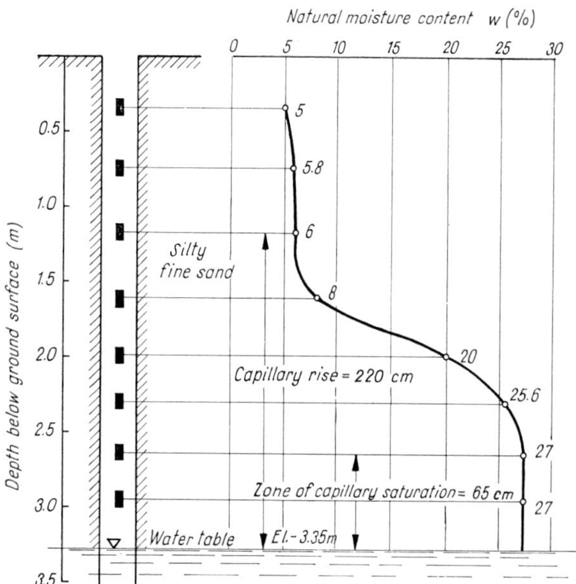

Fig. 214. Variation of water content above the free water table in the field

lary moisture distribution was developed by RÉTHÁTI (1957).

A really true picture of the capillary properties of the soil can be obtained, under favourable conditions, by taking undisturbed samples at regular intervals from the strata located above the water table and determining the phase compositions of them. A plot of degree of saturation against depth below the ground surface will permit the maximum capillary head $h_{c\,max}$ and the capillary saturation head h_{cs} to be determined with sufficient accuracy. The result of such an investigation is shown in *Fig. 214*.

6.4.3 Rate of capillary rise

In many practical problems it may be of importance to know to what height water would rise by capillarity within a given time. The saturation of the subgrade of road pavements, the effect of rising ground water table on the physical properties of the soil, frost action etc. can only be studied with knowledge of the rate of capillary rise in the soil. Let us discuss therefore the time-height of capillary rise relationship and the velocity of capillary flow.

First let us consider the case of a dry sand, assuming, for the sake of simplicity, that capillary water exists only in a zone of complete saturation extending upward to a sharply defined boundary above which the sand is perfectly dry. The movement of water is assumed to be governed by *Darcy*'s law. The water proceeds vertically upward and at a time t the line of saturation stands at a height z above the free water surface. Assuming that at the top of the zone of saturation the vertical component of the surface tension is constant, the pressure head with respect to the free water surface is equal to $h_c - z$, where h_c is the maximum capillary rise. Hence the hydraulic gradient may be written

$$i = \frac{h_c - z}{z}.$$

The discharge velocity v, i.e. the vertical component of the average velocity of the water, is given by the differential coefficient dz/dt. Using the relationship between the seepage velocity v_s and the discharge velocity v derived in Section 6.2.1 we have

$$v_s = \frac{v}{n} = \frac{dz}{dt}.$$

By Darcy's law

$$\frac{dz}{dt} = \frac{k}{n}\frac{h_c - z}{z}. \qquad (135)$$

Integrating the differential equation,

$$h_c - z - h_c \log_e (h_c - z) = \frac{k}{n} t + C.$$

Using the boundary condition that $z = 0$, when $t = 0$, the integration constant is obtained as

$$C = h_c(1 - \log_e h_c).$$

By substituting and solving the equation for t, we have

$$t = \frac{nh_c}{k}\left(\log_e \frac{h_c}{h_c - z} - \frac{z}{h_c}\right). \qquad (136)$$

For small values of z/h_c the above relationship can be replaced with sufficient accuracy by a quadratic parabola having the equation

$$t = \frac{n}{2kh_c} z^2 \quad \text{or} \quad z = \sqrt{\frac{2kh_c}{n} t}. \qquad (137)$$

The assumption used in the preceding derivation that the zone in which capillary rise takes place is completely saturated is never strictly fulfilled in natural soils, since, as was pointed out earlier, there always exists a zone of partial saturation in the capillary fringe above the water table. In soils, where the capillary "tubes" are extremely variable in size, the water, while filling practically all the voids to a certain height dependent on the size of the widest pores, may rise much higher in the narrow ones. Thus a three-phase zone is formed in which the moisture distribution is as was shown in Figs 210 and 211. In this zone the relative water-permeability tends to decrease rapidly with growing intrusion of air, and thus the laws governing the movement of water in three-phase systems apply. (See Section 6.2.5.) The value h_c, assumed to be a constant in Eq. (136), also increases as the water rises higher and higher, since above a certain height it is no longer the average pore diameter but the diameter of the active water-conveying voids that determines the rate of capillary rise. Variation of the two quantities—relative permeability and limiting height of capillary rise—affects the velocity of capillary flow in opposite ways. A theoretical relationship taking all these factors into account, however, still remains to be found.

If we are not particularly concerned with the moisture distribution within the zone wetted by capillarity and only seek the advance of the wetting front, an empirical formula suggested by JÁKY (1944) is very useful. He found the following relationship for the rate of capillary rise

$$h = at^b \qquad (138)$$

wherein,

h = height of capillary rise

t = time elapsed from the beginning of capillary rise

a and b = empirical constants.

Equation (138) describes with great accuracy the capillary rise of water in the partially saturated zone. In the saturated zone, i.e. at an early stage of capillary rise, the deviations from the measured values are more pronounced.

In logarithmic form Eq. (138) becomes

$$\log h = \log a + b \log t.$$

The plot of this in a log–log system is a straight line. By differentiation with respect to t, the velocity of capillary rise is

$$v = \frac{dh}{dt} ab\,t^{b-1} = b\frac{h}{t}. \tag{139}$$

The applicability to practical problems of Eqs (138) and (139) is rather limited since they have been derived from tests conducted on perfectly dry samples. In real soils there is always a given initial three-phase state to be considered, and reliable information can only be obtained from field measurements and observations. A comparison of laboratory and field data then provides the basis on which we can predict the changes in soil moisture that are likely to occur as a consequence of some interference due to engineering activities.

For the determination of the rate of capillary rise in two-layered or multilayered systems a graphical method has been developed by RÉTHÁTI (1953).

The coefficient of permeability (k) and the height of capillary rise (h_k) are inversely proportional quantities. In fine-grained soils k is small and h_k is large, just opposite to the case of coarse-grained soils. It is therefore a plausible assumption that there must exist a relationship between k and h_k. For capillary rise in sands, TAYLOR (1948) found the following limits:

$$\frac{0.7}{\sqrt{k}} < h_s < \frac{2.4}{\sqrt{k}} \tag{140}$$

where h_s is expressed in cm and k in cm/s. The formula gives good approximate values for the height of capillary saturation h_{cc}. It is not valid for clays, and cohesive soils in general, since the movement of the water invading a dry clay no longer obeys *Darcy*'s law.

6.4.4 Variation of soil moisture in the field

Under field conditions the variation of soil moisture in the layers located near the ground surface is greatly affected by precipitation, temperature, weather conditions, type of soil, elevation of the water table, slope of the ground surface, vegetation and many other factors.

Soil water is also subject to the action of various forces of which gravity and the surface tension of the water are the most important. The movement of water by gravity and capillary action can, as we have seen earlier, be treated mathematically. But natural soils are never truly homogeneous and the capillary conditions are not as simple as in a laboratory test. Theoretical investigations, in spite of the useful information they provide, cannot replace field observations in solving practical problems.

The various processes that may take place in the top zone of the subsoil are illustrated in *Fig. 215*. The soil is simulated by a bundle of vertical capillary tubes with their lower ends immersed in the ground water. At first the water in the pipe stands at a height h_c above the free water surface. If water is later allowed to enter the system from above (case *a*, Fig. 215), it will be drawn into the tubes by capillary action, and between the two zones filled with water there will be entrapped air over a distance of h_a. At the air–water boundary the pressure in the air is equal to the algebraic sum of the pressure in the water and the pressure exerted by the surface tension on the air. If water is continuously supplied from above, the entrapped air together with the water is forced to move downward in the pipe until it escapes into the free water. If the supply of water ceases before the air has been able to escape, the top of the water–air–water zone comes to rest at a height $h_c + h_a$ above the free water surface (*b*, Fig. 215), and henceforth the air

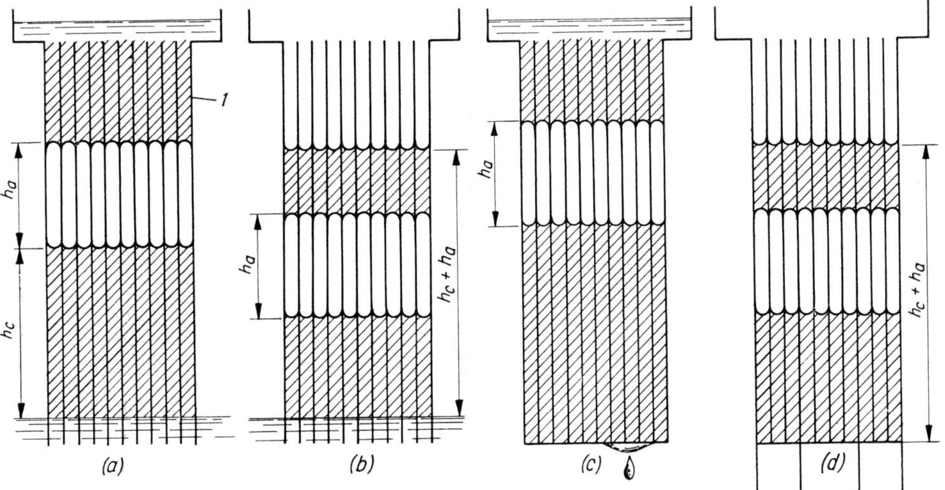

Fig. 215. Processes occurring in the upper layer of the soil

remains permanently entrapped between the two zones of capillary water.

If the bottoms of the capillary tubes are not immersed in water (case *c*), the water flows out freely below in large drops. If the water supply from above is disrupted, the water continues for a while to drain out of the tubes. But as soon as the water level has dropped to the height $(h_c + h_a)$ above the bottom of the tubes, the movement ceases and only the water held by capillarity remains in the tubes.

Case *d* in Fig. 215 represents a fine sand layer underlain by a coarse-grained gravel in which there is no capillary action.

Figure 216 shows some examples of the distribution of soil moisture under field conditions. Parts *a* and *b* show a layer of fine silty sand containing a lense of coarse sand and gravel. In the top zone exposed to the seasonal changes of air humidity and temperature and to the action of roots, the permeability of the soil is much higher than at greater depths. In this zone the distribution of soil moisture changes with the season from the continuous to semi-continuous to discontinuous states. Below this zone discontinuous soil moisture may exist only in pockets of coarse sand or gravel and then only temporarily. In dry weather, the distribution of soil moisture is as shown in Fig. 216*a*. There is a zone of capillary saturation above the water table and another perched on top of the gravel lens. Note that in the gravel lens itself there is no continuous water. A paradoxical situation arises in that saturated zones of soil occur above a three-phase zone. This resembles a waterlogged sponge which, if placed onto a layer of dry, coarse-grained material, e.g. gravel, retains by capillarity all the water contained in it. But a more absorbent material, e.g. blotting paper, will vigorously suck in water from the sponge.

During heavy rainstorms water enters the soil, and alters the former moisture distribution in many ways (Fig. 216*b*). Wherever continuous capillary water existed previously, the level of the saturated zone rises considerably. The amount of the semi-continuous water also greatly increases and a portion of the water accumulates at the bottom of the lens, forming temporarily a second free water surface. As the rainfall ceases, the excess water drains into the ground water and the original conditions shown in Fig. 216*a* become gradually restored. Entrapped air bubbles incidentally formed during the rainstorm will in subsequent dry periods migrate towards the free surface and eventually escape into the atmosphere.

Figure 216*c* shows a stratified soil profile in which layers of fine silty sands alternate with thin seams of coarse sand or gravel. Such stratification is typical of flood plain deposits laid down during alternating high-water and low-water seasons. Owing to the regular occurrence of coarse-grained seams, the amount of water that the silty sand can permanently hold by capillarity greatly increases. The explanation for this is that a zone of continuous capillary water may be formed above every layer of gravel just as in the analogous capillary-tubes model shown in Fig. 215*d*. If the sequence of fine-grained and coarse-grained layers is such that the vertical distance between any two successive layers of gravel is smaller than the height to which water could rise in the intermediate layer of silty sand, then the entire mass of silty sand may become permanently saturated by capillarity to practically any height above the water table. Thus, once the entire stratified subsoil

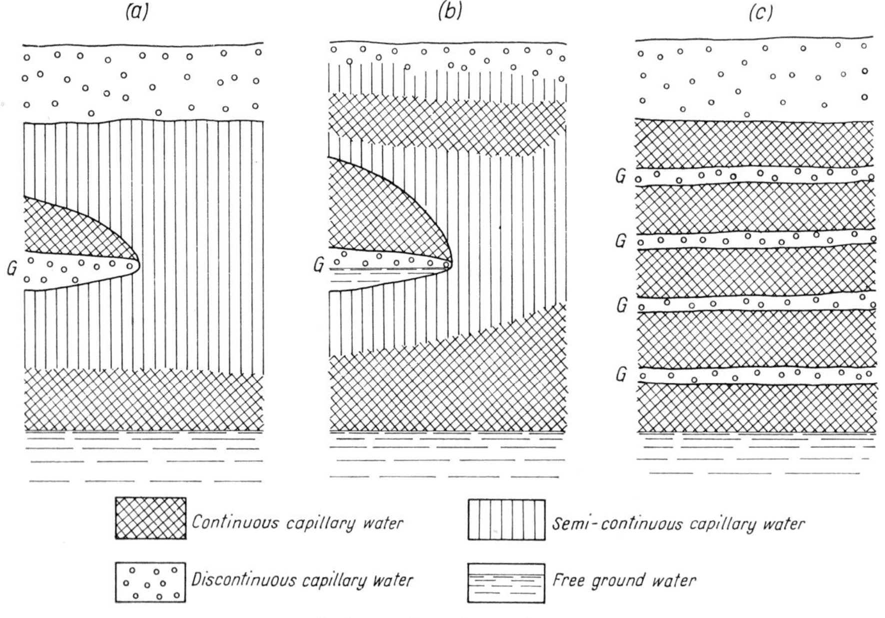

Fig. 216. Distribution of moisture in the soil

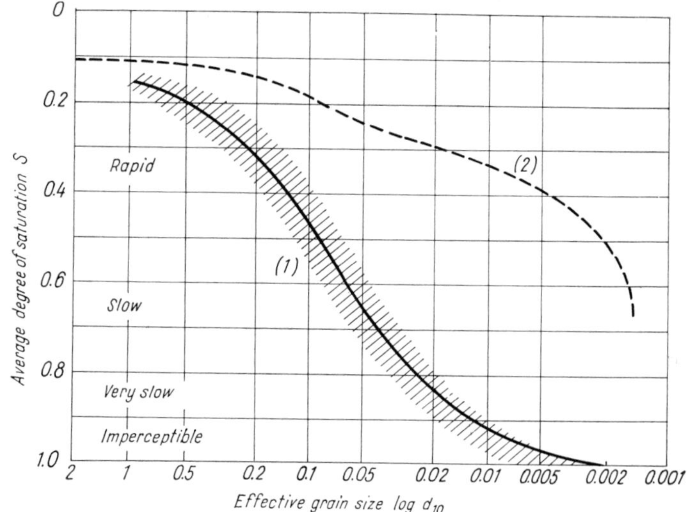

Fig. 217. Drainage of sand due to gravity: 1 — results of field measurements; 2 — laboratory tests

has become saturated through infiltration, the water can no longer be removed from the fine sand layers by gravity. On the other hand, a portion of the voids in the gravel may periodically contain air. During heavy rainstorms gravitational water may accumulate in the gravel layers, forming secondary ground water levels.

The phenomena described here play an important role in judging the frost-susceptibility of the subsoil for the design of road pavements.

6.5 Drainage of saturated soils by gravity

The first extensive laboratory investigations on the drainage of sand by gravity were made by King (1899). A sand column 2.5 m high, was initially saturated and then allowed to drain under the influence of gravity. After drainage for 2 1/2 years in laboratory, the distribution of moisture within the sample was measured. The results of *King*'s experiments were similar to those shown earlier in Fig. 210. The finer the grains of a soil, the smaller is the amount of water that can be drained out of the ground by gravity. On the basis of field measurements, Terzaghi (1948) presented the empirical relationship shown by curve *C* and the shaded area in *Fig. 217*. In the figure the average degree of saturation *S* that can be attained at all after a long period of drainage is shown as a function of the effective grain size d_{10}. Owing to the surface tension of water there will be, even after prolonged drainage, a certain amount of water held in the soil which can no longer be removed by gravity. Curve *2* represents (after Lebedeff, 1928) an empirical relationship between *S* and d_{10}, obtained by laboratory tests.

In connection with drainage by gravity the important question arises of the rate at which the process takes place. This problem can be dealt with on the simplifying assumption that in a sand mass the drained zone is separated by a distinct boundary from the zone of capillary saturation.

Consider a homogeneous horizontal sand stratum of thickness *h* underlain by a highly pervious coarse gravel (*Fig. 218*). Let the surface of the ground water be level with the ground surface. Drainage of the sand stratum is to be accomplished

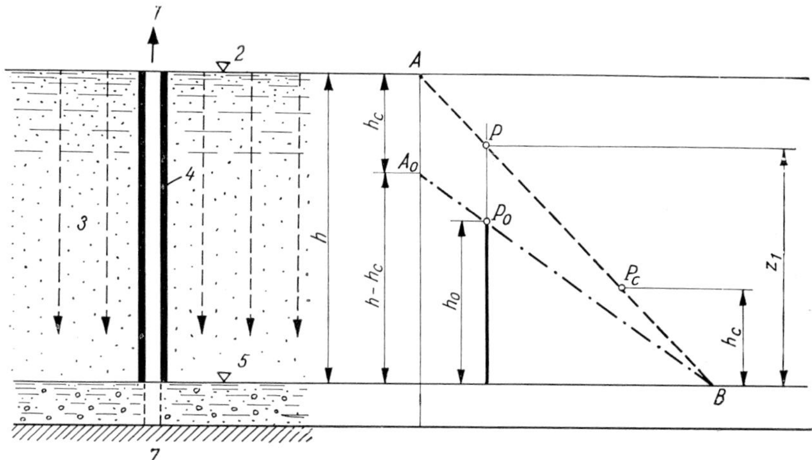

Fig. 218. Drainage of a sand layer in the vertical direction: 1 — pumping; 2 — original water table; 3 — sand; 4 — casing; 5 — lowered water table; 6 — gravel; 7 — impermeable layer

by pumping water out of the gravel through a row of open-end solid steel tubes driven down to the upper surface of the gravel so as to ensure that the water draining out of the sand moves in the vertical direction. As soon as drainage through pumping has started, the surface tension of the water begins to act with the result that at the point just below the ground surface the piezometric head immediately drops to the capillary head h_c. Thus the initial hydraulic head, considering that at the bottom of the sand stratum the pressure is zero, is reduced to $h - h_c$. Hence the initial hydraulic gradient is

$$i_0 = \frac{h - h_c}{h}. \tag{141}$$

The value i_0 can be interpreted in Fig. 218 as the slope of line $A_0 B$ which in turn represents the piezometric levels along the vertical plane AB at $t = 0$. By proportionality, the initial hydraulic head at a depth $h - z_1$ below ground surface (point P) is

$$h_0 = (h - h_c) \frac{z_1}{h}.$$

Because of the capillary tension, the neutral stress in the pore water is negative everywhere in the sand stratum. When $t = 0$, the neutral stress at the ground surface is

$$u_0 = h_c \gamma_w$$

and at a depth $h - z_1$ below ground surface

$$u_0 = h_c \gamma_w \frac{z_1}{h}$$

which is represented by the distance PP_0 in the figure.

As the drainage proceeds, the upper level of the saturated zone drops continuously, but the piezometric head at the changing boundary remains constant and negative and equal to the capillary head h_c throughout. Thus when the top of the saturated zone is at a height z above the lowered free water surface, the hydraulic head is $z - h_c$, and the hydraulic gradient

$$i = \frac{z - h_c}{z}. \tag{142}$$

When the line of saturation is lowered to the height h_c (point P_c) above the bottom of the sand stratum, i becomes zero, and the process of drainage is complete. The velocity, at which the line of saturation descends from an elevation z toward the final elevation h_c, is $-\mathrm{d}z/\mathrm{d}t$. The amount of water draining into the gravel across a unit area in unit time is given by

$$v = -\frac{\mathrm{d}z}{\mathrm{d}t} n(1 - S). \tag{143}$$

In the formula $n(1 - S)$ is that portion of the voids, in unit volume of the soil, in which the water will eventually be replaced by air, and S is the degree of saturation after completion of drainage. This final S value can either be estimated on the basis of Fig. 217 or determined experimentally. By *Darcy*'s law

$$v = ki.$$

Substituting Eqs (142) and (143) gives

$$k \frac{z - h_c}{z} = -\frac{\mathrm{d}z}{\mathrm{d}t} n(1 - S). \tag{144}$$

The general solution of the differential equation is

$$t = \frac{n(1 - S)}{k} [h_c - z - h_c \ln (h_c - z)] + C.$$

To determine the constant, we use the boundary condition that $z = h$ when $t = 0$. Hence we have

$$t = \frac{h_c n(1 - S)}{k} \left[\ln \frac{h - h_c}{z - h_c} - \frac{z}{h_c} + \frac{h}{h_c} \right]. \tag{145}$$

Equation (145) gives the progress of drainage as a function of time. The amount of water that has passed through a unit area of the bottom of the sand in time t is $(h - z)\, n(1 - S)$. The ratio, in percent,

$$D\% = 100 \frac{h - z}{h - h_c} \tag{146}$$

may therefore be regarded as the degree of drainage.

Figure 219 shows, using the data of a numerical example, the relationship between the degree of drainage and the time elapsed since the start of the drainage.

For coarse-grained soils in which the capillary rise is negligible ($h_c \sim 0$), Eq. (144) reduces to

$$k = -\frac{\mathrm{d}z}{\mathrm{d}t} n(1 - S)$$

whence

$$-\frac{\mathrm{d}z}{\mathrm{d}t} = \frac{k}{n(1 - S)} = \text{constant}.$$

Integrating

$$z = h - \frac{k}{n(1 - S)} t.$$

The degree of drainage becomes

$$D\% = 100 \frac{h - z}{h} = 100 \frac{kt}{n(1 - S)\, h}. \tag{147}$$

In Fig. 219 this expression is represented by a straight line which intercepts the time axis at

$$t = \frac{n(1 - S)}{k} h.$$

Fig. 219. Degree of drainage versus time

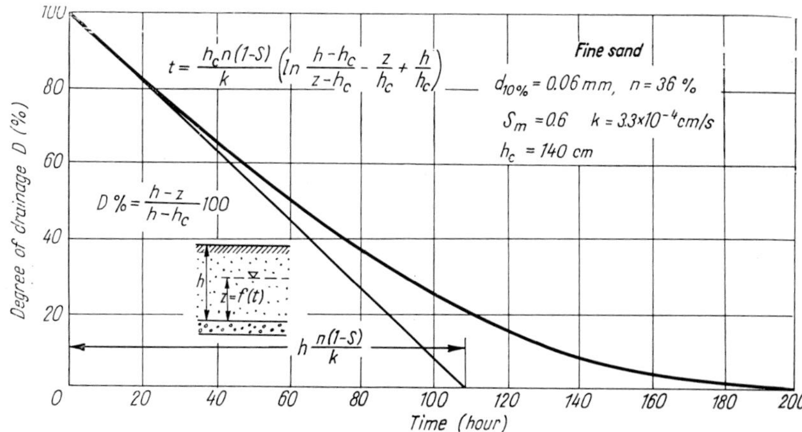

The curve computed by Eq. (146) lies above this straight line at every value of *t*. This indicates the twofold effect of capillary action on drainage: firstly, it reduces the amount of water that can drain out of the soil by gravity, secondly, it retards the process.

6.6 Movement of water in saturated soils due to loading (consolidation)

6.6.1 Basic concepts

We saw in Ch. 5 that an increase in external load induces neutral stresses in the pore water, both in saturated and partially saturated soils. Differences in excess pore pressure due to load give rise to a hydraulic gradient in the soil, thereby initiating a flow of water. As water is being squeezed out of the soil under the load, the soil skeleton is simultaneously compressed. Drainage and compression proceed at the same rate but in opposite directions. In the following we shall examine the laws that govern this kind of flow of water known as *consolidation*. The fundamental considerations previously adopted for the flow of water by gravity will still hold with the one assumption of the incompressibility of the soil skeleton omitted.

First we consider the case that the soil is completely saturated, and the state of stress along any vertical section in the soil mass under load is the same. This implies that the water can drain out only in the vertical direction. This is referred to as one-dimensional consolidation. The mathematical solution presented in the next section will first furnish the neutral and effective stresses as functions of time. Knowing these, we can compute the rate of consolidation.

The process of consolidation and the factors influencing it can be clearly demonstrated by means of *Terzaghi*'s mechanical model (1922). It consists of a cylindrical vessel that contains a series of pistons separated by springs (*Fig. 220*). The space between the pistons is filled with water, and the pistons are finely perforated. Let a pressure p per unit of area act on the top of the uppermost piston. The length of the springs at the first instant of loading is unchanged because sufficient time has not yet elapsed to allow the escape of the water through the perforations. The springs do not carry a load until their length decreases, and the entire load must at first be carried by the water. The initial excess hydrostatic pressure in the water is therefore equal to the superimposed load per unit of area: $u = p = h\gamma_w$. The water in the piezometric tubes rises to the same elevation above the original water level in the vessel: $h = p/\gamma_w$.

After a certain time t, some water has already been squeezed out through the perforations, from the upper compartments, but in the lowest compartment the amount of water is still unchanged. Because of the decrease in volume of the upper compartments, the upper springs have been slightly compressed as an indication that they already carry a portion of the applied load, while

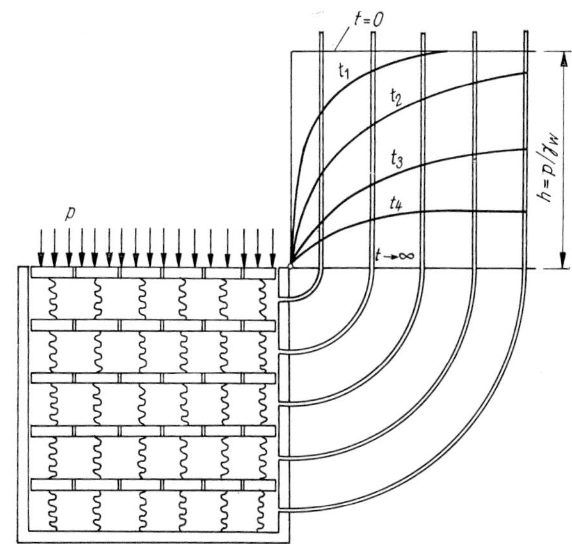

Fig. 220. Mechanical model to explain the process of consolidation

the pressure in the water has been reduced accordingly. At this stage the water in the piezometer tubes stands at the levels shown by the curve t in Fig. 220. If only a short time t_1 has elapsed, the curve merges into the initial horizontal line at elevation h. Let the corresponding total compression of the set of springs be denoted by Δh_1. Any curve which connects the water levels in the piezometer tubes at a given time is called an *isochrone.* After a time t_2, the water levels are located on the curve t_2. Finally, after a very long time the excess hydrostatic pressure becomes very small, since the total amount of excess water corresponding to the increase in load has been squeezed out whereupon the system comes to rest. The excess pressure everywhere in the water is then zero, and the total load is carried by the springs. Let the final compression be denoted by Δh.

For the above model the time required to reach the final compression obviously depends on the speed with which the water can escape through the fine bores, in other words, on the opening of the bores. Furthermore, if the springs were less stiff, so that the same compression would be produced by a smaller load, more water would have to be squeezed out in order to allow a given state of consolidation.

A measure of the degree to which consolidation has progressed is given by the ratio

$$\varkappa\% = 100 \frac{\Delta h_t}{\Delta h} \qquad (148)$$

called the degree of consolidation. Δh_t is the compression after an elapsed time t, and Δh is the final compression which can, theoretically, be reached only after infinite time.

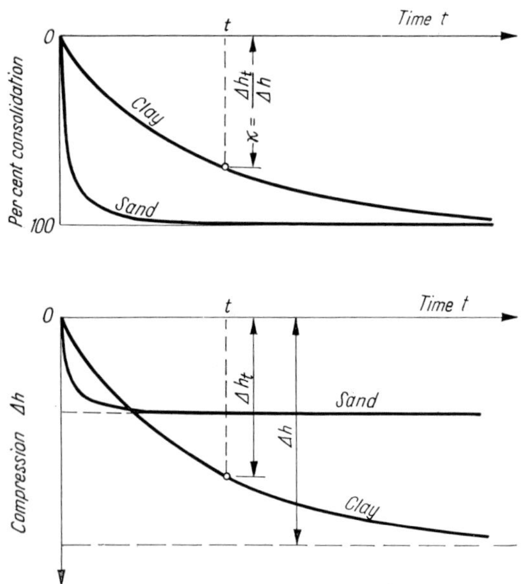

Fig. 221. Consolidation curves for sand and clay

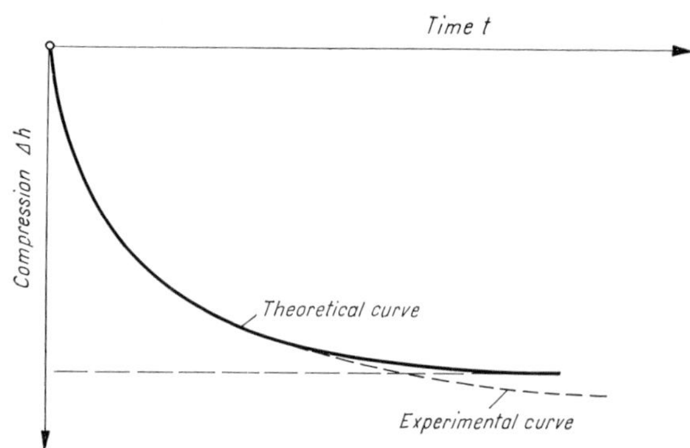

Fig. 222. Consolidation curve; primary and secondary consolidation

In a mass of clay subject to loading the process of consolidation is analogous to that in the mechanical model. The applied load is at first carried entirely by the pore water, and the resulting increase in neutral stress causes no compression. With the lapse of time, however, the water is gradually squeezed out of the voids. The soil skeleton, analogous to the set of springs, is compressed accordingly. An increasing portion of the load is transferred to the soil skeleton, until the excess pore pressures dissipate completely and the final compression has been reached. In the mechanical model described above the springs represent the skeleton of a saturated soil, and the water between the pistons represents the water filling the voids of the soil. The opening of the perforations is analogous to the permeability of the soil; the compressibility of the springs to the compressibility of the soil. The more compressible the soil, the longer the time required for complete consolidation, and the more permeable the soil, the more rapid the consolidation.

Typical time-consolidation curves for sand and clay are shown in *Fig. 221.* The conspicuous difference in the character of the two curves is due to the permeability of the sand being higher, and its compressibility much lower than that of the clay. Consequently, the compression of the sand, under constant load, takes place almost instantaneously, while that of the clay may take a very long time (Fig. 221*a*). In addition, the absolute value of the final compression is, normally, much higher for clays than for sands (Fig. 221*b*).

The rate of consolidation of the piston—and—spring model can be computed on the basis of the laws of hydraulics. Such computations would lead to a relationship between the degree of consolidation $\varkappa$ and the elapsed time similar to that indicated by the solid line in *Fig. 222.*

The consolidation curve obtained experimentally in laboratory compression tests usually do not quite fit the theoretical relationship but have a shape shown by the dashed line in Fig. 222. Up to a degree of consolidation of about 80% the two curves run close together, but after that the experimental curve, instead of approaching a horizontal asymptote, continues on a gentle slope. This difference between the two curves is due to what is called the secondary time effect. This additional time lag in compression is connected with the frictional resistance to slipping between the grains undergoing a process of reorientation under the load. A probable explanation for this lies in the very high viscous resistance to deformation of the layers of adsorbed water surrounding the soil particles.

Because of the great importance of the consolidation in the study of the settlement of structures and the stability of foundations and earthworks, the process of primary consolidation, which means the gradual drainage of the pore water out of the soil due to an increase of load, has long been the subject of intensive research. A treatise published by TERZAGHI in 1923, one of the earliest fundamental works on Soil Mechanics, was devoted to this subject and gave the first rigorous theoretical solution to the problem of consolidation, opening up a new field of research in Soil Mechanics.

6.6.2 Theory of one-dimensional consolidation

The basic relationship can be derived on the following assumptions.

1. The voids in the soil are completely saturated with water.
2. Both the water and the individual soil grains are incompressible.
3. The movement of water in the consolidating layer obeys *Darcy*'s law, and the coefficient of permeability is constant.
4. The time lag of the compression is due exclusively to the slow draining of water out of the voids.
5. The compressible layer is laterally confined. At every stage of consolidation, the neutral or the effective stresses are the same at every point of a horizontal section through the layer. Flow of water occurs therefore only in the vertical direction.
6. An increase in the effective stress from the initial value $\bar{\sigma}_0$ to $\bar{\sigma}(\Delta\bar{\sigma} = \bar{\sigma} - \bar{\sigma}_0)$ causes a compression Δh in the thickness of the layer. The compression curve is replaced by a straight line for this internal, thus the compression can be given as

$$\frac{\Delta h}{h} = \varepsilon = \frac{\Delta\bar{\sigma}}{M} \tag{149}$$

where M is the modulus of compressibility.

At the instant of its application, the load induces only neutral stresses in the pore water (saturated soil, the pore pressure coefficient B is equal to 1). As time goes on, an increasing portion of the pore water is squeezed out, whereupon the effective stress in the soil skeleton increases. At any point in the consolidating layer, the value u of the excess neutral stress at a given time is equal to $h\gamma_w$. As long as the applied load remains constant, the sum of the effective and neutral stresses is also constant in every point of the consolidating layer throughout the process of consolidation.

$$\Delta\sigma = \Delta\bar{\sigma} + u\,. \tag{150}$$

Since $\Delta\sigma$ is constant, it will be sufficient to consider the variation of u only.

Figure 223 shows the section of a compressible layer of clay located between two layers of sand. The water is free to escape through both the upper and lower surfaces of the consolidating layer and it moves only in the vertical direction. The total consolidation pressure $\Delta\sigma$ is assumed to be uniform everywhere in the clay layer.

If we insert piezometer pipes at points 1 to 5 on a vertical, similar to those in the model shown in Fig. 220, the water in the pipes will rise to heights corresponding to the excess neutral stresses, induced by the consolidation pressure, at those points. If the pipes are arranged in such a manner that the horizontal distances 1 — 2′, 1 — 3′, etc., are equal to the corresponding vertical distances 1 — 2, 1 — 3, etc., then the line connecting the water levels in the standpipes at a given time is an *isochrone*, which represents the variation of the excess pressure head with the depth.

At time $t = 0$, the intitial or "zero" isochrone is represented by the horizontal line d — e located at an elevation $\Delta\sigma/\gamma_w$ above the top surface of the consolidating layer. As has already been discussed in connection with Fig. 220, the consolidation of a layer starts at the free surface of drainage and gradually proceeds toward the interior. Thus, at an early stage of consolidation, the neutral stresses at the central part of the layer are still unchanged i.e. equal to the total consolidation pressure $\Delta\sigma$, while those at the boundaries have already diminished to zero as shown by the isochrone C_1. At a later stage the neutral stresses have decreased at all depths, but least at the centre of the layer (isochrone C_2). Finally, after a very long time, all the excess neutral stresses

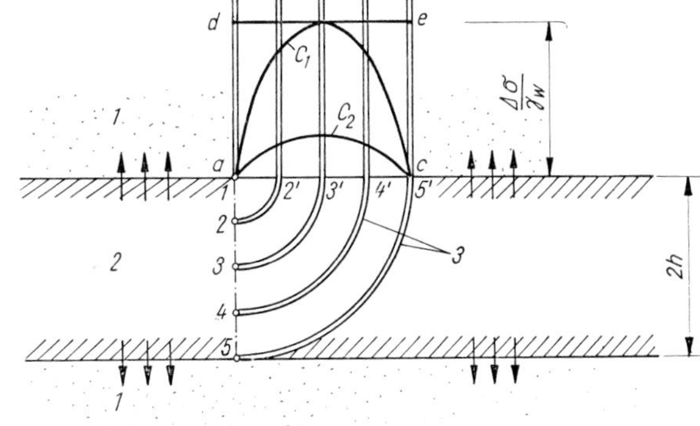

Fig. 223. Isochrones to visualize the neutral stresses acting in the consolidating layer:
1 — sand; *2* — clay; *3* — piezometer tube

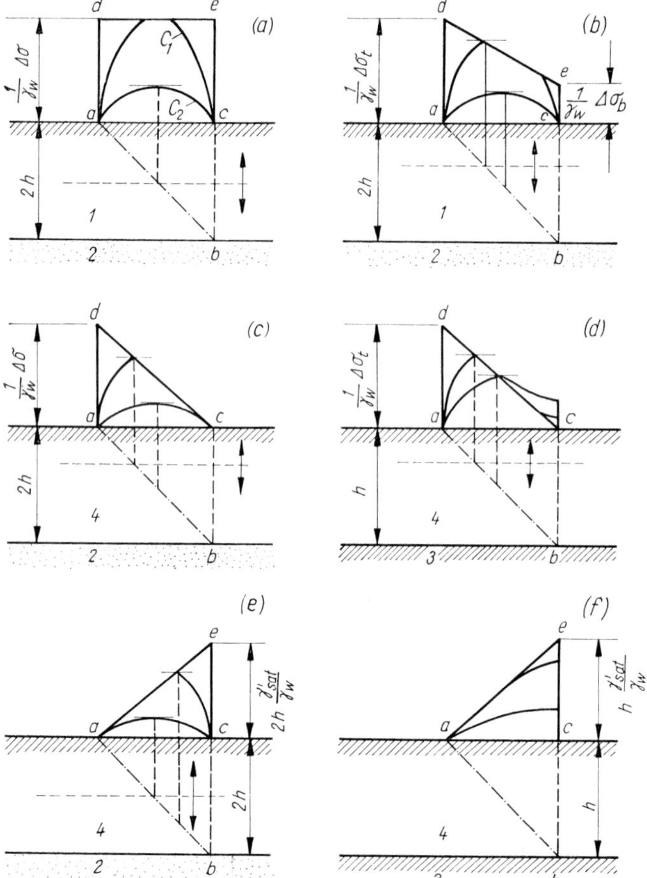

Fig. 224. Isochrones for different boundary conditions: *1, 4, 3* — clay; *2* — sand

have completely dissipated. The final isochrone corresponding to $t = \infty$ is represented by the horizontal line $a - c$.

The case illustrated in Fig. 223 represents only one of the possible boundary conditions, namely, that the layer is free to drain through both of its boundary surfaces. Such a layer is called an *open* layer and its thickness is denoted by $2H$. If the water can escape through only one surface, the layer is termed *half-closed*, and its thickness is H.

Further particular cases of consolidation are illustrated in *Figs 224b* to *f*. If the consolidating layer is thick in comparison with the width of the loaded surface area, the consolidation pressure on a vertical section is no longer uniform but decreases according to the laws of pressure distribution in a continuous medium. Assuming that the decrease of the pressure with depth is linear, the initial isochrone is represented by the line $d - e$ in Fig. 224*b*. The consolidation pressure at the top of the layer is $\Delta\sigma_t$, and that at the bottom is $\Delta\sigma_b$.

If the consolidating layer is very thick, the total pressure $\Delta\sigma$ at the bottom of the layer may be assumed to be zero. For an open layer the isochrones are shown in Fig. 224*c*. In a half-closed layer (Fig. 224*d*), flow of water occurs at first toward both the upper and the lower boundaries causing a temporary swelling of the clay in the lower part of the layer. At a later stage, all excess water is forced to leave the layer through the upper boundary.

Figures 224*e* and *f* represent the cases of hydraulically placed fills. Here the only force acting on the layer is the weight of the soil itself. The free water surface is at the top of the layer. In the case *e* the consolidating fill rests on a permeable sand stratum, hence the layer is open. Case *f* corresponds to a half-closed layer. The consolidation pressure in either case is due to the submerged weight of the soil and, hence, increases from zero at the top to $H\gamma_b$ at the bottom of the layer (γ_b is the submerged unit weight of soil). The initial isochrones (lines $a - e$) and the final isochrones (lines $a - c$) are the same for both cases, but at intermediate stages the isochrones may be very different in shape, and therefore the rate of consolidation is also different for the open and for the half-closed layer.

The differential equation of one-dimensional consolidation for the simple case of a uniformly loaded open layer can be derived as follows (*Fig. 225*).

Consider a thin horizontal slice with the consolidating layer thickness of dz. By *Darcy*'s law, the velocity of flow $v = ki$. The hydraulic gradient i at the elevation of the thin slice is equal to $-\,\mathrm{d}h/\mathrm{d}z$, where d$h$ denotes the drop in hydraulic head across the thickness of the slice. The negative sign indicates that the flow of water takes place upwards, i.e. against the positive direction of the z-axis. During the process of consolidation, the relationship between the hydraulic pressure, the neutral stress increment and the effective stress increment may be written

$$-\gamma_w \Delta h = -\Delta u = \Delta\bar{\sigma} \,. \tag{151}$$

Hence the hydraulic gradient

$$i = -\frac{1}{\gamma_w}\frac{\partial u}{\partial z} \,. \tag{152}$$

The quantity of water that flows per unit time

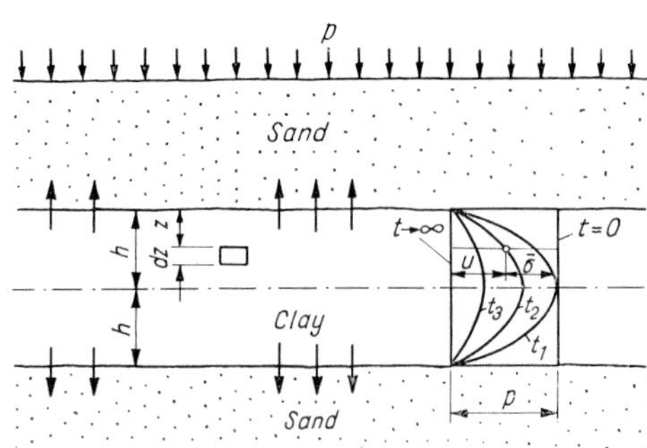

Fig. 225. Vertical section of consolidating layer

through unit area of a horizontal plane located at a depth z below the ground surface is

$$q = Fv = ki = -\frac{k}{\gamma_w}\frac{\partial u}{\partial z}. \qquad (153)$$

For a horizontal plane located higher by an elementary distance dz the rate of flow per unit cross-sectional area is

$$q + \mathrm{d}q = -\frac{k}{\gamma_w}\left(\frac{\partial u}{\partial z} + \frac{\partial}{\partial z}\frac{\partial u}{\partial z}\mathrm{d}z\right).$$

Therefore,

$$\mathrm{d}q = -\frac{k}{\gamma_w}\frac{\partial}{\partial z}\frac{\partial u}{\partial_z}\mathrm{d}z. \qquad (154)$$

Hence the quantity of water squeezed out of unit of volume becomes

$$\frac{\mathrm{d}q}{\mathrm{d}z} = -\frac{k}{\gamma_w}\frac{\partial^2 u}{\partial z^2}. \qquad (155)$$

Making use of the assumption that the quantity of water squeezed out is equal to the decrease in volume of the layer we can write

$$-\frac{k}{\gamma_w}\frac{\partial^2 u}{\partial z^2} = \frac{\partial \varepsilon}{\partial t} \qquad (156)$$

and

$$-\frac{k}{\gamma_w}\frac{\partial^2 u}{\partial z^2} = \frac{1}{M}\frac{\partial \Delta\bar{\sigma}}{\partial t} = -\frac{1}{M}\frac{\partial u}{\partial t}.$$

Finally we have

$$\frac{kM}{\gamma_w}\frac{\partial^2 u}{\partial z^2} = \frac{\partial u}{\partial t}. \qquad (157)$$

The constant part of the above expression may be replaced by a single coefficient c called the coefficient of consolidation. Hence Eq. (157) becomes

$$\boxed{c\frac{\partial^2 u}{\partial z^2} = \frac{\partial u}{\partial t}.} \qquad (158)$$

The above differential equation of one-dimensional consolidation is analogous to the differential equation of heat conduction in an isotropic slab with parallel boundary planes of constant temperature.

From the fundamental relationship

$$u + \bar{p} = p$$

it follows that the effective stresses $\bar{p}$ in a consolidating layer are also governed by Eq. (158).

To solve the differential equation we must find a function $u = f(z,t)$ that satisfies both Eq. (158) and the given boundary conditions.

For the case of one-dimensional consolidation of an open layer under a constant uniform load $\Delta\sigma$ (Fig. 223), the initial and boundary conditions are as follows.

$$\left.\begin{array}{lll} \text{When } z = 0, & u = 0. \\ \text{When } t = 2h, & u = 0. \\ \text{When } t = 0, & u = \Delta\sigma, \\ \text{and as } t \to \infty, & u = 0 \text{ for all values} \\ & 0 \leq z \leq 2h. \end{array}\right\} \qquad (159)$$

Assume, after *Bernoulli*, that the required function $u = f(z, t)$ may be expressed as a product of some function of z and some function of t, i.e.

$$u = F(z)\,\Phi(t). \qquad (160)$$

Equation (158) may then be written

$$c\Phi(t)\,F''(z) = F(z)\,\Phi'(t), \qquad (161)$$

or

$$\frac{F''(z)}{F(z)} = \frac{\Phi'(t)}{c\Phi(t)}.$$

Since the left-hand side contains only functions of z and the right-hand side contains only functions of t, equality for all values of z and t is possible only if each of the two sides is equal to a constant. Let this constant be denoted by $-A^2$.

Hence from the left-hand side of Eq. (161) we obtain

$$F''(z) = -A^2 F(z).$$

The solution of this is

$$F(z) = C_1 \cos Az + C_2 \sin Az.$$

The right-hand side of Eq. (161) gives

$$\Phi'(t) = -A^2 c\Phi(t).$$

Integrating,

$$\Phi(t) = C_3 e^{-A^2 ct}.$$

Thus the general solution of the differential equation becomes

$$u = (C_4 \cos Az + C_5 \sin Az)\,e^{-A^2 ct}. \qquad (162)$$

Particular solutions for concrete cases can be derived by making use of the given boundary conditions. For the case in question—open layer plus uniform pressure distribution throughout the depth of the consolidating layer—the boundary conditions were given under (159). The first condition is satisfied if $C_4 = 0$, and the second one is satisfied if $2A = n\pi$, wherein n is any integer. This leaves

$$u = C_5 \sin\frac{n\pi z}{2h}\,e^{-n^2\pi^2 ct/4h^2}. \qquad (163)$$

C_5 is an arbitrary constant, and can assume any integer value. Therefore an infinite series of the form

$$u = B_1 \sin \frac{\pi z}{2h} e^{-\pi^2 ct/4h^2} + B_2 \sin \frac{2\pi z}{2h} e^{-4\pi^2 ct/4h^2} + \dots$$

$$+ \dots B_n \sin \frac{n\pi z}{2h} e^{-n^2\pi^2 ct/4h^2} + \dots =$$

$$= \sum_{n=1}^{\infty} B_n \sin \frac{n\pi z}{2h} e^{-n^2\pi^2 ct/4h} \qquad (164)$$

is also a solution. Here B_1, $B_2 \dots$, B_n are constants.

In order to satisfy the third boundary condition, the constants B_n are determined so that

$$u_0 = \sum_{n=1}^{\infty} B_n \sin \frac{n\pi z}{2h}.$$

The terms of the right-hand side series constitute a family of orthogonal functions. The constants may be readily determined with the aid of the following set of standard definite integrals.

$$\int_0^{\pi} \sin mx \sin nx \, dx = 0$$

and

$$\int_0^{\pi} \sin^2 nx \, dx = \frac{\pi}{2},$$

wherein m and n are unequal integers.

Putting $\pi z/2h$ for x gives

$$\int_0^{2h} \sin \frac{m\pi z}{2h} \sin \frac{n\pi z}{2h} \, dz = 0$$

and

$$\int_0^{h} \sin^2 \frac{n\pi z}{2h} \, dz = h.$$

If both sides of Eq. (164) are multiplied by

$$\sin (n\pi z/2h) \, dz$$

and integrated between 0 and $2h$, all terms in the series except the nth term will assume the form of the first definite integral above and vanish; whereas the nth term will be of the form of the second definite integral.

Hence

$$\int_0^{2h} u_0 \sin \frac{n\pi z}{2h} dz = B_n \int_0^{2h} \sin^2 \frac{n\pi z}{2h} dz = B_n h$$

and

$$B_n = \frac{1}{h} \int_0^{2h} u_0 \sin \frac{n\pi z}{2h} dz.$$

Substituting this value in the solution of the differential equation, Eq. (164), gives

$$u = \sum_{n=1}^{\infty} \left(\frac{1}{h} \int_0^{2h} u_0 \sin \frac{n\pi z}{2h} dz \right) \left(\sin \frac{n\pi z}{2h} e^{-n^2\pi^2 ct/4h^2} \right). \qquad (165)$$

By introducing the dimensionless quantity

$$T = \frac{ct}{h^2} = \frac{k}{\gamma_w m_v} \frac{t}{h^2} \qquad (166)$$

called the time factor and integrating we get

$$u = \sum_{n=1}^{\infty} \frac{2u_0}{n\pi} (1 - \cos n\pi) \left(\sin \frac{n\pi z}{2h} \right) e^{-\frac{1}{4} n^2 \pi^2 T}.$$

When n is an even number, $1 - \cos n\pi = 0$. When n is odd, $1 - \cos n\pi = 2$. Therefore it is convenient to let $n = 2N + 1$. The above expression then becomes

$$u = \frac{4}{\pi} u_0 \sum_{N=1}^{\infty} \frac{1}{2N+1} \cdot$$

$$\cdot \sin \left[\frac{(2N+1)\pi z}{2h} \right] e^{-(2N+1)^2 \pi^2 T/4}. \qquad (167)$$

From this solution for the neutral stresses the effective stresses can also be readily obtained. By use of Eq. (167), we can compute the stresses at any depth within the consolidating layer and any time during the process. The derivation was based on the assumption of an open layer, but the results are easily applicable to a half-closed layer as well, with the only modification that the

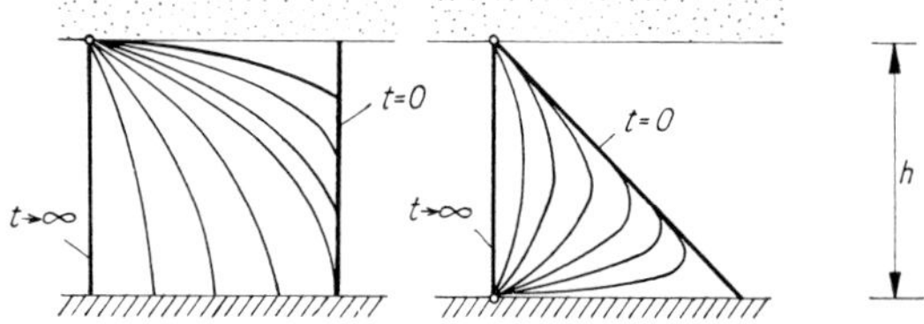

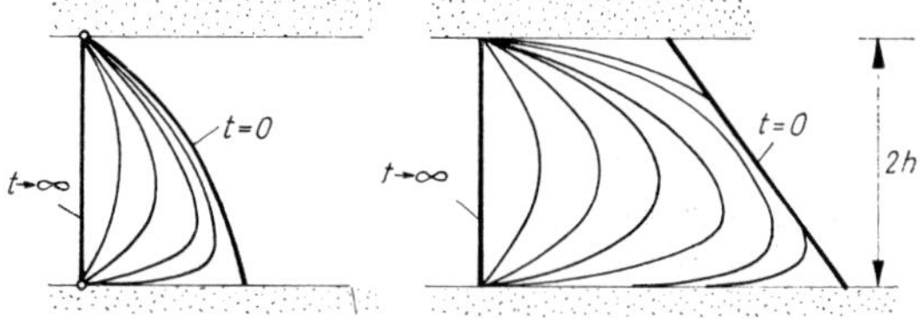

Fig. 226. Isochrones to given initial and boundary conditions

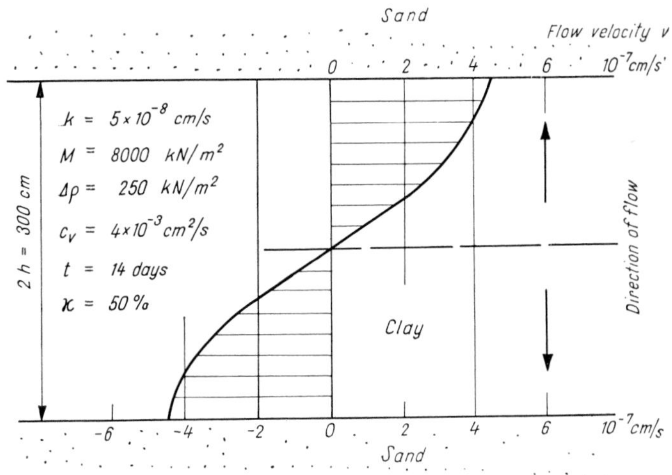

Fig. 227. Velocity distribution in the vertical section of the consolidating layer

vertical distance h used in the solution must mean the total thickness of the layer.

From Eq. (167) the isochrones for any time t can also be determined. Examples for some initial and boundary conditions are presented in *Fig. 226*. The zero isochrones for these cases, and others, are summarized later in Fig. 229. Finally, *Fig. 227* gives a numerical example of the distribution of the velocity of water on a vertical section within a consolidating layer.

6.6.3 Computation of rate of consolidation

As was shown in the preceding derivation, the process of consolidation can be described in terms of the dimensionless time factor

$$T = \frac{ct}{h^2} = \frac{k}{m_v \gamma_w} \frac{t}{h^2}. \tag{166}$$

The solution giving the neutral stresses at any depth z and at any time t [Eq. (167)] can be expressed in the general form

$$u = f(z, T). \tag{167a}$$

As a next step, we determine the relationship between the time factor and the degree of consolidation $\varkappa$. To this end, let us consider the compression of the consolidating layer. The effective pressure acting on an elementary slice with thickness dz at time t is $\bar{\sigma} = \sigma_1 - u$. The compression of this elementary slice is

$$\mathrm{d}\,\Delta h = m_v(\sigma_1 - u)\,\mathrm{d}z. \tag{168}$$

Thus the total compression of the layer after time t may be written

$$\Delta h_t = \int_0^h m_v(\sigma_1 - u)\,\mathrm{d}z = m_v(\sigma_1 h - \int_0^h u\,\mathrm{d}z). \tag{169}$$

Substituting u from Eq. (167) and integrating,

$$\Delta h_t = m_v \sigma_1 h \left[1 - \frac{8}{\pi^2} \sum_{N=0}^{\infty} \frac{1}{(2N+1)^2} \mathrm{e}^{-(2N+1)^2\pi^2/4T}\right]. \tag{170}$$

After a very long time ($t = \infty$) the final compression of the layer becomes

$$\Delta h = m_v \sigma_1 h$$

and the corresponding degree of consolidation in percent is

$$\varkappa\% = 100 \frac{\Delta h_t}{\Delta h} = 100 f(T). \tag{171}$$

It may be seen that the function $f(T)$ is equivalent to the bracketed expression in Eq. (170). This means that the degree of consolidation depends only on the boundary conditions and the time factor. The function $f(T)$, which also depends on the boundary conditions, has been determined for many cases of loading, and only special conditions warrant an elaborate computation on the basis of Eq. (170). The function $\varkappa = f(T)$ for specified conditions can once and for all be evaluated and presented in the form of graphs or tables. With the aid of these, almost every problem of one-dimensional consolidation likely to be met in practice can be readily solved so that only the coefficient of the time t in the expression of the time factor needs to be determined. Such a graph presenting the solutions for the loading conditions illustrated by Fig. 224 is shown in *Fig. 228*.

For every open layer with thickness $2h$ the relationship between $\varkappa$ and T is represented by the curve C_1, regardless of the slope of the zero isochrone. Curve C_1 applies therefore to the cases shown in Figs 224*a*, *c* and *e*. If the zero isochrone is horizontal, which indicates a uniform pressure distribution with depth in the consolidating layer,

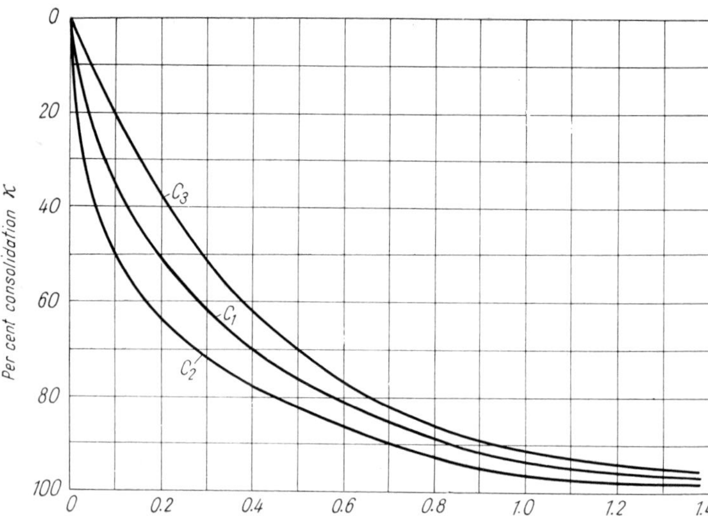

Fig. 228. Consolidation ratio versus time factor

curve C_1 also represents the process of consolidation for a half-closed layer with thickness h. The following numerical example illustrates the use of the graph.

Example. A layer of clay with thickness $2h = 3.0$ m is located between two permeable layers of sand (open layer). The coefficient of permeability of the clay is $k = 5\times10^{-10}$ m/s and its modulus of compressibility is $M = 8000$ kN/m². How much compression will the clay layer experience under a load increment of $\Delta p = 250$ kN/m², and how much time will elapse before the settlement becomes 20, 50 and 80% of its final value?

First we calculate the total settlement:

$$\Delta h = 2h\frac{\Delta p}{M} = 3.0\frac{250}{8000} = 0.094 \text{ m}.$$

According to Eq. (157), the coefficient of consolidation is

$$c = \frac{kM}{\gamma_w} = \frac{5\times10^{-10}\text{ m/s}\times8\times10^3\text{ kN/m}^2}{10\text{ kN/m}^3} = 4\times10^{-7}\text{ m}^2/\text{s}.$$

From $T = ct/h^2$

$$t = T\frac{h^2}{c} = T\frac{1.5^2\text{ m}^2}{4\times10^{-7}\text{ m}^2/\text{s}} = 5.62\times10^6\,T\text{ s}$$

and since

$$1\text{ month} = 30\times24\times60\times60 = 2.59\times10^6\text{ s}$$

therefore

$$t = 2.16\,T\text{ months}.$$

Table 24. Time factor and degree of consolidation

Time factor T	Degree of consolidation for the cases shown in Fig. 229				
	$\varkappa_1$	$\varkappa_2$	$\varkappa_3$	$\varkappa_6$	$\varkappa_1-\varkappa_2$
0.004	0.071	0.008	0.135	0.009	0.063
0.008	0.101	0.016	0.186	0.019	0.085
0.012	0.124	0.024	0.223	0.029	0.100
0.020	0.160	0.040	0.279	0.048	0.120
0.028	0.189	0.056	0.322	0.067	0.133
0.036	0.214	0.072	0.356	0.085	0.142
0.048	0.247	0.096	0.398	0.112	0.151
0.060	0.276	0.120	0.433	0.138	0.156
0.072	0.303	0.144	0.462	0.163	0.159
0.100	0.357	0.197	0.516	0.218	0.160
0.125	0.399	0.244	0.554	0.265	0.155
0.167	0.461	0.318	0.605	0.338	0.143
0.20	0.504	0.370	0.638	0.388	0.134
0.25	0.562	0.443	0.682	0.460	0.119
0.30	0.613	0.508	0.719	0.523	0.105
0.35	0.658	0.565	0.752	0.588	0.093
0.40	0.698	0.615	0.780	0.627	0.083
0.50	0.764	0.700	0.829	0.708	0.064
0.60	0.816	0.765	0.866	0.773	0.051
0.80	0.887	0.857	0.918	0.861	0.030
1.0	0.931	0.913	0.950	0.915	0.018
2.0	0.994	0.993	0.996	0.993	0.001
∞	1.000	1.000	1.000	1.000	0.000

Reading off the T values for $\varkappa = 20$, 50 and 80% respectively on curve C_1, we get 0.031, 0.19 and 0.57. Hence the corresponding times are $t = 0.067$, 0.41 and 1.23 months.

Conversely, it may be required to find the degree of consolidation after a given time. If $t = 1$ month the time factor becomes

$$T = \frac{t(\text{months})}{2.16} = \frac{1}{2.16} = 0.462$$

and the required degree of consolidation is

$$\varkappa = 74\%.$$

For a half-closed layer with thickness $h = 1.50$ m the same results would be obtained.

If the zero isochrone for a half-closed layer has the shape of a triangle with pressure decreasing from p at the top to zero at the bottom of the layer as shown in Fig. 224*d*, the relationship between $\varkappa$ and T is given by the curve C_2 In the case of a hydraulic fill, in which the pressure is zero at the surface and it increases linearly to p_b at the impervious base (Fig. 224*f*), the curve C_3 applies. For intermediate cases we can interpolate between the curves.

The time factor can also be obtained from *Table 24*, which contains corresponding values of T and $\varkappa$ for six different conditions of loading. The subscripts from 1 to 6 refer to the pressure distribution diagrams illustrated, for a half-closed layer, by *Fig. 229*. For the cases 1, 2, 3 and 6 $\varkappa$ values as functions of T can be obtained directly from the table. For the cases 4 and 5, $\varkappa$ values are calculated by the following formulae:

$$\varkappa_4 = \varkappa_1 - \frac{1 - \xi}{1 + \xi}(\varkappa_1 - \varkappa_2) \tag{172}$$

and

$$\varkappa_5 = \varkappa_1 - \frac{\xi - 1}{\xi + 1}(\varkappa_1 - \varkappa_2), \tag{173}$$

wherein ξ denotes the ratio of the pressure at the free surface of drainage to that at the impervious boundary. To facilitate the numerical work, the differences $\varkappa_1 - \varkappa_2$ are also given in Table 24. So for every open layer with any type of linear pressure distribution the $\varkappa_1$ values apply. Note that in an open layer h means the half-thickness of the layer.

A mathematical analysis has shown that up to a degree of consolidation of $\varkappa = 52.6\%$ the curve C_1 can be replaced by a parabola having the equation

$$T = \frac{\pi}{4}\varkappa^2;$$

substituting ct/h^2 for T, we get

$$\varkappa = \sqrt{\frac{4C}{\pi h^2}}\sqrt{t}, \tag{174}$$

i.e. the degree of consolidation changes in direct proportion to the square root of the time. If $\varkappa$ is greater than 52.6%, a fair approximation to the theoretical curve C_1 can be attained by the formula

$$T = -0.9332 \log(1 - \varkappa) - 0.0851.$$

As we have seen, the consolidation theory described in the preceding section is based on a number of simplifying assumptions that are often satisfied only in part. For that reason, no matter how complicated mathematical approach we use, the results obtained by calculation can be regarded merely as a crude estimate giving at best only the order of magnitude. A further major discrepancy between theory and reality is due to the secondary time effect. (See Section 6.6.1 and also Vol. 2.)

The results of computations of rate of settlement are not even approximately correct unless the assumed hydraulic boundary conditions are satisfied in reality. If a bed of clay contains pervious seams of sand that are connected to water-bearing strata adjacent to the clay, the process of consolidation will be substantially accelerated. In contrast with this, pockets and lenses of sand and silt entirely surrounded by the clay have no draining effect. But it may prove extremely difficult to ascertain, on the basis of test borings, whether or not these seams of sand and silt are interconnected. In case of such uncertainties the consolidation theory can be used only for determin-

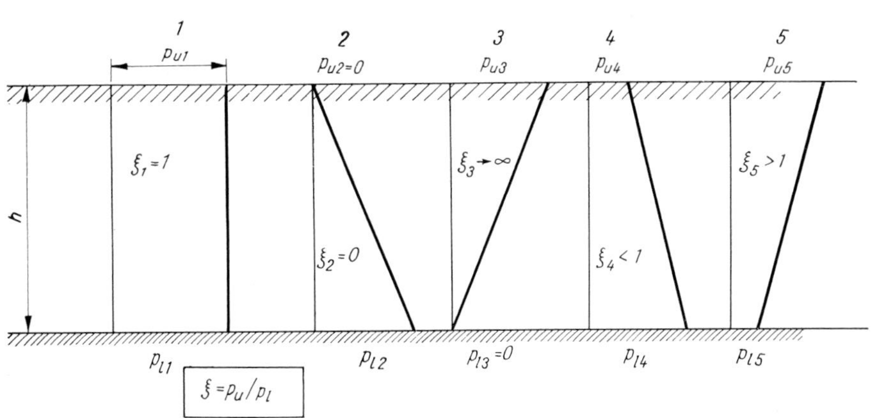

Fig. 229. Zero isochrones for the cases listed in Table 27

ing an upper and a lower limiting value of consolidation. The real time rate of settlement can then be determined only by means of direct measurements on the completed structure.

The differential equation of one-dimensional consolidation has been derived on the assumption that the total load causing settlement is applied to the soil instantaneously. In reality, however, the stage of consolidation under constant load is normally preceded by a period during which the load is gradually increased to the final value while consolidation is progressing simultaneously. Since the consolidation due to a small increment of load is independent of both the consolidation already in progress and the consolidation due to any subsequent load increment, the degree of consolidation during the loading period can be determined to any desired degree of accuracy by superposition. An approximate method advanced by *Terzaghi* is as follows.

Let us consider for example the compression of an open layer of clay with thickness $2h$ under a load increasing linearly during the construction period from zero to a final value p_1. The loading diagram is as shown by the broken line Oab in *Fig. 230*. The consolidation curve C_1 (dashed line) in the lower part of the figure corresponds to the instantaneous loading. As a practical approximation it may be assumed that the degree of consolidation at some time t when the load equals p is the same as that which would have resulted in time $t/2$ under a constant load p throughout. At time $t/2$, the degree of consolidation due to an instantaneous load p_1 would be $\varkappa'$. Thus for the real case the degree of consolidation at time t is given by

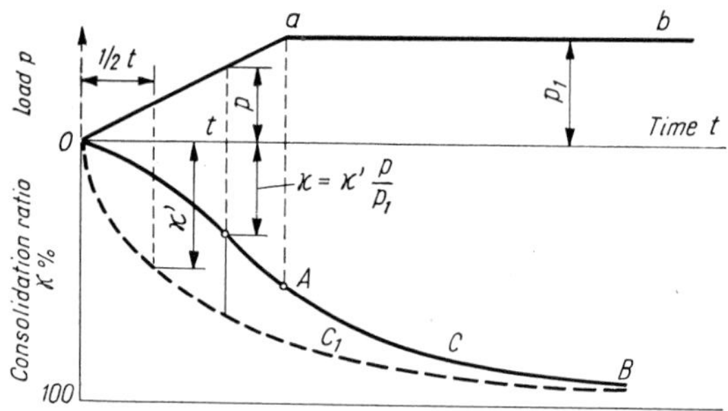

Fig. 230. Construction of consolidation curve for linearly increasing load

$$\varkappa = \varkappa' \frac{p}{p_1}.$$

By plotting the $\varkappa$ value against t as abscissa and repeating the procedure for several values of t, we can construct the part OA of the real consolidation curve C (solid line). Beyond point A the curve C is identical to the consolidation curve that would have resulted from an instantaneous load p_1 applied at time $t_1/2$. The part AB of the curve C is therefore obtained by simply transferring the instantaneous curve C_1 by an amount of $t_1/2$.

6.6.4 Three-dimensional consolidation

In the consolidation problem considered in Section 6.6.2 we assumed that the state of stress was the same in any vertical section and that flow was only in the vertical direction. This case was termed the *one-dimensional* consolidation. In the following we shall briefly deal with the *three-dimensional* case, when the flow occurring through an element of soil has three different velocity components in the coordinate directions x, y, z. Let these components at a point with coordinates (x, y, z) be v_x v_y and v_z (*Fig. 231*). The flow velocities at the faces of the element, at some small distances $d_x/2$ and $d_z/2$ from that point, are also shown in the figure.

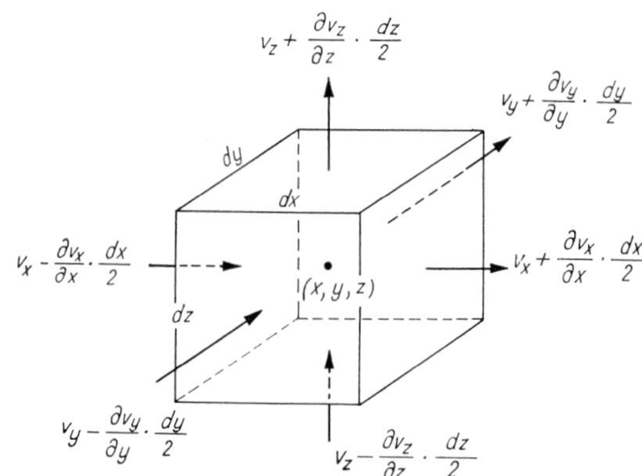

Fig. 231. Seepage velocities on the sides of the elementary cube

The flow per unit of time into the element may be written

$$q = \left(v_x - \frac{\partial v_x}{\partial x}\frac{dx}{2}\right) dy\,dz + \left(v_y - \frac{\partial v_y}{\partial y}\frac{dy}{2}\right) dx\,dz + \left(v_z - \frac{\partial v_z}{\partial z}\frac{dz}{2}\right) dx\,dy. \tag{175a}$$

The flow out of the element is

$$q' = \left(v_x + \frac{\partial v_x}{\partial x}\frac{dx}{2}\right) dy\,dz + \left(v_y + \frac{\partial v_y}{\partial y}\frac{dy}{2}\right) dx\,dz + \left(v_z + \frac{\partial v_z}{\partial z}\frac{dz}{2}\right) dx\,dy. \tag{175b}$$

In the following we use the same basic assumptions as in the discussion of one-dimensional consolidation, notably that

(1) the consolidating layer is saturated with water;
(2) the water and the solid particles of soil are incompressible;
(3) the volume change is small as compared with the initial volume of the soil.

The change of volume can then be expressed as the difference between the flow out of the element and that into the element: $q' - q$. According to the third assumption

$$\frac{\partial V}{\partial t} = \left(\frac{\partial v_x}{\partial x} + \frac{\partial v_y}{\partial y} + \frac{\partial v_z}{\partial z}\right) dx\,dy\,dz. \tag{176}$$

If V_s is the total volume of the solid particles contained in a volume of soil V, and e is the void ratio, then

$$V = V_s(1 + e) = dx\,dy\,dz\,.$$

Differentiating with respect to t gives

$$\frac{\partial V}{\partial t} = V_s \frac{\partial e}{\partial t}\,.$$

Considering that

$$V_s = \text{constant} = \frac{1}{1+e}\,dx\,dy\,dz\,,$$

the rate of change of volume becomes

$$\frac{\partial V}{\partial t} = \frac{1}{1+e}\,\frac{\partial e}{\partial t}\,dx\,dy\,. \tag{177}$$

Equating the right-hand sides of Eqs (176) and (177) gives

$$\frac{\partial e}{\partial t} = (1+e)\left(\frac{\partial v_x}{\partial x} + \frac{\partial v_y}{\partial y} + \frac{\partial v_z}{\partial z}\right). \tag{178}$$

Assuming further that

(4) *Darcy*'s law holds, and
(5) its validity can be extended to anisotropic soils;

the three velocity components are obtained as follows.

$$\left.\begin{aligned} v_x &= \frac{k_x}{\gamma_w}\,\frac{\partial u}{\partial x}\,,\\ v_y &= \frac{k_y}{\gamma_w}\,\frac{\partial u}{\partial y}\,,\\ v_z &= \frac{k_z}{\gamma_w}\,\frac{\partial u}{\partial z}\,. \end{aligned}\right\} \tag{179}$$

If k_x, k_y and k_z are independent of the coordinates x, y and z (6th assumption: homogeneous medium), then by differentiating the first expression under (179) with respect to x, the second one with respect to y and the third one with respect to z, and substituting in Eq. (178), we get

$$\frac{\partial e}{\partial t} = \frac{(1+e)}{\gamma_w}\left(k_x \frac{\partial^2 u}{\partial x^2} + k_y \frac{\partial^2 u}{\partial y^2} + k_z \frac{\partial^2 u}{\partial z^2}\right). \tag{180}$$

Let us denote the initial value of the effective stress by $\bar{\sigma}_0$, the total stress increment by $\Delta\sigma$ and the initial neutral stress due to $\Delta\sigma$ by u_0. At time t, the effective stress is $\bar{\sigma}$ and the neutral stress is u. Hence

$$\bar{\sigma}_0 + u_0 = \bar{\sigma} + u\,, \tag{181}$$

or

$$\bar{\sigma} = \bar{\sigma}_0 + u_0 - u\,.$$

We shall now consider the case that the total stress increment $\Delta\sigma$ is applied instantaneously (7th assumption). u_0 is therefore independent of time. Differentiating Eq. (181) with respect to t gives

$$\frac{\partial \bar{\sigma}}{\partial t} = -\frac{\partial u}{\partial t}\,. \tag{182}$$

It follows from Eq. (181) that any increase in σ will give rise to an equal decrease in u. Thus

$$\frac{\partial e}{\partial \bar{\sigma}} = -\frac{\partial e}{\partial u}\,. \tag{183}$$

According to Eq. (149)

$$a_v = -\partial e/\partial \bar{\sigma}\,.$$

Substituting in to Eq. (183),

$$\frac{\partial n}{\partial u} = a_v\,.$$

Since

$$\frac{\partial e}{\partial t} = \frac{\partial e}{\partial u}\,\frac{\partial u}{\partial t}\,, \tag{184}$$

we may write

$$\frac{\partial e}{\partial t} = a_v \frac{\partial u}{\partial t}\,. \tag{185}$$

Combining Eqs (180) and (185) leads to

$$\frac{\partial u}{\partial t} = \frac{1+e}{a_v\,\gamma_w}\left(k_x \frac{\partial^2 u}{\partial x^2} + k_y \frac{\partial^2 u}{\partial y^2} + k_z \frac{\partial^2 u}{\partial z^2}\right). \tag{186}$$

Equation (186) is the three-dimensional counterpart of Eq. (158). It can be simplified by substituting the coefficient of consolidation

$$c_v = \frac{k(1+e)}{a_v\,\gamma_w}\,.$$

Equation (186) then reads

$$c_{vx}\frac{\partial^2 u}{\partial x^2} + c_{vy}\frac{\partial^2 u}{\partial y^2} + c_{vz}\frac{\partial^2 u}{\partial z^2} = \frac{\partial u}{\partial t}\,. \tag{186a}$$

Solutions to the above equation have been developed for some very special cases by Carslaw (1935) and Barron (1948). However, exact solutions for boundary conditions commonly encountered in engineering practice are rather rare. Numerical methods using the finite element method have been worked out by Scott (1963).

The theory of consolidation is useful in calculating the rate of settlement for structures. Practical methods of settlement calculations will be given in Vol. 2.

6.7 Movement of soil water due to electric current

In recent years, soil response to electric current has aroused a growing interest among researchers mainly on account of the wide scope of promising prospective applications. But the study of electro-osmosis and other electrochemical phenomena has so far yielded results that are contradictory in many respects.

The phenomenon called *electro-osmosis* became known in physics as early as the beginning of the 19th century (REUSS, 1808). Essentially this term refers to the flow of water under the action of a direct current form the positive electrode toward the negative electrode (*Fig. 232*). The cause of electro-osmosis is the potential between two adjacent media. As is well known, an electric potential difference exists at the interface of two different substances. The two bodies in contact may be likened to a plate capacitor. The substance with the greater dielectric constant is positively charged. The described phenomenon is produced also when water is brought into contact with soil. In a capillary tube (*Fig. 233*) for example, there is an electrostatic attraction acting on the electrons of both substances within the electric double layer adjacent to the liquid–solid interface. This gives rise to a potential difference known as the corpuscular electrokinetical potential. The electric double layer acts as a capacitor. If a potential difference is produced by means of electrodes between the two ends of a capillary tube, the electrically charged fluid particles begin to move under the action of the electrical force field. The migration of the water particles normally occurs from the positive electrode to the negative one; since the water, having a high dielectric constant, carries positive charges. This process can be produced also in interconnected capillary systems such as soils, and can be used for drainage of soil

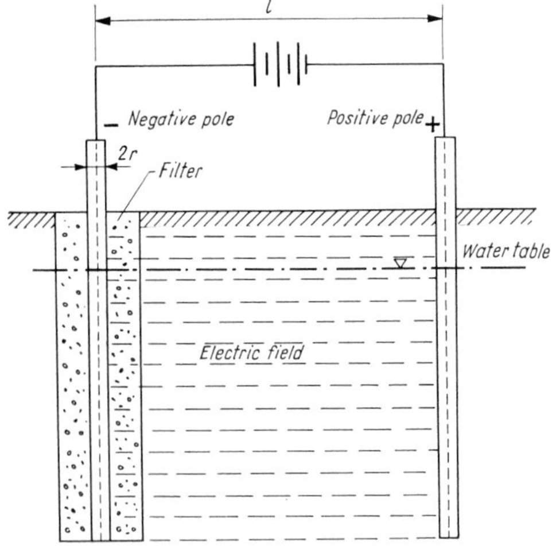

Fig. 232. Electrically activated well

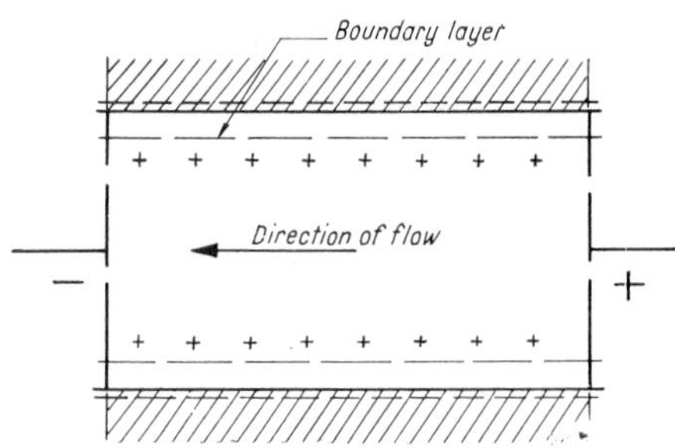

Fig. 233. Liquid flow in a capillary tube caused by electrical potential drop

or, generally, for conveying liquids through fine-grained porous media by electric current.

HELMHOLTZ, PERRIN and SMOLUHOVSKI found that the mean velocity v at which water flows under the action of electric current through a cylindrical tube is given by the formula

$$v = \zeta \frac{\varrho D}{4\pi\eta} I . \tag{187}$$

Herein

ζ = corpuscular electrokinetical potential
ϱ = specific resistance of the liquid
D = dielectric constant of the liquid
η = kinematic viscosity of the liquid
I = mean current density across the capillary tube.

If, in addition to the flow of water produced by electric current, there is also a hydraulic head between the two ends of a capillary tube, the two kinds of effect are superimposed.

Considering that in the above formula the product ϱI means the field intensity (i.e. the potential drop per unit of length of the pipe), the quantity of water per unit time that flows by electro-osmosis under a potential difference V between two electrodes having a surface area A and spaced at a distance l is given as

$$q = vF = \zeta \frac{D}{4\pi\eta} \frac{V}{l} F = k_E EF .$$

For a surface area of $A = 1$ the velocity expression becomes

$$v = k_E E . \tag{188}$$

It is analogous to *Darcy*'s law; k_E (cm²/volts sec) is called the *coefficient of electro-osmotic permeability*. k_E means the velocity of flow under a unit electric potential gradient ($E = 1$ volt/cm) and zero hydraulic gradient ($i = 0$). In contrast to the coefficient of permeability k, the coefficient k_E varies little with the soil type. For fine-grained soils it is of the order of $k_E = 5 \times 10^{-5}$ cm²/volts sec (L. CASAGRANDE, 1969).

The electro-osmosis is greatly influenced by the cations (Na, Ca, Cl etc.) adsorbed on the clay particle surfaces. It may also be affected by the exchange capacity of clays, by the material of the electrodes and by various salts dissolved in the ground water. Investigations by SHUKLA (1953) have shown that the treatment of clays by an electric current results in a decrease in the liquid limit and the plastic limit and an increase in permeability. L. CASAGRANDE (1947) found that a remoulded clay may regain its natural cloddy or mosaic structure under the action of the electric current.

A marked electro-osmosis occurs only if the soil is fine-grained enough so that the total wetted particle surface is comparatively large, since the electrokinetical potential is proportional to the specific surface of the soil. If the pores are relatively large, the galvanic effect causing electrolysis becomes predominant, and the electric current passed through the soil can no longer be used for drainage by electro-osmosis.

6.8 Flow of water by heat action

Some early observations (PLATTEN, 1909; BOUYOUCOS, 1915) suggested that temperature differences in the soil also could induce flow of water. For example the distribution of soil moisture beneath a road pavement was found to have an entirely different pattern from that in the adjacent soil masses. Large buildings exert a heat insulating effect with the result that moisture accumulates in the underlying soil. Infiltration of precipitation water, hindered evaporation etc. may account for such anomalies but another factor not at all negligible is the migration of water under a thermal gradient. This process is known as *thermo-osmosis*. HABIB and SOEIRO (1953 and 1957) gave experimental evidence of thermo-osmosis and established some basic relationships. They used the experimental setup shown in *Fig. 234*. A cylindrical specimen 11 cm high and 25 cm² in cross-sectional area was placed between a hot and a cold water bath. The tested soil was a silt from Orly ($w_L = 34\%$, $w_P = 19\%$). The water passing through the sample contained potassium iodide with some radioactive iodine. Before the test the moisture distribution, and hence the distribution of the radioactivity, were uniform throughout the sample. During the test the sample was subjected to a constant thermal gradient, and after 8 days the distributions of the moisture content and the radioactivity were determined again. The moisture content after the test showed a decrease on the hot side and an increase on the cold side, whereas the radioactivity was increasing with increasing temperature. This proves that the soil moisture moves, in the form of water vapour, from the hot to the cold side where it collects as condense water while the water in liquid phase moves in the opposite directions (*Fig. 235*). This is a probable explanation of the fact that in early winter before frost sets in, a steady decrease in ground temperature gives rise to intense fog formation. If both phases are free to move, a dynamic equilibrium is established. If either of the phases is restricted in its movement, a static equilibrium results. The latter condition arises, for example, at low water contents of the soil, when the liquid water exists in an adsorbed state; or in a saturated soil with no air-filled voids open for vapour transfer. A typical result of the cited experiments is shown in *Fig. 236*. The measure of water and vapour transfer depends on the initial moisture content of the soil. There is a certain degree of saturation at which

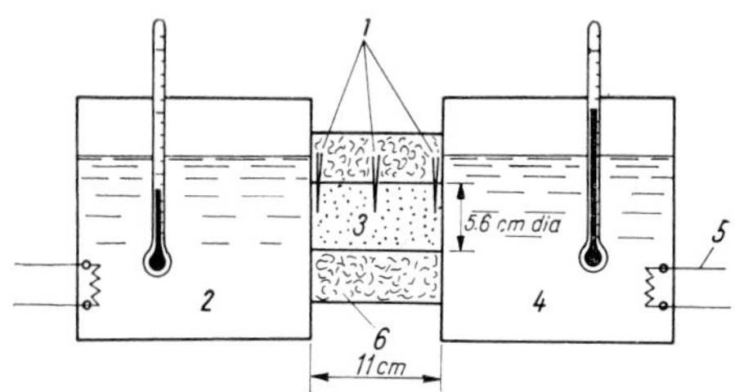

Fig. 234. Test arrangement for the investigation of thermo-osmosis:
1 — thermoelectric couples; 2 — cold bath; 3 — soil sample; 4 — warm bath; 5 — heating coils; 6 — heat insulation

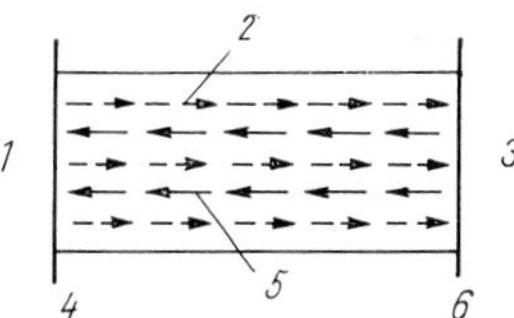

Fig. 235. Movement of water vapour and water in soils:
1 — warm side; 2 — water vapour; 3 — cold side; 4 — moisture content drops; 5 — water; 6 — moisture content increases

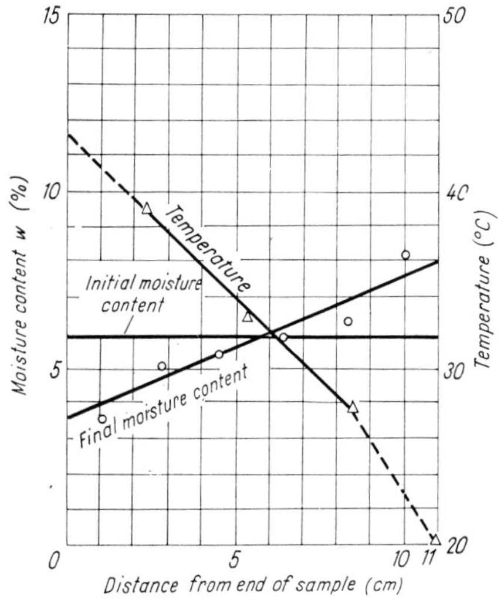

Fig. 236. Distribution of moisture content under a thermal gradient

the intensity of the migration of vapour toward the cold side is a maximum.

The results of the experiments permitted some conclusions to be drawn concerning the accumulation of soil moisture below buildings and road pavements. If the surface temperature is lower than that at greater depths of the soil, the water moves upwards, and it moves downwards under reverse conditions. If the cycles of the seasonal fluctuation in soil temperature show a marked asymmetry with much longer periods of cooling than warming, the moisture tends to accumulate in the upper zones of the soil. Should the free evaporation be prevented by an insulating layer, e.g. a road pavement, the wetting of the subsoil will steadily increase from year to year.

The temperature of the soil is primarily dependent on the amount of solar radiation reaching the earth. But the water may, owing to its higher specific heat, also convey a substantial amount of heat to the soil either through precipitation or through condensation. Another source of heat is the exothermic decomposition of organic compounds in the soil. Data for the specific heat and the heat conductivity of the constituent parts of the soil are given in *Table 25*. Specific heat is

Table 25. Thermal properties of soil constituents

Soil constituent	Specific heat kcal/kg °C	Thermal conductivity cal/°C cm s
Air		5.6×10^{-5}
Water	1	140×10^{-5}
Quartzite	0.19	1600×10^{-5}
Sand	0.20	80×10^{-5}
Clay	0.23	500×10^{-5}

defined as the amount of heat, in calories, required to raise by one centigrade the temperature of one gram of a substance. It can be seen that the specific heat of the mineral constituents is only about one-fifth of that of the water. It is therefore largely owing to its moisture content that the soil is capable of storing heat.

The heat conduction in soil can be described for the one-dimensional case by the *Fourier* equation. According to this

$$\lambda \frac{\partial^2 T^0}{\partial z^2} = c\gamma \frac{\partial T^0}{\partial t}. \tag{189}$$

Herein

T^0 = temperature at time t and at depth z, in °C

λ = thermal conductivity, i.e. the amount of heat that flows under a temperature gradient $\partial T^0/\partial z = 1$ in unit of time through unit of cross-sectional area, in g cal per hour per cm per °C

c = specific heat.

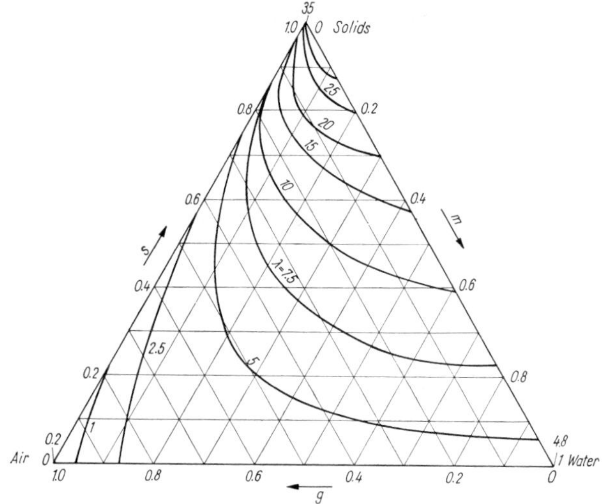

Fig. 237. Coefficient of thermal conductivity of a soil as a function of the phase composition

The variation of the thermal conductivity λ with the phase composition of the soil is shown in the form of a triangular diagram, in *Fig. 237*. The diagram clearly demonstrates that heat flows primarily through the solid grains and that the high air content is responsible for an increased heat insulation.

A practical application of Eq. (189) is in the study of frost penetration into the soil. We may examine, for instance, for a given variation of surface temperature, to what depth the frost penetrates into the soil and what fluctuations in soil temperature may be expected. Assuming for example a sinusoidal variation of surface temperature, the fluctuation of soil temperature at greater depths will be as shown for different times by the curves in *Fig. 238*.

Below the freezing temperature of water, frost phenomena occur in the soil. These are also thought to be due, in part, to the movement of

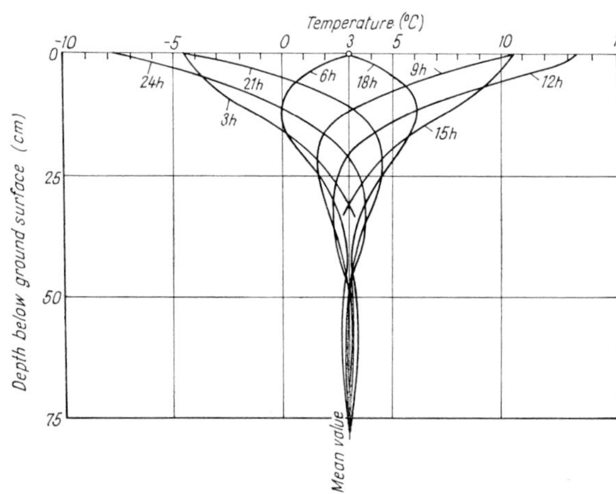

Fig. 238. Changes in soil temperature due to periodic temperature changes on the surface

soil moisture. The thermodynamic processes that take place in the soil during a freezing period are, however, not yet fully understood. The cause of moisture movement induced by frost action and of the formation of ice lenses is attributed by many to the fact that the water in the fine capillaries is, under the action of the forces of adsorption, subjected to a considerable pressure and its freezing point drops below 0°C. According to another concept (Ruckli, 1950) the cause of this moisture movement lies in the crystallization power of the growing ice crystals. This force attracts the supercooled water and the adsorbed water surrounding the solid particles. Frost action in soil and its practical implications will be discussed in Vol 2.

Recommended reading

Andrei, S. (1967): Apa in paminturile nesaturate. Editura Technica, București.

Baver, L. D. (1948): Soil Physics. J. Wiley & Sons, New York.

Bear, J., Zaslavsky, D., Irmay, S. (1968): Physical principles of water percolation and seepage. Published by the United Nations Educational, Scientific and Cultural Organisation, Paris.

Bernatzik, W. (1947): Baugrund und Physik. Schweizer Druck- und Verlagshaus, Basel.

Bölling, W. H. (1972): Sickerströmungen und Spannungen in Böden. Anwendungsbeispiele und Aufgaben. Springer Verlag, Wien New York.

Cedergren, J. (1966): Seepage. J. Wiley & Sons, Inc., New York.

Dallavalle, J. M. (1948): Micromeritics. The Technology of Fine Particles. New York.

Harr, M. E. (1966): Foundations of Theoretical Soil Mechanics. McGraw-Hill Book Co., New York.

Highway Research Board (1958): Water and its Conduction in Soils. Spec. Report 40. An International Symposium.

International Association for Hydraulic Research (1972): Fundamentals of transport phenomena in porous media. Elsevier Publishing Co. Amsterdam.

Kovács, Gy. (1972): A szivárgás hidraulikája (Hydraulics of seepage). Akadémiai Kiadó, Budapest.

Koženy, J. (1937): Hydraulik.

Lambe, T. W.–Whitman, R. V. (1969): Soil Mechanics. J. Wiley & Sons, Inc., New York.

Mansur, Ch. I.–Kaufman, R. I. (1962): Dewatering. (Ch. 3 in Foundation Engineering. *Leonards, G. A.* (editor). McGraw-Hill Book Co., Inc. New York.

Pavlovski, N. N. (1956): Motion in ground water (in Russian). Izdat. Akad. Nauk SSSR. Moscow–Leningrad.

Polubarinova-Kochina, P. Ya.: (1952): Teoriya dvizheniya gruntovikh vod. Moscow.

Rode, A. A. (1952): Pochvennaya vlaga. Izd. Akad. Nauk USSR, Moscow.

Scheidegger, A. E. (1957): The physics of flow through porous media. Univ. Toronto Press, Toronto.

Symposium on Permeability of Soils. American Society for Testing Materials. Spec. Techn. Publ. No. 163. Philadelphia, 1955.

Taylor, D. W.: (1948): Fundamentals of Soil Mechanics. J. Wiley & Sons, New York, chapters 69–9.

Ven Te Chow (editor) (1964): Handbook of Applied Hydrology. McGraw-Hill Book Co., New York.

De Wiest-Roger, J. M. (1965): Geohydrology. J. Wiley & Sons, Inc., New York–London–Sydney.

Zunker, F. (1930): Das Verhalten des Bodens zum Wasser. Handbuch der Bodenlehre von E. Blanck. Springer, Berlin, Bd. 6.

Chapter 7

Soil strength

7.1 General discussion

One of the most important tasks in the application of soil mechanics to engineering problems is the study of soil behaviour under load. In the design of structures the engineer relies upon the laws of applied mechanics and he determines the stresses and strains in the structural elements on the basis of a few physical characteristics of the construction materials used. Steel is unique in this respect in that it exhibits, within a well-defined stress regime, such simple mechanical properties that permit a straightforward application of the theoretical analysis to practical problems. In the case of any other material structural design is based on a series of simplifying assumptions which are more or less deviating from reality. The same applies to soils. As far as soils are concerned, the theory of elasticity has a rather limited scope, therefore, problems of stability and strength such as bearing capacity, earth pressure on retaining walls, safe angle of slope, etc., are usually solved by *limit-state stability analysis*. This means that e.g. in the study of load-bearing capacity we determine, regardless of the soil deformations, the ultimate pressure that causes a slip failure beneath the foundation, and the "allowable soil pressure" is then obtained by safety considerations. Similarly, for slopes we determine the critical angle at which the shearing forces due to the dead weight of the mass of soil adjacent to the slope surpass the shearing resistance of the soil, and a slide occurs.

In this chapter we study the internal resistances that allow the soil to withstand, up to a certain limit, shearing forces, and discuss the laws that govern failure criteria in soils. These laws will find practical applications in the study of earth pressure, stability of slopes and problems of load bearing capacity of soils.

The term shearing strength means the maximum resistance of soil to shearing stress. If external forces surpass this internal resistance, a failure occurs.

The "strength" of a material, a simple and familiar concept by itself, is rather difficult to define exactly, especially for soils. The difficulty arises from the many ways in which soils may fail. Failure may take the form of an abrupt brittle rupture as well as of a plastic flow with large and continuous deformations. Such differences in soil behaviour are illustrated by typical stress-strain curves in *Fig. 239*. Curve *1* represents the brittle failure with a definite ultimate strength value. In contrast to this, Curve *2* exhibits no marked rupture value but it approaches a rather vaguely defined vertical asymptote. Curve *3* does not even have a vertical asymptote, and failure cannot be defined at all. Curve *4* shows the not uncommon case that the peak and ultimate strength values are different.

Failure itself may take place in two different ways. In the first instance failure conditions are satisfied only on a single surface known as surface of sliding or surface of rupture whereas the rest of the soil mass remains in the elastic state. In the second case an entire mass of soil, or a part of it bounded by a surface of sliding, is in a state of rupture. Failure conditions prevail at every point within the mass, and at least two intersecting surfaces of sliding pass through every point (*Fig. 240*).

The deformations of large soil masses are mainly due to the relative displacements between the

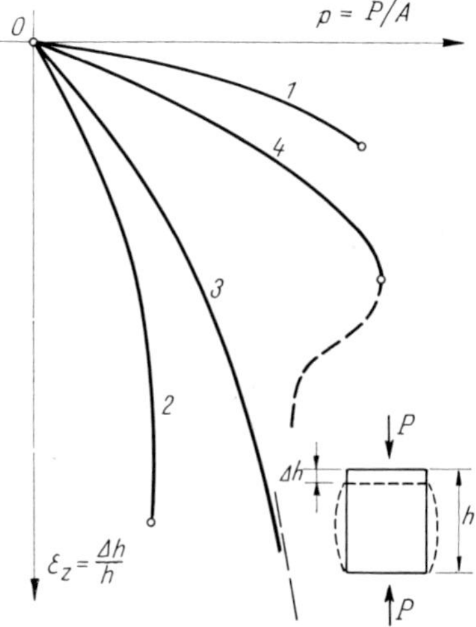

Fig. 239. Modes of failure in soils:
1 — brittle failure; *2* — deformation failure; *3* — there is no proper failure; *4* — maximum failure load and its ultimate value are different

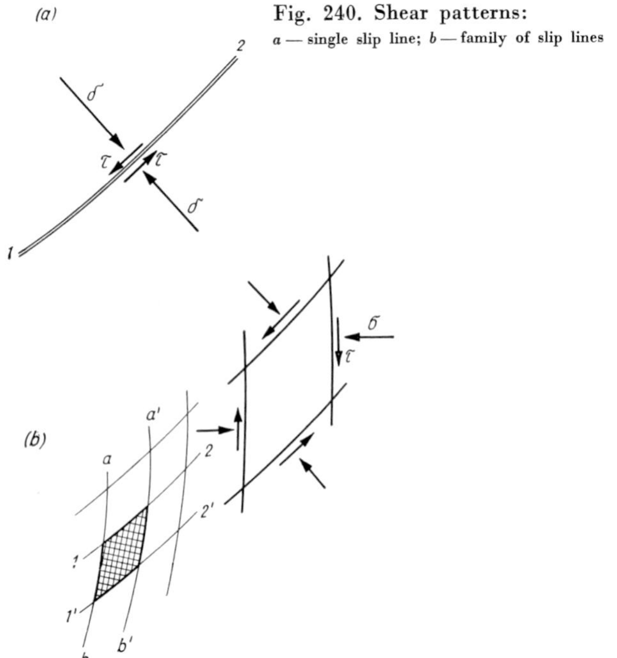

Fig. 240. Shear patterns:
a — single slip line; *b* — family of slip lines

particles. By strength of earth masses is, therefore, primarily meant the shearing strength. If only a single surface of sliding exists (*a*, Fig. 240), shear deformations are confined to that surface. In the second case (*b*, Fig. 240) deformations develop within the entire mass in state of rupture. Shearing strength was previously defined as the ultimate shearing stress on the surface of sliding. For the second case of failure, however, this surface is difficult to exactly define, and failure criteria must therefore be given in terms of the principal stresses that can yet be mobilized. This will be discussed in detail in the following section.

7.2 Calculation of stresses

This section presents a revision of formulae familiar from the theory of strength of materials, which are necessary for the calculation of the stresses in loaded soil masses. First we assume that the axes of a rectangular coordinate system are parallel to the directions of the major and minor principal stresses, and we determine the stresses that act on an arbitrary plane $a - a$ making an angle α with the plane of the major principal stress (*Fig. 241*). Let us consider an elementary prism cut out of the soil mass. We assume that in the direction parallel to that of the intermediate principal stress the soil mass is of infinite extent. Hence on every plane section parallel to that shown in Fig. 241 the state of stress is the same (*plane state or strain*). Let one side of the elementary prism coincide with the plane $a - a$. The remaining two sides are parallel to the principal stress planes. Let us dissolve the elementary force acting on plane AB into normal and tangential components, σ ds and τ ds, respectively. The plane $a - a$ is oriented at angle α, measured counter-clockwise, to the plane of the major principal stress. Stresses are considered positive when compressive. For equilibrium

$$\sigma_3 \sin \alpha \, ds - \sigma \sin \alpha \, ds + \tau \cos \alpha \, ds = 0$$

$$\sigma_1 \cos \alpha \, ds - \sigma \cos \alpha \, ds - \tau \sin \alpha \, ds = 0 .$$

Solving this for σ and τ,

$$\left.\begin{aligned} \sigma &= \frac{1}{2}(\sigma_1 + \sigma_3) + \frac{1}{2}(\sigma_1 - \sigma_3) \cos 2\alpha \\ \tau &= \frac{1}{2}(\sigma_1 - \sigma_3) \sin 2\alpha . \end{aligned}\right\} \quad (190)$$

Introducing the expressions

$$\sigma_0 = \frac{\sigma_1 + \sigma_3}{2} \quad \text{and} \quad R = \frac{\sigma_1 - \sigma_3}{2} \qquad (191)$$

and eliminating α from Eqs (190) gives

$$1 - \frac{\tau^2}{R^2} = \frac{(\sigma - \sigma_0)^2}{R^2}$$

whence

$$(\sigma - \sigma_0)^2 + \tau = R^2 . \qquad (192)$$

Equation (192) describes a circle whose centre, in the (σ,τ) coordinate system, lies on the σ-axis at distance σ_0 from the origin and whose radius R is equal to $(\sigma_1 - \sigma_3)/2$ (*Fig. 242*). For any angle α the stresses σ and τ can readily be determined from this graphical representation known as the *Mohr circle* (Mohr, 1882). As is seen in the figure, any point on the circle such as P represents the stress components σ and τ on the plane oriented at angle α to the plane of the major principal stress. The resultant stress p is represented by the distance OP, because $p^2 = \sigma^2 + \tau^2$.

Once the *Mohr* circle is known, the stress components σ and τ on any arbitrary plane $a - a$ can be graphically determined by means of the polar point O_p. The construction is illustrated by *Fig. 243*. Let *1 — 1* and *3 — 3*, respectively, be the planes of the major and minor principal stresses. Given the principal stresses σ_1 and σ_3, we can construct the *Mohr* diagram (Fig. 243*a*). If we draw a line through σ_1 parallel to plane 1 — 1, it will intersect the circle at the pole O_p. Next we draw a line through the pole parallel to the direction $a - a$. It will make an angle α with line 1 — 1. The central angle contained by the

Fig. 242. Mohr's circle

Fig. 241. Calculation of stresses in the plane state of deformation

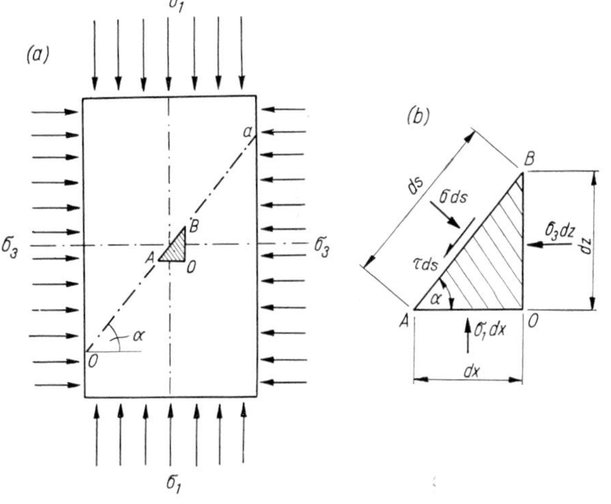

σ-axis and the line aC is, therefore, 2α, i.e. the coordinates of point a supply the required stresses σ and τ.

Figure 244 shows the stresses σ and τ as functions of angle α. The maximum shear stress will occur on planes lying at 45° to the principal stress directions. The angle δ which the direction of the resultant stress on a plane makes with the normal

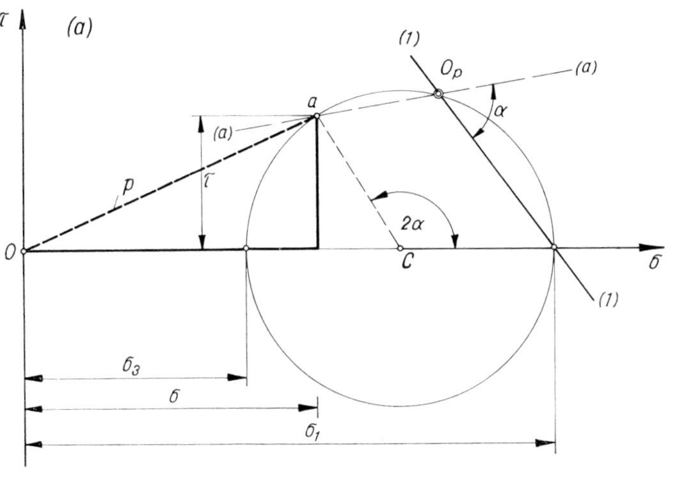

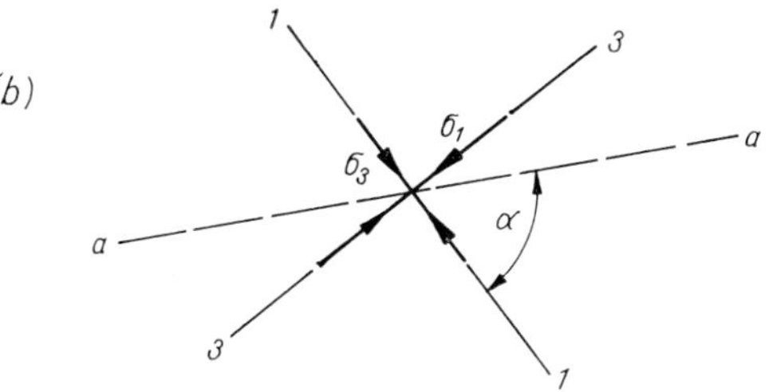

Fig. 243. Determination of stresses pertaining to a given direction with the aid of the pole

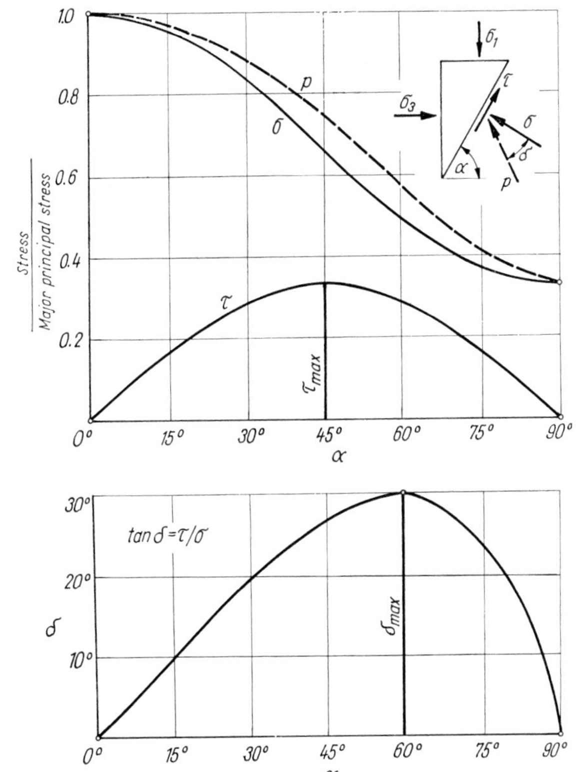

Fig. 244. Variation of normal and shear stresses with the slope angle of the elementary surface

to the plane is calculated by the expression

$$\tan \delta = \frac{\tau}{\sigma} = \frac{(1-k)\sin 2\alpha}{(1+k)+(1-k)\cos 2\alpha} = \frac{K \sin 2\alpha}{1+K\cos 2\alpha}. \quad (193)$$

Herein k is the principal stress ratio

$$k = \frac{\sigma_3}{\sigma_1}$$

and

$$K = (1-k)(1+k).$$

$\tan \delta$ values for various angles α and k-values are given in *Fig. 245*. Largest values of $\tan \delta$ occur when

$$\alpha = \alpha_0 = \frac{1}{2}\cos^{-1} K = \frac{1}{2}\cos^{-1}\frac{1-k}{1+k}. \quad (194a)$$

Then

$$(\tau/\sigma)_{max} = \cot 2\alpha_0. \quad (194b)$$

If in a two-dimensional problem (plane state of stress) the stress components parallel to the horizontal (x) and vertical (z) coordinate axes, (σ_x, σ_z, τ_{xz}), are known, it is possible to determine the magnitude and orientation of the principal stresses by plotting the *Mohr* diagram. For this purpose first we plot points (σ_x, τ_{xz}) and (σ_z, τ_{zx}); one of them will be above and the other below the positive σ-axis (D and D' in *Fig. 246*). A line drawn through D and D' will intersect the σ-axis at the centre of the *Mohr* circle. The circle can now be constructed, its radius being equal to the distance CD. The points at which the circle cuts the σ-axis, represent the magnitude of the major and minor principal stresses σ_1 and σ_3. The central angle between line DD' and the positive σ-axis is twice the angle the major principal stress makes with the direction of the vertical normal stress σ_z. As can easily be read off from the figure,

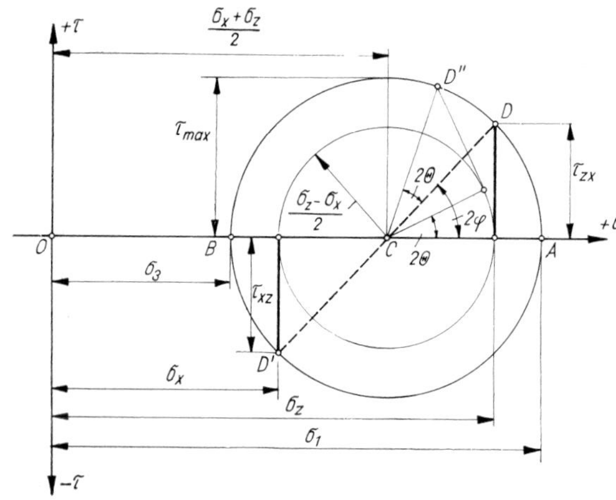

Fig. 246. Determination of the principal stresses and of the principal directions

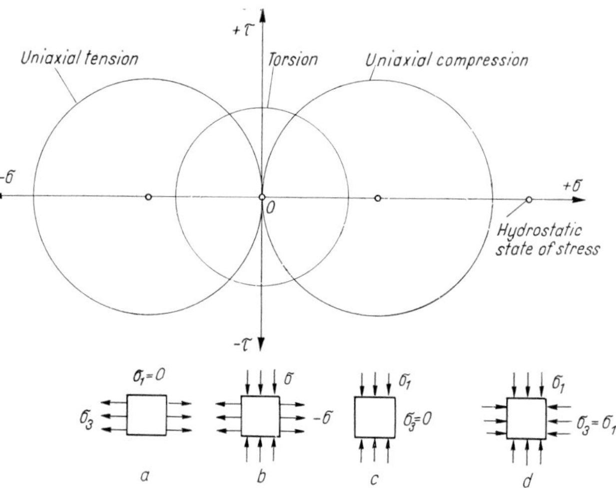

Fig. 247. Mohr's circles for special states of stress

$$\left.\begin{aligned} \sigma_{1,3} &= \frac{\sigma_z+\sigma_x}{2} \pm \sqrt{\left(\frac{\sigma_z-\sigma_x}{2}\right)^2 + \tau_{xz}^2}, \\ \tau_{max} &= \frac{\sigma_1-\sigma_3}{2} = \sqrt{\left(\frac{\sigma_z-\sigma_x}{2}\right)^2 + \tau_{xz}^2} \end{aligned}\right\} \quad (195a)$$

and

$$\left.\begin{aligned} \left.\begin{matrix}\sigma_z\\ \sigma_x\end{matrix}\right\} &= \frac{1}{2}(\sigma_1+\sigma_3) \pm \frac{1}{2}(\sigma_1-\sigma_3)\cos 2\varphi, \\ \tau_{xz} &= \frac{1}{2}(\sigma_1-\sigma_3)\sin 2\varphi. \end{aligned}\right\} \quad (195b)$$

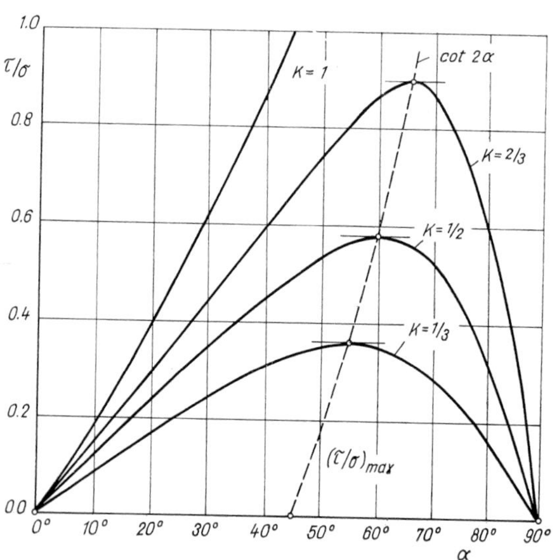

Fig. 245. Variation of the angle of obliquity with the slope angle of the elementary surface

The stress components (σ, τ) that act on an arbitrary plane oriented at angle Θ to the vertical are given by

$$\left.\begin{aligned}\sigma &= \frac{\sigma_z + \sigma_x}{2} + \frac{\sigma_z - \sigma_x}{2}\cos 2\Theta - \tau_{xz}\sin 2\Theta,\\ \tau &= \frac{\sigma_z - \sigma_x}{2}\sin 2\Theta + \tau_{xz}\cos 2\Theta\,.\end{aligned}\right\} \quad (195c)$$

As was shown before, the centre of the Mohr circle lies on the σ-axis, and in the general case when $\sigma_1 \neq 0$ and $\sigma_3 \neq 0$, its radius is different from zero. When the major and minor principal stresses are equal $(\sigma_1 = \sigma_3)$, the Mohr circle reduces to a point, and the normal stress in every direction is a constant. This special case is called the *hydrostatic state of stress*. In a uniaxial state of stress there is only one principal stress that is different from zero (uniaxial tension or compression), therefore the Mohr circle is tangential to the vertical τ-axis (*Fig. 247*).

7.3 Failure criteria

One of the most important problems in strength theory is to define the state of stress in which internal stresses due to external loads in a solid body are increased to an extent sufficient to overcome the internal resistance of the material whereupon "failure" occurs, i.e. the body actually ruptures or else it undergoes very large permanent deformations. The stress at which failure occurs is referred to as limiting stress. It is, of course, the combined effect of all stress components acting at a given point that produces failure conditions. However, one is often faced with almost insurmountable difficulties to exactly simulate actual and often very complex stress conditions experimentally. It has therefore become necessary to determine strenght properties by relatively simple tests and, using the results obtained, predict the behaviour of materials subjected to composite states of stress by theoretical means. This approach to strength problems has inspired a number of failure theories of which the *Mohr* failure theory is the foremost in civil engineering applications. In the light of recent research, *Mohr*'s concept does certainly not mean the last word in strength theory, and in many problems it has proved inadequate in describing the true behaviour of materials. Yet, for engineering purposes, it has become a very useful and dependable tool in juding, by strength computations, the danger of failure in solid bodies under general stress conditions.

Failure condition may be expressed, in a general way, as a functional relationship between the three principal stresses:

$$f(\sigma_1, \sigma_2, \sigma_3) = 0\,. \quad (196a)$$

Mohr's concept defines the limiting condition as follows. If the *Mohr* circles are plotted to represent states of stress at failure, a common envelope can be drawn to these circles. Its shape depends on the material. Supposing a *Mohr* circle intersects the *Mohr envelope*, it would mean that the corresponding state of stress is beyond the limiting state, but this is impossible. A *Mohr* circle, in turn, that lies entirely below the envelope indicates that failure condition has not yet been reached. An essential assumption in the *Mohr* failure theory is that the failure condition is independent of the intermediate principal stress. Hence Eq. (196a) reduces to

$$f(\sigma_1, \sigma_3) = 0\,. \quad (196b)$$

If we adopt the *Mohr* representation of state of stress, then Eq. (196b) gives the equation of the Mohr envelope. It may be written in the form

$$\tau = f(\sigma)\,. \quad (196c)$$

The two expressions for failure condition [Eqs (196b) and (196c)] are shown diagrammatically in *Fig. 248*. In soil mechanics a simplified form of the *Mohr* failure theory is used, which assumes, after *Coulomb*, that the failure relationship $\tau = f(\sigma)$ is linear. According to the early concept of Coulomb, failure of a material occurs when

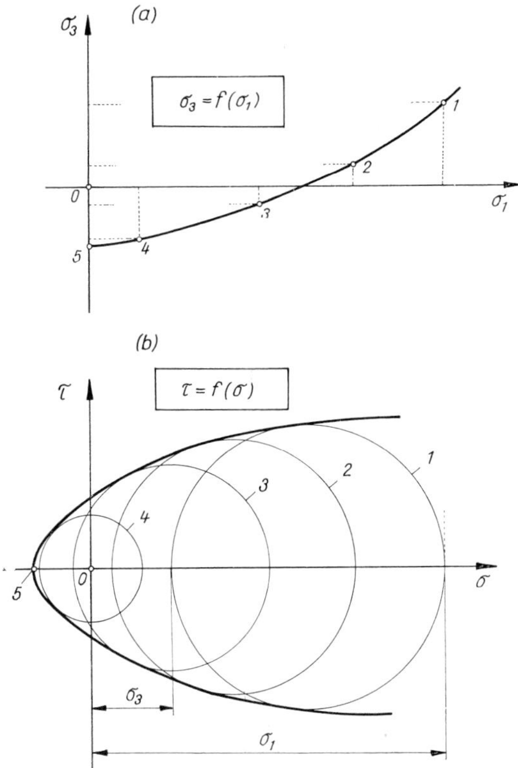

Fig. 248. Failure conditions:
a — relationship between major and minor principal stresses in the case of failures
b — envelope of Mohr's circles; *1–5* Mohr's circles

the shear stress on a surface element exceeds the internal frictional resistance and cohesion available between the particles of that material. At failure, the following linear relationship exists between normal and tangential stresses acting on a surface element *ds*.

$$\tau = \sigma \tan \Phi + c\,. \tag{197}$$

A surface, at every point of which the normal and tangential stresses satisfy the failure condition expressed by Eq. (197) is called a *failure surface* of *slip surface*. In the formula, Φ is the *angle of internal friction* and is the *cohesion* per unit of surface area.

According to the Coulomb theory the shearing strength is made up of two parts:

1. Internal friction: $\tau_1 = \sigma \tan \Phi$, which is proportional to the normal stress acting on an element of the slip surface and is characterized by the friction coefficient $\tan \Phi$.

2. Cohesion: $\tau_2 = c$, which according to Eq. (197) is independent of the normal stress. It can also be interpreted as the shearing strength at zero normal stress.

Later in this chapter it will be pointed out that Eq. (197) represents a very simplified approach to the problem of shearing resistance. There are many other factors, not included in the simple linear relationship, that influence the shearing strength of a given soil. Besides, it is extremely difficult, if not impossible, to determine experimentally numerical values of the two components of shearing strength separately, the more so since an arbitrary separation of soil strength into two parts implicit in the Coulomb equation does not truly reflect the actual stress behaviour of soils. Nevertheless, subject to certain limitations and modifications, the *Coulomb* equation is even today a very useful and widely used aid in the analysis of limiting equilibrium conditions in soils.

The general *Coulomb* relationship together with two special cases are illustrated in *Fig. 249*. In the general case both Φ and c are different from zero. When $\Phi = 0$, the shearing strength is constant and independent of the normal stress. Lastly, when $c = 0$, the shearing strength is entirely due to frictional resistance. Thus at zero normal stress the shearing strength also becomes zero. These straight lines represent special cases of the *Mohr–Coulomb* envelope. According to *Coulomb*'s theory failure occurs if the Mohr circle representing the state of stress at a point is just tangent to the *Coulomb* line. It also follows that failure does not occur along the planes on which the shearing stress is maximum (i.e. planes lying at 45° to the principal stress directions), but on planes where the resultant stress makes the largest possible angle with the normal to the elementary surface.

The *Coulomb* failure criterion may be expressed in terms of principal stresses as follows. Using notations shown in Fig. 249

$$\frac{\sigma_1 - \sigma_3}{\sigma_1 + \sigma_3 + 2k} = \sin \Phi\,.$$

After rewriting

$$\sigma_3 = \sigma_1 \tan^2 (45° - \Phi/2) - 2c \tan (45° - \Phi/2)\,. \tag{198a}$$

When $\Phi = 0$, the relationship becomes

$$\sigma_3 = \sigma_1 - 2c\,. \tag{198b}$$

When $c = 0$,

$$\sigma_3 = \sigma_1 \tan^2 (45° - \Phi/2)\,. \tag{198c}$$

If the σ-axis indicates also the direction of the plane on which the major principal stress acts, then the lines ***aP*** give the direction of the slip surface for the three cases shown. For the general case (*a*, Fig. 249), this angle is equal to $45° + \Phi/2$.

The *Mohr* failure theory represents one of the general methods of defining the state of failure. It is based on the assumption that failure occurs solely due to slip. Failure is independent of the deformation characteristics and the *Poisson* number of the material, and strains due to tension or compression and the respective strength values are not necessarily equal. These are the essential

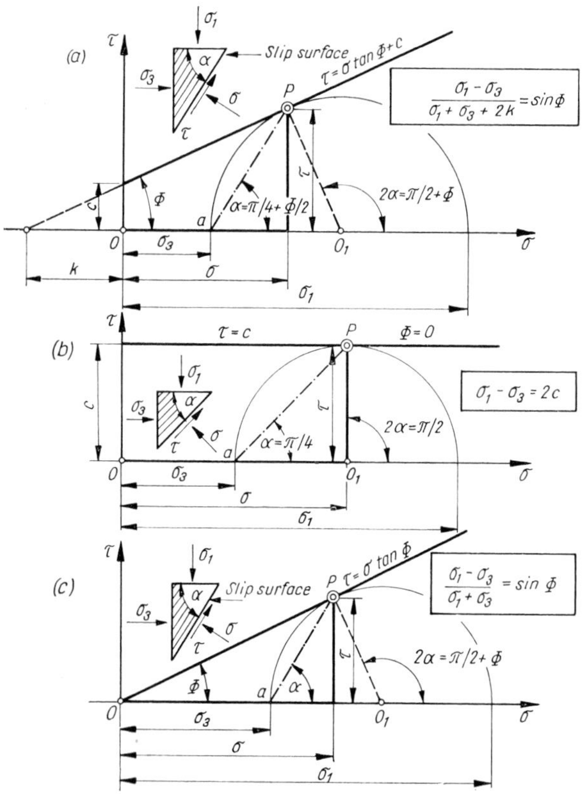

Fig. 249. Coulomb's failure condition in the general and in two particular cases

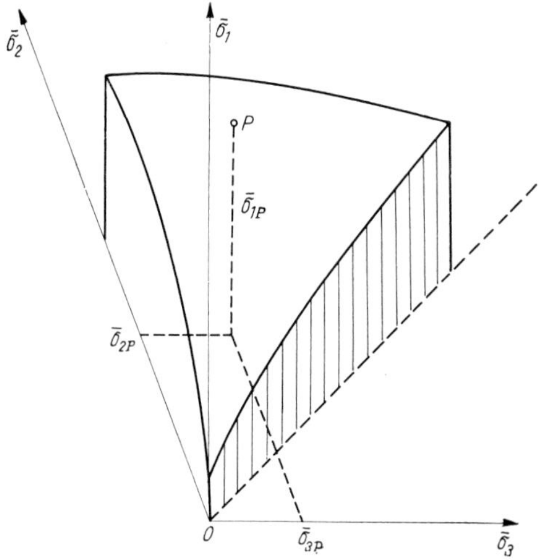

Fig. 250. Mohr's condition of failure for three-dimensional stressed state

features that make the *Mohr* theory particularly suited to the study of soil strength. Only one question remains to be answered, namely, what kind of normal stress (whether total or neutral or effective) is responsible for shearing resistance. As LEONARDS (1960) pointed out, the experiments of BISHOP and ELDIN (1950), RENDULIC (1937) and HENKEL (1958) gave conclusive evidence that shear strength is related to effective normal stress.

The Mohr failure theory can be generalized to the three-dimensional state of stress (HENKEL, 1958). The general failure criterion [Eq. (196a)] is represented in the $(\bar{\sigma}_1, \bar{\sigma}_2, \bar{\sigma}_3)$ coordinate system by a curved surface (*Fig. 250*). In this representation special states of stress are characterized by certain lines or points. In special cases the three-dimensional failure criterion may be depicted by a circular cone or a circular cylinder or a regular hexagonal pyramid. The space diagonal which makes an angle of $\alpha = \cos^{-1}(1/\sqrt{3}) = 54°44'$ with each of the coordinate axes represents the hydrostatic state of stress $(\bar{\sigma}_1 = \bar{\sigma}_2 = \bar{\sigma}_3)$. Finally, the three-dimensional failure criterion may also be represented in the form of a triangular diagram as shown in *Fig. 251*. Plotted on the sides are the ratios of the principal stresses to the first invariant of the state of stress or octahedral normal stress $(\sigma = \sigma_1 + \sigma_2 + \sigma_3)$. Here point 0 indicates the hydrostatic state of stress. The shaded hexagonal area bounded by lines that correspond to the failure state represents the domain of physically possible states of stress. Such a plot is particularly useful in visualizing variations of state of stress. Starting, for example, from the hydrostatic state of stress (point 0), we may increase the major principal stress σ_1. In this way, following any prefixed course, we may reach the boundary representing failure. But failure condition can be produced in many different ways, e.g. by reducing the minor principal stress, or by using any combination of changes in principal stresses. Any of these processes is described by some line in the triangular plot. The equivalent line in the usual *Mohr* diagram is called the *stress path.* This is obtained as follows. The state of stress is continuously changed according to some prefixed procedure, and the resulting *Mohr* circles

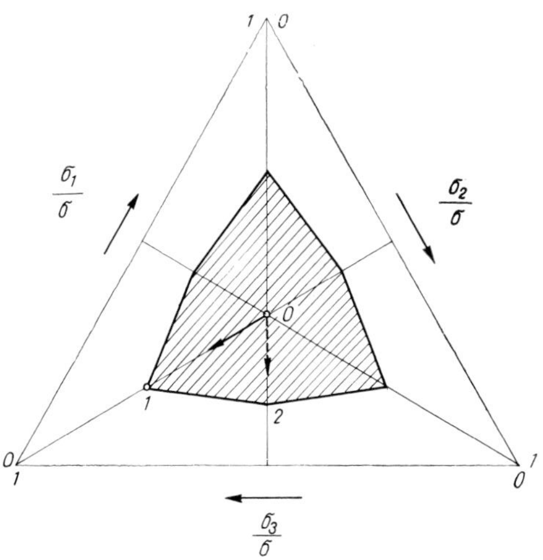

Fig. 251. Failure condition in triangle diagram

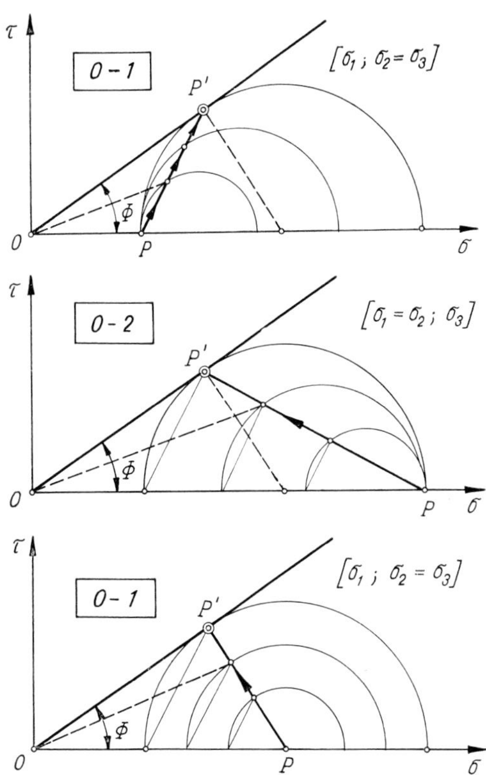

Fig. 252. Stress paths

at consecutive stages are plotted. The stress path is defined as the locus of a point that marks on each circle the resultant stress acting on the potential slip surface. Several variants for the stress path are shown in *Fig. 252*. The corresponding lines are indicated also in Fig. 251.

7.4 Physical causes of friction and cohesion

To move two solid bodies which are in contact relative to each other from their rest position, a force is required that has a component tangential to the surface of contact and acting in the direction of motion. The resistance which must be overcome before motion takes place is called *friction*. It is important to realize that friction is a *passive* force, which is mobilized only to the extent determined by the magnitude of the active force. The active force can be increased until the physically possible upper limit of the passive force is reached. At that limit the friction is completely mobilized and a continuous motion ensues. On the basis of practical experience and experimental evidence we assume that the ultimate frictional resistance is directly proportional to the force normal to the contact surface, i.e. the *Coulomb* friction law holds which states

$$T = N \tan \varphi = Nf. \tag{199}$$

The available theories on the inner nature of friction are often contradictory. Already in the last century there were researchers (RENNIE, 1829, 1861; MARIN, 1833) who disputed whether the coefficient of friction, f, was really independent of the area of the contact surface, of the magnitude of the normal force and of the velocity of motion.

Coulomb's concept of friction reflected a purely mechanical approach. According to him friction is accounted for by the fact that the surface of contact between two solid bodies is never perfectly smooth but rather rough showing hollows alternating with corners and higher spots. In order to permit motion the interlocking superficial irregularities must be sheared off. But as is well known, two solid bodies cannot be brought into direct contact, because a water film adsorbed on the surface prevents them from actually touching each other even under pressure. If we can manage to remove this adsorption layer in some way, e.g. by heating or applying very high pressure, the two bodies will be fused together over the surface of contact. In this case, the passive force to be overcome to permit motion is the shearing resistance of the fused zone. This phenomenon was first observed by W. B. and T. K. HARDY (1919). After having been slid on one another, two glass plates with smooth and chemically clean surfaces showed long pointed, wedge-shaped scars with a sharp-edged oval spot in the centre resembling, under the microscope, a droplet of molten glass (*Fig. 253*). The coefficients of friction measured under such conditions were extremely high: from 0.84 to 1.10 between glass and glass, and 0.74 between steel and steel. The quoted numerical values were obtained from tests conducted in a sealed, dry, and dust-free chamber. Under ordinary test conditions the f values were much lower and showed a wide scatter. This indicated that the presence of an adsorbed film on the interface may substantially reduce friction, and the shearing resistance measured

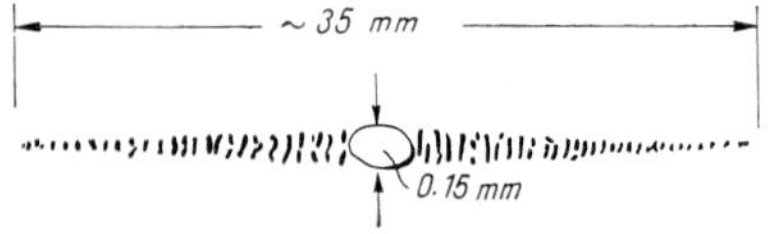

Fig. 253. Scratches on a plane plate after surface friction test

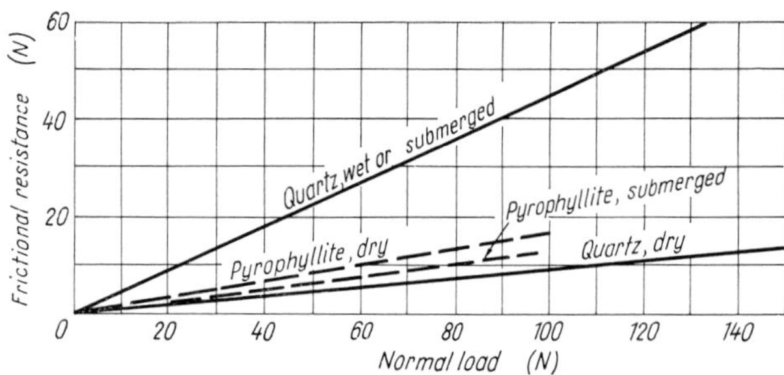

Fig. 254. Relationship between normal stress and sliding resistance for quartz and pyrophyllite

was, in fact, that of the film. Tests with various lubricants showed the following relationship between the coefficient of friction, f, and the molecular weight of the lubricant, M

$$f = b - aM.$$

Here a depends on the chemical properties of both the solid body and the lubricant, whereas b depends only on the properties of the lubricant. The lubricating effect was found to be practically independent of the quantity of the lubricant. Liquids are classified as active or inactive according to their effect on the coefficient of friction. Those in the first group are grease and heavy lubricating oils. Water, alcohol, petrol, etc. belong to the second group.

For centuries, oil as a lubricant has been known and used to reduce friction between metallic surface. The effect of water as an antilubricant is not widely known. Since earlier experiments on friction had been carried out with glass, metals and various other materials but not with earth minerals, TSCHEBOTARIOFF and WELCH (1948) studied the friction between fixed mineral particles and polished surfaces of the same minerals in the dry, wet and submerged states. The relationship obtained between the tangential force causing slip and the normal force was as shown in *Fig. 254*. It is seen from the figure, that there is a pronounced difference between the values measured in the dry and wet states. Experimental results were influenced even by a low air humidity, probably because of the formation of an adsorbed film on the interfaces. For quartzite the coefficient of friction was found to be 4.5 times as high in the wet state as in the dry state. For calcite the ratio was 2.5. For both hydrophilic minerals, however, the coefficients of friction measured in slightly wet or in submerged state were almost the same. Thus, the increase in frictional resistance relative to that in the dry state cannot be accounted for by the effect of surface tension but it is due to changes in the physical state of the water in the adsorbed layer.

In the case of hydrophobic minerals, the coefficient of friction was slightly reduced by the action of water. Since most minerals are hydrophilic, water exerts no lubricating effect on them; instead, it acts so as to increase surface friction.

The *Coulomb* law expresses a linear relationship between normal pressure and frictional resistance. TERZAGHI (1925) attempted to give a physical explanation for it. He considered contact between two curved surfaces. He distinguished between total and net contact areas. The former can be calculated by the *Herz* formulae which state that the total contact area depends on the radius of curvature and is directly proportional to the 2/3 power of the normal pressure. The net contact area is composed of small scattered spots within the total area. Its magnitude is obtained, according to *Terzaghi*, by dividing the total pressure by the yield stress of the softer substance. Full cohesion is acting on the net contact area, and in order to produce motion, the shearing strength of the material itself must be overcome. If f denotes the coefficient of friction on the interface, N is the normal thrust, A_n is the net contact area, and τ is the shearing strength of the material, then the ultimate resistance to motion T may be written

$$T = Nf = A_n \tau_0$$

whence

$$f = \frac{A_n}{N} \tau_0 .$$

As the practical evaluation of the net contact area by the described method is rather uncertain, the above formula is mainly of theoretical value. Nevertheless, it explains why the frictional resistance should be related to the normal pressure.

After the normal load has ceased to act, the resistance to lateral displacement reduces to the force necessary to shear away a single layer of molecules. This resistance is known as zero friction. The normal value of the coefficient of friction occurs only under higher pressures. For very low normal pressures the relationship between the normal and tangential forces is no longer linear. In general, the total resistance to slip is made up of two effects: *adhesion* and *friction*, the latter of which is dependent on the normal pressure. In ordinary friction tests, no appreciable adhesion is observed, but in the case of an assemblage of microscopic particles it may have significant values.

The coefficient of friction may, in addition, depend on the rate of displacement. It is a common experience that the friction at rest (static friction) is much higher than the friction during motion (kinetic friction). The explanation for this is that under static loading conditions the adsorbed film becomes highly compressed at the points of contact, giving rise to increased density and shearing strength. During motion, provided it takes place at a sufficiently high rate so as not to permit compression, the adsorbed film retains its larger thickness and offers only a relatively lower frictional resistance.

The overall size of the surfaces sliding on one another is also an important factor. Recent research has shown that the coefficient of friction decreases with increasing sliding surface area.

The fact that friction at rest is higher than its kinetic value can be easily demonstrated by the simple setup shown in *Fig. 255.* A weight placed on a sloping plane is to be moved up the slope by means of a screw coupled with a spring. If we rotate the handle at a constant rate, the motion produced will not be uniform at all, but jerky and intermittent. Initially, the applied force will be compensated by the extension of the spring before any displacement of the block could take place. As soon as the block starts moving, the friction is reduced to its kinetic value with the result that the block moves ahead a somewhat greater distance than would correspond to the original tension in the spring.

A similar phenomenon can be observed when e.g. we want to recover a pile driven into the ground: even if we apply a constant pull-out force, the resulting movement will be intermittent.

According to the preceding discussion, the coefficient of friction greatly depends on the properties of the film located between the surfaces in contact. This film is responsible also for the phenomenon of adhesion. If the solid grains are large in relation to the thickness of the adsorption hull, then the effect of adhesion is negligible. Because of the relatively small number of points of contact between the grains, the contact pressures due to load are high; the film is therefore greatly compressed, and pressure is practically transferred from grain to grain. In such coarse-grained materials the internal friction becomes the predominant factor in the shearing resistance, while the adhesion is negligible. In an assemblage of fine grains the number of contact points per unit of area is greatly increased, and the adsorption films, acted upon by relatively small pressures, will be much thicker. Thus the shearing resistance of the film will decrease. On the other hand, its adhesion comes gradually into play. This explains why fine-grained soils exhibit as a rule small internal friction but considerable cohesive strength.

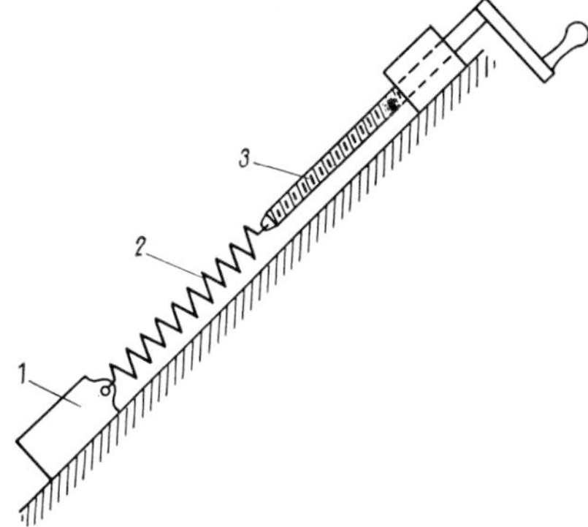

Fig. 255. Setup to demonstrate the difference between frictional resistances at rest and in motion:
1 — friction block; *2* — spring; *3* — screw

Cohesion as was just discussed can be derived, in part, from the adhesion of the adsorption films. Another factor that also accounts for cohesive strength in fine-grained soils is the capillary attraction caused by the surface tension of water pulling the grains together. (Contact moisture, Ch. 3.)

In the literature one often finds views that attempt to distinguish between two kinds of cohesion, namely, true and apparent cohesion. The true cohesion is believed to be caused by molecular attraction between the grains, whereas the apparent cohesion is due to the surface tension of water. This concept holds, furthermore, that the apparent cohesion is bound to vanish as soon as the soil is submerged in water. The true cohesion, on the contrary, is likely to diminish only after prolonged exposure to water. Experimental data showed, however, that at a distance of 10^{-6} cm the molecular attraction becomes negligibly small. Besides, the cohesive property of solid crystalline bodies is of an electrical nature. X-ray analysis has revealed that e.g. the crystals of common salt have a space lattice structure made up of cubes having sides 2.8×10^{-8} in length, with positive Na and negative Cl ions placed at alternating corners of the cubes. The force that keeps the structure in equilibrium is the electrostatic attraction between opposed charges. Similarly, electric phenomena are responsible for the cohesive strength of metals which, in the physical sense, are supercooled liquids. Cooling results in shrinkage initiating thereby a process of crystallization, during which immense hydrostatic stresses are developed. This produces cohesion.

Between individual soil particles, cohesion in the strict physical sense is non-existent. We can also easily show by the following approximation that in soils the cohesion due to intergranular gravity forces must be subordinate to other physical effects.

Let us consider a cubic array (see Fig. 54) of uniform spheres 0.0002 mm in diameter. The number of spheres n over a width of 1 cm is $10/0.0002 = 5 \times 10^4$. We shall consider the gravity forces between two adjacent rows of spheres only, neglecting the attraction exerted by more distant spheres since it is inversely proportional to the square of the distance. Thus we calculate the force by which one sphere attracts six adjacent spheres. The distance between centres of touching spheres is d, and between two diagonally placed spheres is $d\sqrt{2}$. Reducing the problem to that of attraction due to gravity between two masses, m_1 and m_2, we obtain the force F exerted by a sphere on six adjacent spheres as

$$F = f\frac{m_1 m_2}{d^2} + 4f\frac{m_1 m_2}{(d\sqrt{2})^2} .$$

For equal masses $(m_1 = m_2 = m)$

$$F = 3f\left(\frac{m^2}{d^2}\right) .$$

f is the gravitational constant (by the *Eötvös* experiment $f = 6.67 \times 10^{-8}$ cm³/gs²). Substituting $d = 0.0002$ mm and taking 2.8 for the specific gravity of solids gives

$$m = \frac{3}{4}\pi d^3 \varrho = \frac{3}{4}\pi \times 2^3 \times 10^{-5} \times 2.8 = 5.3 \times 10^{-14}\,\mathrm{g}\,.$$

Hence

$$F = 3 \times 6.67 \times 10^{-8} \times \frac{5.3^2 \times 10^{-28}}{2^2 \times 10^{-10}} = 1.40 \times 10^{-24}\ \mathrm{dyne}\,.$$

The number of spheres over an area of 1 cm² is $(5 \times 10^4)^2 = 2.5 \times 10^9$.

The total force they exert, i.e. the gravitational attraction is

$$\sum F = 2.5 \times 10^9 \times 1.40 \times 10^{-24} = 3.5 \times 10^{-15}\ \mathrm{dyne/cm^2} \approx 3.5 \times 10^{-16}\ \mathrm{N/m^2}\,.$$

If we consider that even soft saturated fine-grained soils may exhibit a cohesive strength of 1000 to 2000 N/m², it becomes clear, that the "true cohesion" of these soils can by no means attributed to gravitational attraction.

7.5 Measurement of shearing strength

7.5.1 Testing methods

The basic principle of the experimental determination of the shearing strenght is that a soil specimen is subjected to a comparatively simple and controlled state of stress, and the *Mohr* circle, or circles, at which failure occurs are recorded. The common tangent to the circles so obtained furnishes the *Coulomb* line which is characterized geometrically by the angle of internal friction and the cohesion intercept. The commonly used testing methods are as follows.

Direct shear test: a soil specimen subjected simultaneously to a constant vertical normal load and a horizontal shearing force is caused to fail through shear by increasing the horizontal force while the transversal strain in the plane of shear is prevented; *unconfined compression test*: a cylindrical specimen is subjected to an axial force which is increased to the failure point while the specimen is free to strain in the horizontal direction;

triaxial compression test: a cylindrical specimen is first subjected to a constant all-around confining pressure and is then caused to fail by altering the state of stress according to some preset procedure.

The *Mohr* representation of these states of stress is shown in *Fig. 256*.

In the determination of shearing strength the neutral stresses have a decisive role, since that portion of the shearing strength which is dependent on normal pressure can be generated only by effective stresses. With respect to the effect of the neutral stresses, we distinguish between *drained* (or open) and *undrained* (or closed) systems.

In a *drained system* excess water is allowed to freely move out of a mass of soil under the action of neutral stresses that may build up during loading. The boundary surfaces are comparatively pervious, thus permitting consolidation to take place. The neutral stresses will change with time, and the shearing strength will change accordingly. The strength behaviour of the soil will, of course, be different in saturated and partially saturated states, since consolidation in these two cases takes place in different ways.

In an *undrained system* the drainage of water is prevented. It is therefore only in partially saturated soils that the neutral pressure changes during loading because of the compression, or solution in water, of the pore air. In a saturated soil the state of stress does not change with time under conditions of no drainage.

Whether in a given case the system should be regarded drained or undrained depends not only on permeability, but also on the rate at which load is applied to the soil. For it is quite possible, even if the boundary surfaces are pervious, that loading changes so rapidly that there will be insufficient time for the moisture content to adjust itself to changes of state of stress. In such an instance a mass of soil will virtually behave as an undrained system. Shear tests performed under undrained or drained conditions, respectively, are often referred to as "quick" of "slow" tests. Essentially, the difference between the two types of test is that in the first instance the system is generally undrained, in the second it may be considered either drained or undrained depending on the rate and conditions of loading.

We should further differentiate between the two cases when in a given soil mass the neutral stress is independent of, or dependent on, the total normal stress. The first case is simpler in that the pore pressure depends solely on the position of the point considered. This is the case e.g. when a mass of soil with no surcharge acting on it is at rest, and either the water table is stationary or a steady flow of water takes place in the soil. Once a steady flow has developed and continues unchanged in time, the seepage pressure is uniquely determined by the position of the point in question. Under such conditions no change in volume occurs, or at least it is negligibly small.

In the second case, and this is more general, changes in the effective normal stress or in the shearing stress give rise to a tendency to volume changes in the soil mass. In an undrained system, this results in a pore pressure build-up. The magnitude of the induced pore pressure depends on the extent of the change in stress, on the value of the coefficient of consolidation, and lastly, on the distance of the point under consideration from the boundary surface of free drainage. The effective and neutral stresses are, in this case, essentially time-dependent even if the total stress remains constant throughout. Such conditions arise when e.g. the load of a structure is imposed on natural soil strata, or water is suddenly removed from the face of a slope (sudden drawdown), or a cut is quickly excavated. In the last example, the change

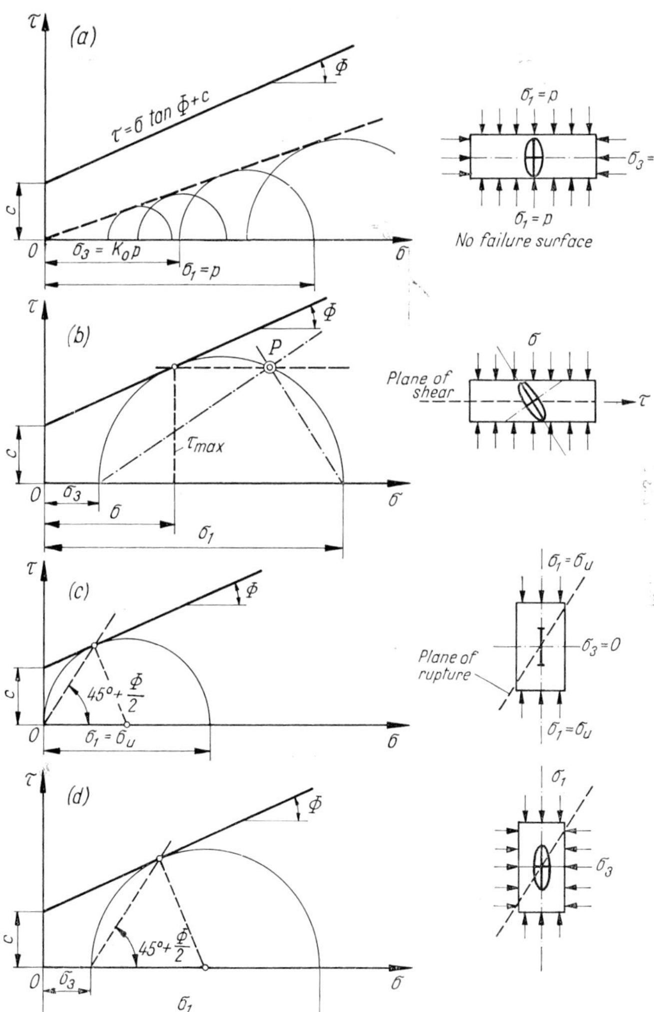

Fig. 256. Mohr's circle and stress ellipse for the tests used to determine the shear strength of soils:
a — confined compression test; *b* — direct shear test; *c* — unconfined compression test; *d* — triaxial compression test

in pore pressure is due to the expansion of the soil during unloading.

In order to obtain appropriate strength values experimentally for any of the above mentioned cases, we must produce a state of stress causing failure under test conditions that simulate as closely as possible the actual conditions under which the soil will be stressed. The relationship between the normal and shearing stresses at failure, i.e. the *Coulomb* line, can then be readily plotted. Formerly, it was common practice to plot the results of the shear tests in terms of the total stress employed. This method has the disadvantage of leading to different strength parameters, Φ and c, depending on the magnitude of the neutral stresses built up during the test. In present practice, the second method is preferred when we plot shear stresses at failure against effective normal stresses, and use Φ and c values thus obtained for the purpose of stability calculations. Using this method, we are able to take the effect of neutral stresses on shearing strength into consideration and to predict, in a given stability problem, the variation of the factor of safety with time. Our guiding principle will be therefore that the shearing strength should be expressed as a function of the effective normal stresses. The corresponding shear strength parameters will be denoted by Φ' and c'; they are often called the true angle of internal friction and true cohesion of the soil. Thus the failure law becomes

$$\tau = (\sigma - u) \tan \Phi' + c'. \tag{200}$$

Failure condition may be produced in many different ways in the various types of test. But experience shows that although the deformations measured in different types of test may differ greatly, the Φ' and c' values are fairly constant.

In connection with the measurement of shear strength we have to add three general remarks.

The first one concerns the initial state of stress of the soil. Subsurface strata are normally acted upon by the overburden pressure, but occasionally they have been subjected, in their past history, to heavy preloading. If the ground surface is horizontal, the vertical and horizontal stresses, respectively, are the major and minor principal stresses. The two are as a rule very different. If a sample taken from a soil stratum that has consolidated under natural conditions is subjected to either a shear test, unconfined compression test or triaxial test, the initial state of stress in the test will, in general, not be identical to the state of stress that prevailed in nature. For example, in a triaxial test we start from a hydrostatic state of stress. As we have seen previously, neutral stresses are caused by changes in the state of stress. Therefore, if we work with different initial states of stress, the neutral stresses induced during the test and, hence, the resulting shearing strength will be different, too. For this reason, we must always consider the initial state of stress of a neutral mass of soil before we use experimental data for practical calculations.

The second remark relates to the definition of failure condition. As seen before, the *Coulomb* line is obtained as the common envelope of *Mohr* circles representing failure. But it is often difficult to decide what should be regarded as failure condition. If a soil tends to expand during shear, its strength will decrease. If the soil compacts, the strength will increase. Sometimes the soil undergoes the stages of both compaction and expansion in the course of shear. Before test results are evaluated, it is advisable to state what the criterion of failure should be.

Finally it should be pointed out that according to the failure theories used in Soil Mechanics only the deviator stress, $(\sigma_1 - \sigma_3)$, is of significance in defining the condition of failure, whereas the intermediate principal stress has no influence on it. A thorough study of the problem reveals that this cannot be so, but available experimental data are still contradictory (JANBU, 1957; PELTIER, 1957).

7.5.2 Direct shear test

The principle of the oldest and simplest method of measuring the shearing resistance is illustrated in Fig. 256*b*. Also shown in the figure is the *Mohr* circle representing the state of stress at failure. A sketch of the practical form of the apparatus, called *shear box*, is shown in *Fig. 257*. It consists of a square metal box split across the middle into two parts, one fixed, the other movable in the horizontal direction. The soil specimen is held between two indented porous stones or metal grilles. During the test, a vertical confining load is first applied to the specimen through a rigid top block, then a gradually increasing horizontal shearing force is applied through the upper movable part until the sample fails. Both the horizontal shear displacement and the vertical deformation are measured by precise dial gauges. The loading causes the void ratio of the specimen to decrease. If the voids are filled only with air, compression occurs almost instantaneously, whereas in water-saturated soils consolidation may take a very long time. The test procedure may be of the following three kinds depending on the purpose of the test.

1. The specimen is first consolidated under the vertical confining load. The horizontal load is then increased stepwise allowing again sufficient time for the specimen to consolidate before the next load increment is applied (*slow test*). In this case the water content of the specimen can always adjust itself to the magnitude of the shearing force.

2. After the specimen has consolidated under the vertical load, the horizontal force is applied in small increments but at such short intervals that the moisture content of the sample remains practically unchanged (*consolidated quick test*).

3. Both the vertical and the horizontal forces are applied at such a quick rate that there is insufficient time for consolidation to take place. As a consequence, the moisture content of the

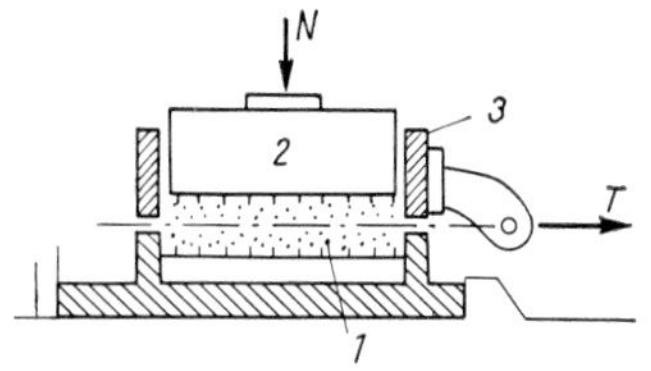

Fig. 257. Sketch of shear box:
1 — soil sample; *2* — loading piece; *3* — moving frame; *4* — shear plane

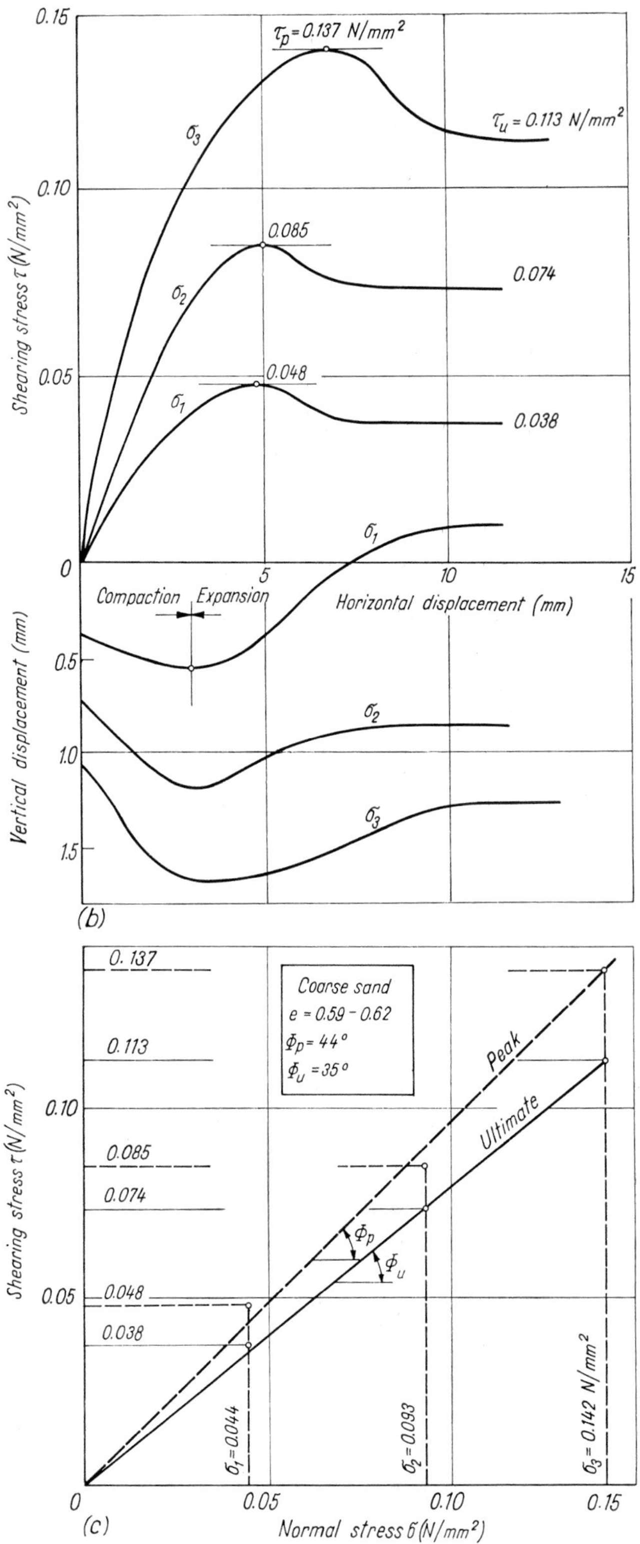

Fig. 258. Results of direct shear tests

specimen remains almost the same throughout the test (*quick test*).

The direct shear test may, furthermore, be carried out in the following alternative ways:

(a) The shearing force is applied in steps.

(b) The shearing force is increased continuously with simultaneous recording of shear displacement (*stress-controlled shear test*).

(c) The two parts of the shear box are moved away in relation to each other at a constant rate and the shearing force necessary to maintain continuous motion is recorded (*strain-controlled shear test*).

The usual size of the sample is 6 cm × 8 cm, or 9 cm × 12 cm, with a thickness of 2 cm. For direct shear tests on coarse-grained gravelly soils or stone chippings large shear boxes, 30 cm², are used.

The results of direct shear tests are plotted as shown in *Fig. 258*. For various constant normal loads the horizontal forces that cause the specimen to fail in shear are measured and the corresponding stress values calculated. On the left hand side of Fig. 258 the shearing resistance and the vertical expansion or compaction of the sample are plotted against the horizontal displacement. By reading off the ultimate shear stress values from the stress–strain curves and plotting them against the corresponding normal stresses held constant during each test we obtain the *Coulomb* line as a result (Fig. 258*c*). Hence, the shear strength parameters, Φ and c, can be determined graphically.

7.5.3 Unconfined compression test

A commonly used method of measuring the cohesive strength of soils is the unconfined compression test, mainly because it is quick and easy to perform and supplies, by relatively simple means, information from which e.g. the variation with depth of the bearing capacity in a soil profile can reliably be estimated. In principle the test is identical to the compression, or crushing, test used in materials testing. But in contrast to the cube test standardized in the testing of timber and concrete, the unconfined compression test is carried out on cylindrical specimens with a diameter–height ratio of 1 :1.5. The reason for this is that a reliable compressive strength value can only be measured if a surface of failure develops during the test. As theoretically the plane of failure lies at $45° + \Phi/2$ to the horizontal, the minimum diameter—height ratio is $1 : (45° + \Phi/2)$. This condition is satisfied by the standard ratio of 1 : 1.5. The specimens are usually 60 mm high by 40 mm in diameter.

The test procedure is very simple. The apparatus consists of a hand-operated or mechanized loading

Fig. 259. Apparatus for performing unconfined compression test

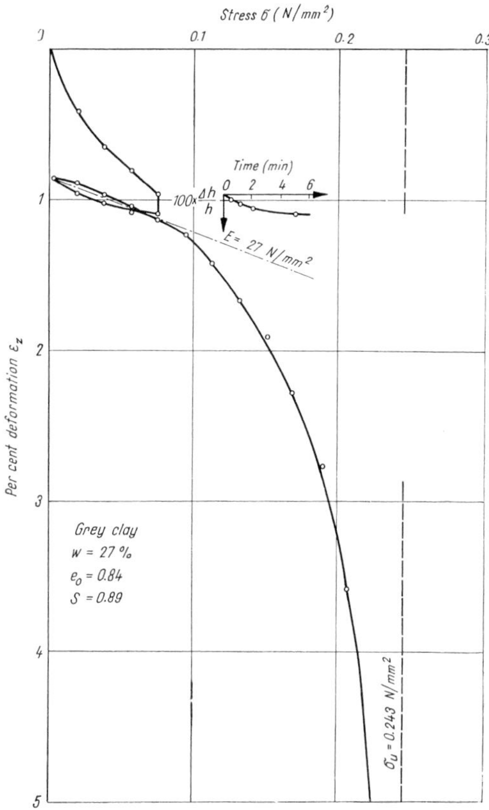

Fig. 260. Unconfined compression test; load-deformation diagram

machine sufficiently precise in the range of relatively small loads normally encountered in soil testing, and a dial gauge to measure axial displacement.

In order to prevent the specimen from drying out through evaporation, especially in slow tests, it is placed inside a plexiglass cylinder which is sealed on the top with a rubber membrane. Saturation air humidity inside the cell is ensured by placing wetted cotton wool around the base. The specimen is tested between flat and smooth platens so that the interference from surface friction is minimized. It is essential that the two ends of the specimen are absolutely parallel and the load is centric. The assembled apparatus is shown in *Fig. 259*.

A typical stress/strain curve for the unconfined compression test is shown in *Fig. 260*. It may be useful to completely unload and reload the specimen at an early stage of the test. During unloading the soil does not behave as a perfectly elastic material in that a portion of the total deformation will be irreversible. In a quick test the load increments are applied at a constant time interval, usually 30 s. Before unloading, however, we wait until the deformation under the last load increment is complete. This normally takes several minutes. This process is represented by the short vertical section of the stress/strain curve in Fig. 260. The small diagram inserted in the figure shows the corresponding time/deformation curve. As the load is further increased and approaches failure, the deformation will rapidly increase, cracks develop and eventually the sample fails. Typical examples of failed specimens are illustrated in *Fig. 261*. In hard, brittle materials failure occurs along a single or only a few clearly defined rupture planes. Soft plastic clays usually fail by bulging with two families of conjugate slip surfaces developing at the advanced stage of failure. The relevant stress/strain curves are shown in *Fig. 262*. The deformation at failure varies from 3 to 8% for stiff, brittle clays, 8 to 14% for semi-rigid clays and 14 to 20% for plastic clays.

The Mohr representation of unconfined compression was previously shown in Fig. 256. Since only one of the principal stresses, viz. the vertical stress σ_z, is different from zero, the *Mohr* circle is tangential to the τ-axis. At failure, it is also tangential to the *Coulomb* line. From Fig. 256, the cohesion intercept is

$$c = \frac{\sigma_u}{2} \tan\,(45° - \Phi/2)\,. \qquad (201)$$

The unconfined compression strength is very useful in judging the consistency of soils. A correlation between soil consistency and unconfined compressive strength is given in *Table 26*.

Table 26. Consistency of clays characterized by the unconfined compressive strength

Consistency	Unconfined compressive strength kN/m²
Very soft	<25
Soft	25–50
Medium	50–100
Hard	100–200
Very hard	200–400
Extraordinarily hard	>400

As soils exhibit a non-linear relationship between compressive stress and strain, their behaviour under compression cannot be characterized by a single constant, the *Young* modulus, as in the case of an elastic material. In practice, the follow-

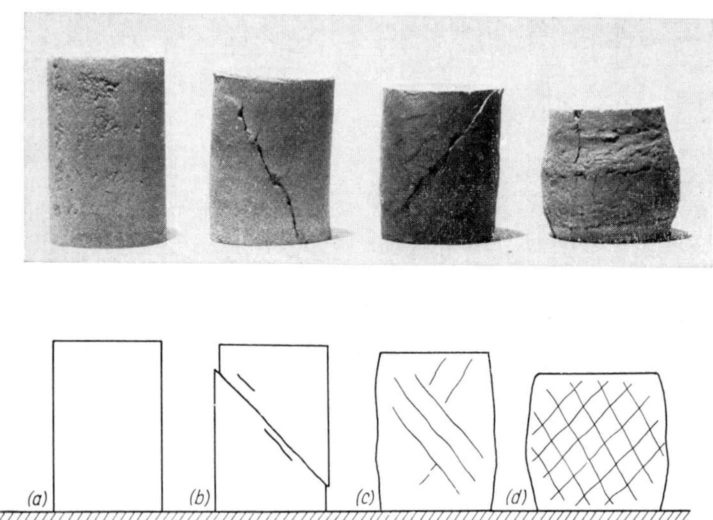

Fig. 261. Characteristic failure patterns:
a — specimen before testing; *b* — brittle failure; *c* — plastic failure; *d* — plastic flow

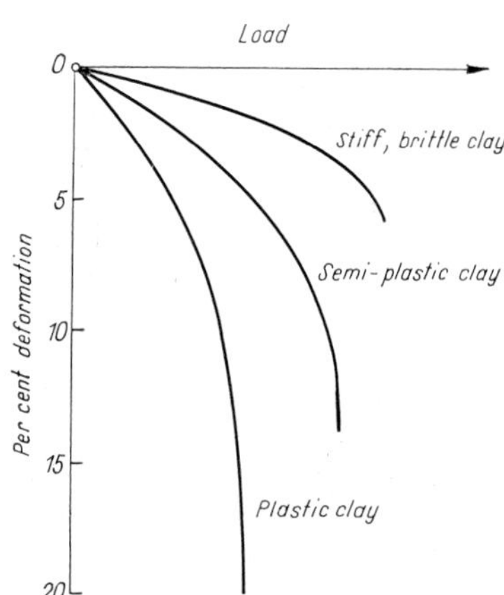

Fig. 262. Characteristic deformation curves at different consistencies

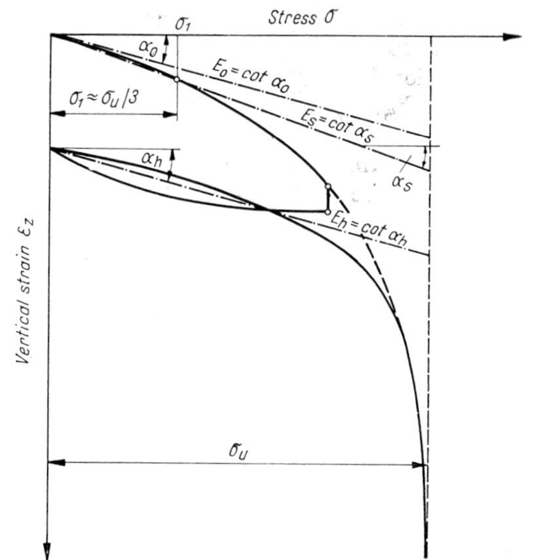

Fig. 263. Determination of different deformation moduli from the load-deformation diagram

ing characteristics of the stress/strain curve are used to express the elastic properties of soils: (1) The slope of the initial tangent to the curve; (2) the slope of the chord between the origin and a point located on the curve at a prefixed stress value (usually one-third of the ultimate strength); (3) the slope of the hysteresis loop obtained by unloading and reloading the test specimen (*Fig. 263*).

If a clay specimen is first subjected to unconfined compression in its undisturbed, natural state, then remoulded and tested again with its water content held constant, it usually exhibits much lower strength and modulus of elasticity in the remoulded state. This loss in strength is associated with changes in the clay structure. Remoulding has a twofold effect: it temporarily disturbes the oriented arrangement of the molecules in the adsorbed layers and destroys the stable flocculated structure of the clay formed during sedimentation. The soil will later regain a portion of the strength lost through reorientation of the molecules, (thixotropy, see Ch. 3), but the remaining loss in strength due to structural destruction is irreversible.

The compressive strength obtained by the unconfined compression test is not truly representative of the *in situ* strength of a soil. First of all, the unconfined compressive strength as measured in this test is greatly affected by the size of the specimen and by the test conditions, especially the rate of loading. In addition, there are always inevitable disturbances during the sampling operation. For these reasons, the unconfined compressive strength should primarily be regarded as an index property suitable to characterize soil consistency.

7.5.4 Triaxial compression test

The most common and versatile test used to determine the soil strength is performed by a *triaxial pressure cell.* A cylindrical specimen of soil is tested under an axially symmetrical state of stress. An axial stress σ_1 is applied to both flat ends of the specimen and a uniformly distributed lateral confining pressure to the sides. The state of stress is thus defined as $(\sigma_1, \sigma_1 = \sigma_3)$.

During the test we measure the stresses, the axial displacement and the neutral stresses induced in the pore water. By increasing the deviator stress, $(\sigma_1 - \sigma_3)$, the specimen can be loaded to failure. From a series of tests, *Mohr* circles representing failure can be plotted and the *Coulomb* envelope constructed, and hence the angle of internal friction and the cohesion can be determined.

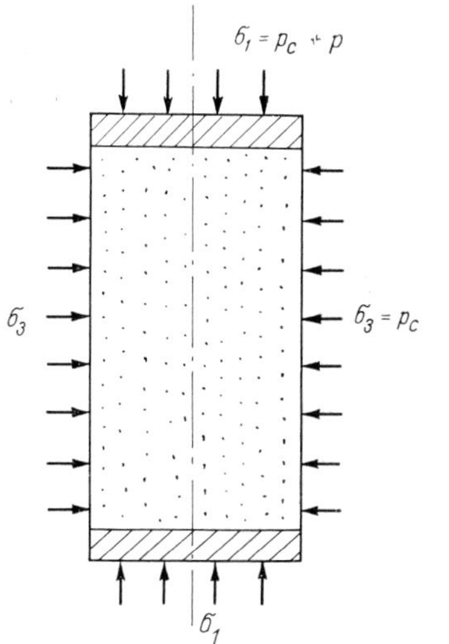

Fig. 264. Principle of triaxial compression test

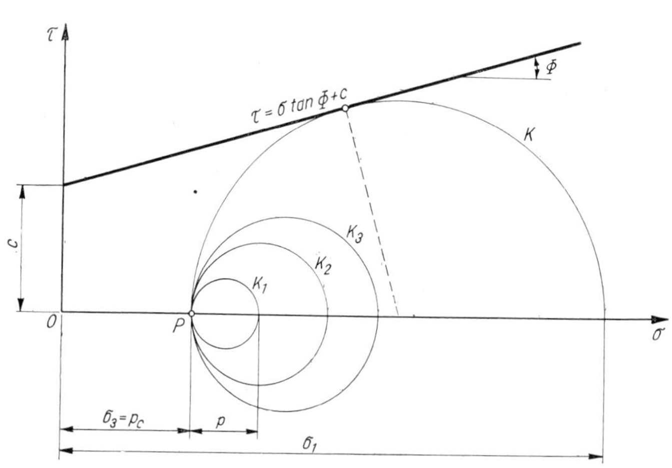

Fig. 265. Stress circles for different stages of testing specimens in triaxial compression

The principle of the triaxial cell is illustrated in *Fig. 264.* A cylindrical specimen of soil which is enclosed in a thin rubber membrane is placed inside the cell. The specimen is first subjected to a constant all-around pressure p_c. Water is allowed to escape through a porous disc inserted into the base of the cell, thus permitting the dissipation of the neutral stresses. In this initial state of stress, called isotropic compression, the *Mohr* circle reduces to a point. (Point P in *Fig. 265.*) The axial stress is then increased by some value p. The altered state of stress is defined as ($\sigma_1 = p_c + p$, $\sigma_2 = \sigma_3 = p_c$). The deviator stress, p, is then gradually increased, while the confining lateral stress is held constant, until the specimen fails in shear. At the ultimate stage the corresponding *Mohr* circle (K) just touches the *Coulomb* line. Also shown in the figure are the *Mohr* circles, K_1, K_2 and K_3, representing intermediate stages of loading. By performing several tests at different initial confining pressures, a series of *Mohr* circles is obtained, whence the shear strength parameters, Φ and c, can graphically be determined.

If it is required to express the shearing strength in terms of effective normal stresses, then the pore pressures must be continuously measured during loading. It is also important to measure the changes in the volume of the specimen. A complete plot of triaxial test data is given in *Fig.* 266.

As was pointed out previously, time effect is another important factor to be considered in the determination of shearing strength. Accordingly, the triaxial test may be conducted in several ways.

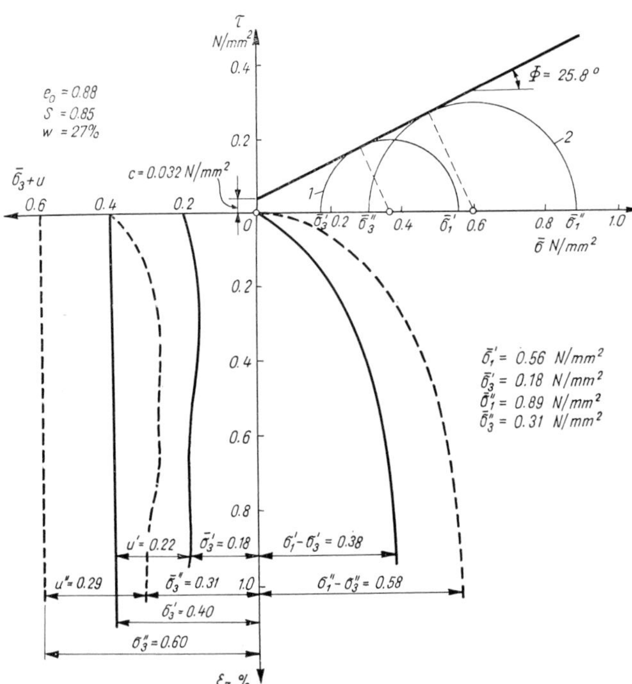

Fig. 266. Presentation of results of triaxial compression test

Usually, we start by applying a constant hydrostatic pressure to the specimen and then wait until consolidation is complete. If the valve in the drainage line (Fig. 1) is closed while the axial load is being increased, then the moisture content of the specimen cannot change during the test. Such condition has been termed in connection with the direct shear test a ***consolidated quick test.*** The testing apparatus is so designed that with the outlet valve shut no change in moisture content can occur inside the specimen regardless of the rate of axial loading. Failure, too, occurs at the initial moisture content. If, on the other hand, the outlet valve is open while the axial load is being applied at a sufficiently slow rate to permit the dissipation of pore pressure, the water content within the specimen can continuously adjust itself to the acting load (*slow test*).

In both cases, the specimen fails in shear along rupture planes. It is impossible, however, to directly measure the stress components acting on a slip surface. We can only measure, by means of manometers and spring type dynamometers, the principal stresses, σ_1 and σ_3, acting at the instant of failure.

7.5.5 Tensile strength of the soil

It seldom occurs that we need to know the tensile strength of the soil. Nevertheless, special problems of stress computation, the design of road pavements, or the stability analysis of slopes may call for the determination of the tensile strength.

The test could be performed on X-shaped specimens similar to the briquettes used in the testing of Portland cement. But the cutting and trimming of such specimens is rather difficult; moreover, the state of stress in the specimen is not simple axial tension, as assumed, because the central part with uniform cross-section is usually too short. Better results can be obtained with an apparatus in which a cylindrical specimen is placed between glass plates. In this way a firm grip can usually be attained, the adhesion between glass and clay being greater than the tensile strength of the soil.

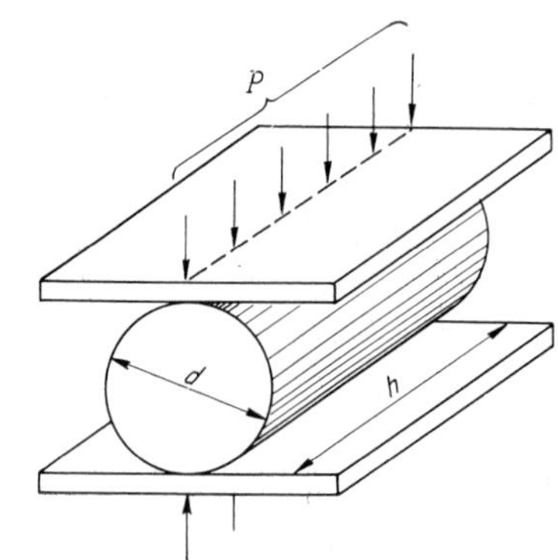

Fig. 267. Split testing of cylindrical specimen

By performing a simple axial tension test and compression tests (unconfined or triaxial) on the same soil and plotting *Mohr* circles respresenting failure, we may obtain the *Coulomb* envelope and determine the angle of internal friction and the cohesion of the soil.

For stiff clays or stabilized soils, which are likely to fail in brittle fracture, the tensile strength can be best determined by a test adopted from concrete testing known as splitting test or Bra-

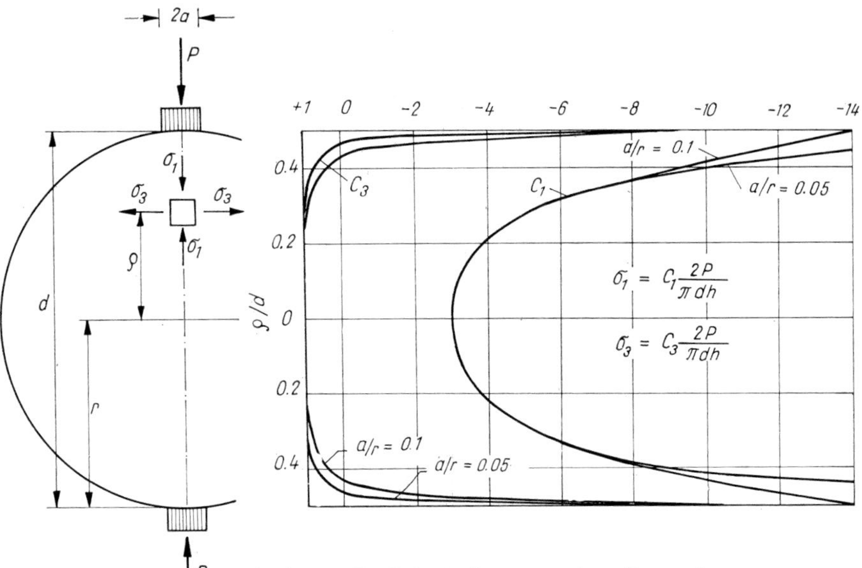

Fig. 268. Principal stresses in split testing

silian test. A cylindrical specimen laid on the side is placed between two rigid metal platens as shown in *Fig. 267*. The load P is gradually increased until the specimen fails. In the vertical section of the specimen, a composite state of stress develops with the major principal stress σ_1 being compressive and the minor principal stress σ_3 being tensile. The principal stresses are given by

$$\sigma_1 = C_1 \frac{2P}{\pi \, dh}$$

and

$$\sigma_3 = C_3 \frac{2P}{\pi \, dh} .$$

The variation of the coefficients C_1 and C_3 along the vertical section is as shown in *Fig. 268*.

By running a splitting test and an unconfined compression test on the same soil, we obtain two Mohr circles belonging to failure. The common tangent to them will be the *Coulomb* envelope from which the angle of internal friction Φ and the cohesion c can be determined. Analytically, Φ and c are given by the following expressions (Nagyváti, 1958):

$$\tan\left(45° - \frac{\Phi}{2}\right) = \sqrt{\frac{\sigma_3}{\sigma_u - 3\sigma_3}}$$

$$c = \frac{\sigma_u}{2} \sqrt{\frac{\sigma_3}{\sigma_u - 3\sigma_3}} .$$

7.6 Shear strength of cohesionless soils

7.6.1 Factors affecting internal friction

The shear strength of sands and gravels is made up, in general, of two components: firstly, the frictional resistance to the relative movement between the particles which in turn is partly sliding friction, partly rolling friction; secondly, the structural resistance due to the interlocking of the particles. These two effects are illustrated in *Fig. 269*. In the case of a loosely packed sand, to start shear it is only necessary to make the particles slide upon one another. The small slip

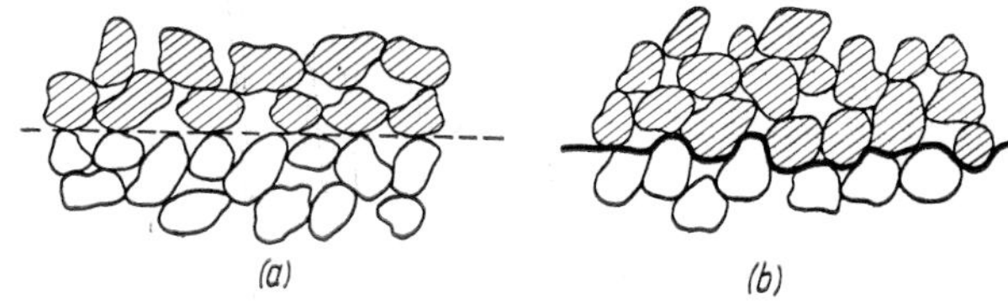

Fig. 269. Shear resistance in granular soils due to (a) — surface friction and (b) — interlocking

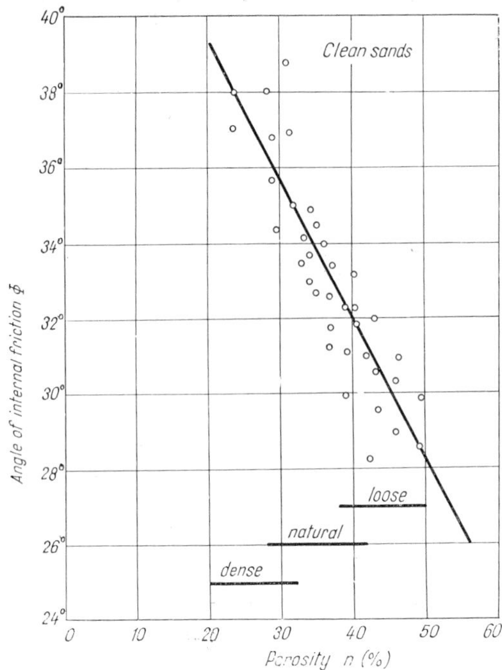

Fig. 270. Angle of internal friction of silt-free sands versus initial void ratio

surfaces lie approximately in the same plane. In the second case (dense sand) a shear displacement is possible only if the particles do not merely slip along the plane of shear but also move upwards and roll over one another.

The effect of the interlocking of the particles is especially significant in dense sands. Their angle of internal friction as determined by the direct shear test depends greatly on porosity. To illustrate this point, *Fig. 270* shows an empirical relationship, compiled from a large number of direct shear tests, between friction angle values and initial porosity. The plotted data represent clean sands containing no particles smaller than 0.02 mm in size.

That the angle of internal friction decreases rapidly with increasing amount of silt, or even of fine sand, in a coarse-grained soil is clearly demonstrated by *Fig. 271*, in which measured Φ values for medium dense sands (n = 30 to 40%) are plotted against the percentage of fractions smaller than 0.1 mm. It can be seen that up to about $S_{0.1} = 10\%$ the effect of the fine fractions is negligible. For higher S values a rather steeply descending straight fit can be made and e.g. a silt content of 20 to 25% reduces the friction angle from its value of 35° obtained for clean sand to 28 to 30°.

Typical results of direct shear tests on dense and loose sands are shown in *Fig. 272*. For loose sands the Coulomb straight fit holds, while for dense sands the failure envelope is a slightly curved line.

From Fig. 269 it is clear that the angle of internal friction is also influenced by the shape of the grains and by the uniformity of the grain-size distribution. Some representative values of Φ are given, after TERZAGHI and PECK (1948), in *Table* 27.

Table 27. Angle of internal friction; limit values for sand (Terzaghi and Peck, 1948)

	Round grains, uniform grain-size distribution	Angular grains, well graded
Loose sand	28.5°	34°
Dense sand	35°	46°

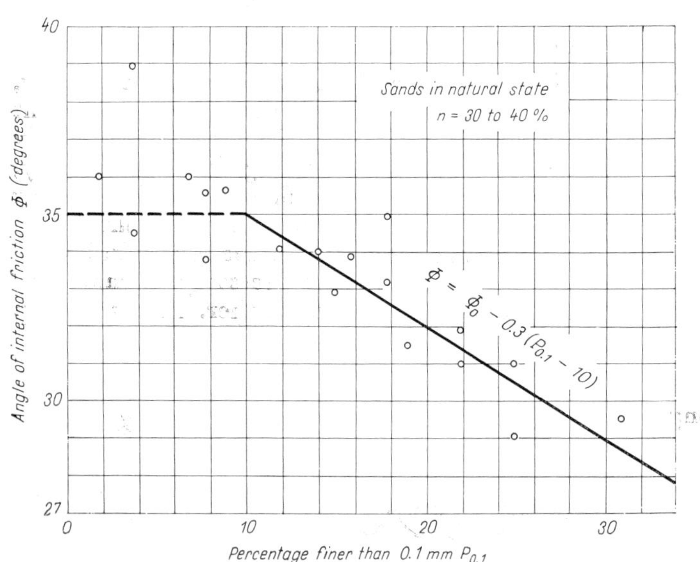

Fig. 271. Decrease of the angle of internal friction with increasing silt content

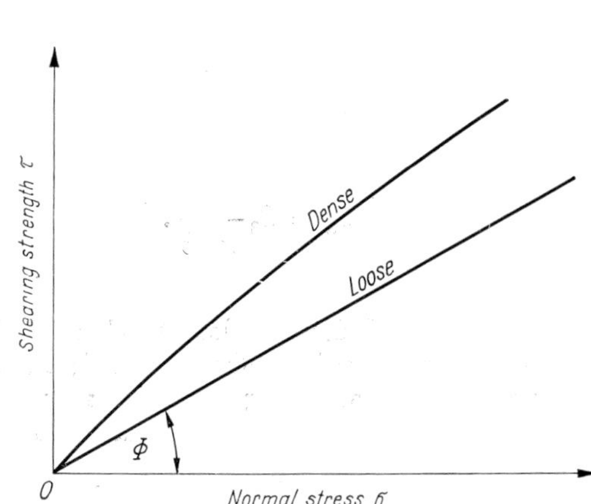

Fig. 272. Shear strength of dense and loose sand

There are several empirical formulae in use by means of which the various factors affecting the shearing strengths of granular soils can be taken into consideration. The following formula, suggested by LUNDGREN (1960), has proved very useful for estimating the friction angle of sands and gravels

$$\Phi = 36° + \Phi_1 + \Phi_2 + \Phi_3 + \Phi_4. \qquad (202)$$

Herein the constant term, 36°, represents the friction angle of an "average" sand, and Φ_1 to Φ_4 are correction terms. Grain shape is taken into account by the term Φ_1 as follows:

Angular grains	+1°
Subangular grains	0°
Rounded grains	−3°
Well-rounded grains	−6°

The correction for grain size, Φ_2:

Sand	0°
Fine gravel	+1°
Medium and coarse gravel	+2°

The correction for uniformity of grading, Φ_3:

Poorly graded soil, small *u* value	−3°
Medium uniformity	0°
Well-graded soil	+3°

The correction for density, Φ_4:

In loosest packing of soil	−6°
At medium density	0°
In densest packing	+6°

According to the above values, the angle of internal friction ranges between 20° and 48°. The most important factor is density. Numerical data can be derived from experiments by CHEN (1948). He found that for any given sand the following relationship holds:

$$\Phi° + a \log e = \text{constant}. \qquad (203)$$

In the expression *e* is the void ratio, and *a* is a constant with an average value of 60°.

To take the effect of density into consideration, WINTERKORN (1960) suggested on the basis of analogies the following formula

$$\tan \Phi = \frac{C}{e - e_{min}}. \qquad (204)$$

e is the void ratio, *C* and e_{min} are material constants. KÉZDI (1966) presented a theoretical verification of Eq. (204) by the following reasoning. If we load a granular mass, it will in general undergo both a decrease in volume and angular distortions. Pure compression is caused by hydrostatic pressure, while angular distortion is due to shearing stresses. Let us assume as a first approximation a linear relationship between the volume change and the applied hydrostatic pressure σ. If V_0 is the volume in the densest state of the mass, then

$$\frac{V - V_0}{V_0} = \frac{\Delta V}{V_0} = \frac{\sigma}{E}$$

and

$$\sigma = \frac{E}{V_0}(V - V_0) = C_1(V - V_0),$$

E is the modulus of elasticity.

Under the action of shear stresses shear deformations develop in the loaded mass. The relationship between the angular distortion β, ($\beta = 90° - \alpha$), and the applied shear stress τ can be expressed on the basis of fluid mechanics analogies as

$$\tau = -\frac{\partial F}{\partial \beta}.$$

F is the free energy of an element of the granular mass with respect to an adjacent element. This free energy is, in the general case, made up of two parts: a static energy, which is a function of the equilibrium array of the particles, and a thermal energy. For the problem considered, this part is negligible. The potential energy is proportional to the difference in height, $h - h_0$, resulting from the movement of the particles. h_0 is the minimum value of *h*. For the sake of simplicity, let us assume that the relationship $h = h(\alpha)$ can be replaced by a straight line. Hence $\partial F/\partial\beta$ = constant and also the shear stress τ = constant = C_2. Therefore, the coefficient of friction can be written as

$$\tan \Phi = \frac{C'}{V - V_0} = \frac{C}{e - e_{min}}. \qquad (204)$$

A better approximation can be obtained if we adhere to the actual stress-compression relationship. Then the expression of the coefficient of friction becomes

$$\tan \Phi = C \frac{\sqrt[3]{e - e_{min}}}{\exp\left(\frac{e - e_{min}}{k} - 1\right)}, \qquad (205)$$

where *k* is a material constant (KÉZDI, 1964).

As we have just seen, the shear strength of granular soils is mainly the result of rolling friction and structural resistance. For a given density, the shear strength becomes

$$\tau = \bar{\sigma} \tan \Phi,$$

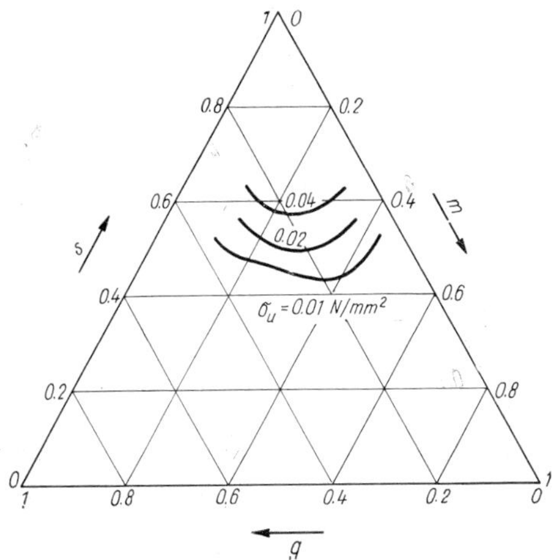

Fig. 273. Unconfined compressive strength of fine sand, depending on the phase composition

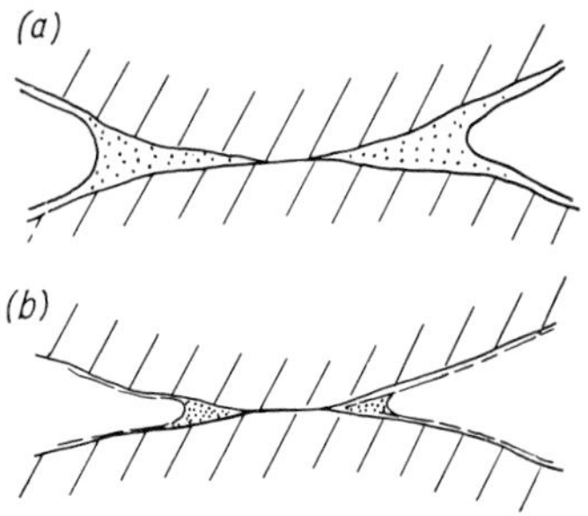

Fig. 274. Effect of small floating solid particles on the cohesion of granular soils

where $\bar{\sigma}$ is the effective normal stress acting on the plane of shear. The angle Φ' is generally regarded as a constant, although its value is slightly influenced by the magnitude of the shear stress and of the intermediate principal stress and by the velocity of shear motion.

Moist sands often possess some cohesion due to capillarity effect (Section 3.2). Its value depends, besides grain size and grading, primarily on the degree of saturation. *Figure 273* shows, in a triangular plot, the relationship between the unconfined compressive strength of a fine sand and the volumetric ratios *s*, *m* and *g*. Below and above certain values of *g* the unconfined compressive strength is seen to be zero. This indicates that both drying out and saturation with water may cause the cohesive strength of the sand to vanish.

It sometimes occurs in sands that the grains are bound together at the points of contact by some kind of natural binder (e.g. calcareous sands). This again leads to cohesive strength within the sand. A similar effect accounts for the often high compressive strength of certain well-graded soil mixtures used for the purpose of "mechanical soil stabilization". Such mixed materials always contain very fine clay-size particles.

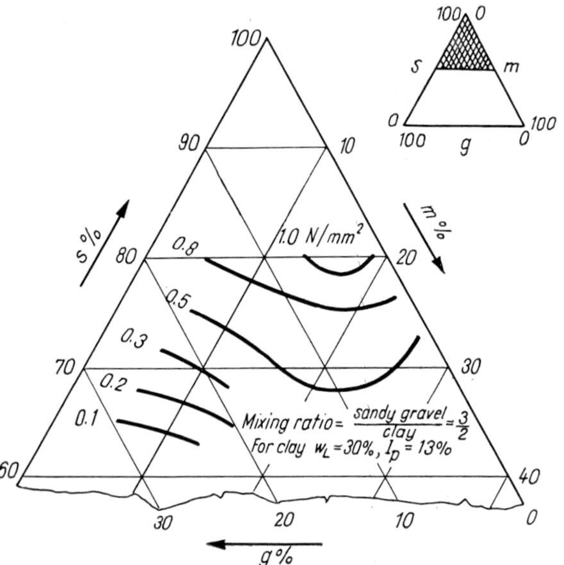

Fig. 275. Unconfined compressive strength of mechanically stabilized soil mixture depending on the phase composition

If the soil becomes too wet, the fine particles get loose and float in the pore water which fills the interstices between the coarser grains. If, on the other hand, the soil dries out, the fine particles tend to pack into the smallest possible space as shown in *Fig. 274.* Having reached their tightest packing, they do not permit any further contraction of the menisci. Stresses develop which, though they may considerably increase as drying continues, act practically on the same surface area throughout the process of desiccation. Therefore the interparticle attraction due to capillarity

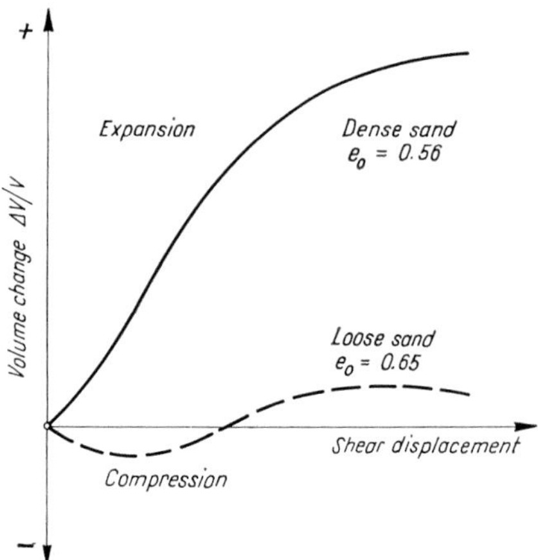

Fig. 276. Volume change of sand during shear

will greatly increase leading to an equally increased structural strength. This type of soil strength depends on the properties of the binder, on the strength of the mentioned bonds, furthermore, on the density and moisture content of the soil mixture. Representative test results are given in *Fig. 275*, where unconfined compressive strength values are plotted as a function of the phase composition of the mixture.

7.6.2 Volume change during shear

If we observe the volume changes of a sand sample during a shear test, we notice that a dense sand tends to dilate, whereas a loose sand will decrease in volume. The effect of shear to the volume of the sand is illustrated in *Fig. 276.* Let us assume first that the sand is dry so that the effect of the neutral stresses can be disregarded. If we plot the shear displacement and the volumetric strain as functions of the shearing stress (*Fig. 277*), we find that in dense sands the shearing resistance attains a peak value, τ_{max}; with continuing shear displacement it drops back to a lower ultimate value τ_1 and remains at that constant level during further shear (curve *1*). As

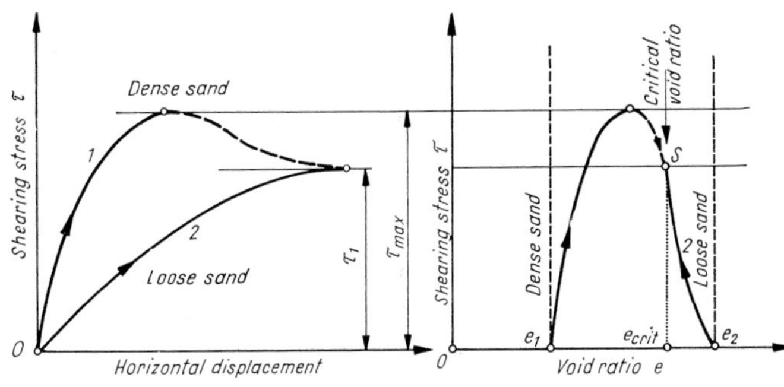

Fig. 277. Shear tests with loose and dense sand:
a — shear displacement; *b* — changes in void ratio during shear

the shear stress decreases from the peak value, the expansion of the sand gradually ceases, to become a decrease in volume until the sand eventually reaches a certain density, marked by point *S* in Fig. 277*b*, at which continuous shear displacement at a constant shear stress becomes possible.

This density lies between the two extreme values representing the densest and the loosest states of the sand. It is called, after CASAGRANDE (1936), the *critical density*. Its value depends chiefly on the uniformity of the material. The more uniform the sand, the lower is its critical density.

In the loose state, most sands are liable to decrease in volume when subjected to shear under constant normal load. The shear stress gradually increases until it reaches the ultimate value τ_1. Thereafter it remains constant even at further shear displacement. In the final stage the void ratio of the sand must correspond to the critical density. It follows that the stress–strain curves representing dense sand (curve *1* in Fig. 277) and loose sand (curve *2*) must merge at the critical density (point *S*).

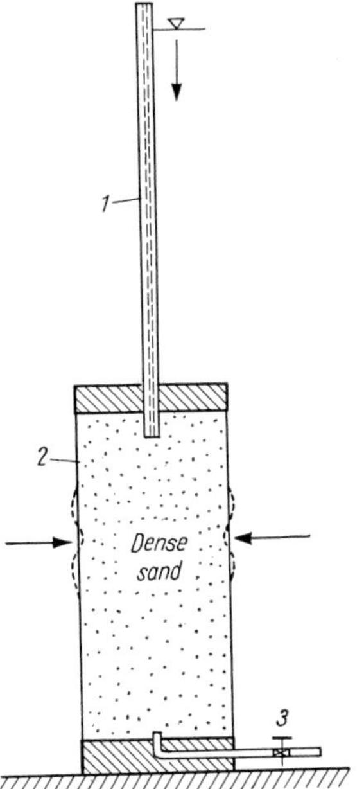

Fig. 278. A simple device to demonstrate loosening of dense sand during shear:
1 — standpipe; *2* — rubber membrane; *3* — drain valve

The fact that a dense sand dilates under shear can readily be demonstrated by the simple setup shown in *Fig. 278*. Sand is compacted into a rubber membrane (*2*) so as to form a cylindrical specimen. The specimen is then saturated with water through the base inlet (*3*) until the excess water overflows at the top of the standpipe (*1*). Valve (*3*) is then closed. If we now apply a sudden strong pressure by two fingers to the rubber membrane, the specimen will bulge because of the large shear deformations taking place in it. The volume of the specimen, i.e. the total volume of the void spaces, will increase and, as a result, the water level in the standpipe will suddenly drop. If we compact the specimen again, the water will rise to its original level.

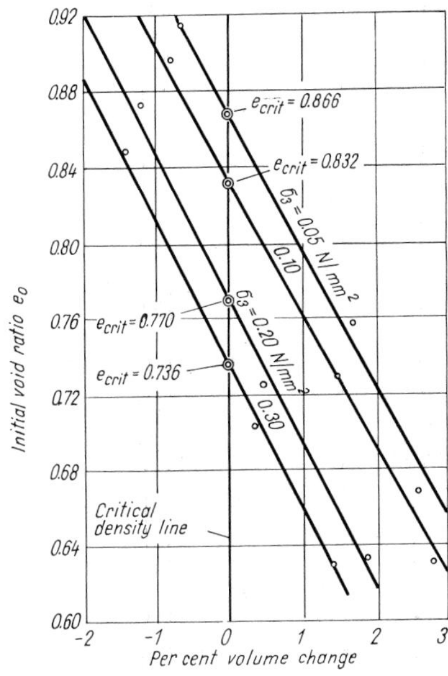

Fig. 279. Determination of critical density

Numerical values for the critical density can be obtained by triaxial tests. Using the apparatus shown in Fig. 153, we measure the volume change at failure of the specimen. If we run series of triaxial tests with several constant confining pressures, σ_3, and then plot the volume changes at failure against the initial void ratios, straight lines will result (*Fig. 279*). The points at which these lines intersect the vertical line pertinent to zero volume change represent the critical void ratios. e_{crit} is seen to decrease with increasing confining pressure, σ_3. It follows that at very high lateral pressures the void ratio after shear failure is always smaller, whereas at very low pressures it is always larger than the initial value. At intermediate σ_3 values expansion or contraction will occur depending on whether the initial void ratio is smaller or larger than the critical void ratio.

It is chiefly on account of the volume changes accompanying shear that the shearing strength of granular soils cannot be treated as a simple phenomenon. The main difficulty is not so much to measure the true friction angle Φ', in the fundamental equation, $\tau = \bar{\sigma} \tan \Phi'$, as to determine the effective normal stress $\bar{\sigma}$. Should there be any restraint that prevents volume changes to take place during shear, neutral stresses will develop, which must be measured so that $\bar{\sigma}$ can be calculated. The mechanism of volume changes is very complex. In the literature we often find conflicting views concerning this problem. This controversy is probably due to the fact that the experimental data upon which theoretical considerations are based are inherently biased by the testing apparatus and testing technique.

Let us examine in the following the shearing strength of water-saturated sands.

In the pore water, if both the sand mass and the ground water are at rest, there exists a hydrostatic pressure controlled only by the depth below the free water table. External loads are transferred from grain to grain. In sands, the effect of the adsorbed water film is negligible; water-saturated sands exhibit therefore no cohesion. If we run drained shear tests on a saturated sand, then, with the porosity held constant, we obtain friction angles lower by only 1 to 2° than in the case of a dry sand. This indicates that the water has no lubricating effect and the angle of internal friction remains practically the same. Consequently, all the properties of sand that are derived from friction such as the principal stress ratio at failure or the earth pressure coefficient at rest also remain unaltered. But as soon as shear deformations accompanied by volume changes occur, the conditions will be essentially changed in that, owing to the incompressibility of water, any volume change must initiate a movement of water. If a saturated sample of sand is first loaded under drained conditions until consolidation is complete, and then is sheared under undrained conditions, a portion of the stress induced by shear will be carried by the pore water. In such cases the shear strength depends on whether the sand has a tendency to expand, or to contract, during shear, i.e. whether its actual void ratio is below or above the critical void ratio. If the porosity tends to decrease, excess hydrostatic pressures build up in the pore water with the result that the effective normal stress acting on the potential slip surface is reduced and the sample fails at much smaller shear stresses than it would under drained conditions. If the sample shows a tendency to expand (actual porosity is below critical density), the opposite effect will occur.

Figure 280 shows typical results of shear tests on saturated sands. The solid lines represent drained tests, with shear stresses given in terms of effective normal stresses. The slope angle is Φ' and there is no cohesion intercept. If the soil is first allowed to consolidate under a given vertical load and sheared thereafter under undrained conditions, then any resulting volume change will induce neutral stresses in the pore water. As we have just seen (Fig. 279), the critical density depends on the hydrostatic consolidation pressure σ_3. The higher this pressure, the lower is the critical void ratio. As in sands the volumetric strain caused by isometric compression is very small, the void ratio along the solid lines *1* in Fig. 280 can be regarded as practically constant. There must exist then a critical normal stress σ_{crit} such that at stresses larger than this the critical void ratio of the sand is smaller than the void ratio consistent with line *1*. This means that in the stress regime $\sigma > \sigma_{crit}$ the sand has a tendency to contract during shear, but this is prevented under undrained conditions thereby giving rise to excess pore pressures. If we plot the results of the shear test in terms of total normal stresses, the shearing strength τ for values $\sigma > \sigma_{crit}$ will be smaller than would be obtained in a drained test. If on the other hand $\sigma < \sigma_{crit}$, the critical void ratio is larger than the actual void ratio. The soil has now a tendency to dilate, which, when suppressed, induces negative pore pressures in the soil. The effective normal stress will increase and so will the shearing strength τ. The failure envelope obtained from a consolidated—undrained test will therefore be similar to the dashed line *2* shown in Fig. 280*a*. In the stress regime $0 < \sigma <$

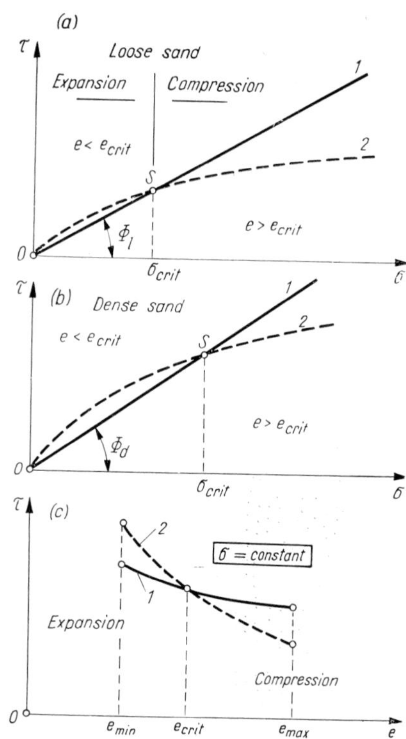

Fig. 280. Results of drained (1) and undrained (2) shear tests on saturated samples of sand:

a — loose sand; *b* — dense sand; *c* — relation between shearing resistance and initial void ratio for saturated samples of sand after complete consolidation under given pressure

$< \sigma_{crit}$ line *2* runs above line *1* and at stresses $\sigma > \sigma_{crit}$ below it. If e just equals e_{crit}, the drainage conditions have no influence on the shearing strength, since neither expansion nor contraction take place and the pore pressures remain unaltered during the shearing process.

If we perform the above described test on dense samples of sand, the resulting failure envelopes (Fig. 280*b*) will show a similar character except that the slope of line *1* and the critical normal stress σ_{cr} will be considerably larger than they would be in the case of a looser sand.

In Fig. 280*c* the void ratios of saturated samples, consolidated under a given constant pressure σ, are plotted as abscissae, and shear strength values as ordinates. Drained shear is represented by the solid line *1* and the undrained case by the dashed line *2*. For a given consolidation pressure the critical void ratio is constant. If the actual void ratio of the sand is smaller than e_{crit}, then shearing causes the expansion of the sand, or, in the undrained case, it induces negative pore pressures. Consequently, the effective normal stress increases and the shearing strength will also be greater than it would be in the drained case at the same values of σ and e. If $e > e_{crit}$ the shearing causes expansion. In an undrained system this is accompanied by a build-up of excess pore pressures. A positive neutral stress reduces the effective normal stress with the result that the shearing strength will be smaller than it would be in the drained case at unchanged values of σ and e. Finally, at the critical density when the volume of the sand neither increases nor decreases during shear, the drained and the undrained tests yield the same result.

A study of the critical density is essential in the stability analysis of earth dams consisting of very fine sand, especially if the dam is acted upon by a water pressure on one side only. As we have seen, the shear strength of saturated sands depends not only on the angle of internal friction and the normal stress due to the overburden pressure acting on the potential surface of sliding but also on the density of the sand and the rate of increase of the shearing stresses. It is the actual value of the density which indicates whether for a given normal stress, the sand is in a state above or below the critical density, i.e. whether negative or positive pore pressures are more likely to occur during shear. The rate of increase of the shearing stress must also be taken into consideration as the main factor controlling the degree of the drainage of the pore water, i.e. whether in a given case the system should be regarded as drained or undrained.

Prevented tendencies toward volume changes are also responsible for what is known as the liquefaction of saturated sands.

Loose saturated sand deposits may be very troublesome during construction works. They may under certain conditions completely lose their shear strength and behave like viscous fluids. In such a state, the sand literally floods the bottom of an excavation, it is incapable of carrying loads and causes slides even on gentle slopes, etc. For long such phenomena were believed to be due to a peculiar kind of soil, called "quicksand", an evil dreaded by many engineers engaged in subsurface works in low-lying lands, e.g. the Great Hungarian Plain. Unquestionably, grain shape and grading, i.e. the type of soil, must have a bearing on the susceptibility to liquefaction of the soil. Yet, the main cause is found in the changes in the pore pressure due to load. It is an established fact that every granular soil can be brought into fluid state by seepage pressure under a sufficiently high gradient. On the other hand, there are certain sand deposits with a marked tendency to spontaneous liquefaction. Present information available on the subject does not yet permit the straightforward laboratory determination of this property. What we know for certain is that such soils show extremely high porosity after laboratory sedimentation; their grains are well-rounded and they contain a certain amount of silt-size particles. Their effective diameter d_e is smaller than 0.1 mm, their uniformity coefficient is smaller than 5 and their natural porosity is at least $n = 44\%$. However, the listed properties are necessary but not satisfactory conditions.

The liquefaction of saturated sands can be explained as follows.

If a sudden shock or dynamic effects cause rapid local deformations within a loosely packed saturated sand mass, then at the first instant the induced stresses are entirely carried by the pore water. Since very little time has passed for the consolidation to take place, the soil mass practically behaves as an undrained system. As a result a local breakdown of the hitherto stable solid skeleton takes place and some of the grains deprived of their supports begin to float in the pore water surrounding them. Since the overall value of the total stress has not changed appreciably, the sudden increase of the neutral stress must result in an equal decrease in the effective stress, leading to a serious reduction of the shearing strength. Locally even a complete loss of the bearing capacity of the soil may occur, and the stresses formerly carried by the affected zone of the soil mass will be transferred to the adjacent stable parts. In a similar manner the process of stress transfer may spread vehemently until the entire soil mass is turned into a thick suspension or viscous fluid in which individual solid grains float in the pore water. Note that the process takes place very rapidly so that the system remains practically 'undrained' throughout allowing no time for the dissipation of excess pore pressure. A similar phenomenon is caused by the seepage pressure of water percolating through the soil. The resulting complete loss of bearing capacity of the soil is called *quicksand condition.*

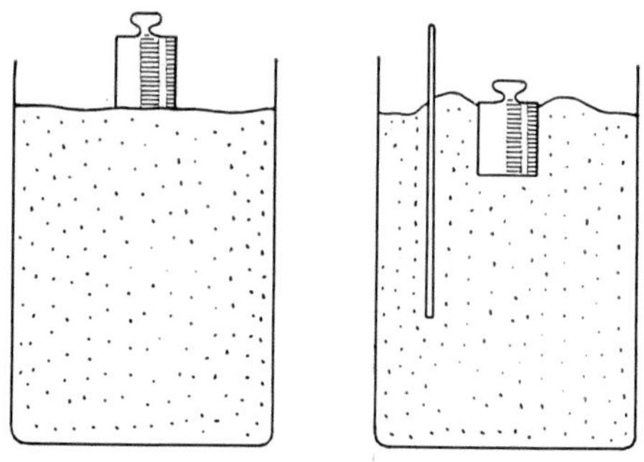

Fig. 281. Demonstration of quick sand condition

The liquefaction of a saturated loose sand can be demonstrated after CASAGRANDE (1938) by the following simple experiment (*Fig. 281*). Dry sand is poured gently in a water-filled container so that a loosely packed saturated sample is obtained. The top surface of the sample is just covered with a thin layer of free water. If we carefully place a weight on the top of the sample, it will sink only little into the sand, as the skeleton formed by the solid grains possesses a certain load bearing capacity. If we now plunge a thin rod into the sample, the weight will suddenly sink below the surface. Such a local deformation which is small but sudden enough to make the sample behave as if it were undrained is sufficient to trigger a rapid spreading of stress transfer and cause an almost immediate liquefaction of the sand.

The stability of sands is of great importance in the design of embankments and dams. Contrary to the widespread belief that a sand mass is always stable as long as its slope angle is smaller than the natural angle of repose of the material, there is ample evidence to show that dams built on fine sand, and judged safe, have failed through sudden liquefaction and complete loss of strength of their material. Because of the sudden and hardly predictable character of the failure, such events often had disastrous consequences. The triggering causes are numerous. In fact anything that causes a local overstressing, beyond the elastic limit, within the sand mass leading to plastic deformations may produce liquefaction. Earthquake, pile driving, sudden local increase of surcharge load, rapid excavation of cuts, ground water fluctuations, rainstorm effect on fills, rapid drawdown in dams, mechanical vibrations, etc. can be mentioned as typical causes.

In this connection an interesting case was described by BERNATZIK (1948). After the first world war, a war memorial was erected at Campione by the Lake of Lugano, on an artificial peninsula made of sand fill. Following a high water period, the lake level suddenly sank; this initiated an intense ground water flow towards the lake, and the sand fill slid into the lake and also the memorial sank to such a depth that every attempt to recover it was futile.

Returning now to the stress–strain behaviour of granular soils, let us determine quantitatively the effect of volume change on the angle of internal friction.

In the case of the direct shear test, the change in the volume of the sample can be simply expressed as a change of the initial height h by an amount of Δh. Let us introduce the following notation

τ — shearing stress on the failure plane
$\bar{\sigma}$ — effective normal stress on the failure plane
$\Delta\tau$ — shearing stress increment
Δs — shear displacement due to a shearing stress increment $\Delta\tau$
A — cross-sectional area of the shear plane.

The work done as the shearing stress τ is being increased by $\Delta\tau$ can be assumed to be composed of two parts: one expended to actually produce volume change. Hence we get

$$\left(\tau + \frac{\Delta\tau}{2}\right) A\,\Delta s = \Delta W + \bar{\sigma} A\,\Delta h\,.$$

ΔW represents that portion of the total work which has been used up for shear. This can be expressed as

$$\Delta W = \tau' A\,\Delta s$$

where τ' is the stress required to produce shear independent of volume change within the sample. Substituting in the first equation and neglecting the term infinitesimal of higher order gives

$$\tau'\,\Delta s = \tau\,\Delta s - \bar{\sigma}\,\Delta h$$

or

$$\frac{\tau'}{\bar{\sigma}} = \frac{\tau}{\bar{\sigma}} - \frac{\Delta h}{\Delta s}\,.$$

In the state of failure τ is equal to the ultimate shearing strength τ_s, further,

$$\frac{\tau_s}{\sigma} = \tan\Phi\,.$$

Hence the angle of shearing resistance Φ_r can be expressed in terms of the shearing stress τ' as

$$\tan\Phi_r = \frac{\tau_s'}{\bar{\sigma}} = \frac{\tau_s}{\bar{\sigma}} - \frac{\Delta h}{\Delta s} = \left(\tan\Phi - \frac{\Delta h}{\Delta s}\right)_{\max}.$$

Δh is positive when the sand is dense and an increase in volume occurs during shear; at the same time Δs is small. For a loose sample, Δh is negative and Δs is large. Φ_r depends only slightly on the porosity and should, therefore, be considered much more a material property than the friction angle Φ. Comparative values of Φ and Φ_r for a medium sand are given as functions of the initial porosity of the sample in *Fig. 282*. Φ_r is controlled chiefly by the coefficient of particle-to-particle

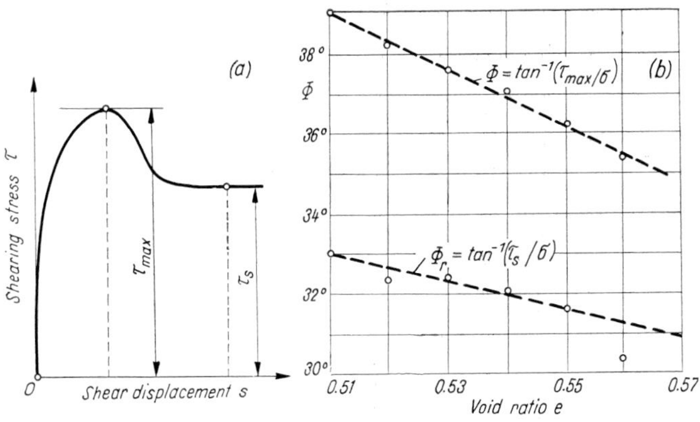

Fig. 282. Components of the angle of internal friction due to surface friction and interlocking

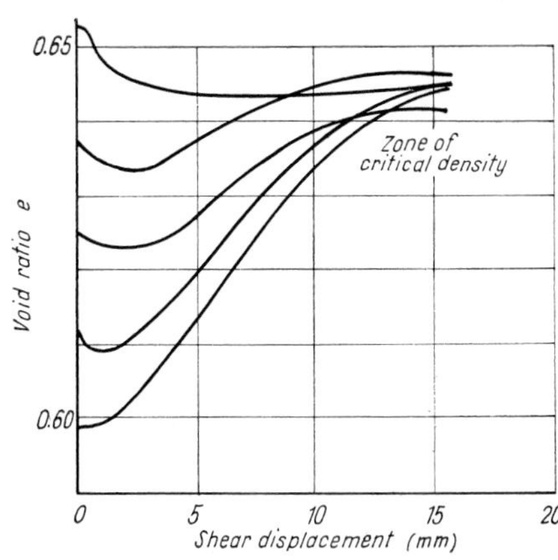

Fig. 284. Variation of void ratio during shear in tests with different initial void ratios at constant normal stress

friction. A generalized application of the above described analysis to problems of three-dimensional, axially symmetrial state of stress was given by KÉZDI (1963).

The two effects controlling the shearing resistance of sands, namely the structural resistance and the particle-to-particle friction, can best be separated by calculating friction angle values from peak shearing resistance, Φ_{max}, and then from ultimate shearing resistance, Φ_u, and plotting them against the initial void ratios as shown in *Fig. 283.* The difference $(\Phi_{max} - \Phi_u)$ is chiefly due to structural resistance, whereas Φ_u is predominantly due to sliding friction between the grains. For practical purposes, it seems appropriate to choose Φ_u to represent the shearing strength of the sand.

The volume change, or more correctly the tendency to volume change, is of great significance as the main factor controlling the build-up of pore pressures within a sand during shear. It also greatly influences the mechanism of failure in such problems as earth pressure on retaining walls, passive resistance of the soil to external loads or the load bearing capacity of the soil under foundations etc. If in an initially dense sand the shearing stress has somewhere reached its peak value τ_{max} and a definite surface of failure has once developed, then shear motion will continue on the surface already available, even though the shearing stress reduces later to some lower value and the resultant stress makes an angle smaller than the peak friction angle Φ_{max} with the normal to the sliding surface. There is no reason why failure should start again along some other surface. In contrast to this, in a loose sand a local increase of the shearing stress causes a decrease in volume at that point. But since the compacted soil will possess greater shearing strength, shear will cease at that place to start again somewhere nearby where the sand has still remained loose. That is why in loose sands we cannot think of a general shear failure along definite surfaces of rupture; instead, the failure assumes the form of *local shear failure.* (See Vol. 2.)

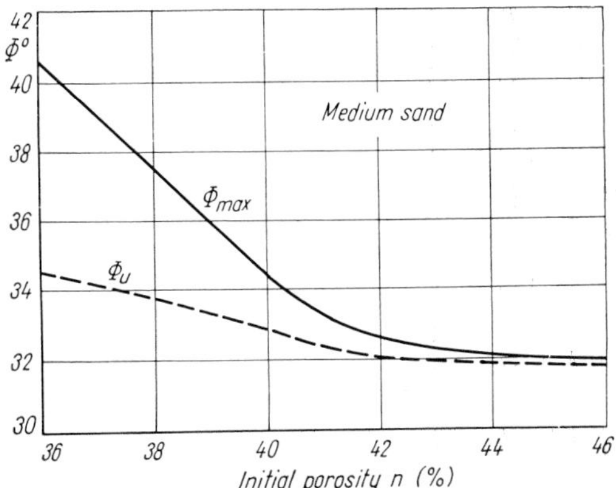

Fig. 283. Angle of internal friction as determined by using peak and ultimate values of shear strength versus initial porosity

A comprehensive treatment of the general normal stress-shearing strength–void ratio relationship was given by ROSCOE, SCHOFIELD and WROTH (1958). From direct shear tests they obtained the results shown in *Fig. 284.* Here the

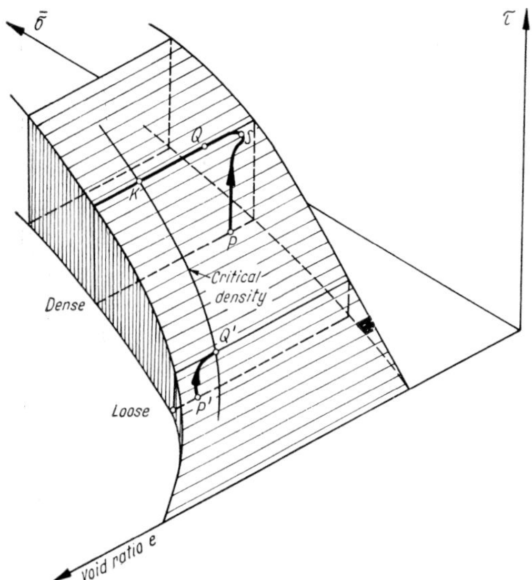

Fig. 285. Relationship between effective normal stress, shearing strength and void ratio at failure

development of the critical state can be clearly recognized. If we plot the same results in a three-dimensional ($\bar{\sigma}$, τ, e) diagram, the picture shown in *Fig. 285* is obtained. With the aid of this graphical representation, we can exactly trace the progress of a direct shear test. For example, the sample is first consolidated under a hydrostatic pressure whereupon its void ratio decreases. This state is represented by point P. With the effective normal stress kept constant, increasing shearing stresses are next applied to the specimen until it fails by shear. In the course of the straining, the specimen is first slightly compressed (points S), then it continuously dilates (an intermediate state is indicated by point Q) before it eventually fails (point K). During the shear motion the void ratio increases with increasing shearing stresses until the critical void ratio is reached. The foregoing statements refer to dense sand. If the initial void ratio is high, only a compaction of the sand takes place during shear.

7.6.3 Some other factors influencing the shear strength of granular soils

In the following section we discuss some further effects on soil strength that are of direct consequence to practical engineering problems.

Let us examine first the role of the normal stresses. As was shown previously, the shear strength is made up, in part, of frictional resistance to relative sliding of the grains and, in part, of structural resistance due to interlocking. The first part may be assumed to be proportional to the normal stress. As for the second part, linearity holds only for high normal stresses and when the soil is loose to medium dense. At small normal stresses the particles are relatively free to move and the interlocking effect is more pronounced. Consequently, the initial portion of the failure envelope is not straight but bulging as shown in *Fig. 286*. The prolongation backwards of the straight section normally passes through the origin 0. The convex section $\overline{Oa}$ extends to a normal stress of 20 to 30 kN/m². This peculiarity of the *Coulomb* envelope must be taken into consideration, for example in model tests on earth pressure or anchor resistance in which, owing

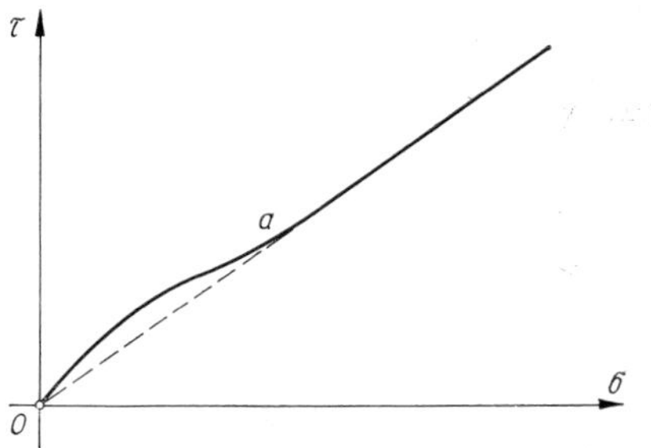

Fig. 286. Failure line at low normal stresses

to the usual small dimension of the apparatus, the normal stresses acting on a slip surface seldom exceed the value of 10 to 20 kN/m². *Figure 287a* shows the origin of the shearing strength in granular soils. The angle of internal friction is given as a function of the initial porosity; the components are due to true interparticle friction, to particle rearranging, and to dilatancy. If we wish to consider the effect of the normal stress, particularly in the case of low σ values, Fig. 287*b* is obtained, where Φ is given as a function of the initial porosity and of the normal pressure. These figures give a deeper insight into the mechanism of failure for granular soil.

It has been mentioned earlier that in the conventional treatment of soil strength the effect of the intermediate principal stress is disregarded. Let us now look more closely into this problem. If we adopt the triangular plot, according to Fig. 251, of the failure criterion and run shear tests in which soil samples are strained to failure in a general state of stress ($\sigma_1 > \sigma_2 > \sigma_3$), then we have a picture of the influence of the intermediate principal stress on the shear strength. But the triangle of principal stresses used in Fig. 251 can be interpreted also in a different way. Let point P in *Fig. 288* represent a given state of stress in the three-dimensional coordinate system (σ_1, σ_2, σ_3). The plane that passes through point P and is normal to the space diagonal ($\sigma_1 = \sigma_2 = \sigma_3$) is described by the equation

$$\sigma_1 + \sigma_2 + \sigma_3 = 0 .$$

In this representation, the position of a point such as P on the plane is uniquely determined by its distance from the point P', in which the space diagonal pierces the plane. In terms of stresses, point P' corresponds to the hydrostatic portion of a general state of stress, whereas the deviator stress is depicted by the vector $\overline{PP'}$. This plane can also be conveniently used, similarly to the triangular diagram in Fig. 251, to plot the failure criterion on it.

In this space coordinate system, the failure criterion $\sigma_3/\sigma_1 = (1 - \sin \Phi)/(1 + \sin \Phi)$ is depicted by a hexagonal pyramid having its apex at the origin 0. The intersection of a particular plane, ($\sigma_1 + \sigma_2 + \sigma_3 =$ constant), with this failure surface is a hexagon. For given parameters σ_1 and Φ, the hexagonal intersections are as shown in Fig. 251.

The representation, according to Fig. 251, of the states of stress raises the important question, whether, and to what extent, the failure condition in the soil is influenced by the sequence of the variations in the state of stress through which failure is achieved (or in terms of the graphical representation, by the course along which an internal stress point P reaches the perimeter of the hexagon representing failure). For example, the path $\overline{01}$ in Fig. 251 depicts a triaxial test such

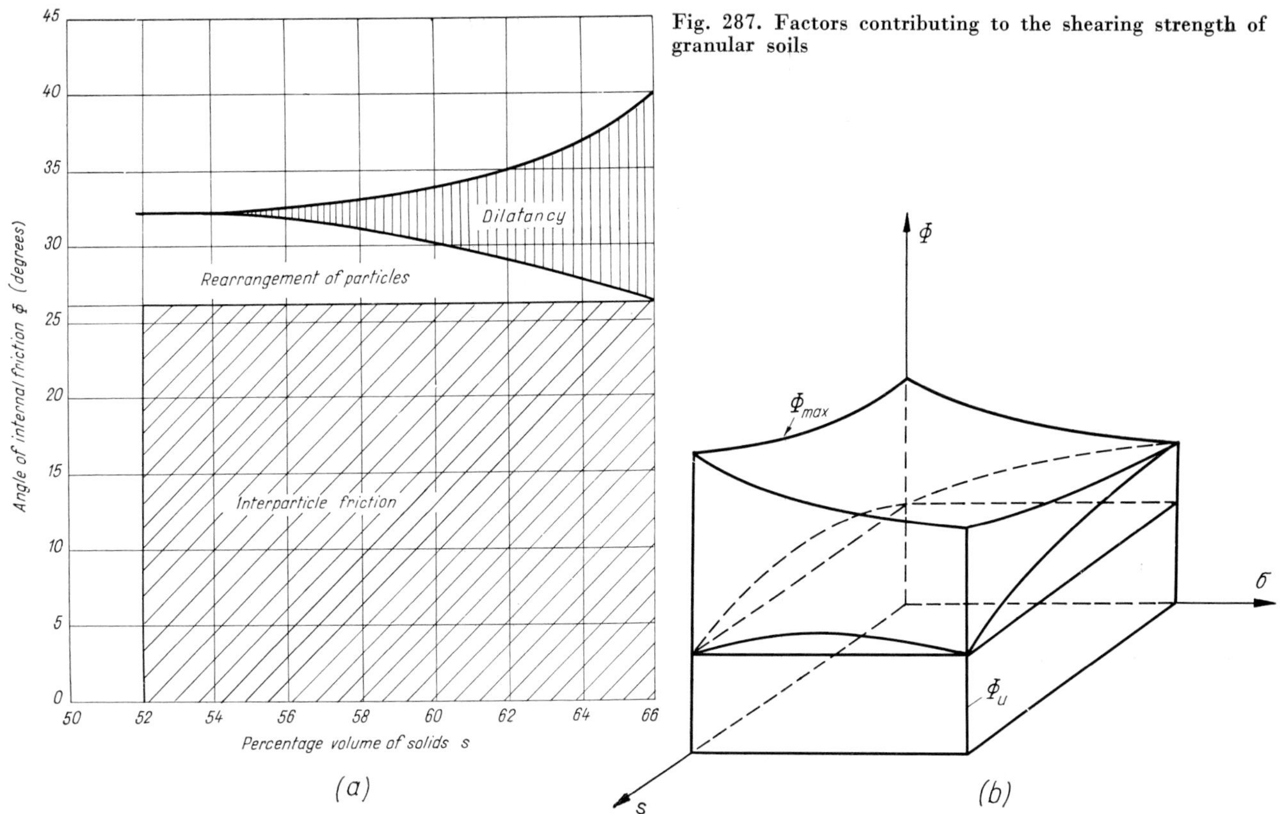

Fig. 287. Factors contributing to the shearing strength of granular soils

that the specimen is first subjected to a hydrostatic state of stress $\sigma_1 = \sigma_2 = \sigma_3$, then the axial stress is increased until the specimen fails (point 1). Successive stages of stress variations along the line $\overline{O1}$ are shown in the usual *Mohr* representation in Fig. 289. If we locate that point on every stress circle where the angle between the stress direction and the normal to the pertinent plane

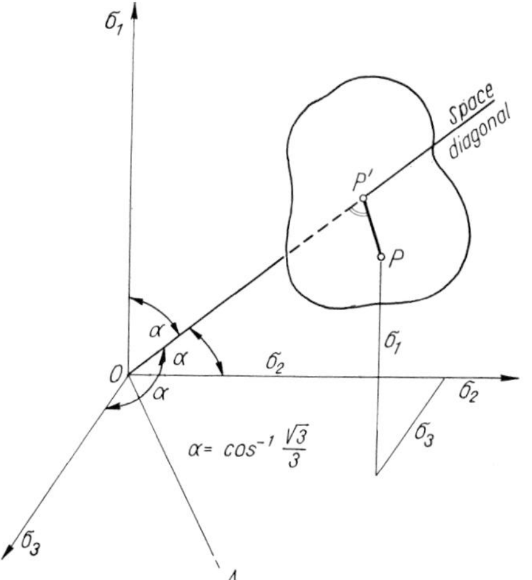

Fig. 288. Representation of the state of stress in three-dimensional coordinate system

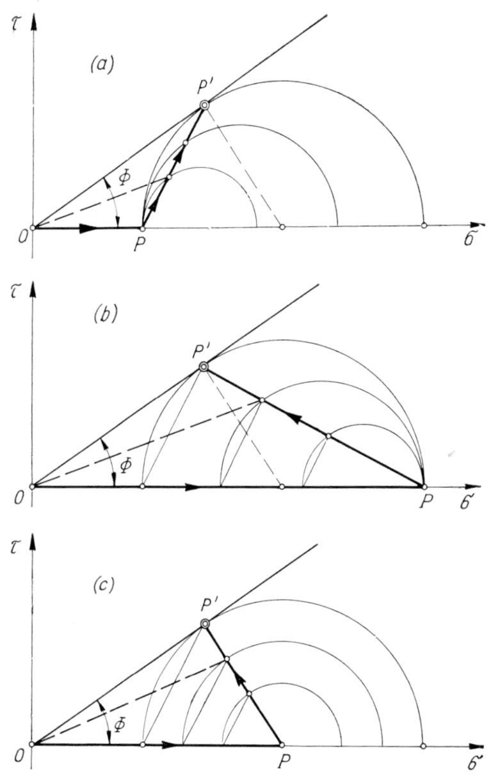

Fig. 289. Different modes of approaching failure

is the greatest, and we connect these points, we get the vector curve corresponding to the stressed states during the pertinent test. The type of test can be characterized by this curve. In the different cases which are shown in Fig. 289 the shear strength parameter (the angle of integral friction) is approximately the same, although the vector curves are quite different. The dashed lines show the resulting stresses in an intermediate stage. In case *b* the test starts from a hydrostatic state, then the stresses $\sigma_2 = \sigma_3$ are decreased until failure occurs. Another possibility to reach failure is shown in the diagram *c*; in this case the mean value of the principal stresses was kept constant.

The generalized Coulomb failure criterion in the form presented in Fig. 251 needs experimental verification to become useful for practical application. Unfortunately, the information about failure tests performed under general stress conditions is rather scarce in the literature. Experiments by KJELLMAN (1936), KIRKPATRICK (1959), BISHOP and ELDIN (1953), HABIB (1963) and MALISHEV (1954) should be mentioned in this connection. Some of the quoted results are plotted in one sector of the stress plane in *Fig. 290*. The data given by KJELLMAN and KIRKPATRICK coincide fairly well with the hexagonal failure envelope or are located only slightly off the line 1 2, regardless of the manner in which failure was reached in the individual tests. HABIB found that compression and expansion tests on sands performed according to the paths *01* and *02* respectively lead to substantial differences. The friction angle obtained from the *02*-test was lower

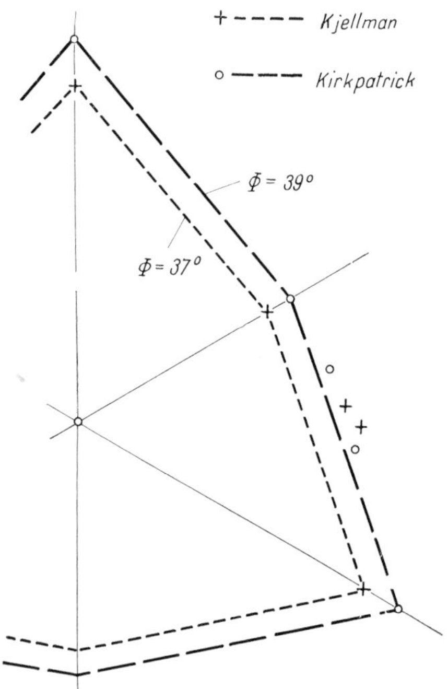

Fig. 290. Results of strength tests plotted in triangular diagram

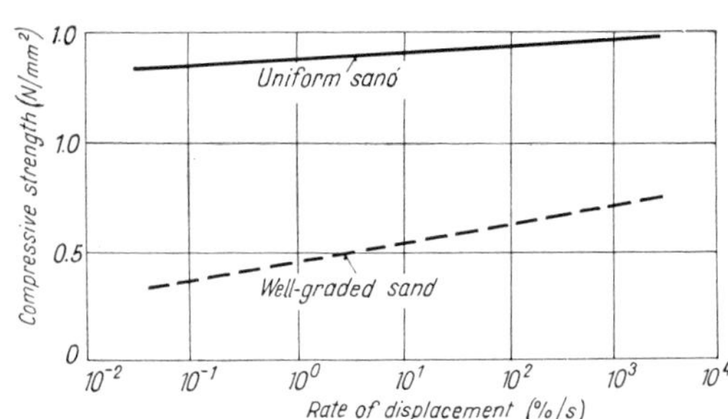

Fig. 291. Effect of rate of displacement on the shear strength of saturated sands

by some 7° than that supplied by the *01*-test. Such discrepancy might be, however, attributed to differences in testing techniques. Available experimental data appear to substantiate the conclusion that in sands the failure condition is independent—at least for practical purposes,—of the intermediate principal stress.

Tests performed under some general state of stress ($\sigma_1 > \sigma_2 > \sigma_3$) as e.g. the experiments by KJELLMAN and KIRKPATRICK give actually somewhat higher calculated friction angle values than those obtained from (σ_1, $\sigma_2 = \sigma_3$)-tests, the difference increasing with increasing density of the soil. But by neglecting this difference we are on the safe side. The finding that the intermediate principal stress is seemingly without influence on the failure of sands can probably be explained by the fact that granular materials do not generally possess a structure with an inherent tendency towards anisotropy which in turn would greatly influence development of the state of stress causing failure. If we produce such anisotropy on purpose e.g. by tamping a laboratory sample in layers into the mould, then the effect of the stress path to failure will be more pronounced. In sands, the void ratio at the instant of failure turns out to be the decisive factor governing the shearing strength. Variations in the state of stress prior to failure should be considered only to the extent resulting in changes in the void ratio.

A final remark concerns the influence of the rate of loading on the failure process. This effect is rather difficult to investigate experimentally. According to WHITMAN (1957) the effect is small in dry sands. In a saturated sample of sand, non-uniform volume changes take place during triaxial compression, which induce pressure gradients in the pore water giving rise to the transfer of a portion of the pore water inside the sample. A certain portion of the total applied deviator stress is thus used up to maintain this flow with the result that the shearing strength is increased. WHITMAN (1957) obtained the experimental data shown in *Fig. 291*. The stress–strain behaviour of sands under dynamic loading will be dealt with in Vol. 2.

Fig. 292. Shearing strength in undrained system; saturated cohesive soil

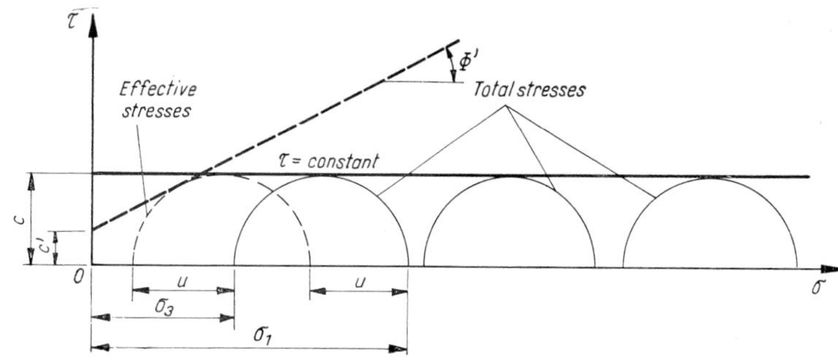

7.7 Shearing strength of cohesive soils

7.7.1 Drained and undrained shearing strength

The shearing strength of cohesive soils is one of the basic, and perhaps most disputed problems of soil mechanics. This is best evidenced by a succession of International Conferences, six in the last two decades, specifically devoted to this topic, besides the regular Conferences of ISSMFE where cohesive soil strength is also constantly being discussed in a number of papers. Whilst these conferences helped a great deal to clarify the fundamentals, a comprehensive synthesis of the problem would, at present, seem premature. The following section is confined therefore chiefly to the practical aspects of the question.

Let us restate the basic difference between drained and undrained shear. As was previously mentioned, the shear strength parameters defined in terms of effective normal stresses can be regarded as practically constant. It follows that the pore water pressures play an important role, since they must be known first to permit the calculation of effective stresses.

In an undrained system the excess pore water cannot escape during the application of normal and shearing loads. If the soil is completely saturated, the deviator stress $(\sigma_1 - \sigma_3)$ causing failure will be independent of the initial hydrostatic pressure imposed on the soil. The volume change of the soil will be negligible. The soil behaves as a perfect plastic medium, i.e. it exhibits a constant shearing strength which is independent of the total normal stress (*Fig. 292*).

Thus, for saturated, undrained conditions, the evaluation of shearing strength in terms of total normal stresses invariably leads to $\Phi = 0$ and $\tau = c = \frac{1}{2}(\sigma_1 - \sigma_3)$. It is only in this case that the $\Phi = 0$ analysis of the stability of earth masses strictly applies (Jáky 1944, 1948; Skempton, 1948).

If we measure the pore pressure during loading, then it becomes possible to calculate the effective stresses. However, as experiments have shown, in saturated clays both the major and the minor effective principal stresses $(\bar{\sigma}_1 = \sigma_1 - u;\ \bar{\sigma}_3 = \sigma_3 - u)$ are independent of the hydrostatic portion, σ_3, of the superimposed stress. Hence, only a single effective stress circle (shown by the broken line in Fig. 292) can be drawn for any confining pressure, σ_3, and therefore Φ' and c' cannot be determined in this manner either.

If the triaxial specimen is only partially saturated, the deviator stress pertinent to failure will increase with increasing confining pressure and the failure envelope is no longer a horizontal line. This is due, in part, to the compression and, in part, to the solution in the pore water of the entrapped air. The more advanced the compression and solution in water of the pore air with increasing superimposed stresses, the more the solids–water–air system approached the two-phase state. From the instant the soil has reached complete saturation, the situation is identical to that previously discussed and the envelope to the circles of total stresses merges into a horizontal asymptote as shown in *Fig. 293*. In the range of low normal stresses the failure envelope is a curved line. A constant angle of internal frinction Φ does not exist, as the slope angle of the tangent to the

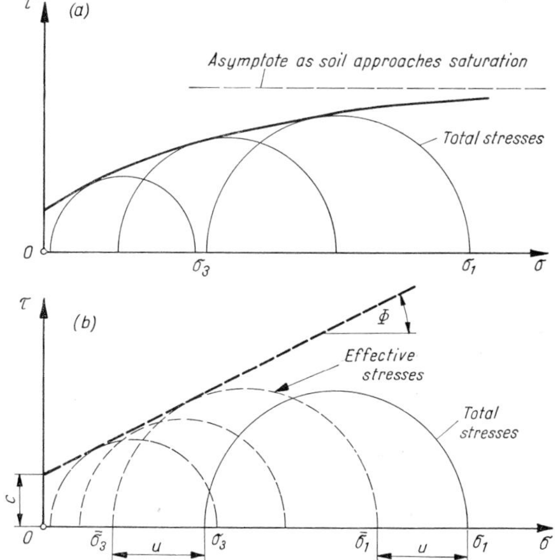

Fig. 293. Mohr circles in undrained system; unsaturated cohesive soil:
a — envelope of circles of total stress; b — envelope of circles of effective stress

envelope varies with increasing total normal stress σ.

If, on the other hand, the pore pressures are measured during the test, it is possible to plot effective stress circles. Experience shows that the envelope to these circles is again a straight line. With the neutral stresses known, the variation of the pore pressure parameter B can also be determined.

If a soil increases in volume, when compressed under increasing stress in one direction (dilatant soil), the difficulty again arises of distinctly defining the state of failure. This is so because the shearing strength characterized by the parameters c' and Φ' is fully mobilized, i.e. the maximum effective deviator stress, $(\sigma_1 - \sigma_3)_{max} = (\bar{\sigma}_1 - \bar{\sigma}_3)_{max}$, is reached, at comparatively small deformations. Any apparent further increase in effective deviator stress is only the consequence of a decrease in pore pressure with continuing expansion of the soil.

The strength behaviour assumes a different pattern, if a soil which has already consolidated in a given state of stress, i.e. excess pore pressures have completely dissipated, is afterwards subjected to an increasing shearing load at constant water content (undrained condition) so that neutral stresses build up in the pore water. Let us assume that a saturated soil specimen is allowed to consolidate under a given confining pressure ($\sigma_1 = \sigma_2 = \sigma_3 = p$). If this consolidation pressure p is gradually increased, the water content (or the void ratio) changes as shown in the top diagram in *Fig. 294.* (Compression curve. See Section 1.8.) At each pressure p, after consolidation is complete, the specimen is caused to fail by increasing the axial stress σ_1, while the free escape of the water is prevented. Such condition was previously termed undrained shear. It was also stated (Fig. 292) that the obtained shearing strength, expressed in terms of total normal stresses, was independent of the applied confining pressure: $\Phi = 0$, $c = \frac{1}{2}(\sigma_1 - \sigma_3)$. But the value of c must depend on how large the consolidation pressure p has been at which each shear test is to be performed. Experiments show a linear relationship between the consolidation pressure p and the undrained shear strength c (Fig. 294*b*). Provided that the soil is normally consolidated, the pore pressure parameter A at failure is $A_f = 1$ for any p value. In the case of a saturated soil the parameter B also equals unity. Hence the neutral stresses can be computed and the shear strength plotted as a function of the effective normal stress. The resulting failure envelope is a straight line, which passes through the origin of the coordinates. (*Coulomb* line, Φ'; $c' = 0$.) The shearing strength of cohesive soils is thus found to be a linear function of the consolidation pressure.

Experience shows that the shearing strength as determined by the above described procedure is somewhat higher than the value obtainable by field tests. The probable explanation is that the actual state of stress in which natural soil deposits consolidated was not hydrostatic, the lateral pressure being smaller than the vertical overburden pressure.

A soil which had consolidated under a pressure p_a and was relieved afterwards of a portion of the consolidation pressure (e.g. p_a was reduced to p_b, Fig. 294*b*), is said to be overconsolidated or preloaded. The undrained shear strength measured in this state is larger than that obtained with a normally consolidated soil. The cause of this is that as a consequence of preloading the pore pressure parameter at failure A_f has changed considerably. It may have reduced from unity to zero or even become negative. In the case of an overconsolidated soil, the shearing strength plotted against the effective normal stress also will be increased (Fig. 294*d*). Φ' is slightly higher than with a normally consolidated soil and, there is a cohesion intercept c'.

In a drained system the excess pore water can always freely flow out of the soil. Typical examples of the drained condition are the deposits of sand

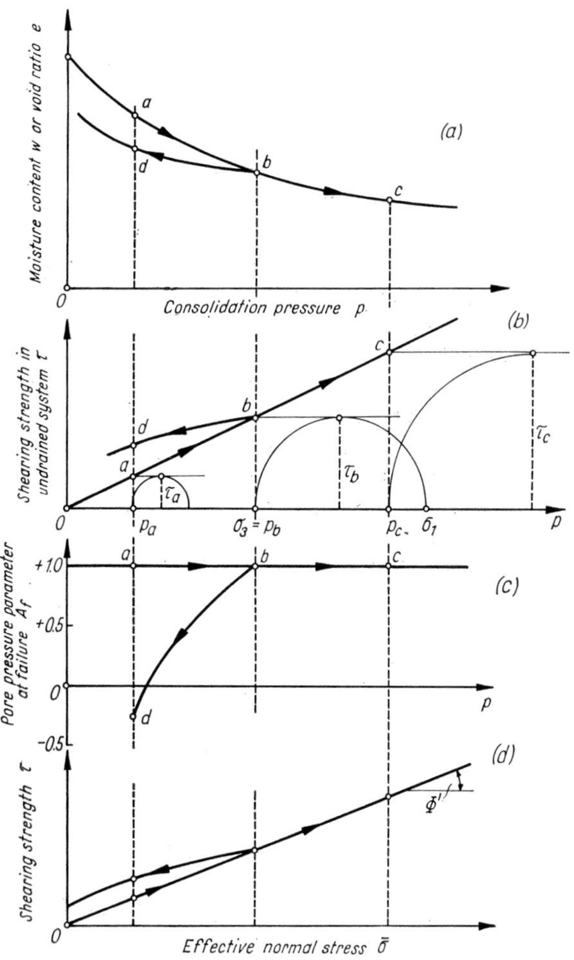

Fig. 294. Consolidated soil; shearing in undrained system: *a* — moisture content or void ratio versus consolidation pressure; *b* — shear strength versus consolidation pressure; *c* — pore pressure parameter at failure A_f; *d* — shear strength of normally consolidated and overconsolidated soil

Fig. 295. Coulomb's lines in drained system:
a — normally consolidated clay; *b* — overconsolidated clay

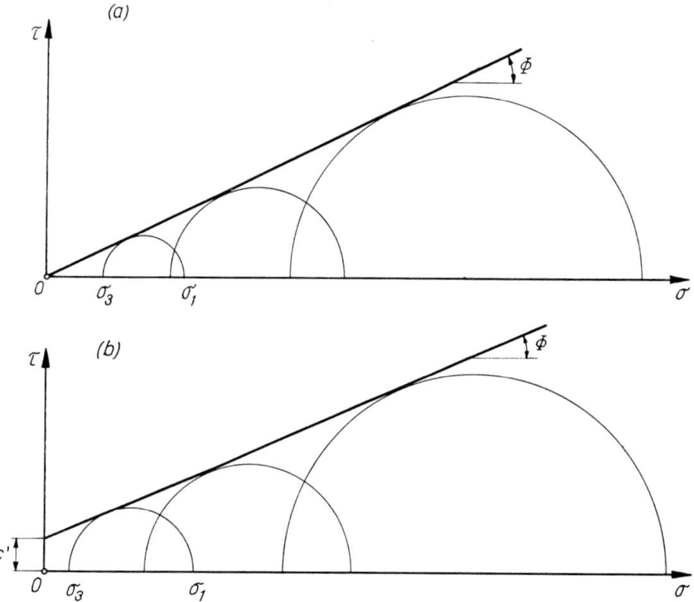

and gravel with pervious boundary surfaces. Owing to the high permeability of these soils, the pore water can easily move in the soil as directed by the volume changes due to loading, and excess pore pressures do normally not develop. A clay stratum, on the other hand, can only be regarded as a drained system when the loading proceeds at a very slow rate so that the consolidation can always follow the change of stress without a time lag.

Let us now consider the drained shear test on a saturated clay sample of liquid consistency. First we consolidate the specimen under an all-around pressure p, then we slowly increase the deviator stress making sure that the effective stress is equal at all times to the applied total stress. In this manner, we can obtain Mohr circles representing failure in terms of effective stresses. In the case of a normally consolidated clay, the failure envelope passes through the origin. If the soil is overconsolidated, there is always a cohesion intercept, the value of which depends on the magnitude of the preconsolidation stress (*Fig. 295*).

If we run a series of drained shear tests on an overconsolidated clay so that the specimens are first gradually unloaded, then reloaded up to, and beyond, the preconsolidation stress, then the shearing strength for reloading will be between the strength values measured with normally consolidated and unloaded specimens. Typical test results are shown in *Fig. 296:* first loading, unloading and reloading respectively produce the curves of virgin compression (*0 — 1*), expansion (*1 — 2*) and recompression (*2 — 3*) shown in Fig. 296*a*. The corresponding relationship between shear strength and effective normal stress is given in Fig. 296*b*. A common feature of both diagrams is that the branches of unloading and reloading form a hysteresis loop. Shear tests performed in practice normally give a straight line representing the reloading section of the rupture line, since clays are always more or less overconsolidated under natural conditions.

The phenomenon that a clay may acquire cohesive strength through preloading is associated with the presence of adsorbed water on clay particle surfaces. During consolidation, the ordinary water is being gradually squeezed out of the voids, whereas the amount of adsorber water remains practically constant, and as the void ratio decreases and the particles are brought to a closer contact, the higher viscous resistance of the

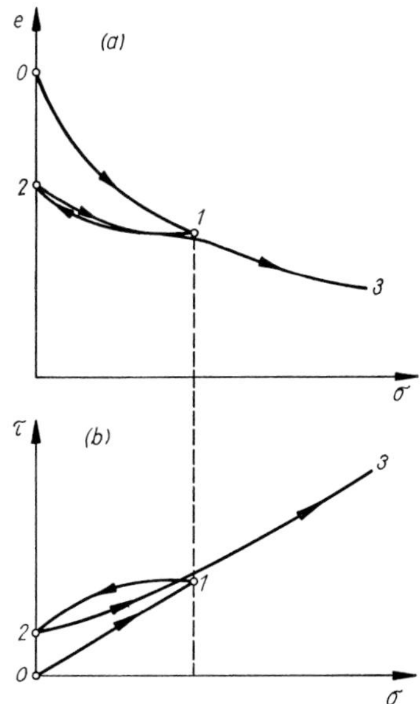

Fig. 296. Results of direct shear tests: first loading, decompression and recompression

adsorbed layer gradually comes into play resulting in a higher cohesion.

For a given soil, the shear strength parameters Φ' and c' expressed in terms of effective normal stresses are practically independent of the test procedure by means of which they have been determined. This fact again strongly corroborates the generally accepted principle that strength and stability problems associated with soils should whenever possible be treated on the basis of effective stresses.

In the case of heavily overconsolidated clays and sands the drained shear test gives slightly higher Φ' and c' values than those obtained with normally consolidated soil, because the volume changes due to shear consume extra work, and failure occurs at relatively small strains. In comparison, the same soil when tested under undrained conditions will exhibit zero volume change, hence, its angle of internal friction, which is after all a measure of its resistance to shear strain, becomes smaller.

Comparing the behaviour of heavily overconsolidated and normally consolidated clays, we observe that it is not so much the value of the shearing resistance as the deformations and volume changes during shear that are greatly affected by the direction and sequence of the changes in stress.

It is clear from the foregoing discussion that for clays the usual shearing strength characteristics cannot be regarded as material constants, since they depend on numerous factors, some of which are not yet fully understood. The same applies, to even a greater extent, to the deformation characteristics of soils.

In the past it was customary, probably after Coulomb's concept of shear failure, to sharply separate the shearing strength of a clay into a cohesive and a frictional resistance. The view has also been advanced that shearing loads are at first resisted solely by cohesion, and the frictional resistance comes into play only when cohesion has been fully mobilized (Winkler, 1871). According to existing knowledge, we take shearing strength as an entity, which may well depend, besides on soil type, on the stress history of the soil, pore pressure, density, water content, the rate of loading, drainage conditions and many other factors, yet any arbitrary separation of it into two components as suggested by the usual formulae can only serve practical purposes.

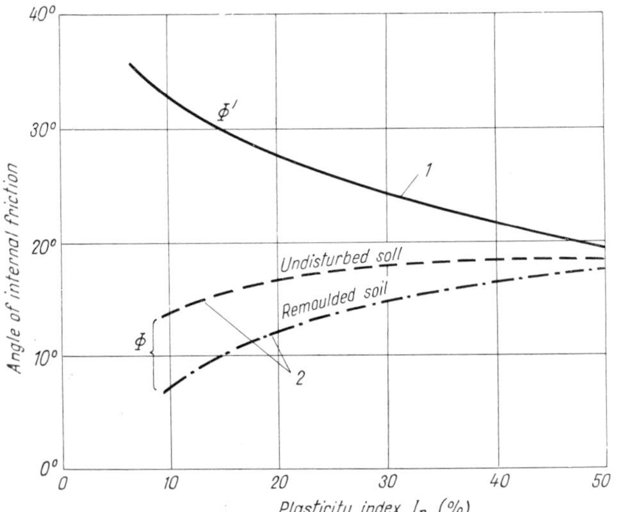

Fig. 297. Statistical relationship between plasticity index and angle of shearing strength according to Terzaghi (1955):
1 — shearing in drained system; *2* — consolidated soil, shearing in undrained system

The shearing strength of clay is essentially influenced by the *cations* adsorbed on the surface of mineral particles. A clay with a given mineral composition may exhibit strongly varying shearing strength, when its originally adsorbed cations are replaced by others. For example, the shearing strength of a clay containing adsorbed Fe and Al cations is about twice as much, at constant amount of free water, as it would be in the case of Na cations. This variation is due to the differences in the thickness of the adsorbed water films. Such peculiarities of clay structure must not be overlooked when one attempts to correlate e.g. liquid limit to shearing strength. Because of the mentioned physicochemical effects, dependable correlations can only be obtained with soils of common geological origin.

Many such empirical formulae have been suggested in the literature. However, they can serve a useful purpose only if the conditions under which the shear strength parameters have been determined are exactly stated.

Empirical relationships between plasticity index and friction angles obtained in drained shear test and in consolidated undrained test, respectively, are shown after Terzaghi (1955) in *Fig. 297*. Φ' decreases, whereas Φ increases with increasing plasticity index. The difference between the two kinds of friction angle tends to diminish as I_p increases, because the volume change during shear and the corresponding change in neutral stress are comparatively smaller with soils of high plasticity. At a plasticity index of about $I_p = 50\%$ the curves practically merge.

Similar statistical relationships between shear strength and plasticity index were given for Hungarian soils by Kopácsy (1953) and Szilvágyi (1955).

7.7.2 Factors affecting the shearing strength

In connection with the study of stresses in two-phase and three-phase systems of cohesive soils (Ch. 5) we have derived the equations of stresses for clays with dispersed structure [Eq. (58)] and with flocculated structure [Eq. (59)]. We also recall the curves of Fig. 110 from which it is clear that when the particles are in close contact, or at least the distance between them is very small, the interparticle forces of repulsion, per unit of area, are negligible as compared with the forces of attraction, and the more so with the total stress. Hence, Eq. (59) reduces to the form

$$p_s = \frac{\sigma_v}{\varphi_s} \tag{206}$$

i.e. to obtain the interparticle pressure we simply divide the stress due to interparticle attraction by the solid-to-solid contact area ratio ($\varphi_s = F_s/F$). This pressure p_s, on the other hand, is according to the concept already mentioned in Ch. 5 (Bowden and Tabor, 1950) just equal to the yield stress of the material at the point of contact. It follows that in a clay with flocculated structure the average actual stress between the particles is constant and independent of the applied external stress. An increase in the latter will not cause an increase in p_s. All the same, another effect, namely the increase of the total contact area under the action of shearing forces can be observed also in clays.

As is inferred from the above analysis, the shearing strength of a flocculated clay cannot be altered by merely increasing the external load or the effective normal stress. Increased shearing strength can only be achieved through an increase in the number of points of contact between the particles. This is the cardinal difference between the strength behaviour of cohesive and granular soils. It also follows from the foregoing statements that every effect that results in an increase of the number of interparticle contacts will lead to increased shearing strength. Repeated loading (provided that the pulsating stress value is only a fraction of the yield stress of the material), long-lasting loading, heat effect etc. may bring about such changes in clay structure. These will all cause, at the same time, an increase in soil stiffness. Of two clays the one containing finer particles, other properties being the same, will exhibit higher strength owing to the larger number of contact points per unit of area. That is why e.g. a clay with high a montmorillonite content shows greater strength than a kaolinite clay.

Let us examine in the following what practical consequences concerning the shearing strength of clays can be drawn from the physicochemical laws governing the formation of clay structure. In general, it can be stated that the shearing strength of a cohesive soil depends on the soil structure, on the void ratio (i.e. the average distance between the grains) and on the rate of shear deformation. Other factors not mentioned here act through these main factors.

With respect to soil structure, we distinguished between dispersed and flocculated clays (see Section 3.4.1). The behaviour of these clay types can be characterized by their shear deformation lines. The stress–strain diagram (see *Fig. 298*) is a smooth curve for the dispersed structure (*a*); with increasing shear strains, the shear stress continuously increases. For a flocculated clay (*b*), there is a marked maximum in the shear strength. The causes of this behaviour were discussed in Section 3.4.1. The effect of remoulding on the strength behaviour of clays is characterized by the term *sensitivity*, meaning the ratio of the peak shearing strength to the ultimate strength of the same clay.

The structure of a flocculated clay, the average distance between the particles, the void ratio within the shear zone and the number of contact points all depend on what stresses acted upon the soil in the past. The stress history also greatly influences the shearing strength of the soil. Under the action of increasing effective normal stresses, the number of contact points between the particles will also grow with the lapse of the time. This explains that in flocculated clays just as in granular soils the shearing strength is a function of the effective stresses. But the similar behaviour is, as we have seen, the outcome of differing mechanisms.

The shearing strength is greatly influenced, as was mentioned previously, by the void ratio measured in the shear zone and by the structure of the soil. Effective normal stresses acting on a soil mainly cause changes in the void ratio but do not affect appreciably the soil structure. Shearing stresses, on the other hand, primarily alter the structure by orienting the particles. In an oriented state, the soil can withstand normal stresses only to a lesser extent, therefore the compression of the soil when acted upon simultaneously

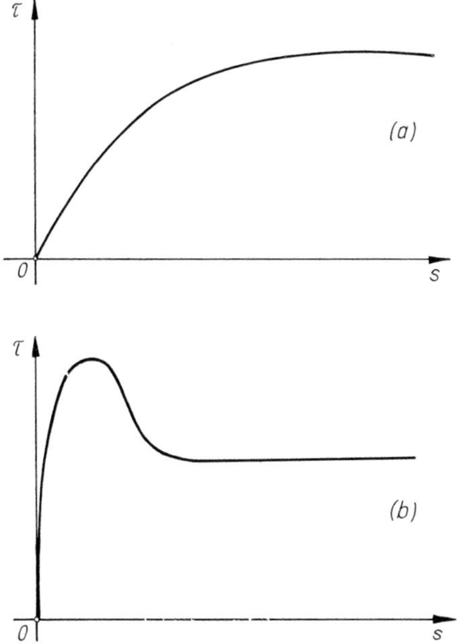

Fig. 298. Shear displacement curves:
a — maximum and ultimate values are identical; *b* — after a critical displacement the movement continues even at reduced shear stress

by shearing and effective stresses will be larger than under the action of normal stresses alone. It is well known that clays cannot be effectively compacted by mere static normal loading, but if shearing stresses are applied concurrently, much higher compaction can be attained. This effect is illustrated after SCOTT (1963) in *Fig. 299*, where the void ratio is plotted against the effective normal octahedral stress

$$\bar{\sigma} = (\bar{\sigma}_1 + \bar{\sigma}_2 + \bar{\sigma}_3)/3 = J_1/3$$

(J_1 is the first stress invariant). For a hydrostatic state of stress ($\bar{\sigma}_1 = \bar{\sigma}_2 = \bar{\sigma}_3$) and random structure the compression curve *1* is obtained. Curves *2* and *3* represent again compression under a hydrostatic state of stress but for increasingly higher degrees of orientation of the particles. Note that for each curve the soil structure can be regarded as practically unchanged.

Assume that a clay with a dispersed, randomly oriented structure has consolidated under a given all-around pressure $\bar{\sigma}$. Its final equilibrium state is represented by point *P* on curve *1*. If this clay is later subjected to shear, e.g. by simultaneously increasing $\bar{\sigma}_1$ and reducing $\bar{\sigma}_2$ and $\bar{\sigma}_3$ in such a manner that the mean effective principal stress remains unchanged, i.e.

$$\bar{\sigma} = \frac{1}{3}(\bar{\sigma}_1 + \bar{\sigma}_2 + \bar{\sigma}_3) = \text{constant},$$

then its structure will be altered and its void ratio decreased as indicated by the line *PQR*. Let point *Q* represent the state in which the shearing resistance is at its maximum (the peak point in Fig. 298*b*). During continuing shear the void ratio soon reaches its final value and retains it from that state on, provided $\bar{\sigma}$ remains constant. The void ratio pertinent to the ultimate condition is analogous to the "critical void ratio" of sands.

Failure may, of course, be produced along many different stress-paths. For example, we increase $\bar{\sigma}_1$ while keeping $\bar{\sigma}_2$ and $\bar{\sigma}_3$ constant. In this case

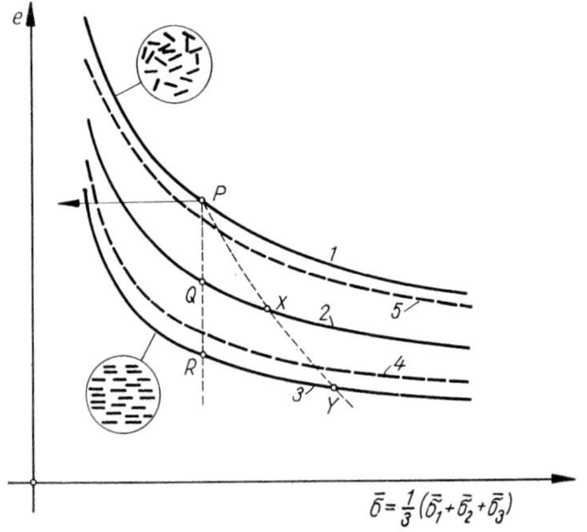

Fig. 299. Changes in the structure and void ratio of clay during shear

the mean effective principal stress acting on the rupture surface is gradually increased. Failure is represented by point *X*, and continuous shear deformation by point *Y*. Failure at constant volume (i.e. at constant void ratio) can only be attained by a vigorous disturbance of the clay structure (e.g. remoulding, kneading). As regards the actual *e* versus $\bar{\sigma}$ curve corresponding to failure, for a soft clay (curve *4*) the failure of which is normally preceded by very large deformations, the curve is likely to approach curve *3;* for a stiff clay (curve *5*) it remains close to curve *1*.

7.7.3 Deformations due to shear

In this section we deal with the peculiarities of the shear deformation of clays. Observations on the behaviour under shear of some soils suggest that appreciable strain occurs only after the shearing stress has exceeded a certain threshold value. In dispersed clays such a threshold is not likely to exist, since there is no direct contact between the mineral particles and the shearing strength is essentially due to viscous resistance. The liquid which fills the voids has no stiffness. The dipolar water molecules do not establish in effect "bonds" but rather create a force field in which individual particles are held and fixed in their relative position by balancing forces, like fine iron particles aligned along lines of force in a magnetic field without necessarily being in contact.

An entirely different situation arises in flocculated clays. The bonds formed at the points of contact between the particles offer a certain elastic resistance and yield only to a critical shearing stress which can be then regarded as the threshold value. Past this limit, the stress–strain relationship assumes a different character though ultimate failure may still be far ahead. A similar effect can be observed when in a dispersed clay matrix there are embedded large grains whose behaviour is predominantly governed by mass forces. Under the action of shearing forces, these grains may come into real contact and develop an additional resistance to shear. The initial shearing load necessary to overcome this is again observed as a "threshold" value. But in this instance, the threshold value cannot be expressed in terms of shearing stress; instead, it is defined by the criterion whether or not an initial shear strain is followed, under constant shearing stress, by sustained slow deformation. As regards clays, the occurrence and the rate of such slow deformations is associated with the consistency of the soil. The rate of strain, at constant water content, is found to be proportional to the applied shearing stress. Characteristic stress versus rate of strain diagrams for different states of consistency are given in *Fig. 300*. It can be seen that there is no conspicuous break in the curves, and the threshold value can be defined at best as the point of intersection of two straight lines fitting the actual curve as is shown e.g. in Fig. 300*b*. The tangent to the final approximately straight section of the actual stress–strain curve is prolonged and the shearing stress τ_0 at which it intersects the τ-axis is the desired threshold value. Whether such a critical stress value, τ_0, really exists, and of what magnitude, is extremely difficult to establish beyond doubt, largely because of the inadequacy of ordinary laboratory measuring technique. It appears that the habit of developing a threshold strength is peculiar to clays having a stress–strain diagram similar to that shown in Fig. 298*b*, such as flocculated clays or clays with embedded coarse grains forming a stiff skeleton.

In the knowledge of the fine structure of clays we can explain the effect of remoulding on their shearing strength. The early concept of *Casagrande*, while still forming an essential element of the whole picture, needs to be completed by taking the various effects due to particle orientation into account. Being a forcible disturbance, remoulding causes substantial changes in the soil structure formed over long geological times. Shearing load of long duration may also seriously interfere with the clay structure by orienting the particles into a more orderly arrangement within the shear zone. These changes are all irreversible, since the original orientation of the particles can no longer be restored by either unloading or the reversal of the direction of shear. The application of shearing stresses invariably results in a higher degree of orientation. Overconsolidated clays are also likely to undergo structural changes under the action of shear. As a consequence, the effect of a previous preloading will be completely obliterated. While preloading usually results in an increase in the number of interparticle contacts and, hence, in the formation of new bonds, shear action tends to destroy these bonds. Therefore, in an overconsolidated clay any shear action that is capable of causing remoulding, will alter the structure of the clay and seriously reduce its shearing strength.

A more detailed picture of the time-shear strain relationship of clays can be obtained from *Fig. 301*.

Assume that a clay specimen acted upon by a given constant normal stress is subsequently subjected to a comparatively small shearing stress. Some time is normally necessary for the resulting deformation to take place before the specimen eventually comes to equilibrium. If we gradually increase the shearing stress, after a while we observe that, with the initial deformation over, the specimen will no more come to a rest. In consequence of the internal structure of the clay, particularly of the viscous behaviour of the water films surrounding the grains, slow deformations begin, the rate of which increases in direct proportion to the shearing stress difference ($\tau - \tau_0$). The time-shearing stress curves approach ascending asymptotes instead of horizontal ones. It follows that if in a cohesive soil the applied shearing stress has exceeded the threshold value, τ_0, the state of stress remains unchanged during continuous shear so long only as the soil can deform unrestricted at a constant rate $\partial\varepsilon/\partial t$. The application of a shearing stress smaller than the ultimate strength

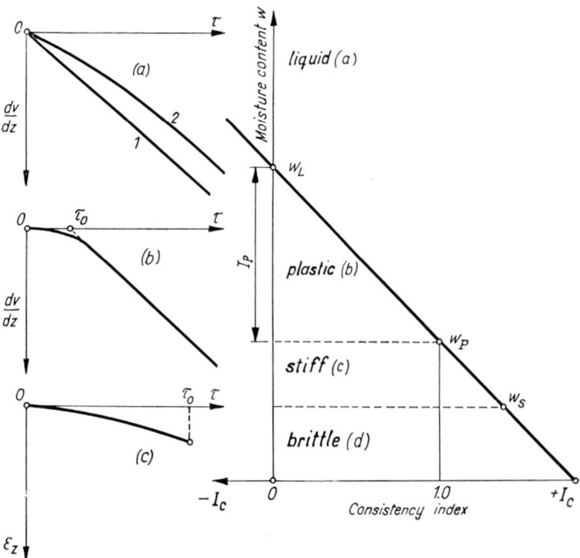

Fig. 300. Shear displacement curves at different consistencies

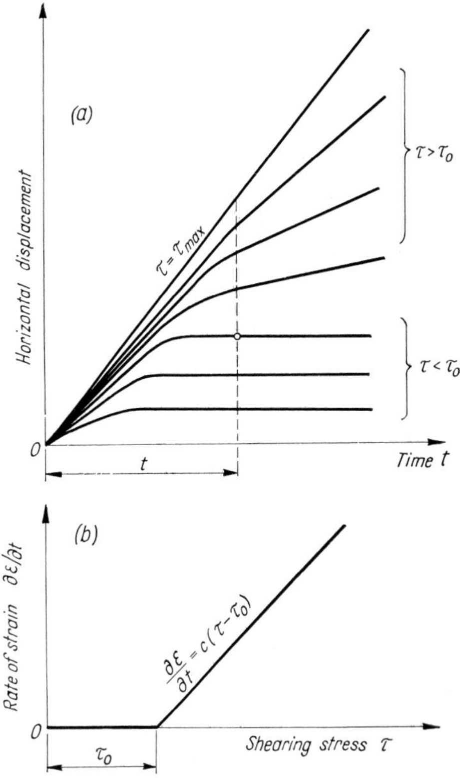

Fig. 301. Time-displacement curves for clay samples subject to increasing shearing stresses

τ_{max} but larger than τ_0 may, through a uniform continuous movement of the retaining structure, seriously endanger the stability of the soil mass, and lead, eventually, to failure. If, on the other hand, the motion is restrained so that the slow deformations and, hence, the shearing stresses within the soil mass cannot fully develop, the share of the load carried by the soil mass is reduced and, for equilibrium, that imparted to the retaining structure is increased.

The above defined threshold value of the shearing strength is called the *fundamental shearing strength* on the grounds that a mass of soil can remain permanently at rest only if the shearing stresses imposed on it are smaller than τ_0. Otherwise, a slow motion known as *creep* occurs, a phenomenon commonly observed on clay slopes and retaining structures supporting cohesive soils (Šuklje, 1970).

A typical form of failure in clays known as progressive failure is as a rule also accompanied by slow deformations. The adjective "progressive" indicates that failure occurs at first at a point or along a limited portion of a line or surface, but spreads rapidly over the rest of the potential surface of failure and reaches the boundary of the soil mass. In the vicinity of the surface of rupture, the shearing stress acting in the clay may reach almost the peak value of the shearing strength, τ_{max}, but on the surface of sliding itself there exists only a reduced resistance to sliding (*Fig. 302*). Thus, at the instant of failure the total shearing resistance available on the surface of sliding is much smaller than could be calculated from the peak strength value. This explains that after peak strength is reached at one point, failure tends to spread over an increasingly larger portion of the sliding surface.

In the previous discussion we have pointed out the great importance of the *time factor* in problems connected with soil strength, be it the role of neutral stresses during consolidation or the slow deformation of clays. In the questions hitherto discussed we have been mainly concerned with the behaviour of soil under static loading, where time effects may also come into play, the more so, the higher the plasticity and the lower the permeability of the soil. But there are quite different

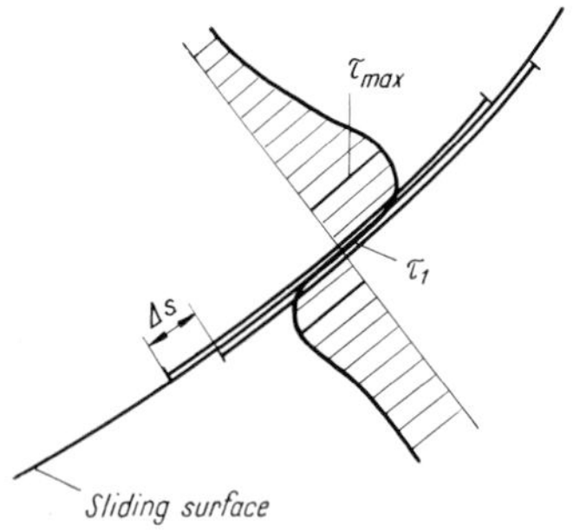

Fig. 302. Progressive failure

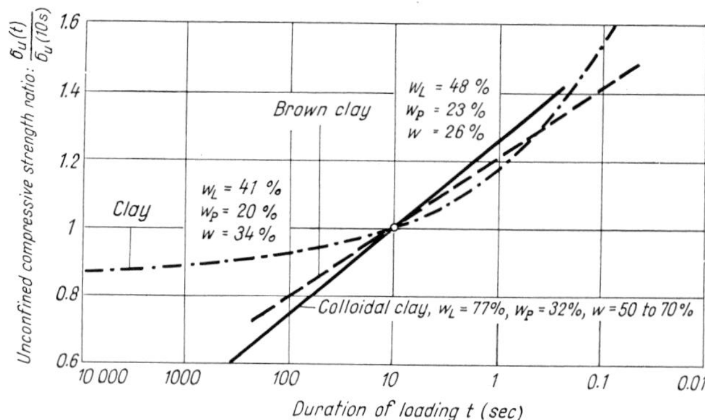

Fig. 303. Effect of duration of loading on the unconfined strength

time effects to be taken into consideration when the loading is of short duration permitting practically no time for consolidation or slow deformation. Soil strength is found to depend greatly on the duration of loading, but so far very little is known of the cause of this. Earthquake, explosion, pile driving can be mentioned as typical cases of rapid loading. Experience has shown that the measured strength is higher, the shorter is the duration of loading. One obvious exception to this rule is, of course, the liquefaction of sands.

The effect of rapid loading on soil strength was investigated by A. Casagrande and W. L. Shannon (1948). Typical test data are shown in *Fig. 303*. The duration of the test is plotted on a logarithmic scale on the horizontal axis, and the ratio of the unconfined compressive strength measured

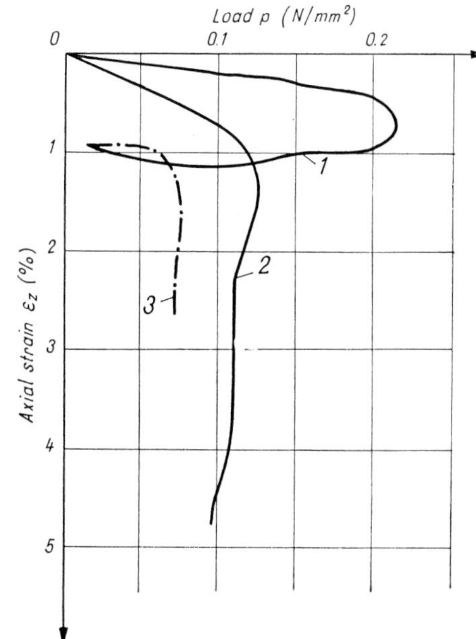

Fig. 304. Load-deformation diagrams for slow and quick loading respectively:

1 — quick transient loading; *2* — slow loading; *3* — slow reloading after transient loading

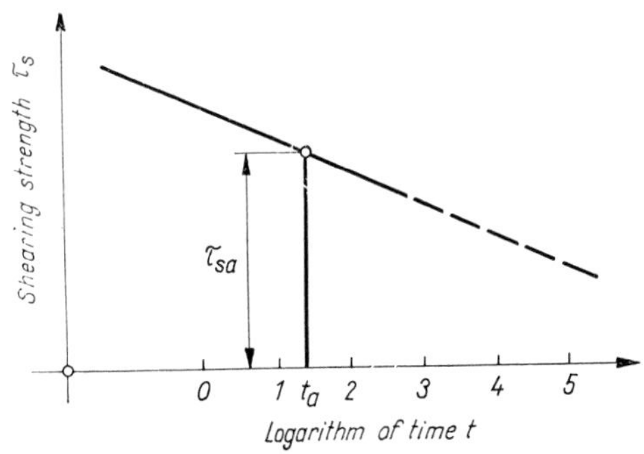

Fig. 305. Shear strength as function of duration of loading

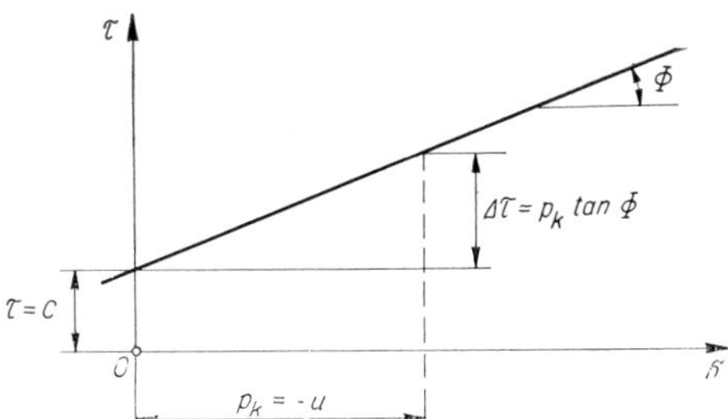

Fig. 306. Increase in shear strength caused by drying

for a given duration to that for a duration of 10s is plotted on the vertical axis. The tested soils exhibited increasing strength, decreasing axial strain and, hence, increasing secant modulus of deformation with decreasing duration of loading. But when a specimen tested first under transient loading was retested under static loading, its compressive strength was lower than that of an identical sample that had not been previously subjected to transient load. Typical stress–strain curves for normal static and rapid transient loadings are shown in *Fig. 304*.

The available experimental data are too scanty to permit as yet a theoretical treatment of this problem, and in judging the soil behaviour under transient loads we have to rely upon the qualitative statements given above.

Another time effect is manifested in the influence of the rate of shear to the shearing strength of the clay. This effect is often disregarded in practice, probably because it is too difficult to handle experimentally. If we perform a compression test at such a very slow rate that absolutely no pore pressure develops in the specimen, then data characteristic of the material itself can be obtained. Results of such tests suggest that the shearing strength obtained at a very slow rate of staining (e.g. 10^{-4} % strain per minute) may drop by as much as 15 to 20% as compared to the value obtained at a speed of 1% per minute. According to the experiments of Casagrande and Wilson (1950) and also of Whitman (1960) the general relationship probably assumes the form given diagrammatically in *Fig. 305*. Hvorslev (1960) found that the relationship can be expressed in the following form

$$\tau_{st} = \tau_{sa}\left[1 - \varrho \log \frac{t}{t_a}\right]. \qquad (207)$$

In the formula ϱ is a soil characteristic, τ_{sa} is the shearing strength measured for a given duration of test of t_a and τ_{st} is the shearing strength for a varying duration t.

The general laws of soil deformations will be discussed in more detail in Ch. 8.

7.7.4 Cohesion of clays

The cohesion of clays, as calculated from the unconfined compressive strength, depends primarily on the consistency, i.e. on the water content, of the soil. When a clay dries, tension (negative neutral stress) develops in the pore water. This tension increases with decreasing water content, and as a consequence, the effective normal stress also increases by an equal amount. Therefore, as the tension in the pore water increases from zero to u_v, it simultaneously produces an effective all-around pressure in the solid skeleton

$$p_k = -u_v.$$

This in turn increases the shearing resistance along any section of the soil by

$$\Delta\tau = p_k \tan \Phi.$$

(See *Fig. 306*.)

For saturated clays ($S = 1$), the unconfined compressive strength is, according to Jáky (1944), an exponential function of the water content. This means that if we represent test data by plotting the water contents on an arithmetic scale

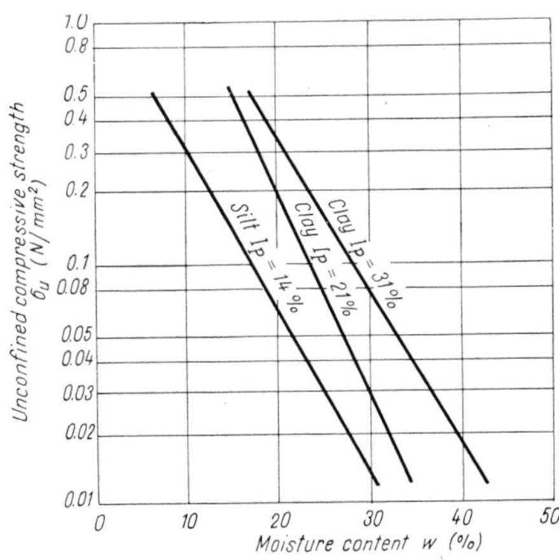

Fig. 307. Relationship between unconfined compressive strength and moisture content in saturated clays

and the corresponding unconfined compressive strength values on a logarithmic scale, straight lines result. This behaviour is analogous to the variation of soil consistency with increasing water content as represented by the "flow lines" (see Fig. 69). Straight lines obtained for various cohesive soils are shown in *Fig. 307*. The empirical relationship can be expressed in a mathematical form as

$$\sigma_u = ae^{cw} \tag{208}$$

$$w = w_0 + c \log \sigma_u \, .$$

If a clay is dried to water contents below the shrinkage limit w_s, air enters the voids of the soil. The tension in the water that remains in the voids produces effective pressures between the grains giving rise to an increased shearing resistance. The water content at which the unconfined compressive strength of a drying soil sample is a maximum depends chiefly on the grain size. This dependence is illustrated by *Fig. 308*, which shows the effect of a decrease of the water content during drying on the compressive strength of the soil. (For water contents smaller than w_s, the degree of saturation can be approximately taken as $S = w/w_s$.)

For a fine clean sand (curve *a* in Fig. 308), maximum compressive strength is obtained at a degree of saturation of about 0.8. (See also Fig. 81.) During further drying, σ_u ultimately reduces to zero. However, if the specimens are moulded with tap water, the impurities contained in it will, with advancing evaporation, adhere to the grains producing a weak bond between them. Thus, even a completely dried sample of sand may exhibit a slight cohesion. Curve *b* in Fig. 308 represents a fine silty sand. As the water content decreases to the shrinkage limit, the compressive strength increases. Below w_s, the strength begins to decrease and drops to its least value at about $S = 0.1$. Thereafter it increases again and reaches a maximum when the soil is absolutely dry.

For clays (curve *c*) the variation of compressive strength with the water content, above the shrinkage limit, is governed by the previously mentioned exponential law. At the shrinkage limit, there is a conspicuous break in the curve. Below w_s the compressive strength increases rapidly as the water content diminishes to zero.

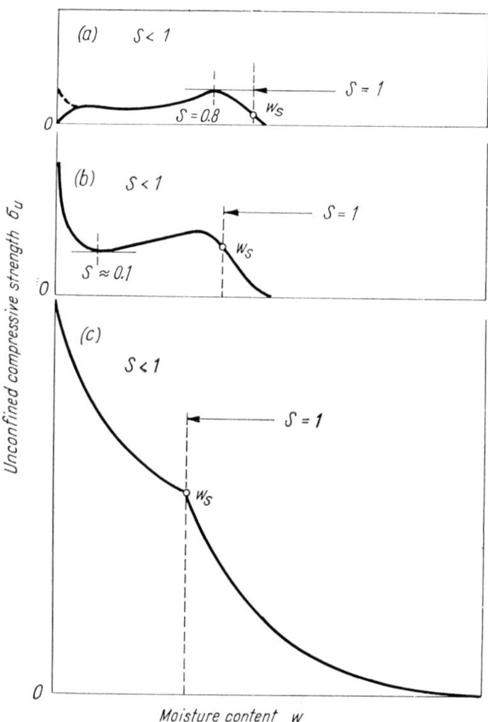

Fig. 308. Relationship between unconfined compressive strength and moisture content

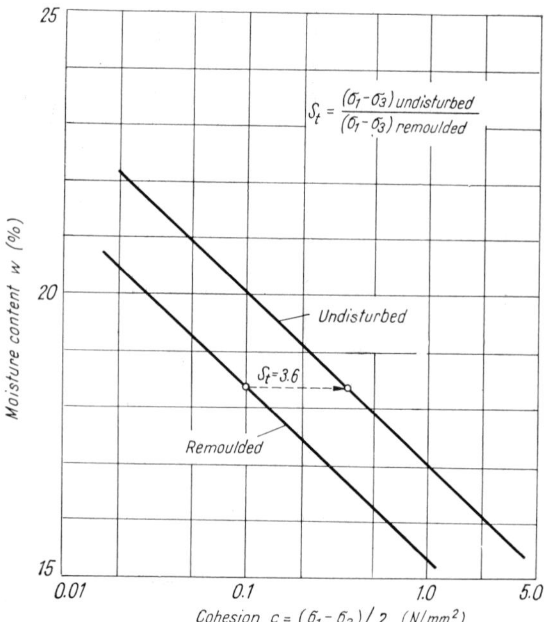

Fig. 309. Compressive strength versus moisture content in undisturbed and in remoulded clay

If we run unconfined compression tests once on undisturbed and once on remoulded specimens of clay at various saturation water contents ($S = 1$), and plot the results in a semi-logarithmic representation, two parallel straight lines result (*Fig. 309*). From these, the sensitivity of the clay (see Ch. 4) can be determined as

$$S_t = \frac{(\sigma_1 - \sigma_3) \text{ undisturbed}}{(\sigma_1 - \sigma_3) \text{ remoulded}} \, .$$

The lines $[\log(\sigma_1 - \sigma_3) - w]$ are parallel to the stress–strain plot representing the compression of a normally loaded clay under increasing all-around pressure (virgin compression curve). By making use of this general relationship, we can determine, from the results of unconfined compression tests at various water contents, the Coulomb failure envelope with respect to effective normal stresses for consolidated—undrained conditions (SCHULTZE and MUHS, 1967, p. 558).

The unconfined compression test is a commonly used method of estimating the cohesive strength of a soil. Since in a compression test performed in the usual manner, the resulting volume change of the specimen is normally negligible, the test conditions can be thought, at first consideration, to be analogous to those of the consolidated undrained shear test. On the other hand, it is a common experience that the Mohr circle of the unconfined compression test, plotted in the usual manner, often does not fit in with the result of a triaxial test. The Coulomb line plotted as a function of the effective stress usually cuts the Mohr circle of the unconfined compression test, but occasionally it also occurs that the unconfined compression test yields the lower strength. The cause of this discrepancy lies in the fact that the "unconfined compression" test does not strictly comply with the conditions of the uniaxial state of stress in that the surface tension in the menisci formed on the surface of the specimen induces negative pore pressures inside the specimen. During the application of the axial load, the pore pressure changes in a way governed by whether the specimen tends to decrease or increase in volume. An increase in the axial load may, therefore, either increase or decrease the initial pore pressure in the specimen; in the former case (increasing u) the net pore pressure at failure may still be negative. The significant u value at failure and, hence, the position of the *Mohr* circle depend on the sign of the tendency to volume change prior to failure. Typical examples of the effect of the neutral stresses on the unconfined compressive strength are illustrated by *Fig. 310.* Diagram *a* shows the representation in terms of total normal stresses.

In this case, the total lateral stress, σ_3, is indeed zero, and the *Mohr* circle touches the τ-axis at the origin. In the diagrams *b* and *c*, the failure envelopes are plotted on the basis of the effective stress principle. Case *b* represents a soil with a tendency to dilate. The initial negative pore pressure due to meniscus effect is further increased during the loading (net $\breve{u}$ is negative), and the resulting effective stress is, according to the formula $\bar{\sigma} = \sigma - \breve{u}$, larger than the applied total axial stress σ_u. In the state of failure, the corresponding *Mohr* circle is tangent to the *Coulomb* line. But if we disregard the effect of the neutral stresses and plot the *Mohr* circle in terms of total stresses, it will intersect the failure line. ($\sigma_1 = \sigma_u$, $\sigma_2 = \sigma_3 = 0$.) In the other case (diagram *c*), when the specimen shows a tendency to contract with approaching failure, the generated excess pore pressure is positive and, in consequence, the effective normal stress decreases. The failure circle touches again the *Coulomb* line. But if we plot the circle as if the state of stress were $[\sigma_1 = \sigma_u, \sigma_2 = \sigma_3 = 0]$, it will be located below the *Coulomb* line. Because of these uncertainties, the unconfined compressive strength is chiefly used only as a characteristic of soil consistency.

We can also conclude from the foregoing statements that the formation of surface menisci due to desiccation can greatly influence the strength of clays even under field conditions.

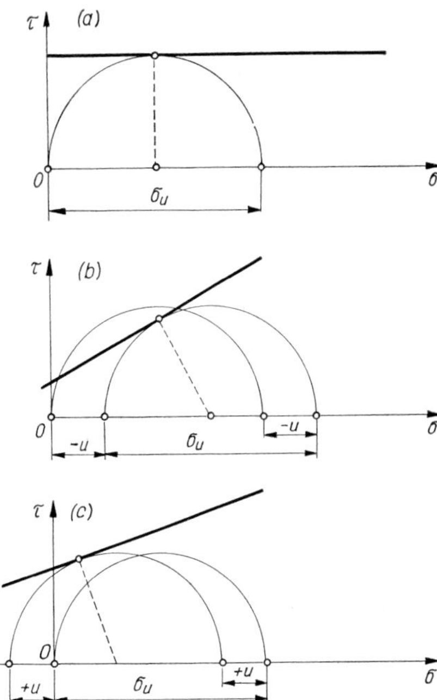

Fig. 310. Effect of neutral stresses on the unconfined compressive strength

The cohesion of a clay can also be determined by a consolidated-undrained test. Here several specimens are first consolidated under a given all-around pressure, σ_c, and then failed under undrained conditions at increasing cell pressures, σ_3. If we plot the results in terms of total stresses, the resulting failure lines will be horizontal. For each initial consolidation pressure there will be a constant shearing strength, $\tau = c$ (*Figs 311a* and *b*). By plotting c as a function of the effective consolidation pressure $\bar{\sigma}_c$, we obtain the relationship shown in Fig. 311*c*. If the consolidation pressures are larger than the maximum past effective stress that ever acted upon the soil, $\bar{\sigma}_p$, (in which case the clay is described as *normally consolidated*), the failure diagram is a straight line which passes through the origin. According to this, when the effective normal stress is zero, the shearing strength of the clay is also zero, since there is no mineral-to-mineral contact between the particles. In nature, however, practically every soil has undergone some kind of preconsolidation, caused e.g. by the overburden pressure or by desiccation, permanent or temporary, in the zone near the ground surface. But preloading,

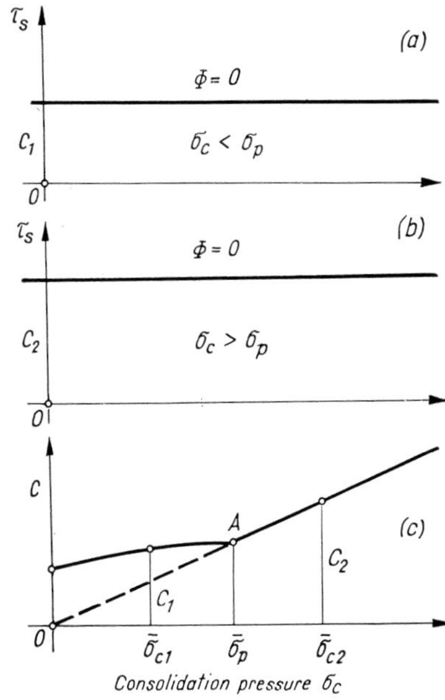

Fig. 311. Shear strength in consolidated, undrained test

as we have seen, creates real bonds between the particles and, thus, produces cohesion. This explains that for consolidation pressures smaller than $\bar{\sigma}_p$ the diagram of c-values in Fig. 311c lies above the failure line obtained for a normally consolidated clay.

The slope of the line OA in Fig. 311c was found to be representative of the make-up of a clay.

Since the strength of cohesive soils is, as we have seen previously, a function of the number of interparticle contacts, it can reasonably be assumed that the finer the grains of which a soil is composed, the greater the number of the potential contacts per unit of area between the grains, hence, the larger the strength increment due to a given consolidation pressure, i.e. the steeper is the line in question. The fineness of the particles is fairly well reflected in the plasticity index and it can be assumed that the slope of the line $\overline{OA}$ also increases with increasing plasticity index. This has been experimentally proved by SKEMPTON (1957). He also found a linear correlation between the plasticity index and the ratio c/σ_c (*Fig. 312*).

It follows from this relationship that for normally loaded natural deposits the consolidated–undrained shearing strength should increase linearly with the depth below the ground surface, since in this case the preconsolidation pressure ($\bar{\sigma}_p$ in Fig. 311c) is equal to the effective overburden pressure at the depth under consideration ($h\gamma$). The validity of this correlation has been satisfactorily confirmed by tests on natural samples of clay taken with meticulous care to preserve them intact.

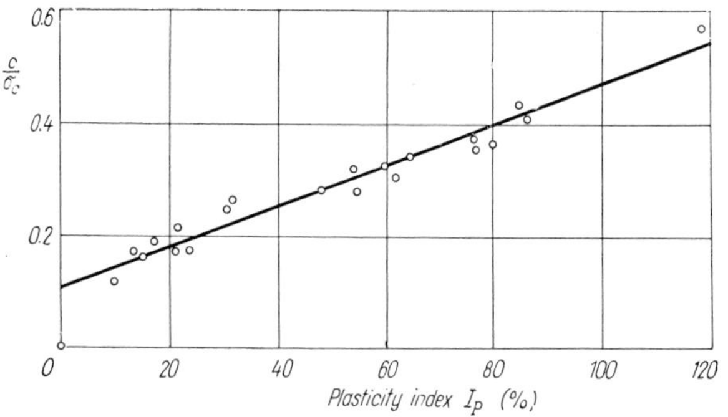

Fig. 312. Relationship between cohesion ratio and plasticity index

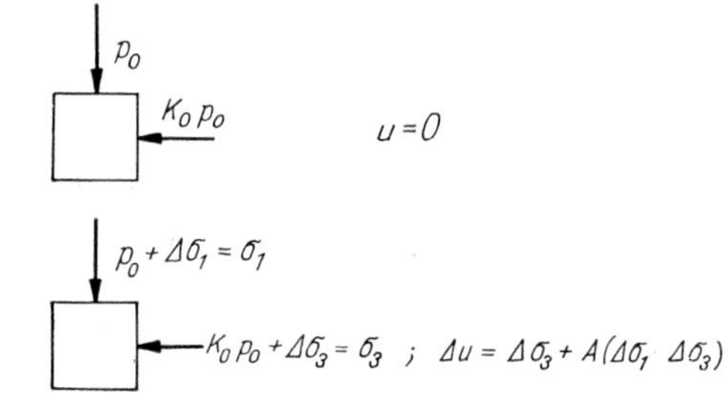

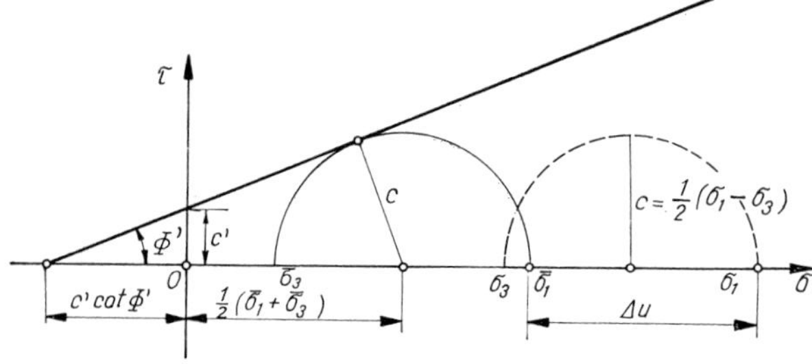

Fig. 313. Undrained shear strength and shear strength parameters as determined on the basis of effective stresses

The undrained shear strength or "cohesion" and the shear strength parameters determined on the basis of effective stresses (Φ', C') are interrelated, as will be seen from the following derivation (LEONARDS, 1960).

The initial conditions together with the failure criterion are illustrated by *Fig. 313*. The vertical principal stress is equal to the overburden pressure ($\sigma_z = p_0 = z\gamma$), and the horizontal principal stress is $\sigma_x = K_0 \sigma_z = K_0 z\gamma$. ($K_0$ is the earth pressure coefficient at rest. See Ch. 9.) The initial excess pore pressure is zero. If we now alter the state of stress by applying principal stress increments of $\Delta\sigma_1$ and $\Delta\sigma_3$, excess pore pressures develop. According to Eq. (75), for a saturated soil ($B = 1$) the excess pore pressure is given by

$$\Delta u = \Delta\sigma_3 + A(\Delta\sigma_1 - \Delta\sigma_3)\,.$$

Hence either total stress circles or effective stress circles can be plotted. Working with effective stresses, we have

$$\sin \Phi' = \frac{c}{\dfrac{\bar{\sigma}_1 + \bar{\sigma}_3}{2} + c\dfrac{\cos \Phi'}{\sin \Phi'}}\,,$$

or

$$c = \frac{\bar{\sigma}_1 + \bar{\sigma}_3}{2} \sin \Phi' + c' \cos \Phi' =$$

$$= \frac{\bar{\sigma}_1 - \bar{\sigma}_3}{2} \sin \Phi' + \bar{\sigma}_3 \sin \Phi' + c' \cos \Phi' =$$

$$= c \sin \Phi' + (\sigma_3 - u) \sin \Phi' + c' \cos \Phi'\,.$$

Hence,

$$c = \frac{c' \cos \Phi' + (\sigma_3 - u) \sin \Phi'}{1 - \sin \Phi'}\,.$$

From Fig. 313

$$\sigma_3 - u = K_0 p_0 + \Delta\sigma_3 - u = K_0 p_0 - A(\Delta\sigma_1 - \Delta\sigma_3)$$

and

$$c = \frac{1}{2}(\sigma_1 - \sigma_3) = \frac{1}{2}(\Delta\sigma_1 - \Delta\sigma_3) + \frac{1}{2}(1 - K_0)\, p_0$$

whence

$$\Delta\sigma_1 - \Delta\sigma_3 = 2c - (1 - K_0)\, p_0$$

and

$$\sigma_3 - u = p_0[K_0 + A(1 - K_0)] - 2Ac\,.$$

Substituting in the expression of c gives

$$c = \frac{c' \cos \Phi' + p \sin \Phi[K_0 + A(1 - K_0)]}{1 + (2A - 1) \sin \Phi'}\,. \tag{209}$$

For a normally consolidated clay $c' = 0$, hence Eq. (209) reduces to the form

$$\frac{c}{p_0} = \frac{\sin \Phi'[K_0 + A(1 - K_0)]}{1 + (2A - 1) \sin \Phi'}\,. \tag{210}$$

The ratio c/p_0 is indeed a constant.

7.7.5 General relationship between effective normal stress, shearing strength and water content for saturated clays

RENDULIĆ (1936) already has pointed out that the shearing strength of a saturated cohesive soil can be expressed as a unique function of the water content and of the effective normal stress. HENKEL (1959) and others (ROSCOE, SCHOFIELD, WROTH, 1958; HORN, 1964) have experimentally established this relationship for several soils. *Figure 314* shows the general relationship, after *Horn*, in the form of a function with two variables: $\tau = f(\bar{\sigma}, w)$. *Roscoe* and coworkers used a different representation by plotting the effective deviator stress $(\bar{\sigma}_1 - \bar{\sigma}_3)$ as a function of the mean effective principal stress, $(\bar{\sigma}_1 + \bar{\sigma}_2 + \bar{\sigma}_3)/3$ and of the water content. These representations are based on the assumption that the shearing strength depends only slightly on the stress path along which the stress is applied during shear, therefore the failure surface, $\tau = f(\bar{\sigma}, w)$ is a unique relationship for a given soil. In other words, the position of a point on the limiting surface is independent of the sequence of changes in the state of stress through which failure is eventually attained.

Regardless of whether or not the underlying assumption strictly holds, the graphical representation shown in Fig. 314 is a general, visual means of portraying the shear strength of cohesive soils and the factors influencing it. The failure surface is shown only from a lower limiting water content, w_0, upwards; it is not practical to have specimens saturated below this limit. The plane $(\bar{\sigma}, w)$ for which $\tau = 0$ is used for plotting the compression of a specimen under uncreasing all-around pressures. A consolidated—undrained test is depicted by a stress path in a plane for which $w =$ constant [i.e. a plane parallel to the plane $(\tau, \bar{\sigma})$]. A plane ($w =$ constant) intersects the failure surface in a line which is, thus, representative of the shearing strength at constant water content (i.e. at constant volume). These lines of intersection are found to be straight lines. Their slope angle is called the "*true friction angle*" of the soil. The line in which the failure surface intersects the plane (τ, w) at $\bar{\sigma} = 0$ represents the "*true cohesion*". As shown before, this is an exponential function of the water content. The limiting surface of shearing strength is bounded in space by the curved line Φ' which represents the drained shear strength of the normally loaded material. The projection of the curve Φ' on the plane $(\tau, \bar{\sigma})$ gives a straight line sloping at angle Φ and passing through the origin (SCHULTZE and MUHS, 1967, p. 559).

The advantage of this picture is that it combines all formerly used representations. We may run e.g. drained shear tests on the same clay initially consolidated under increasing all-around pressures. By plotting the results in terms of effective stresses, we obtain a straight line passing through the origin as shown in *Fig. 315*. Initially, the consistency of the clay corresponds approximately to the

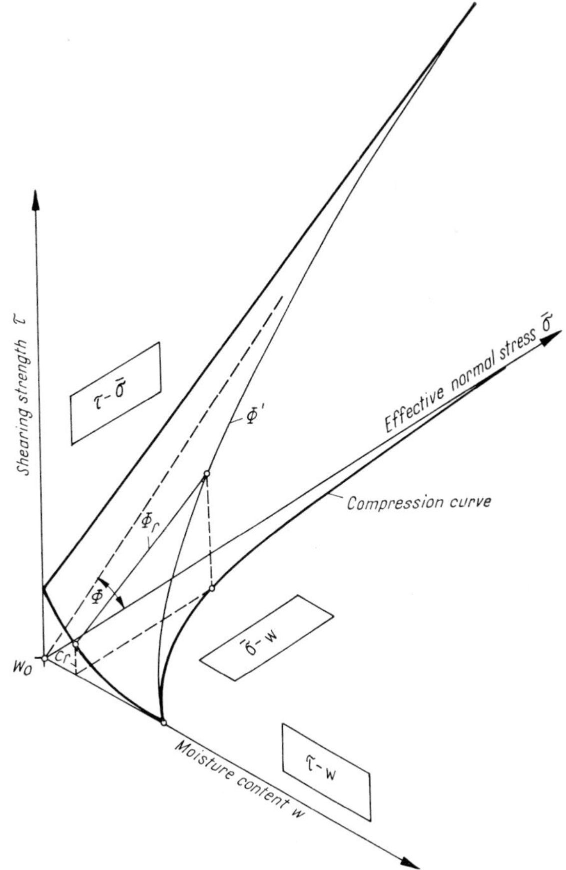

Fig. 314. Diagram of shear strength moisture content and effective normal stress

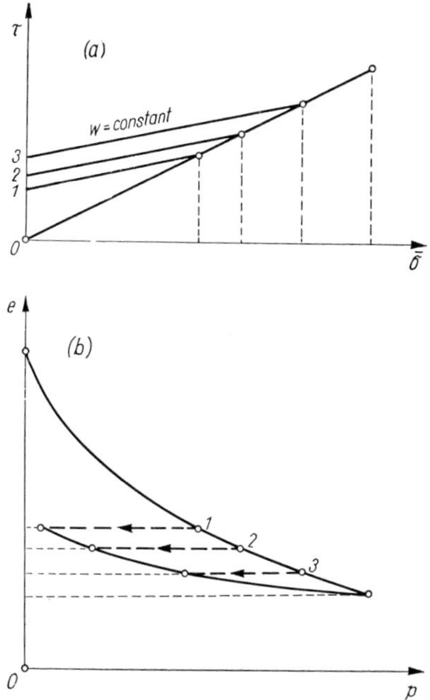

Fig. 315. Shear strength in open system, in terms of effective normal stresses

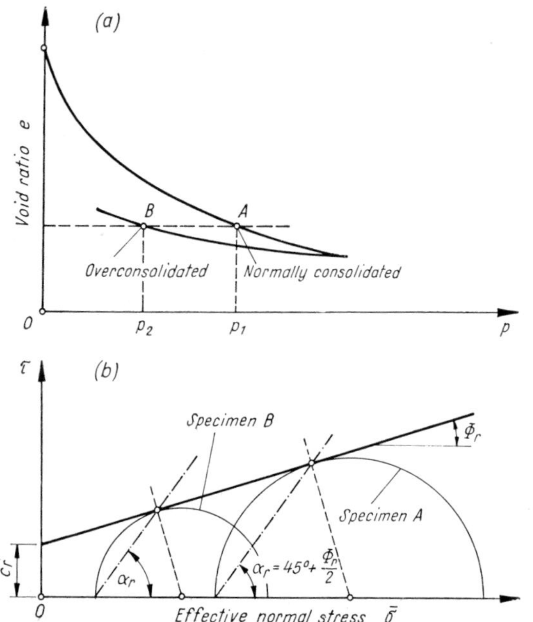

Fig. 316. Determination of true angle of internal friction and true cohesion on two samples having the same moisture content but different overconsolidation load

liquid limit, and every point on the straight line corresponds to a different void ratio. The line itself is an equivalent of the projection of the curve Φ' on the plane $(\tau, \bar{\sigma})$ in Fig. 314. If we now consolidate the clay under different hydrostatic pressures, and then perform shear tests under undrained conditions (i.e. with water contents kept constant) and plot the results in the same diagram, we obtain straight lines in the planes (w = constant). These lines, known as *Krey–Tiedemann* lines, are parallel and differ only in the value of the cohesion intercept. The cohesion is an exponential function of the water content. For different values of $\bar{\sigma}$ = constant, the lines shown in Fig. 315 are obtained.

The slope angles of the lines (w = constant) in Fig. 315*a* should not be regarded as friction angles, since the consistency of the soil is different at every point of a line. It was for this reason that Hvorslev (1937) introduced the concept of a "*true cohesion*", thought to be a unique function of the water content. The "*true friction angle*" characterizes then the rate of increase of the shearing resistance with increasing effective normal stress at a constant water content. These two quantities can be determined by the following procedure (Skempton, 1954). *Figure 316* shows the compression and expansion curves for a saturated clay (see Ch. 8). Owing to the non-elastic behaviour of the soil, the two curves are essentially different. It can be seen from the figure that two states of soil exist in which the water contents are equal but the effective consolidation pressures are different (points A and B). If we perform undrained shear tests on such specimens and measure the pore pressure at the time of failure, we can calculate the effective stresses and plot the corresponding failure circles (Fig. 316*b*). But since the two specimens have equal water contents, their "true" cohesion must also be equal. Hence the difference observed in the shearing strength is due to "true" friction. The failure envelope is obtained as the common tangent to the two circles. Its equation is

$$\tau = c_r + (\sigma - u) \tan \Phi_r . \qquad (211)$$

c_r is the true cohesion and Φ_r is the true angle of internal friction.

From Eq. (211) it follows that under undrained conditions the plane on which a specimen shears makes an angle of

$$\alpha = 45° + \frac{\Phi_r}{2} \qquad (212)$$

with the plane of the major principal stress. A rough estimate of the true friction angle can therefore be obtained by measuring off the angle of inclination α from the failed specimen. According to the experiments of Gibson (1951), if we repeat the *Hvorslev* test at different void ratios, straight lines parallel to that already shown in Fig. 316*b* are obtained. This result suggests that the value of Φ_r depends only slightly on the water content and, within an average range of w, it can be regarded as a constant for a given clay. Measured values of the true friction angle and the true cohesion for several soils are given in *Table 28* (Gibson, 1953, and author's data).

Table 28. True angle of friction and true cohesion of some soils

$\tau = (\sigma - u) \tan \Phi_r + \varkappa p$

Soil type	W_L %	I_p %	Activity	Range of water content	Φ_r degrees	$\varkappa$
Remoulded London clay	74	49	0.98	29÷34	12.5	0.10
Undisturbed clay (Schellhaven)	123	87	1.42	52÷60	18.0	0.10
Clay (Tisza valley)	58	33	0.91	26÷40	21.5	0.03
Lean clay (Tisza)	32	16	0.45	20÷22	27	0.11
Clay of Kiscell (Hungary)	45	23	0.95	25÷28	29	0.02

According to *Hvorslev*, the true cohesion c_r is proportional to the consolidation pressure p on the virgin compression curve and is expressed as

$$c_r = \varkappa p ,$$

where $\varkappa$ is the coefficient of cohesion. Numerical values of $\varkappa$ are also given in Table 28.

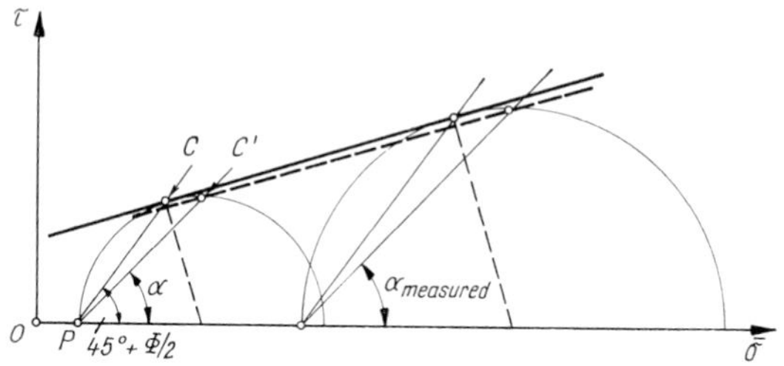

Fig. 317. Correction to the failure for deviation of the sliding surface

The determination of the true friction angle from the inclination α of the rupture plane across the failed specimen gives, according to Whitman (1960), somewhat higher values than the actual Φ_r. The necessary correction can be obtained as follows. From point P on the failure circle, we draw a line parallel to the actual rupture plane (*Fig. 317*). It cuts the Mohr circle at point C'. Having constructed several such points, we can draw the corrected failure envelope (dashed line)

Recommended reading

Bishop, A. W. and Henkel, D. J. (1957): The measurement of soil properties in the triaxial test. Edward Arnold Ltd., London.

Bölling, W. H. (1971): Zusammendrücken und Scherfestigkeit von Böden. Anwendungsbeispiele und Aufgaben. Springer Verlag, Wien—New York.

Brinch Hansen, J. and Lundgren, H. (1960): Hauptprobleme der Bodenmechanik. Springer Verlag, Berlin—Göttingen—Heidelberg.

Casagrande, A. (1936): Characteristics of cohesionless soils affecting the stability of slopes and earth fills. J. Boston Soc. Civ. Engrs. Vol. 23, pp. 13—32.

Henkel, D. J. (1959): The relationships between the strenght, pore-water pressure and volume-change characteristics of saturated clays. Géotechnique, 9, p. 119.

Hvorslev, J. (1937): Über die Festigkeitseigenschaften gestörter bindiger Böden. Ingeniorvideskabelige Skrift. No. 45, Kopenhagen.

Kézdi, Á. (1963): Scherverformungen von Sand. Az Építőipari és Közlekedési Műszaki Egyetem Tudományos Közleményei. Vol. IX, No. 5.

Kézdi, Á. (1966): Grundlagen einer allgemeinen Bodenphysik. VDJ-Zeitschrift, Vol. 108, No. 5.

Lee, I. K. (ed.) (1968): Soil mechanics, selected topics. American Elsevier Publishing Co., Inc. New York.

Poorooshasb, H. B. and Roscoe, K. H. (1961): The correlation of results of shear tests with varying degrees of dilatation. Proc. 5th Int. Conf. of Soil Mech. and Found. Eng. Vol. 1, p. 297, Paris.

Proceedings of the Eur. Conf. on Shear Strength of Soils. Géotechnique. Vol. 1, No. 3, 1949.

Proceedings, ASCE, Research Conf. on Shear Strength of Cohesive Soils. University of Colorado, Boulder, Colorado, 1960.

Proceedings of the Geotechnical Conference, Oslo 1967, on Shear strength properties of natural soils and rocks. Norwegian Geotechnical Institute, Oslo.

Schultze, E. and Muhs, H. (1967): Bodenuntersuchungen für Ingenieurbauten. 2nd edition. Springer Verlag, Berlin—Heidelberg—New York.

Scott, R. F. (1963): Principles of Soil mechanics. Addison-Wesley Publishing Co. Reading, Mass.

Skempton, A. W. (1957): Discussion of "The planning and design on the New Hong-Kong Airport". Proc. Inst. Civ. Engrs., London, Vol. 7, p. 305.

Skempton, A. W. (1961): Effective stress in soils. Concrete and rocks. Conference on Pore Pressure and Suction in Soils. London, Butterworths.

Skempton, A. W. and Bishop, A. V. (1954): Soils. In: Building materials, their elasticity and inelasticity. Edited by Reiner, M., North Holland Publishing Company, Amsterdam.

Symposium on Direct Shear Testing of Soils. American Society for Testing Materials, Spec. Techn. Publ. No. 131, Philadelphia 1953.

Taylor, D. W. (1948): Fundamentals of Soil mechanics. J. Wiley Sons, New York, pp. 13—15.

Tsytovich, N. A. (1963): Mekhanika gruntov. Moscow, Chs. II. 4—II. 5.

Winterkorn, H. F. (1954): Macromeritic liquids. In: Symposium on Dynamic Testing of Soils. ASTM, Spec. Techn. Publ. No. 159.

Chapter 8

Deformation of soils

8.1 Introduction

Every kind of load or stress whether acting inside or on the boundary surface of a soil mass causes strains and displacements in it. These deformations depend on the *in situ* properties of the material, on the applied loads and stresses and, in general, on time. There are cases in which the application of load is followed by deformations occurring at a diminishing rate so that after a while an equilibrium state is attained. In other cases the deformations have a steady character, meaning that their magnitude is indefinite and only the rate of motion has a definite value.

Predicting the magnitude and rate of deformations has always been a difficult task in every branch of engineering science and it is only for certain artificial products such as steel or high-quality concrete that an adequate solution to this question has been found. To certain problems the theory of elasticity can be applied, in others we need to resort to simplifying assumptions as e.g. on anisotropy or on the stress–strain characteristics of the real material. In soil mechanics, the need to calculate soil deformations seldom arises except perhaps for the purpose of predicting the settlement of foundations. In the majority of cases, soils engineering problems are treated on the basis of the theory of plasticity by means of failure state analysis. To such an analysis the soil deformations are completely irrelevant, not that it would at all be feasible to determine them. Therefore, research has always been centered upon soil strength leaving little scope for the study of soil deformations. The reason for this surely lies in the diversity, heterogeneity and anisotropy of soils. It must be stated that at present we do not possess yet a universal theory that would truly describe the stress-behaviour of soil. For this reason, only some of the relatively simple problems of deformation will be considered in this chapter, since the more complicated ones cannot be solved yet numerically. As the deformations due to shear have already been discussed in Ch. 7, this chapter deals mainly with the compression of soil under load. Methods for predicting settlement will be treated in Vol. 2.

We begin our study by writing down the fundamental equations of elastic theory. They will be often referred to in subsequent chapters. In the elastic theory, it is assumed that the general function describing the relationships between stresses and strains contains a linear combination of these same quantities, in other words, that *Hooke*'s law is valid.

Consider an elementary rectangular prism with sides parallel to the coordinate axes (*Fig. 318*). Let a uniform lateral pressure act on two opposite faces of the element. For an isotropic material the strain parallel to the direction of stress is given by

$$\varepsilon_x = \frac{\sigma_x}{E}, \tag{213}$$

where E is *Young*'s modulus of elasticity. It represents the fictitious stress that causes a strain of unity.

The strain occurring in the direction x is accompanied by strains in both transverse directions, y and z, and vice versa. The strains in y and z directions are also assumed to be proportional, in opposite sense, to the applied stress σ_x. Hence

$$\varepsilon_y = -\mu\frac{\sigma_x}{E}; \quad \varepsilon_z = -\mu\frac{\sigma_x}{E}. \tag{214}$$

μ is *Poisson*'s ratio. Its reciprocal value ($m = 1/\mu$) is called *Poisson*'s number.

In the general case (*Fig. 319*), with all stress components, σ_x, σ_y and σ_z acting on the faces of

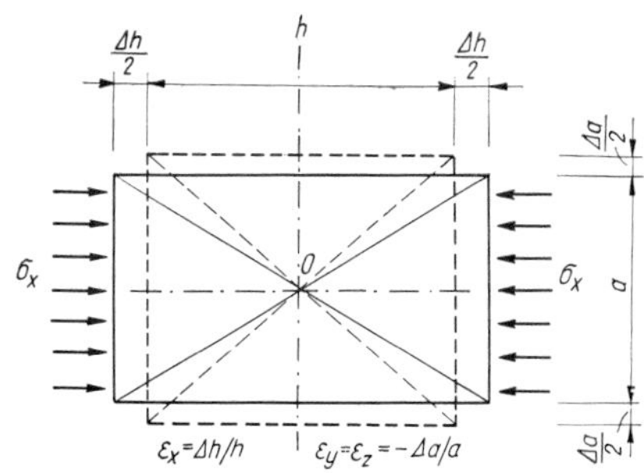

Fig. 318. Axially loaded prism with uniformly distributed load

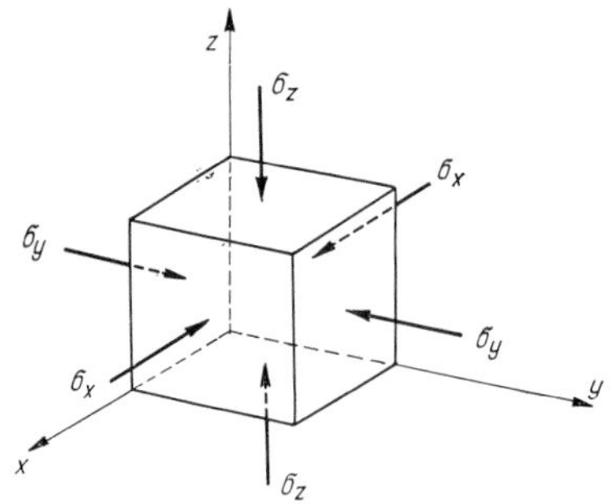

Fig. 319. Stresses acting on the elementary cube

the elementary block, the strains produced by the individual stress components can be obtained from Eqs (213) and (214). By employing the principle of superposition we have

$$\left.\begin{aligned} \varepsilon_x &= \frac{1}{E}[\sigma_x - \mu(\sigma_y + \sigma_z)] \\ \varepsilon_y &= \frac{1}{E}[\sigma_y - \mu(\sigma_x + \sigma_z)] \\ \varepsilon_z &= \frac{1}{E}[\sigma_z - \mu(\sigma_x + \sigma_y)] \end{aligned}\right\} . \tag{215}$$

It must be noted at this point that the application of elastic theory to soils is subject to serious limitations. In the first place, a proportionality between stress and strain does not strictly hold, even when the applied loads are small. The laws expressed by Eqs (213) and (214) can be applied to soils only so long as the deformations are small enough not to essentially influence the applied loads. Another limitation is owing to the apparent lack in soils, particularly in clays, of the property of isotropy on which the validity of Eqs (215) is based.

Beside the just described general, three-dimensional state of stress, in practical applications of soil mechanics we are often concerned with problems in which a soil mass has a very large dimension in one direction and is acted upon by forces that are identical on any plane section perpendicular to that direction. In this case the strain parallel to that same direction must be zero. Such situation is called a *plane-strain condition.* Letting e.g. $\varepsilon_y = 0$, the second equation of Eqs (215) becomes

$$\sigma_y - \mu(\sigma_x + \sigma_z) = 0 .$$

Hence

$$\sigma_y = \mu(\sigma_x + \sigma_z) . \tag{216}$$

In this particular case, the solution of a stress problem only requires the determination of the stress components, σ_z, σ_x and τ_{xz}. Plane-strain condition is typical of most problems of earth pressure, stability of slopes, bearing capacity of long footings, design of sheet-pile walls, etc. Further simplifications of the general stress problem lead to the plane-stress condition in which all stresses are parallel to a given plane; and to axially symmetrical problems.

Finally let us write down the expression of volumetric strain, e, for the elementary block shown in Fig. 319. By neglecting the products of the quantities ε, we have

$$e = \varepsilon_x + \varepsilon_y + \varepsilon_z . \tag{217}$$

Adding Eq. (215) and substituting σ_0 for the sum of three normal stresses, (which sum is called the first stress invariant as it is independent of the system of coordinates), gives the following expression

$$e = \frac{1 - 2\mu}{E} \sigma_0 . \tag{218}$$

Let us examine now more closely the intepretation for soils of the two elastic properties contained in the basic equations, E and μ. For a truly elastic material, the stress–strain diagram is linear for both compression and tension (*Fig. 320*). The proportionality between stress and strain is expressed by the modulus elasticity (*Young*'s modulus) being constant:

$$\cot \alpha = \frac{\Delta\sigma}{\Delta\varepsilon} = \text{constant} = E . \tag{219}$$

This, however, is valid only for perfect elastic, homogeneous and isotropic materials. If the stress —strain diagram is a curved line, the modulus of elasticity is defined as

$$E = \frac{d\sigma}{d\varepsilon} .$$

In such a case, the relationships of the classical elastic theory are only approximations, and the true behaviour of the material can only be described by the laws of non-linear mechanics. In many practical cases, as we shall see, a reasonable approximation can be obtained by assuming, on

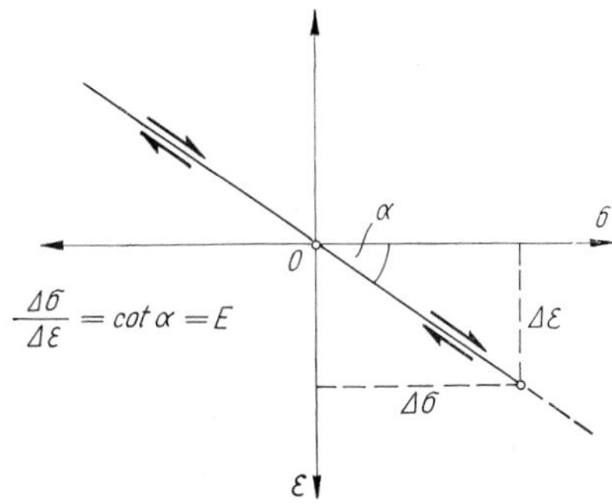

Fig. 320. Definition of the modulus of elasticity

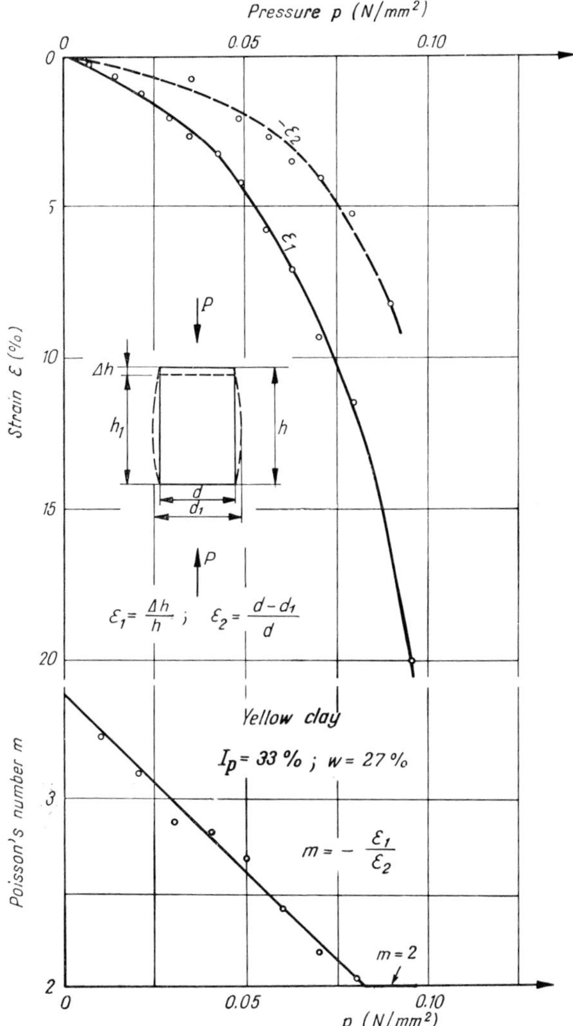

Fig. 321. Determination of Poisson's number by measuring the axial and the lateral deformation of the compressed cylinder

the basis of experimental evidence, that the modulus of elasticity increases linearly with the increase of one of the normal stresses (usually the vertical overburden stress).

A further difficulty in the application of elasticity theory to soils is that different E-values are valid for first loading, unloading and reloading. The deformation of a soil can best be described as elastic when it is subjected to repeated loadings whose upper value is well below that causing failure.

The other elastic property in the relationships of elastic theory is Poisson's ratio, which is defined as the ratio of the lateral strain to the axial strain (see Fig. 318). It has values between zero and 0.5 depending on the material. As is obvious from Eq. (218), the value 0.5 corresponds to a material showing no volume change during loading. The other limiting value, $\mu = 0$, represents a perfect rigid material.

For the direct measurement of *Poisson*'s number, no reliable method has been developed so far. The test results shown in *Fig. 321* were obtained from an unconfined compression test. The specimen was photographed after the application of each load increment, and the lateral strain was simply scaled off from the prints. *Poisson*'s number was then evaluated as the ratio of the axial strain to the lateral strain:

$$m = \frac{1}{\mu} = -\frac{\varepsilon_1}{\varepsilon_3}.$$

As can be seen from the figure, at the early stages of the test m had values of as high as 3 to 3.5, then it decreased almost linearly with increasing load. As failure was approached, m reached a value of 2, which is characteristic of perfect plastic materials. As demonstrated by this test, *Poisson*'s number is not a soil constant in that it depends not only on the physical properties of the material, but also on the magnitude of the load. Some approximate values of m for various materials are given in *Table* 29.

Table 29. Values of Poisson's number

Material	m
Plastic clay, with no volume change	2
Clay (depending on preloading and condition)	2.5–5
Sand (depending on density, load, deformations)	4–6
Hard rocks	7–10
Ideal rigid body	∞

8.2 Deformations during confined compression

8.2.1 Measurement of the compressibility of soil

Let us consider the case that vertical loads due to the weight of a structure are transferred through the foundation to a stratified subsoil. For the sake of the stability and satisfactory performance of the structure, we must know, in advance, how much *settlement* due to the compression of the underlying strata is likely to occur, and whether these settlements, or rather the differential settlements, are likely to cause a detrimental overstressing of the structure or even the formation of cracks. Provided that the loads imposed on the foundation are smaller, within an ample margin of safety, than the bearing capacity of the soil, settlements result predominantly from a compression in the vertical direction, since frictional resistances acting on the base of the foundation and at the boundaries of successive strata are normally large enough

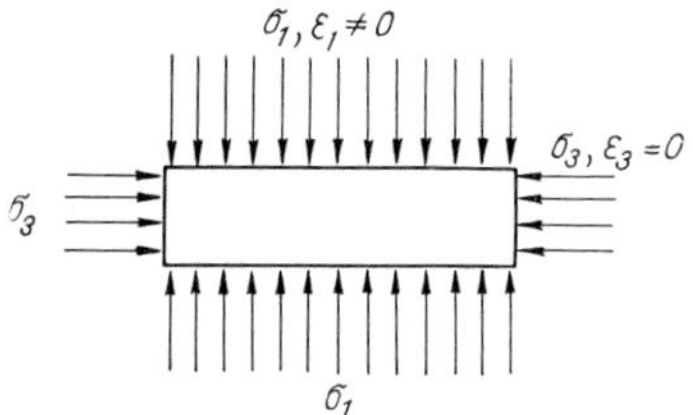

Fig. 322. State of stress in laterally confined compression

to prevent any lateral displacement. The soil mass is then said to be in a state of *confined compression*. The assumption of this simple state of stress greatly facilitates the determination of settlements by means of a compression test in which the lateral strain of the specimen is prevented. The stress–strain behaviour of the soil under such conditions is the subject matter of this section.

The principle of the test is illustrated by *Fig. 322*. A vertical stress σ_1 is applied to the specimen whereupon a vertical compression ε_1 occurs. Since the specimen is laterally confined, the horizontal strain is zero ($\varepsilon_2 = \varepsilon_3 = 0$). But to completely prevent lateral strain, a certain confining stress σ_3 must act on the cylindrical surface of the specimen.

In this test, the ratio of the lateral stress to the axial stress is constant and is called the *coefficient of lateral pressure at rest*, K_0 (see Ch. 9). If we disregard the interference from the side friction, the specimen is in a state of one-dimensional consolidation.

The apparatus in which a state of confined compression can be produced is called an *oedometer*. A usual arrangement of it is shown in *Fig. 323*. The specimen is placed between two porous discs which permit the escape of excess pore water during consolidation. The device is properly sealed to ensure that flow of water occurs only in the vertical direction. The axial compression of the specimen is measured on a dial gauge reading to 1/100 mm.

The size of the specimen used for oedometer tests varies widely with the different types of

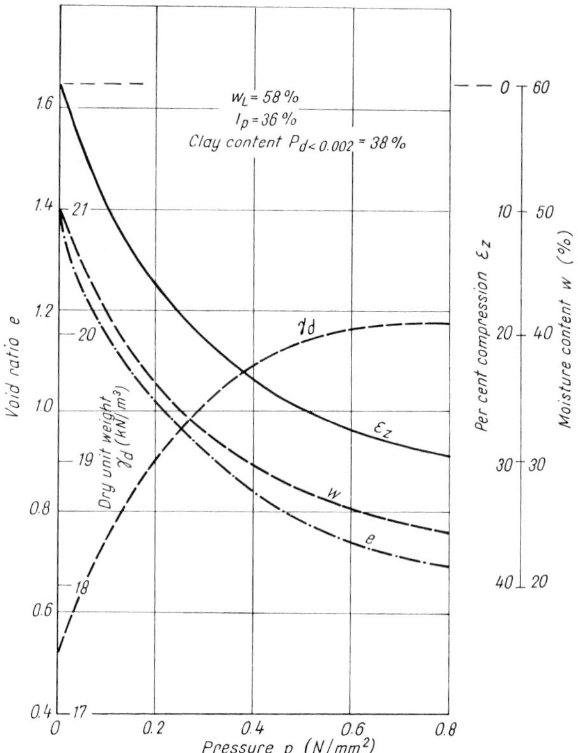

Fig. 324. Typical result of compression test: changes in moisture content, void ratio, bulk density and strain versus vertical pressure

apparatus used by various laboratories. Common sizes vary from 1 to 4 cm in thickness, by 7 to 10 cm in diameter. An argument for the application of smaller sample thickness is that it helps to minimize the undesirable effect of side friction which may markedly hamper free compression with the thicker specimens. Experiments have shown that the disturbance by side friction can be neglected only if the diameter-thickness ratio of the specimen is not smaller than 5 : 1.

The compressive load is increased in steps. Each load is allowed to act until dial gauge readings indicate compression. The duration of the test is at least 5 hours. For saturated clays it is usually

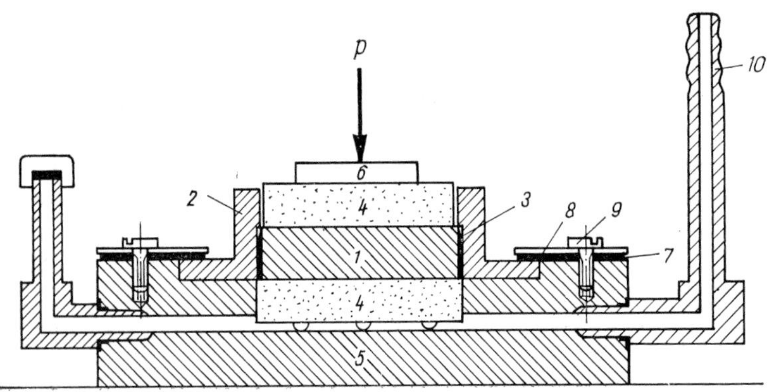

Fig. 323. Apparatus for perfoming laterally confined compression test on undisturbed soil sample: *1* — sample, *2* — ring; *3* — thin-walled ring with cutting edge; *4* — filter stones; *5* — base plate; *6* — load distributing plate; *7, 8, 9* — sealing ring, strap, screw; *10* — standpipe

Fig. 325. Compression curve; relation between (*a*) void ratio and pressure and (*b*) strain and pressure

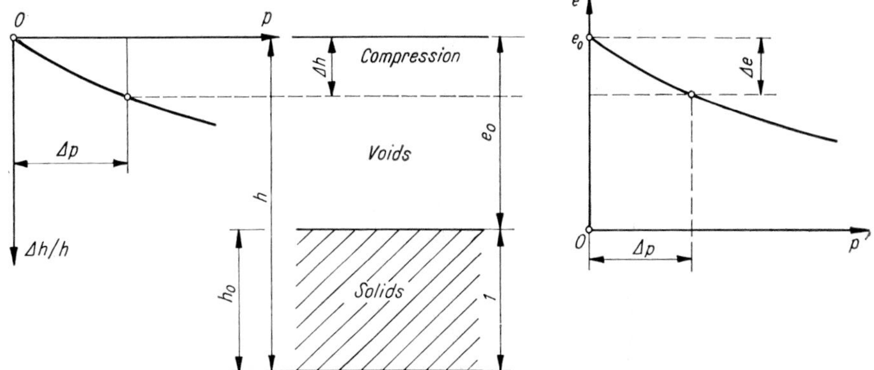

much longer, since compression must keep pace with the slow drainage of water out of the specimen.

Under the action of loading, the specimen decreases in thickness and its void ratio decreases accordingly. As compression continues, excess water is squeezed out of the water-filled pores and escapes through the porous stones. As a result, the water content of the specimen also decreases. Changes in e and w also involve changes in dry density. Now, the changes in the state of the specimen during compression can be expressed in terms of any of the affected properties. By plotting ε, e, w or γ_d against the pressure p, we obtain characteristic curved lines, which start with steep slopes but become gradually flatter as the load increases. *Figure 324* shows the result of an actual compression test on a clay with a plasticity index of $I_p = 36\%$. In practice, the pressure/vertical strain and the pressure/void ratio relationships are most commonly used (*Fig.* 325). From the dial gauge readings (which indicate the decrease in thickness, Δh, of the specimen), the void ratio is calculated by means of any one of the following expressions.

$$e_p = e_0 - \Delta e = e_0 - \frac{\Delta h}{h}(1 + e_0) = \frac{h - h_0 - \Delta h}{h_0} =$$

$$= e_0 - \frac{\Delta h}{h_0} = \frac{V}{V_0} - 1 = \frac{A(h - \Delta h)}{W_0/\gamma_s} - 1 .$$

In the formulas

e_0 = initial void ratio
e = void ratio at pressure p
Δe = decrease in void ratio
h = intial thickness of the specimen
Δh = decrease in thickness
h_0 = fictitious thickness of solids fused into a voidless mass
V = variable volume of the specimen
V_0 = volume of solids
A = cross-sectional area of the specimen
W_0 = dried weight of the specimen
γ_s = unit weight of solids

8.2.2 Some characteristics of the compressibility of soils

The causes contributing to the compression of a soil are manifold. A decrease in void ratio is, in general, the combined effect of the following mechanisms.

1. The grains tend to slide on one another into a tighter packing. This involves, especially with granular soils, a change of the soil structure.

2. Individual soil particles undergo elastic deformations. These in turn are the composite effects of tension, compression, bending and shear due to local stressing of the grains within a granular mass (*Fig. 326*). However, such elastic strains and deformations contribute but a few percent to the total compression. Under excessively large compressive stresses crushing of the grains may also occur. The fragments tend to fill the remaining void spaces, and this again results in a volume decrease. This effect becomes important when a granular soil is subjected to very heavy loads or when it contains a large amount of calcareous shells.

3. For water-saturated fine-grained soils, the dominant factor is the escape of water from the voids. As was explained in connection with Fig. 324, the water content of a saturated compression sample becomes less and less as the pressure is increased, since excess pore water is squeezed out through the pervious boundaries.

If the soil is highly permeable, thus permitting the expulsion of pore water unimpeded or only with a small time lag, compression takes place almost instantaneously upon the application of load. By contrast, in a fine-grained soil with a low

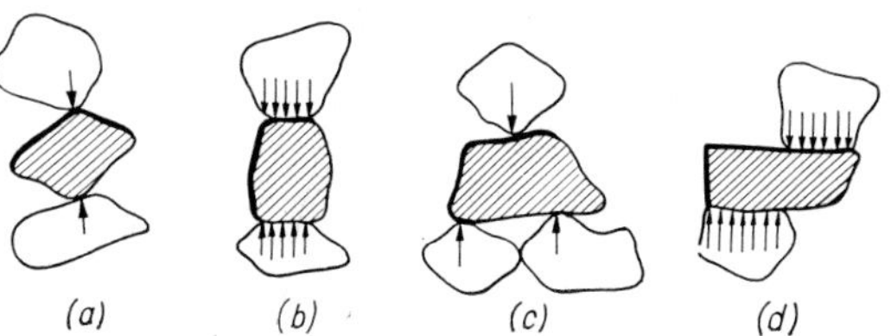

Fig. 326. Soil particles in different states of stress:
a — tension due to lateral expansion; *b* — compression; *c* — bending; *d* — shear

permeability a much longer time is necessary for the water to escape and the process of compression is prolonged in time. This time lag due to the expulsion of water was previously termed consolidation (see Ch. 6).

An idealized compression curve is shown in *Fig. 327*. The void ratio is seen to decrease at a diminishing rate as the compressive load is increased. The laterally confined specimen becomes stiffer and stiffer as the compression continues still further. The branch *AB* of the diagram represents a soil that was never before subjected to compressive loads. This branch of the curve gives a straight line on the semi-logarithmic plot. If the load is first raised to some value represented by point *B* in Fig. 327, and is then gradually removed, an expansion curve (branch *BC*) results. The soil behaves as a semi-elastic material in that the total compression consists of two parts: a permanent deformation, which is due to an irreversible alteration of the soil structure, and an elastic rebound, which is largely due to the recovery of elastic strains within the mineral skeleton (see Fig. 326). The expansion curve is much flatter, especially for clays, than the compression curve. Owing to the usually high content of mica or other flaky mineral particles, clays show in general a much more elastic behaviour.

If after unloading the specimen is reloaded, the new compression curve lies somewhat above the expansion curve. The two intersect just before the maximum stress of the first loading is reached (point *B*), and if the pressure is further increased beyond *B*, the curve merges smoothly into the curve that would have been obtained by uninterrupted first loading (branch *DE*). Between the expansion and recompression curves, a hysteresis loop is formed. Hysteresis loops for different soils differ only in slope and width, but their basic character is the same.

Preloading is a very common event in nature. For example, a clay stratum that originally consolidated under a large pressure may later have been relieved following erosion of the overburden. By performing an oedometer test on such a soil, we can observe that the compression curve consists of two parts: at first, up to about the preconsolidation pressure, it resembles the recompression curve, then it follows the trend of the virgin compression curve. Precompression may also be caused by capillary pressures due to desiccation in the zone near the ground surface. Preloading has an important practical implication. If the pressures imposed by the load of a structure on a soil stratum exceed the value of the preconsolidation pressure, the settlements will be much larger than within the range of precompression.

In sands, compressibility depends chiefly on the initial void ratio, and in clays on the initial water content. Typical compression curves for loose and dense sands are shown in *Fig. 328*. It can be seen that by static compression alone the void ratio of an originally loose sand cannot be reduced, even under high pressures, to the initial void ratio of a dense sand. This is in good agreement with the practical experience that granular soils cannot be effectively compacted by static loading, and only dynamic effects, such as ramming or vibratory compaction, can give satisfactory results.

The compressibility of sands at constant void ratio is the greater, the more rounded the grains and the more uniform the grading. Natural sand deposits are less compressible in the direction perpendicular to the bedding planes than in the

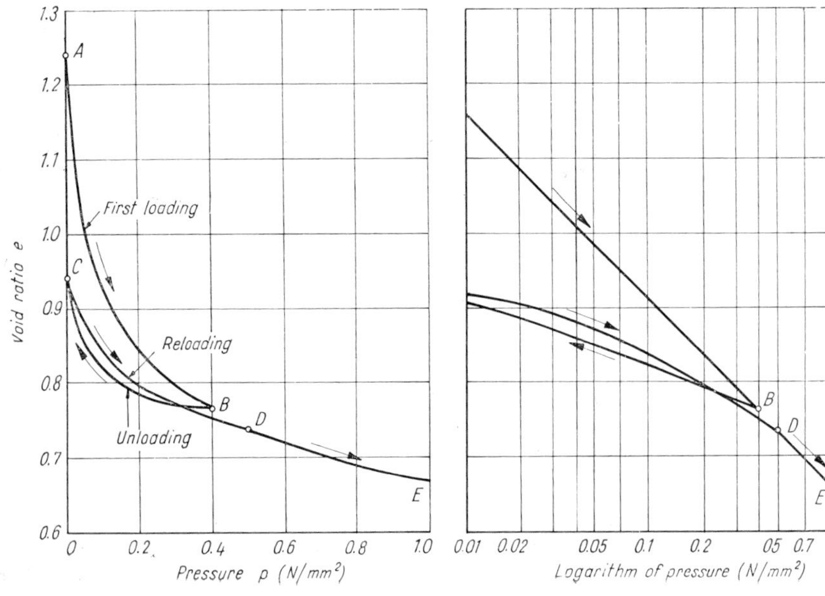

Fig. 327. Idealized compression curve in arithmetic and semi-logarithmic plot

direction parallel to them. A high content of mica greatly increases compressibility. The deformations take place very quickly; the role of the time factor is negligible.

For clays, compressibility depends on the type of clay, on the initial water content, on the stress-history of the soil as well as on the properties of the adsorption complex. Compressibility is the smaller, the stronger the forces binding the exchangeable cations to the surface of clay particles, i.e. the lower the hydration capacity of the cations (SALAS and SERRATOSA, 1953). For example, a Na clay is much more compressible than a Ca clay.

The rate of the application of load also plays an important role. Experience has shown that the total compression for a given load is smaller when the load is applied in very small increments. The probable explanation is that slow loading is more favourable for the formation of interparticle bonds. The bonding forces depend on both the duration and the magnitude of the imposed load. Thus a slow loading leads to increased interparticle bonds and a reduced compressibility. This effect may account in part for the discrepancy between settlements calculated on the basis of relatively rapid compression tests and those actually observed on structures; the latter as a rule are smaller.

We should also point out the great difference between the compressibility of a clay in its natural, undisturbed state, and that of the same clay after its structure has been disturbed. If a clay is remoulded so that its structure suffers irrecoverable injuries, e.g. by pile driving, its compressibility greatly increases. This is due to the fact that clays deposited geological ages ago have a structure of the type shown in Fig. 120 with intrusions of large grains forming a skeleton. Near the contact points between these large grains, the thickness of the water films surrounding the particles is greatly reduced. A load imposed on such a clay will act against the skeleton, this being the stiffer part, and only a small compression occurs. But if remoulding destroys the skeleton of large grains, the water films will grow in thickness, and the so far unconsolidated clay matrix becomes exposed to the load. As a result, the compressibility of the clay becomes much higher.

The theory of elasticity uses a constant stress–strain ratio known as *Young*'s modulus, E, to characterize the behaviour of linearly-elastic materials. For soils, whose behaviour is typically non-linear, the "modulus" analogous to E is not constant and holds only for a limited, narrow range of pressures. Hence, the stress–strain ratio is defined as

$$E_v = \cot \alpha = \Delta p / \Delta \varepsilon \qquad (220)$$

and is called the modulus of compressibility (see *Fig. 329*). If we use the pressure–void ratio curve,

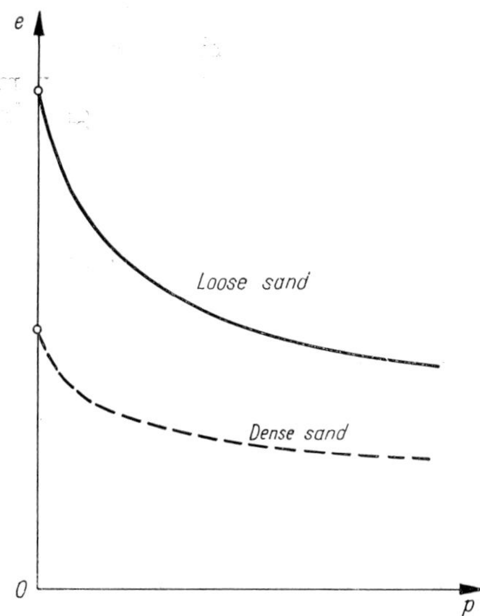

Fig. 328. Compression of loose and dense sand

the corresponding ratio is

$$a_v = \frac{\Delta e}{\Delta p} . \qquad (221)$$

This is known as the coefficient of compressibility. E_v and a_v are interrelated by the following equation.

Since

$$\Delta \varepsilon = \Delta e / (1 + e_0) ,$$

therefore

$$a_v = \frac{1 + e_0}{E_v} . \qquad (222)$$

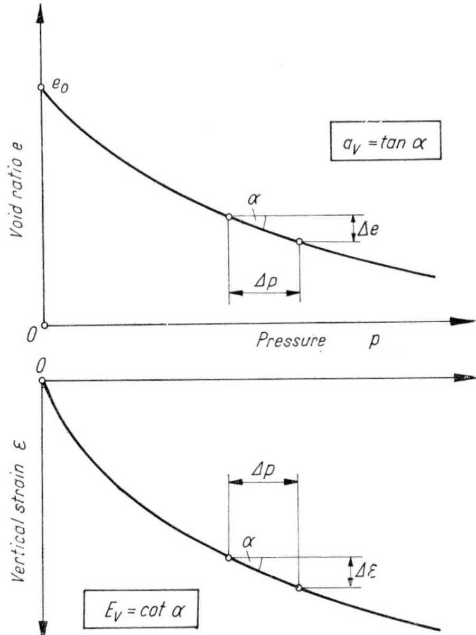

Fig. 329. Coefficient of compressibility and modulus of compression

Both a_v and E_v were found to depend on the pressure. On the basis of the experience that the branches of a compression diagram appear in a semilogarithmic plot as straight lines, TERZAGHI (1925) suggested the following empirical relationship between a_v and p:

$$\frac{1}{a_v} = c(p_0 + p)\,. \tag{223}$$

In the formula, c and p_0 are independent of the pressure and depend only on the initial void ratio and on the soil type. Numerical values of c and p_0 can be obtained from the pressure–void ratio curve by the construction illustrated in *Fig. 330*. a_v represents the slope of the tangent to the (p, e) curve with respect to the horizontal.

Combining Eq. (222) with Eq. (223) gives the following expression for the modulus of compressibility:

$$E_v = c(p_0 + p)\,(1 + e_0)\,. \tag{224}$$

According to Eq. (224), E_v is a linear function of the pressure. This relationship clearly reflects the essential difference between the behaviour of elastic materials and that of soils.

If we take the limiting value of a_v, defined by Eq. (221), as Δp approaches zero and substitute Eq. (223), we obtain

$$\frac{\mathrm{d}e}{\mathrm{d}p} = a_v = \frac{1}{c(p_0 + p)}\,.$$

Integrating,

$$e = \frac{1}{c}\int_0^p \frac{\mathrm{d}p}{p_0 + p} = -c[\log_e(p_0 + p) - \log_e p_0] + C\,.$$

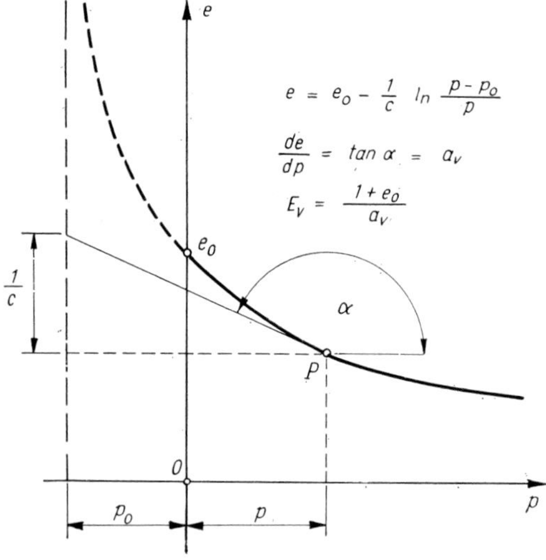

Fig. 330. Characteristics of compression curve

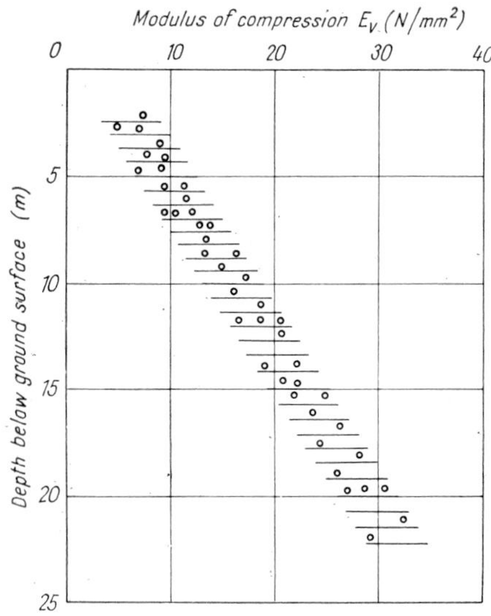

Fig. 331. Empirical relation between modulus of compression and depth

Considering that when $p = 0$, $e = e_0$, we get

$$e = e_0 - \frac{1}{C_1}\log_e\frac{p_0 + p}{p_0} = e_0 - C_c \log\frac{p_0 + p}{p_0}\,. \tag{225}$$

The geometrical meaning of the three constants in Eq. (225) is shown in Fig. 330.

Representative values of C' for various soils are given in *Table 30*.

Table 30. Approximate values of the modulus of compression E_v (N/mm²)

Soil type	Soil condition			
Granular soil	loose		medium	dense
Sandy gravel	30÷80		80÷100	100÷200
Sand	10÷30		30÷ 50	50÷ 80
Non-plastic silt	8÷12		12÷ 20	20÷ 30
Cohesive soils	soft	plastic	hard	very hard
Silty sand	5÷ 8	10÷15	15÷20	20÷ 40
Silt	3÷ 6	6÷10	10÷15	15÷ 30
Lean clay	2÷ 5	5÷ 8	8÷12	12÷ 20
Fat clay	1.5÷ 4	4÷ 7	7÷12	12÷ 30
Organic silt		0.5÷ 5		
Organic clay		0.5÷ 4		
Peat		0.1÷ 2		

For remoulded clays, the compression curve can be expressed by an equation similar to Eq. (225), only the constants have different values

$$e = e_0 - C_c' \log\frac{p_e + p}{p_e}\,. \tag{226}$$

The values of C_c' for different clays can be correlated with the liquid limit. For London clay,

SKEMPTON (1944) found the relationship:

$$C_c' = 0.007(w_L - 10)\,.$$

For clays from the environs of Paris (KÉRISEL, 1956), $C_c' = 0.0035(w_L - 10)$.

For clays of common geological origin, further correlations can be established between the modulus of compressibility, the consistency limits, and the water content. According to KOPÁCSY (1953)

$$E_v = (160 - 2I_p)\, I_C\,,$$

I_C is the consistency index. (See Ch. 2.)

Equation (224) suggests a simple linearity between the modulus of compressibility and the pressure. Since the vertical geostatic pressure in the soil increases proportionally with depth, it can be expected that for a uniformly deposited and normally consolidated clay E_v also increases linearly with depth. This is indeed the case, as is confirmed e.g. by the plot of data for soils from the *Sajó* valley, Hungary (*Fig. 331*).

The compressibility of soils can also be treated on the basis of the analogy between micromeritic and macromeritic liquids. In fluids, the shearing stress that produces a given velocity gradient is proportional to the viscosity of the fluid. The viscosity in turn can be expressed, according to ARRHENIUS and GUZMAN, by the following equation:

$$\eta = A e^{E/RT} \tag{227}$$

where

η = viscosity
A = a constant
E = molar energy necessary to initiate flow
R = gas constant
T = absolute temperature
RT = activation energy

According to BATCHINSKI (1913), the viscosity and the volume V of a liquid are related by the following equation:

$$\eta = \frac{B}{V - V_0} \tag{228}$$

B and V_0 are independent of temperature and of pressure. V is the specific volume of the gas, and V_0 is its possible least value. The difference $V - V_0$ is obviously proportional to the number of the pores in the liquid and is, thus, a measure of liquidity.

If we equate the two equations given for viscosity, Eqs (227) and (228), we obtain the equation of state for macromeritic liquids:

$$V - V_0 = C e^{E/RT}, \tag{229}$$

where $C = B/A$.

According to FRENKEL (1955), E is given by

$$E = E_0 + p\,\Delta V \tag{230}$$

ΔV is the minimum volume of pores. By substituting Eq. (230) into Eq. (229) and combining the constants, we get the following expression for the void ratio:

$$e = a + c e^{-bp}. \tag{231}$$

When $p = 0$, $e = e_0 = a + c$. Hence

$$e = e_0 - c(1 - e^{-bp})\,.$$

With the notation of *Fig. 332*, this becomes

$$\frac{\Delta e}{\Delta e_{\max}} = 1 - \exp(p/p_0)\,. \tag{232}$$

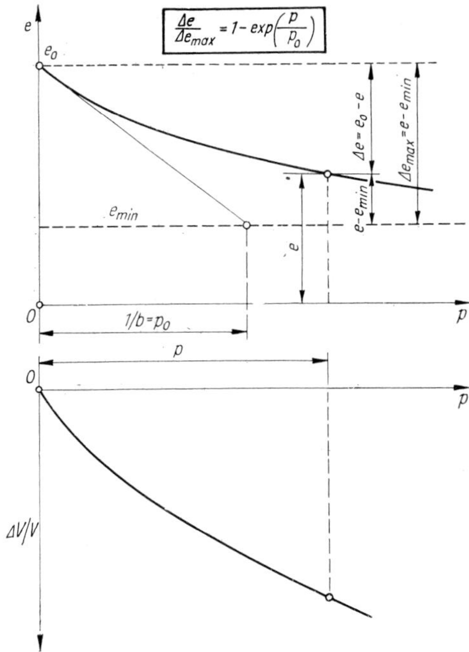

Fig. 332. Compression curve; equation derived on the basis of an analogy with liquids

The diagram in Fig. 332 represents a theoretical relationship between the void ratio and the pressure. Equation (232) is an equivalent of the logarithmic curve defined by Eq. (225) with the extra merit that this one approaches a horizontal asymptote.

8.3 Rheological properties of soils

Rheology is the science dealing with the deformations of plastic bodies. It furnishes the general isothermic equations of state of deformable bodies. The theory of elasticity and the theory of viscous liquids concern themselves with special cases of the general equations of rheology.

In a perfect elastic body the deformation upon load application is almost instantaneous, and there is a unique, though not necessarily linear, relationship between the deformation and the applied load. The deformation is reversible: after removal of the load the body instantaneously recovers its shape and dimensions. It is also reproducible: repeated loadings cause one and the same deformation. For the purpose of mathematical analysis, the stress/deformation relationship must be formulated mathematically, e.g. in the form of a power series. In the elastic theory the terms in higher powers of x are neglected. The simplified condition, linearity, furnishes the definition of the Hookean elastic body. In the general case, the normal and shearing stresses and the corresponding deformations are related through coefficients of proportionality. These need not necessarily be equal in every principal direction.

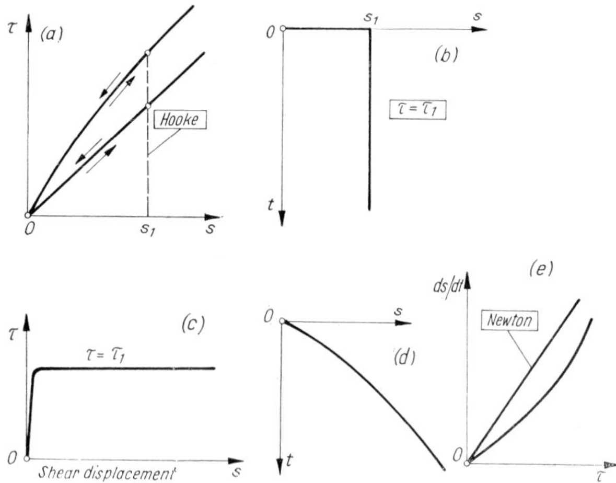

Fig. 333. Deformation characteristics of elastic and viscous materials

In a perfect viscous liquid, a given shearing stress causes a steady, indefinite deformation. There is no unique relationship between stress and strain. But if an acting shearing stress is increased, the deformation continues with a greater speed. Thus, a relationship can be established between the magnitude of the shearing stress and the rate of deformation. If the relationship is linear, we speak of a Newtonian fluid. For a quantitative analysis, the relationship needs to be expressed again in some mathematical form. The constants of the expression are regarded as material characteristics. The kinetic theory of gases derives these constants from the molecular structure of gases. For fluids only partial solutions are available. For soils, the analogy with macromeritic fluids appears to be a promising approach.

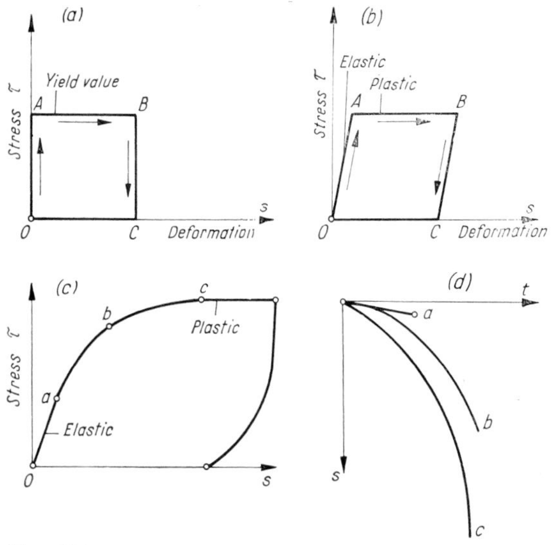

Fig. 334.
a — ideal plastic body; *b* — the plastic flow is preceded by elastic deformations; *c* — general case: behaviour of a plastic body on loading and unloading; *d* — process of deformation in time

The stress-deformation-time characteristics for elastic and for viscous substances are given in *Fig. 333.*

A substance is described as plastic if its deformations up to a certain yield stress are finite and definite but beyond that limit a steady flow occurs. The permanent deformations of a plastic body are independent of the time rate of flow. Once the limit of plasticity is reached, the deformation will continue at a constant rate, while the state of stress remains constant. With incidental small initial deformations neglected, the behaviour of such a body is described as *perfect plastic.*

The general case together with the linearized and the perfect plastic behaviours is illustrated by *Fig. 334.* Also shown in the diagrams is the effect of unloading.

Another case is represented by plastic flow at a stress-dependent rate after a threshold stress value is reached (see Ch. 7). A material exhibiting

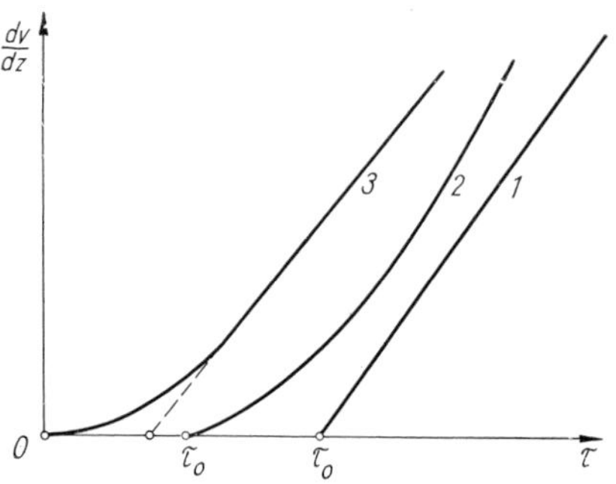

Fig. 335. Behaviour of Bingham-body:
1 — theoretical line; *2* — non-linear behaviour; *3* — approximation of the real behaviour

this property is called a *Bingham* body (*Fig. 335*). In such a material, no flow occurs below the yield stress. Above it, the material behaves as a Newtonian fluid. The rate of shear deformation varies linearly with that portion of the stress which is above the threshold value. In the general case the flow diagram is non-linear.

The preceding discussion presented but a brief review of the diverse stress-deformation-time characteristics of various materials. In order to facilitate quantitative analysis, we used simplified mechanical models expressed in mathematical form to simulate actual behaviour. As far as soils are concerned, their behaviour is a great deal more complex. The deformation components can approximately be described as follows.

Let us consider a mass of soil acted upon by effective stresses. The state of stress is assumed to be homogeneous. We shall examine the changes of some characteristic length. The load-time and deformation-time curves are as shown in *Fig. 336.* If a sudden load is applied, but without dynamic side effects, at first an instantaneous compression occurs (section *01*). This is composed of two

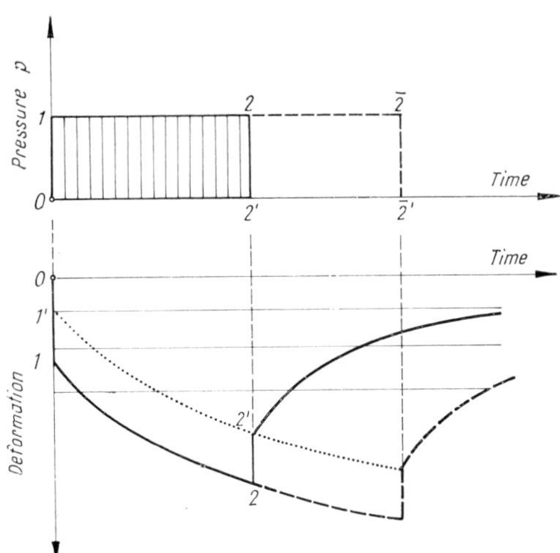

Fig. 336. Deformation during loading and unloading

portions: an instantaneous elastic deformation (*01'*), from which the soil recovers fully and immediately upon removal of load, and a permanent deformation (*11'*). The elastic portion is due partly to elastic strains in the particles (see Fig. 326) and partly to elastic deformations within the soil structure. The permanent portion results from some of the contact pressures between the grains having exceeded the local yield stress of the material. Under steady pressure, the soil mass undergoes a further compression but at a rate gradually decreasing with time (section *1 — 2*). This time-dependent deformation results from the tendency of individual particles to align themselves into a more orderly arrangement matching the applied load, namely one in which the energy of the system is a minimum. In saturated masses of soil, this phase of the deformation is accompanied by movement of the pore water (consolidation, see Ch. 6). At this stage, the soil mass behaves as a non-linearly viscous fluid.

Upon a sudden removal of the load, an instantaneous rebound occurs (section *2 — 2'*), which is approximately equal to the portion *1—1'* of the initial compression. This is followed by a time-dependent expansion resulting from a rearrangement of the particles to match the changed conditions.

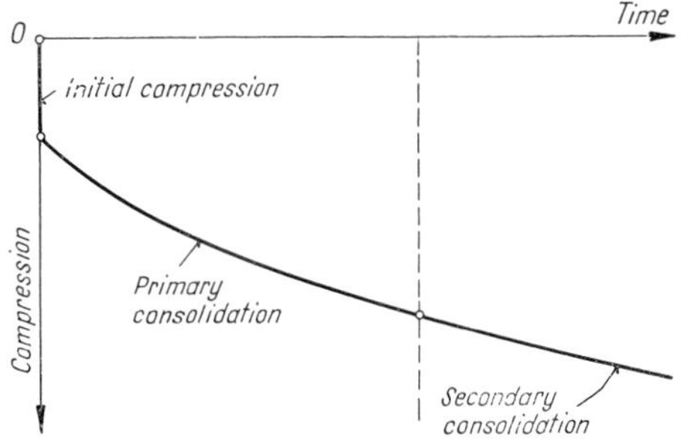

Fig. 337. Three sections in the time-deformation diagram of clays

Scott (1963) explains the above described process as follows. The initially applied steady effective pressure induces stresses in the grains themselves. As a consequence, a viscous flow begins to take place in the pore water and this, in turn, results in relative displacements between adjacent particles. Upon unloading, the viscous deformations that have taken place are incompatible with the tendency of the particles to restore their original position, and as a result, residual stresses remain in some of the grains. When the load is removed, the pore water and the particles themselves undergo viscous deformations in the reverse direction, and the residual stresses gradually cease. This relaxation process tends to restore completely the conditions that existed before the load application. However, a considerable portion of the deformations will not be regained, because of the plastic deformations having taken place at the contact points.

The time rate of these motions is conditioned by the soil type and the initial state of the soil. In coarse-grained soils, gravel and sand, they take place much more rapidly than in clays. Besides, still another difference may be observed. With clays, the final tangent to the curve *1—2* shown in Fig. 336 may be horizontal as well as inclined. Thus, the time-deformation diagrams of clays reflect, in general, the following three characteristic phases of deformation (*Fig. 337*):

— initial compression,
— primary consolidation due to the movement of the pore water and
— secondary consolidation.

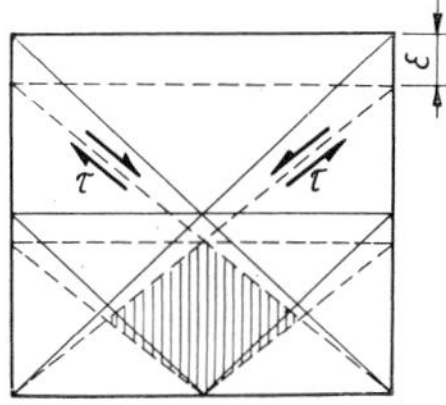

Fig. 338. Shear stresses in one-dimensional consolidation

For the explanation of the secondary consolidation, several concepts have been advanced (Gray, 1936; Buisman, 1941; Haefeli, 1953; Tan Tjong-Kie, 1957, 1967; Suklje 1957, 1967; Murayama and Shibata, 1967 et al.). Compression is always associated with shear strain, even in the case of one-dimensional consolidation, as is shown in *Fig. 338*, and it can be reasonably assumed that the resulting shearing stresses initiate a viscous flow and cause the clay particles to slip relative to one another. The adsorbed water, owing to its relatively higher viscosity, is squeezed out of the pores much more slowly than the "ordinary" pore water. The volumetric change due to the hydrostatic part of the applied state of stress may also be delayed by the viscous properties of the soil skeleton. It seems also possible that a load of long duration tends to weaken the bonds holding the particles together, and causes a slow continuous alteration of the soil structure.

The intensity of the secondary time effect depends on numerous factors, the most important of which are the viscous properties of the solid skeleton, the boundary conditions, the magnitude of loading and the temperature. If a mass of clay is laterally confined, the shear effect during compression is not so marked and the secondary time effect is also less important. On the other hand, under extremely heavy loads, creep phenomena and progressive failure may also come into play. Temperature affects the viscosity of the pore fluid and that of the clay matrix.

The secondary time effect is only seldom of practical importance, and in the majority of problems it can be disregarded. Exceptions in this respect are some organic soils in which slow chemical decomposition may contribute to the secondary compression (ISHII, SHINOHARA, TATEISHI and KURATA, 1957).

The secondary time effect can be studied also in the laboratory. In a semilogarithmic plot the experimental compression/time curve is usually of the shape shown in *Fig. 339*. According to the consolidation theory, at the final stages the curve should turn into a horizontal section. Instead, the experimental curve approaches an inclined asymptote. The compressions due to primary consolidation and to secondary time effect can be separated by the following fitting method devised by *Casagrande*.

The construction aims at the determination of the lines of the zero and 100% primary consolidation. The zero consolidation line can be obtained by considering that in the arithmetic plot the early portion of the primary consolidation curve fits a parabola.

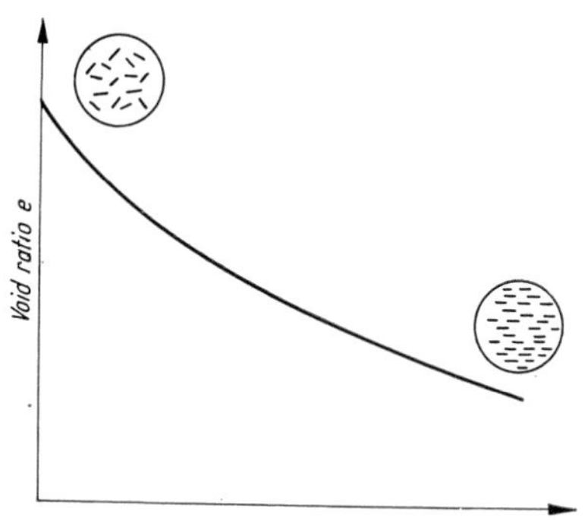

Fig. 340. Change in the degree of orientation during compression test

We choose a point such as A with a time t_1 and another with a time $t_1/4$ and mark off the difference in ordinates between the two points, a, above the consolidation curve at $t_1/4$. If we repeat this procedure with several points such as B, etc. on the early part of the curve, the newly plotted points should be located along a straight line, which is then the zero consolidation line. The ordinate of 100% consolidation is obtained at the intersection of the tangent at the point of inflexion and the final straight section representing secondary time effect.

In the previous discussion (Ch. 6), the time-dependent process of consolidation was treated on the basis of the effective stress concept and of hydrodynamic stress phenomena. But another, rheological, approach is also possible, which permits the mathematical formulation of the compression/time diagram shown in Fig. 336 (KÉZDI, 1966).

In the simplest case when the soil is saturated with water, the alterations in soil structure under the action of an external load are limited to a mere rearrangement of particles during compression. As a consequence, a more oriented soil structure developes. But such a change in the degree of particle orientation does in general not take place instantaneously. It is a time-dependent process, since it requires a certain activation energy to make the particles change their mutual orientation and since the speed of motion is finite. The time-

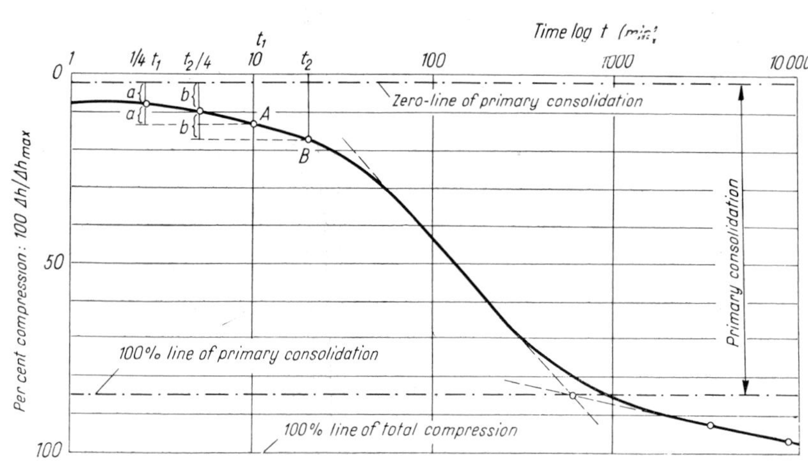

Fig. 339. The logarithm of time fitting method: determination of primary consolidation from laboratory test

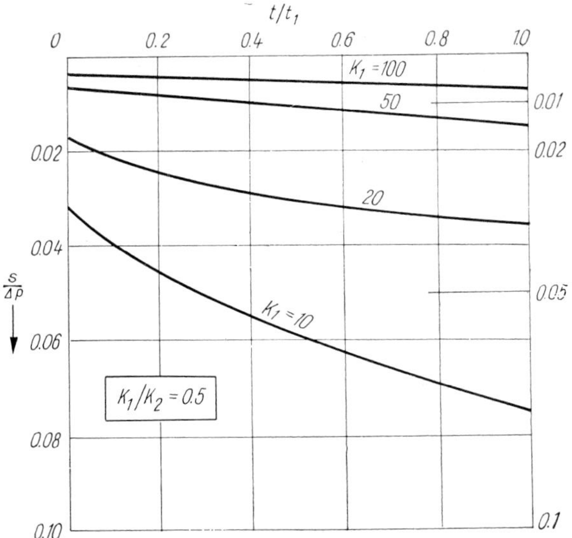

Fig. 341. Compression curve versus ratio K_1/K_2

dependency of compression is therefore primarily due to changes in particle orientation (*Fig. 340*). Analogies with the theory of fluids make it possible to derive, using certain assumptions concerning the degree of orientation, the equation of compression. According to this theory,

$$s = s_0 + s_1 = \Delta p\left[\frac{1}{K_1 + K_2} + \frac{1}{K_1\left(1 + \dfrac{K_1}{K_2}\right)}\right](1 - e^{-t/t_1}). \tag{233}$$

In the formula

s = total compression
s_0 = instantaneous compression
s_1 = time-dependent compression
Δp = pressure increment
t = time
K_1, K_2, t_1 = constants.

The compression/time curve calculated by Eq. (233) is in a very good agreement with the experimental curve shown in Fig. 337. Similar curves for constant ratios K_1/K_2 are shown in *Fig. 341.*

In theoretical rheology, new knowledge is often gained through the study of models devised to simulate the behaviour of the material under consideration. The solution of this, which is a purely mathematical task, will provide the desired relationship which described the behaviour of the material. Such a model was the container with springs and pistons used by *Terzaghi* to demonstrate the process of consolidation. More advanced models which attempt to give a more accurate description of the true behaviour of material are usually made up of three elements: an ideal *spring*, whose deformation is instantaneous and perfectly elastic in conformity with *Hooke*'s law; a *dashpot* with a piston moving in a Newtonian fluid of constant viscosity η so that a linear relationship exists between the stress and the rate of deformation; and finally a friction block or slide plate, which only begins to move when a certain stress (yield stress) is reached (or else experiences an instantaneous permanent deformation at that stress). The basic elements are illustrated by *Fig. 342.* By coupling them in series or in parallel, various composite models can be obtained. For instance, the series coupling of the spring and dashpot elements yields the *Maxwell* model, while the parallel coupling of the same two elements is known as the *Kelvin* model (*Fig. 343*). The elements signify the fundamental rheological state equations. Thus the spring represents a linear elastic behaviour. For this case, the relationship between the volumetric strain and the first stress invariant is given by the following expression:

$$e = \frac{1 - 2\mu}{E}(\sigma_1 + \sigma_2 + \sigma_3)\,. \tag{234}$$

The shear stress and the shear distortion are related by the equation

$$\gamma = \frac{2(1 + \mu)}{E}\,\tau\,. \tag{235}$$

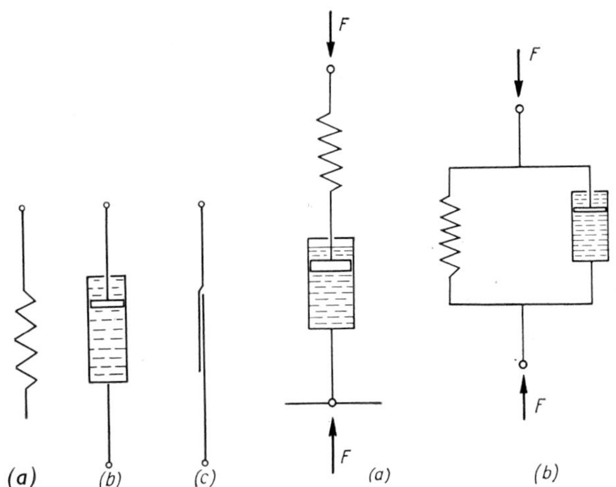

Fig. 342. Elements of rheological models:
a — spring; b — dashpot; c — friction plate

Fig. 343. *Maxwell* (a) and *Kelvin* (b) model

For a uniaxial state of stress,

$$\varepsilon_x = \frac{\sigma_x}{E}\,.$$

The basic reationship for the dashpot is

$$\frac{dv}{dz} = \frac{d\gamma}{dt} = \frac{\tau}{\eta} \tag{236}$$

γ = shear distortion
dv/dz = velocity gradient at right angles to the direction of flow
η = viscosity.

Examples of composite rheological models are shown in *Fig. 344*. Type *a* simulates the behaviour characterized by the deformation/time diagram in Fig. 337. Type *b* depicts the behaviour of the *Bingham* body. Here the force *P* and the displacement of the friction block represent the shear stress and the shear strain, respectively. As the pull *P* is increased, at first only the spring will stretch elastically until the force *P* transferred to the base of the block exceeds the frictional resistance acting there. As *P* is further increased, the frictional resistance on the base remains constant whereas the dashpot becomes active and exerts an increasing viscous resistance to motion.

A typical example of the application of rheological models to the study of soils is the composite model suggested by TAN TJONG-KIE (1957) to demonstrate primary consolidation and secondary time effect (*Fig. 345*). It is an advanced version of the *Terzaghi* consolidation model. A perforated piston P_1 is supported by a combination in parallel of a linear spring S_1 bearing firmly against the base of the vessel and an another spring S_2 coupled in series to a dashpot element which contains a highly viscous fluid. Immediately after the application of load, the piston P_2 does not move appreciably in the dashpot, and the predominant effect is, just as in the basic *Terzaghi* model, the primary consolidation due to expulsion of pore water. With increasing load the "effective" forces in the springs also increase and the dashpot is gradually brought into action. After a certain time, the compression of spring S_1 practically ceases, signifying the end of the hydrodynamic process. From that stage on, further motion is chiefly governed by the series coupling of springs S_2 and the dashpot. The process, which simulates secondary time effect, continues until the applied load is entirely carried by spring S_2. With the aid of this model, TAN TJONG-KIE derived theoretical relationships for the progress with time of the secondary compression and for the magnitude of total compression. He stated that for a given time factor, *T*, the neutral stress calculated on the basis of this model was always smaller than that obtained by the *Terzaghi* model. For comparison, *Fig. 346* gives theoretical curves obtained by the two methods for constant values of *T*.

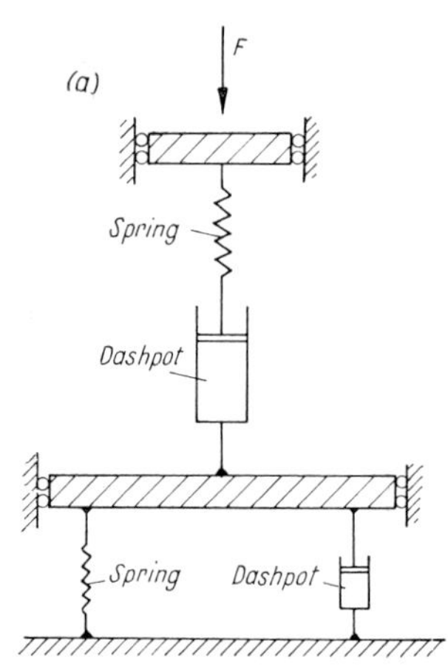

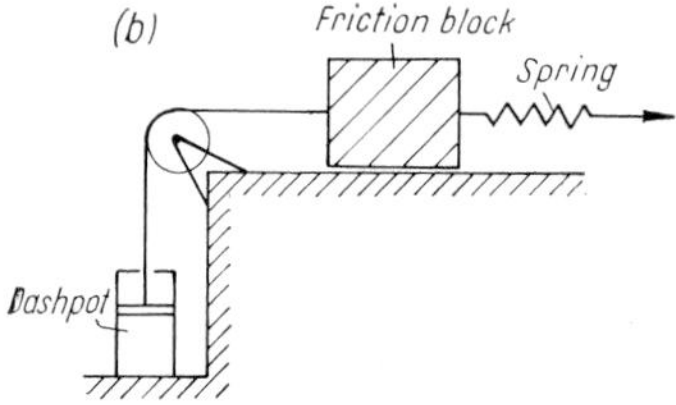

Fig. 344. Combined rheological models

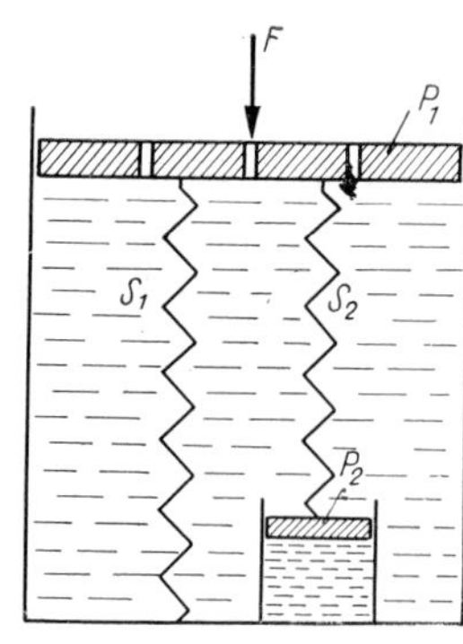

Fig. 345. Model proposed by *Tan-Tjong-Kie* to consider the secondary time effect

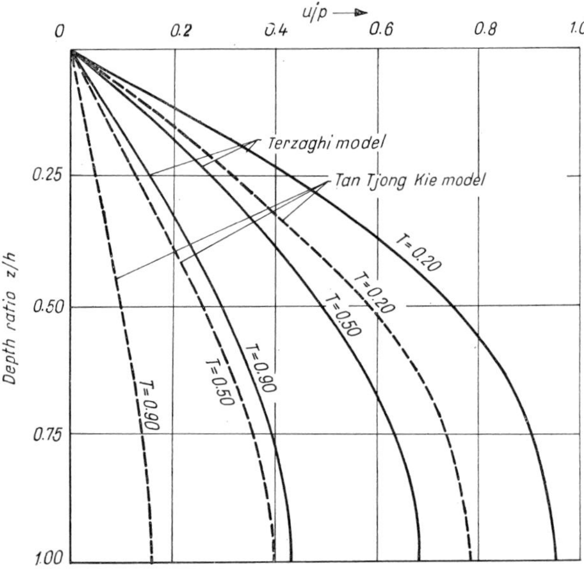

Fig. 346. Isochrones obtained by taking the slow deformation into consideration

8.4 Compression due to repeated loading

There are situations in which an engineering structure transfers repeated loads to the subsoil. This is the case with road pavements, railway ballast, machine foundations, crane runways, bridges, floor slab of lock chambers, etc. The settlements caused by repeated loads cannot be predicted by a simple settlement analysis based on the results of the ordinary confined compression test. Experience shows that, with the magnitude of cyclic load kept constant, an additional settlement results from each load application and that the increment per cycle decreases with increasing number of load repetitions. The compression of the soil depends on the magnitude of the repeated load, and to a great extent on the duration of loading, i.e. on the number per unit of time of the cycles of loading.

The response of soils to sudden loading was studied by A. Casagrande and W. L. Shannon (1948). From experiments they found that for clays the modulus of compressibility under a rapid transient load (0.02 s) was twice the value obtained from a static loading test of 10 minutes duration. For sands, the modulus of compressibility was 30% higher in the rapid loading test.

The author also used repeated loading tests with a rather low frequency but with a great number of repetitions (500 to 1000) to investigate the behaviour of granular soils under cyclic loading. Characteristic test data are shown in *Fig. 347*. In addition to an initial compressive load of $p_0 = 50$ kN/m² a pulsating pressure increment of $\Delta p = 200$ kN/m² was applied to the specimens. After a comparatively small number of cycles the pulsating portion of the deformation became practically elastic (the corresponding modulus, $M = \Delta p/\Delta\varepsilon$, was constant), however, the total permanent deformation continued to increase by a small amount with every reapplication of the load. The total strains plotted against the logarithm of the number of repetitions of loading gave approximately straight lines (Fig. 347*b*). The set of lines represent different initial void ratios. It can be seen that the value of the modulus M decreases considerably with increasing e_0. These experimental results suggest that the compression due to repeated loading should be governed by a logarithmic law. The linear plot in the semilogarithmic representation makes it also possible to extrapolate for a great number repetitions.

In cohesive soils, the effect of repeated loading is different, since here the role of the duration of loading becomes increasingly important depending on the consolidation characteristics of the soil. A detailed experimental investigation of this problem, however, still remains to be done.

8.5 Sudden compression (structural collapse)

If an unsaturated loaded mass of soil that has never been exposed to the effect of water is flooded, usually a sudden compression occurs. This kind of deformation is generally accompanied by a breakdown of the soil structure. This effect is particularly marked in *loess* and in certain granular, cohesionless soil deposits.

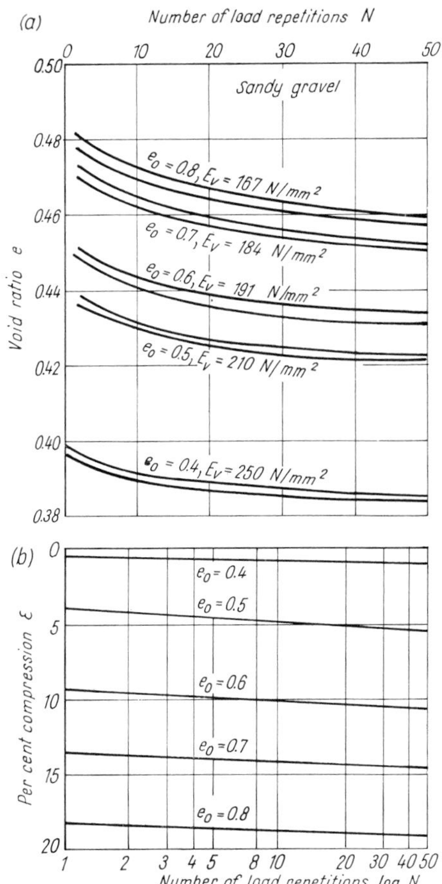

Fig. 347. Compression of sands of different initial density under repetitive loads

In a moist granular soil with a low degree of saturation the capillary water collects in the corners between the grains, forming contact moisture. The surface tension of the water tends to pull the grains together. This effect may prevent the grains from settling into a dense packing. In this manner, a very loose but temporarily stable structure is formed with large hollow spaces in it which are disproportionate to the size of the grains. If such a soil is immersed, the menisci disappear and the apparent cohesion becomes equal to zero. The grains, now free to slide and roll over one another, tend to move under the action of the external load into denser and thus more stable positions. Submersion is therefore invariably followed by a rapid decrease in volume, which is known as *structural collapse*. The magnitude of the sudden compression depends on the grain size, on the uniformity of grading but above all on the initial density of the soil. Experiments by Jáky (1948) showed, e.g., that a column of gravel loaded by a surcharge of $p = 120$ kN/m² underwent the volume decrease shown in *Fig. 348*, as the water in it rose higher and higher. In the loose state, the reduction in volume was twice to three times as much as that in the dense state. *Jáky* also found

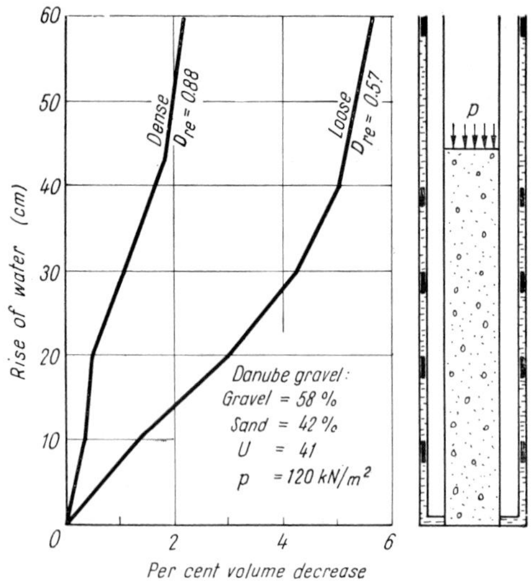

Fig. 348. Sudden compression of sandy gravel due to a rise of water level

that the relationships between per cent structural collapse and relative density for various soils were as shown in *Fig. 349*. He also pointed out that a repeated submersion caused only one tenth to one hundredth of the volume decrease that had occurred during the first submersion, and after 2 to 3 cycles no further collapse was observed. According to the investigations by RÉTHÁTI (1957), a structural collapse can result not only from a rise of the ground water but also from a capillary wetting of the soil.

The above described phenomenon can be very detrimental to fills made of sand. If the sand is placed and compacted in a relatively dry state, the first flooding subsequent to completion of the dam may result in excessive subsidences due to structural collapse to an extent seriously endangering stability.

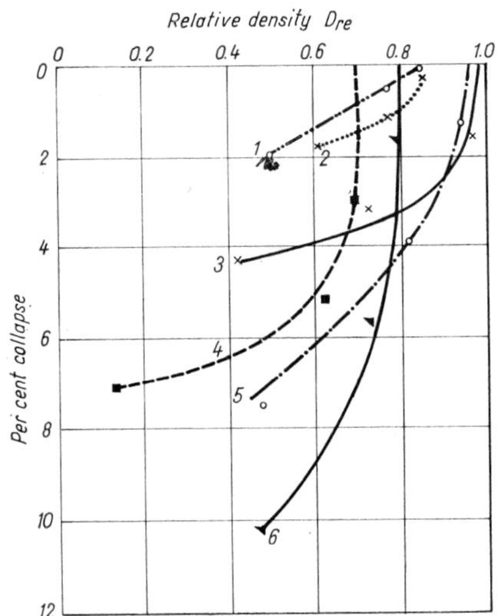

Fig. 349. Sudden compression of granular material as a function of relative density:

1 — fine sand; *2* — coarse sand; *3* — pea gravel; *4, 5* — Danube gravel; *6* — slag

A soil typically susceptible to structural collapse is loess. (ABELYEV, 1951; DENISOV, 1946; *US Bureau of Reclamation*, 1949; SCHEIDIG, 1934.) As discussed before, loess is an aeolian deposit and normally occurs in a partially saturated state. During deposition, the grains adhere to one another, and the fresh bonds gradually increase in strength with time because of the cementing effect of thin films formed between the grains. It is due partly to this cementing effect, partly to the apparent cohesion that the loess develops a peculiar structure with large "macropores" between the grains which are visible even to the unaided eye. Certain chemical substances always present in the floated fine dust and chemical processes occurring at the place of sedimentation also have their share in the formation of bonds between the grains. When the loess is loaded, hair cracks develop in the cementing films and the remaining cohesion is largely due to meniscus effect. Upon submersion, a sudden structural collapse occurs, just as in sands except for the fact that the decrease in volume in loess is much larger than in sands owing to the presence of macropores. In construction works in areas covered with loess the potential danger of structural collapse must be reckoned with and an extensive submersion of the loess must be prevented at all costs.

The cementing film is predominantly calcareous, and it was found (DOMJÁN, 1952) that the higher the lime content of the loess, the larger the amount of structural collapse. Some research workers suggest that clay minerals often present only in traces may also contribute to the formation of interparticle bonds in the loess.

The sudden volume decrease due to the collapse of structure can easily be studied in the laboratory by a special oedometer, devised by VASILIEV (1949). The test is so arranged that the specimen contained in the annular mould can be saturated by gravity from both above and below. During the test, first the natural specimen is consolidated under loads increased in steps. A normal pressure/void ratio curve is obtained (*Fig. 350*). At some prefixed load, after the consolidation is complete, the specimen is wetted through the porous stones with enough water to become fully saturated. A sudden compression can be observed which appears as a vertical line in the pressure/void ratio diagram. During further loading, a more steeply sloping curve is obtained.

The tendency to structural collapse is a maximum when the actual moisture content of a macroporous soil is smaller than its maximum molecular water holding capacity. In soils with much higher water contents, a portion of the structural collapse has already occurred, and a submersion produces only a slight effect. In a macroporous soil with a high clay content, submergence produces two simultaneous but opposite effects; the sudden

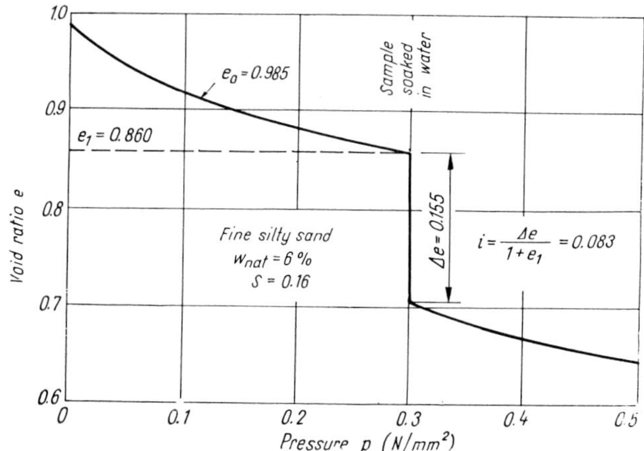

Fig. 350. Compression curve; structural collapse due to wetting

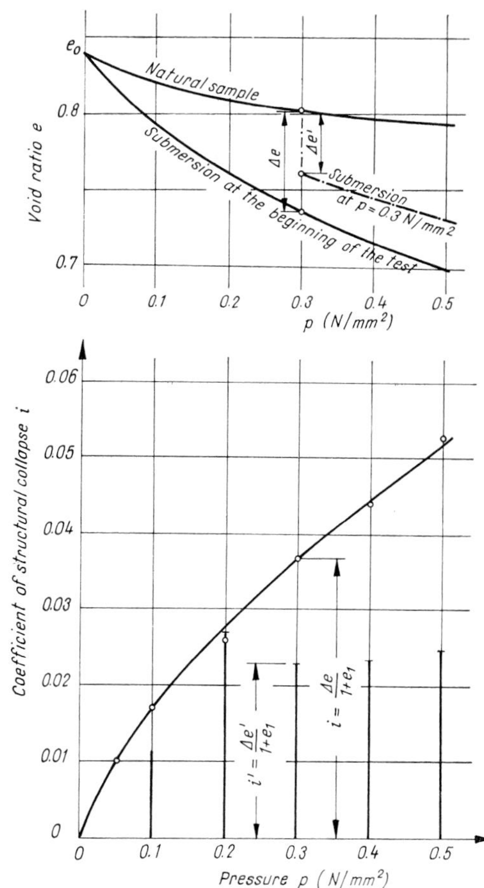

Fig. 351. Coefficient of structural collapse versus pressure

collapse of the structure due to the migration of the macropores and the swelling of the clay fraction. Whether the net volume change is compression or expansion depends on the balance of the two opposing actions.

For a quantitative evaluation of the collapsibility of the soil, ABELYEV (1948) introduced the terms of coefficient of structural collapse and accumulated collapse. The coefficient of structural collapse is defined by the expression:

$$i = \frac{\Delta e}{1 + e_1} \tag{237}$$

wherein

Δe = decrease in void ratio caused by submergence

e_1 = void ratio before submersion (*Fig. 351*).

The accumulated structural collapse, I, is used to characterize a complete soil profile. It is defined as

$$I^{(\mathrm{cm})} = \sum_{i=1}^{n} h_i i_i . \tag{238}$$

h_i is the thickness in cm of the ith homogeneous stratum within the macroporous zone in the profile and i_i is the corresponding coefficient of collapse.

A soil is judged dangerous in respect of structural collapse if $i > 0.02$. DENISOV (1946) used the ratio e_L/e to characterize the collapsibility of a soil. Here e_L is the void ratio corresponding to the liquid limit and e is the natural void ratio. If this ratio is greater than unity, the soil is described as "*collapsible*".

The collapse characteristics of a typical loess deposit flanking the right bank of the Danube in Hungary were investigated by KÉZDI (1954). It was found that maximum structural collapse could be obtained from compression tests on twin specimens, one being natural and the other submerged at the beginning of the test. The volume decrease due to collapse was found to increase with increasing pressure as shown in Fig. 351. The solid curve in the lower diagram shows the relationship between the pressure and the coefficient of structural collapse as determined from the difference in void ratio Δe_1 measured between the two compression curves. The single ordinates

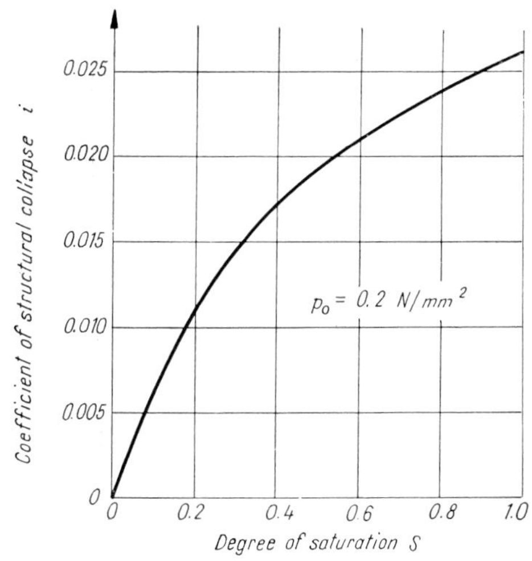

Fig. 352. Effect of quantity of water on the coefficient of collapse

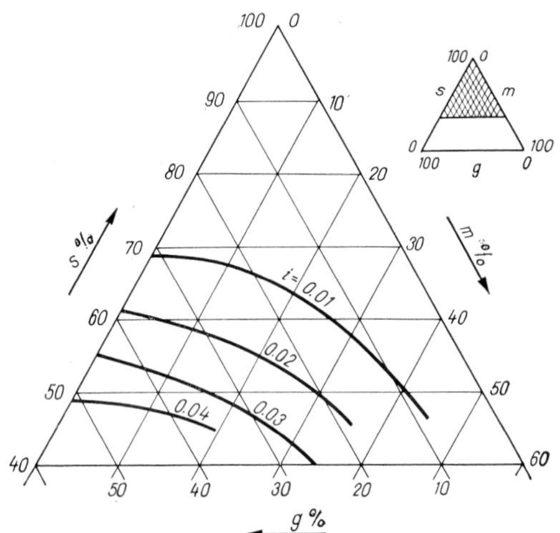

Fig. 353. Structural collapse of loess as a function of phase composition

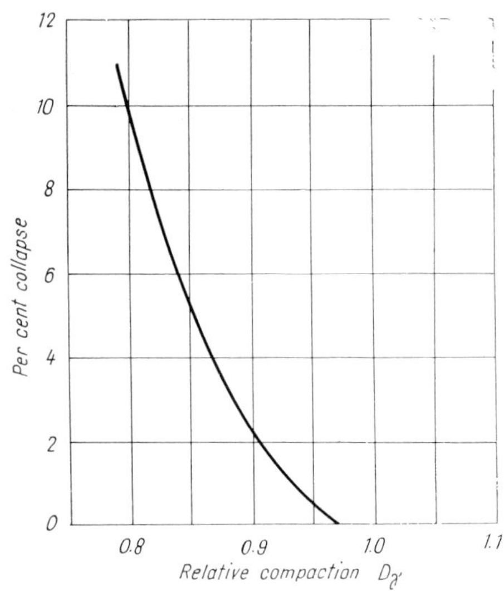

Fig. 355. Collapse in fills at different degrees of compaction

drawn as solid lines represent i values obtained when the sample was first tested in the natural state up to some intermediate load (300 kN/m^2 in the figure) and then submerged while the load was held constant. i values determined by the latter method showed an increase up to a pressure of about 300 kN/m^2, but thereafter remained practically constant. If the amount of water that has penetrated into the loess is insufficient to completely fill the voids, only a slight collapse results (*Fig. 352*). The bulk of the structural collapse usually occurs upon the first saturation, and a repeated submersion produces only a very small further decrease in volume.

The amount of structural collapse for a given mineral composition and grading depends chiefly on the phase composition of a soil. A statistical evaluation of test data available from the mentioned loess plateau by the Danube furnished the empirical relationship between i and the phase composition shown in a triangular plot in *Fig. 353*. For a given phase composition, the coefficient of collapse i varies with the pressure as shown in *Fig. 354*. The reduction in volume is a maximum at a pressure of 300 kN/m^2. In the standard procedure of the collapse test, the submersion of the specimen is therefore effected at $p = 300$ kN/m^2.

A similar phenomenon, subsidence due to collapse of the soil structure, can be observed in man-made fills when the material has been inadequately compacted. However, by thorough compaction this effect can be completely eliminated. This is evidenced by the diagram of *Fig. 355*, where per cent collapse in fills is plotted against relative compaction. For D_γ values higher than 0.8, the collapse is less than 0.02. Above $D_\gamma > 0.95$, practically no collapse is observed.

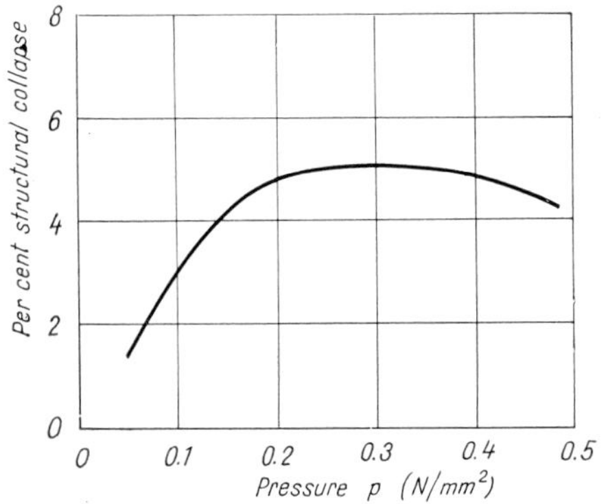

Fig. 354. Structural collapse versus pressure

Recommended reading

CAQUOT, A.–KÉRISEL, J. (1967): Grundlagen der Bodenmechanik. Übersetzt von G. Scheuch. Springer—Verlag, Berlin–Heidelberg–New York.

KRAVTCHENKO, J.–SIRIEYS, P. M. [editors] (1966): Rheology and Soil Mechanics. Springer Verlag, Berlin–Heidelberg–New York.

LAMBE, T. W.–WHITMAN, R. V. (1969): Soil Mechanics. J. Wiley & Sons, Inc., New York.

LEE, I. K. [editor] (1968): Soil Mechanics. Selected topics. American Elsevier Publ. Co. Inc., New York.

LEONARDS, G. A. (1962): Engineering properties of soils. (Ch. 2 in: Foundation Engineering, Leonards, G. A., [editor]) McGraw-Hill Book Inc. New York.

SCHULTZE, E.–MUHS, H. (1967): Bodenuntersuchungen für Ingenieurbauten. 2. Aufl. Springer-Verlag, Berlin–Heidelberg–New York.

SCOTT, R. F. (1963): Principles of Soil Mechanics. Addison-Wesley Publishing Co. Reading, Mass.

SOOS, P. (1967): Das Setzungs- und Verformungsverhalten von Böden. VDJ-Zeitschrift; 109, Nr. 8.

Stress-strain behaviour of soils (1972): Roscoe Memorial Symposium, Cambridge University. G. T. Foulis Co. Ltd., Henley-on-Thames, Oxfordshire, England.

SUKLJE, L. (1969): Rheological aspects of soil mechanics. Wiley-Interscience; London, New York.

Chapter 9

Earth pressure problems

9.1 Introduction

9.1.1 Definitions

One of the most important problems in applied soil mechanics is the determination of forces acting on engineering structures which are in connection or in direct contact with earth masses. Knowledge of these forces is essential for the design of such structures as retaining walls, footings, tunnels etc. In this chapter we shall discuss a particular group of these problems; first we shall give some definitions and classifications and present some general laws and methods that enable us to deal with earth pressure problems.

Earth pressure, in the broadest sense of the word, denotes forces and stresses that occur either in the interior of an earth mass or on the contact surface of soil and structure. Its magnitude will be determined by the physical interaction between soil and structure, and the value and character of absolute and relative displacements and deformations.

Taking this general definition into consideration, one may state that the problem of earth pressure is one of the basic and essential topics of civil engineering. The different cases of earth pressure in connection with engineering structures are illustrated with some examples in *Fig. 356.* The significant difference between all these cases consists in the different absolute and relative movements of the soil and of the structure. Different movements give rise to different stress distributions. For example, a vertical retaining wall can rotate about its toe, it can be displaced horizontally or vertically. Earth pressure acting on flexible sheet piles is a function of the construction methods; under footings, earth pressure forces resist penetration. If we excavate a tunnel in sand, one part of the surrounding soil will be loosened and the supporting structure will suffer rock pressure. All these pressures and deformations will be caused mainly by the weight of the soil itself.

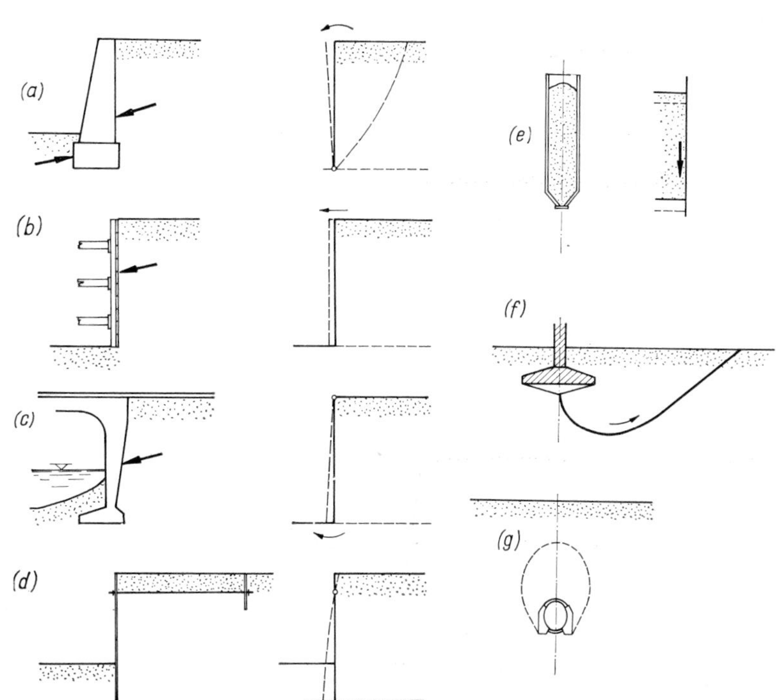

Fig. 356. Occurrence of earth pressure in connection with engineering structures:
a — retaining wall, tilting about the toe; *b* — bracing of cuts; horizontal displacement; *c* — abutment of a frame bridge, tilting about the top; *d* — anchored bulkhead, bending; *e* — silo, vertical movement; *f* — strip footing; ground failure; *g* — tunnel, loosening of the surrounding earth mass

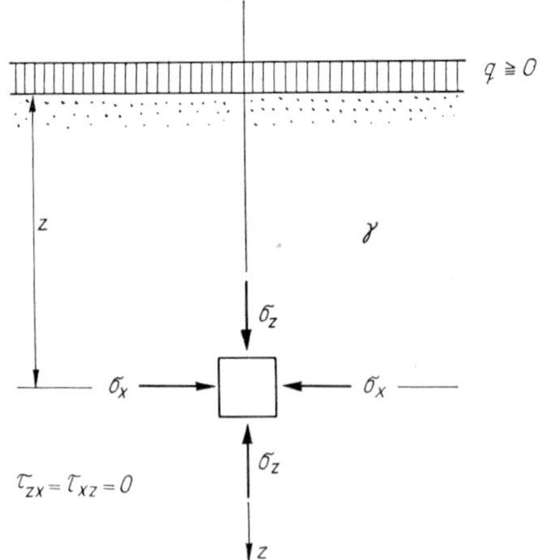

Fig. 357. State of stress in the semi-infinite half space; at rest condition

The fact that the stress field will be determined by the type and measure of the displacements and deformations, makes the problem indeterminate. This indeterminateness prevails also in another meaning. The pressures will be exerted by an assembly of individual particles, but the forces between these cannot be determined (Section 5.1). Soils are dispersed systems, there is never a continuous stress distribution.

Earth pressure problems can be classified into three main groups.

The first group includes cases, where the earth mass is in the state of rest, there are no displacements and deformations. This condition is strictly fulfilled in the half-space at rest; the basic problem in this group consists in the investigation of this state of stress (*Fig. 357*).

In the second group, the horizontal stresses in the earth masses are of primary interest. These are the retaining wall problems, sheet piling, timbering of cuts, silos etc. The retaining structure always suffers displacements and there will be a volume change. Depending on whether this change is a compression or an expansion, we speak of passive and active earth pressure, respectively. A well-known example of this case is the rigid retaining wall, tilting about its toe. If the wall is in the state of rest, we have a problem of the first kind, "*earth pressure at rest*". If the wall is rotated about its toe, a positive rotation (*Fig. 358*) causes expansion in the retained earth mass. In this case, the force acting on the wall will be called the *active earth pressure.* If a force pushes the wall against the retained earth mass, the soil will be compressed and a *passive earth pressure* arises.

There are cases where one part of the earth mass expands and the other one will be compressed. Then one part of the wall surface gets active, another one passive pressure.

The value of the active or passive earth pressure is a function of the displacement. The relationship between the displacement and the acting force, for the examples given in Fig. 358, can be seen in *Fig. 359*. At zero displacement earth pressure at rest will be acting; a positive rotation causes a decrease of the force, because the shearing resistances will be mobilized in the earth mass and they act against the movement. This movement is directed downwards. After a certain amount of displacement, a failure will occur, and a slip surface will be developed, along which the total shearing strength has been developed.

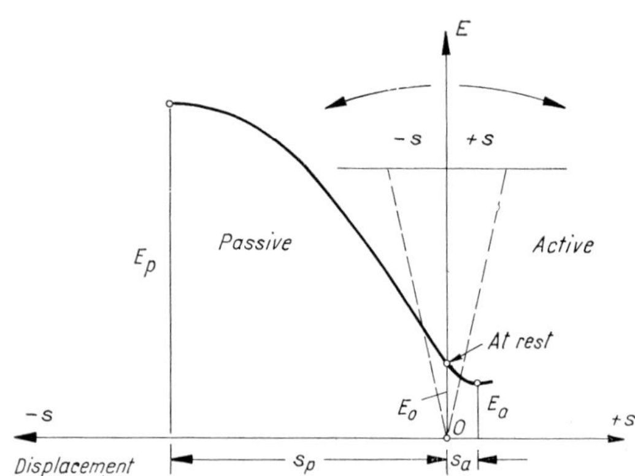

Fig. 359. Relation between wall movement and earth pressure; active and passive case

On the other hand, if we want to push the wall against the earth, we have to mobilize the shearing strength. A continuous negative rotation can be maintained only if we increase the external force. If this force reaches a value which the earth mass cannot withstand, then a failure occurs; again, slip surfaces develop and the shear strength will be mobilized. The active earth pressure and the passive earth pressure both have a limiting value; they belong to the limiting state of equilibrium.

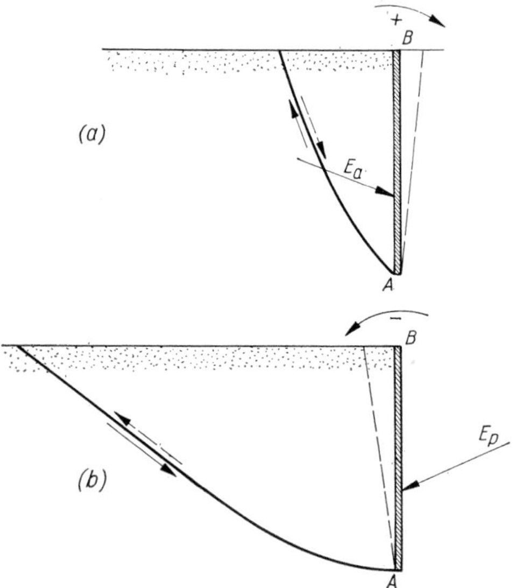

Fig. 358. Active and passive earth pressure

Cases where vertical forces are preponderant belong into the third group: these are the problems of foundation, the investigation of stresses and deformations, and equilibrium and failure state of the soil which is acted upon by vertical forces. Further problems in this group are the vertical stresses in silos, forces acting on a trap door, on tunnels and other embedded structures. There are active and passive cases also in this group, depending on whether an expansion or compression occurs.

In each group we may distinguish between one-, two- or three-dimensional earth pressure problems, with the particular case of axially symmetrical problems in the last group.

9.1.2 Methods to determine earth pressure

Methods which are in more or less general use today to determine earth pressure can be classified into four groups.

1. If we assume or establish a stress-deformation characteristic of general validity for the soil mass, then a method can be developed that furnishes stresses and displacements in every point of the earth mass. For earth pressure problems, only the simplest law of this has been used: that which forms the basis of the Theory of Elasticity. In a two-dimensional case, the two equations of equilibrium and the compatibility equation—assuming the validity of *Hooke's* law—are sufficient to determine the three unknown stresses and also the displacements. However, the soil does not follow *Hooke's* law—the differences are extremely great in the vicinity of failure—, therefore, the use of this method to determine earth pressure is, in general, not justified.

2. Theories of Plasticity determine the stresses in earth masses by making the assumption that the condition of plastic failure is fulfilled either at every point of the mass or along specified surfaces. We have, in the two-dimensional problem, again three equations to determine the three unknown quantities: the two equations of equilibrium and the failure condition. Many important theories belong in this group furnishing solutions to many practical problems.

3. The third method does not analyse stresses acting on volume elements but determines the unknown forces on limiting surfaces by investigating the equilibrium of finite earth masses. Slip surfaces will be assumed or developed in the earth mass; along these surfaces, at every point, the condition of failure is fulfilled. Earth masses bounded by the surfaces of the structure and by slip surfaces are investigated; the static equilibrium, the condition of failure together with the additional requirement that the solution must be kinematically admissible furnish the equations necessary to determine the unknown forces. The stress distribution will remain unknown in this case.

4. The fourth group contains the variation methods. One part of the earth mass behind the wall is considered and it is assumed that the shear strength of the soil is mobilized on the dividing surface, but the earth mass moves as a rigid body. The conditions of equilibrium and the failure conditions are, in this case, not sufficient to determine the earth pressure; the missing equations will be furnished by an extremum condition: the value of the earth pressure has to be a minimum or a maximum. It is essential that the stresses acting on the sliding surface do not enter into the calculations. This group contains e.g. the widely used earth pressure theory of *Coulomb*, assuming a plane surface of sliding, and also the methods of *Rendulić*, *Fellenius* and others.

In this book, two methods of the second group, and a few in the fourth group will be presented and discussed. We have to start, however, with the investigation of the infinite half-space, in the "at-rest" as well as in the plastic conditions. A chapter on the basic relationship between wall movement and earth pressure, and on the general equations will be added to the mainly practical treatment.

9.1.3 Classification of granular media

We have seen in the grouping of earth pressure theories that with one exception, they all use the failure condition, and they all assume sliding surfaces which means that only limiting—failure—cases can be investigated. This is due to the fact that a realistic relationship of stresses and deformations in soil has not been established yet, at least not in a form that could be used to solve practical problems. This fact emphasizes the importance of failure surfaces, and therefore we intend to have a closer look at slip surfaces in this chapter.

We have to distinguish between cases where only one surface of sliding occurs and cases, where whole masses of earth get into the plastic state and the failure condition is fulfilled at every point and not only along a single line. (*Fig. 360;* cf. Section 7.1, Fig. 235.)

In practical problems, the failure condition of Coulomb is used exclusively; it assumes a linear

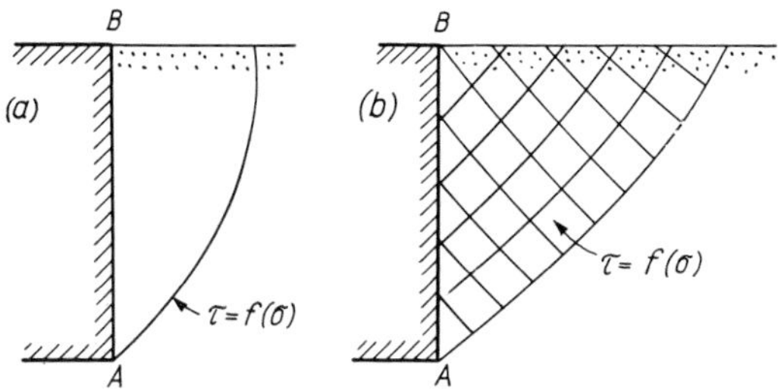

Fig. 360.
a — slip line, "line rupture"; *b* — slip field, "surface rupture"

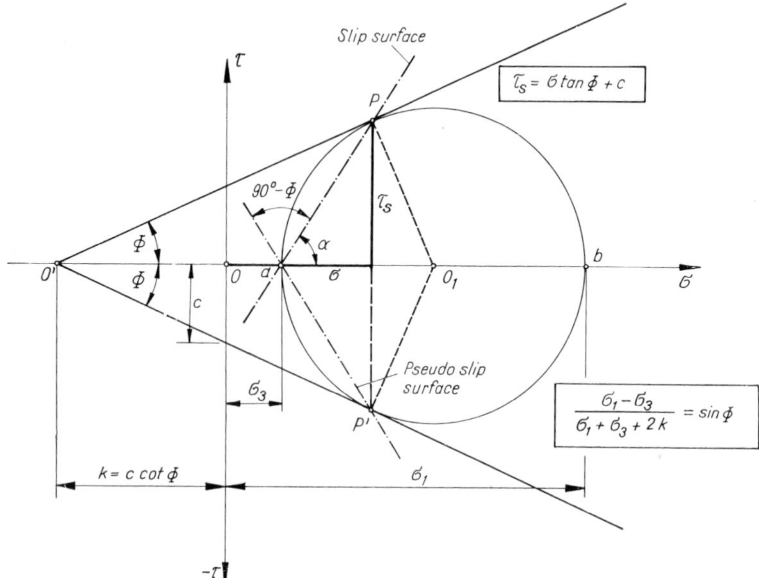

Fig. 361. Condition of failure; direction of slip lines

relationship between normal stress and shearing strength (Ch. 7). It also gives the direction of the slip surface: it makes an angle $\alpha = 45° + \varphi/2$ with the direction of the plane of the major principal stress (*Fig. 361*). It will be given by the line $\overline{aP}$. The line $\overline{aP'}$ gives the direction of the slip line that belongs to the other family; in case of a line rupture, this "pseudo-slip surface" is also present, however, its length is infinitely short.

If we select a wedge-shaped element, one side of which coincides with the sliding surface, and the other two with the principal directions, then we can derive three equations in the two-dimensional case expressing relations between the stresses (*Fig. 362*). We have (see Section 7.2):

$$\left.\begin{aligned} \sigma &= \frac{1}{2}(\sigma_1 + \sigma_3) + \frac{1}{2}(\sigma_1 - \sigma_3)\cos 2\alpha \\ \tau &= \frac{1}{2}(\sigma_1 - \sigma_3)\sin 2\alpha \end{aligned}\right\}$$

and, with $\alpha = 45° + \varphi/2$,

$$\sigma = \frac{\sigma_1 + \sigma_3}{2} + \frac{\sigma_1 - \sigma_3}{2}\sin\varphi ,$$

$$\tau = \frac{\sigma_1 - \sigma_3}{2}\cos\varphi .$$

Besides, the failure condition gives

$$\tau = \sigma \tan\varphi + c ,$$

and thus, Eq. (239) will be obtained:

$$\sigma_3 = \sigma_1 \tan^2\left(45° - \frac{\Phi}{2}\right) - 2c\tan\left(45° - \frac{\Phi}{2}\right); \quad (239)$$

or

$$\frac{\sigma_1 - \sigma_3}{\sigma_1 + \sigma_3 + 2k} = \sin\varphi ,$$

with $\quad k = c\cot\varphi .$

If we wish to consider the weight of the material itself — and it cannot be neglected in every case, because the earth pressure will be produced by heavy earth masses—then another physical characteristic enters into the calculations: the bulk density of the material. So we have three characteristics: bulk density, angle of internal friction, and cohesion. Earth pressure theories consider these as true constants; we shall see later how the values to be introduced into the formulae should be selected. If either none or one or two of them are equal to zero, we get general or special solutions of the system of differential equations. The number of possible variations is eight; these are given in *Table 31.*

Cases 1 to 3 are not earth pressure problems; the first two deal with materials without shearing resistance (ideal gases and liquids), the third with a weightless assembly of individual grains where no normal stresses can act and therefore no shearing resistance can be mobilized. Case 4 is the special plastic body of *St-Venant* and *Tresca;* case 5 is the plastic state analyzed by *Prandtl.* Case 6 represents the granular masses with weight.

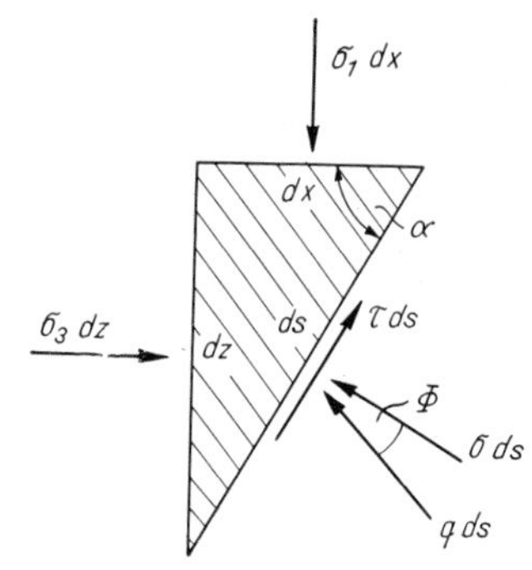

Fig. 362. Equilibrium and failure condition

Table 31. Classification of idealized materials

No.	γ	Φ	c	Remark	Exact solution
1	0	0	0	ideal liquid, processes not involving gravity	+
2		0	0	ideal liquid, processes involving gravity	+
3	0		0	weightless grain assemblies	—
4	0	0		ideal plastic body, without weight	+
5	0			Prandtl's medium	+
6			0	Grain assembly; no cohesion	—
7		0		ideal plastic body, with weight	+
8				general case	—

In the table we indicated the cases for which exact solutions are available. In cases 6 and 8 only the stressed state in the infinite half space can be solved (solution according to *Rankine*); in other cases, we have to use approximate solutions. Table 31 can be considered, from the point of view of failure, as a classification of granular materials which helps to group the respective theories.

9.2 Earth pressure at rest

Let us investigate the state of stress in the half space, which is bounded by a horizontal plane (*Fig. 363*). The half space is filled with homogeneous material; the bulk density is constant. The surface is unloaded, so we have a one-dimensional problem; besides, any vertical plane can be regarded as a symmetry plane. We take a vertical prism having a height z and a normal cross-section with area A. The half space is at rest; there are no displacements and deformations; therefore, the horizontal forces acting on the vertical sides of the prism are in equilibrium and there are no vertical stresses acting there. The weight of the prism is $W = Az\gamma$; therefore, the vertical stress acting at depth z, is equal to

$$\sigma_z = \frac{W}{A} = z\gamma\,. \tag{240}$$

The stress distribution can be given as a triangle. The state of stress in the presence of ground water was treated in Ch. 5. Now we wish to determine the horizontal stresses.

We consider a vertical plane in the half space. Based on the equations of equilibrium, the value of the horizontal stress will remain unknown. However, in this case, we may use the Theory of Elasticity since there are no displacements at all. In the plane state of deformations we have:

$$\left.\begin{aligned} \varepsilon_x &= \frac{1}{E}\left[\sigma_x - \mu(\sigma_y + \sigma_z)\right], \\ \varepsilon_y &= \frac{1}{E}\left[\sigma_y - \mu(\sigma_x + \sigma_z)\right] = 0, \\ \varepsilon_z &= \frac{1}{E}\left[\sigma_z - \mu(\sigma_x + \sigma_y)\right]. \end{aligned}\right\} \tag{241}$$

Then

$$\sigma_y = \mu(\sigma_x + \sigma_z)$$

and

$$\left.\begin{aligned} \varepsilon_x &= \frac{1+\mu}{E}\left[(1-\mu)\,\sigma_x - \mu\sigma_z\right], \\ \varepsilon_z &= \frac{1+\mu}{E}\left[(1-\mu)\,\sigma_z - \mu\sigma_x\right]. \end{aligned}\right\} \tag{242}$$

In the half space $\varepsilon_x = 0$, i.e.

$$\sigma_x = \frac{\mu}{1-\mu}\,\sigma_z = \frac{1}{m-1}\,\sigma_z = K_0\,z\gamma\,. \tag{243}$$

The coefficient of earth pressure at rest, K_0, is therefore a function of the elastic constant m, and a constant itself. Although the relation between stress and deformation is not linear at all, the value K_0 is really a constant for the first loading. This was proved by many tests, among others by TERZAGHI's model tests (1934) with great dimensions. The horizontal stress in the half space is directly proportional to the vertical one, and its distribution is linear with depth; the diagram is a triangle.

Figure 364 represents the stressed state. Diagram *a* shows the linear distribution of σ_z and σ_x, *b* gives the ellipse of stresses and *c* the relationship between the principal stresses. Based on the representation of stresses with *Mohr*'s circles it is

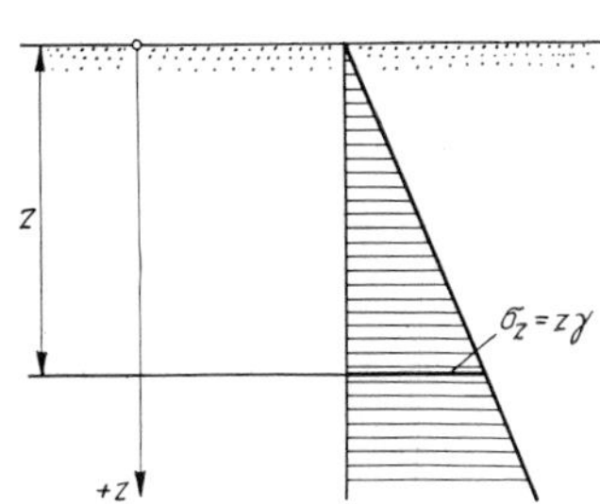

Fig. 363. Semi-infinite half space with horizontal surface

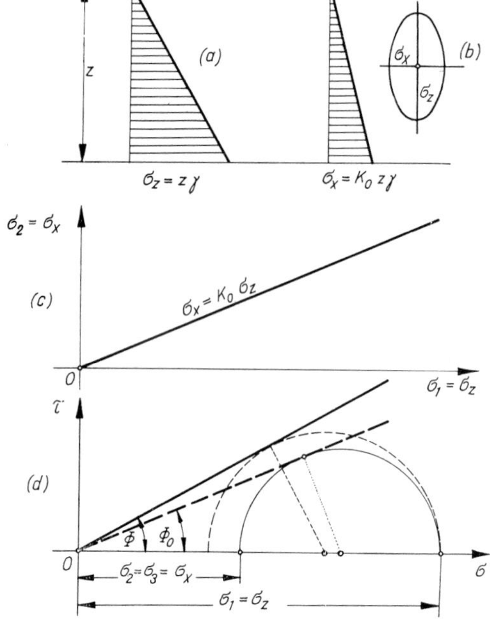

Fig. 364. State of stress in the "at-rest" condition:
a — vertical and horizontal stresses; *b* — stress-ellipse; *c* — relation between the principal stresses; *d* — Mohr's representation of the stressed state

possible to determine the normal and shear stresses for any element. The equations are:

$$\left.\begin{aligned}\tau &= \frac{1}{2}(\sigma_z - \sigma_x)\sin 2\alpha = \frac{1}{2}z\gamma(1-K_0)\sin 2\alpha,\\ \sigma &= \frac{1}{2}(\sigma_z+\sigma_x) + \frac{1}{2}(\sigma_z-\sigma_x)\cos 2\alpha = \\ &= \frac{1}{2}z\gamma[1+K_0+(1-K_0)\cos 2\alpha].\end{aligned}\right\} \quad (244)$$

The resulting earth pressure at rest acting on an oblique plane is given by (*Fig. 365*):

$$E_0 = \frac{h^2\gamma}{2}\sqrt{\cot^2 \alpha + K_0^2}. \quad (245)$$

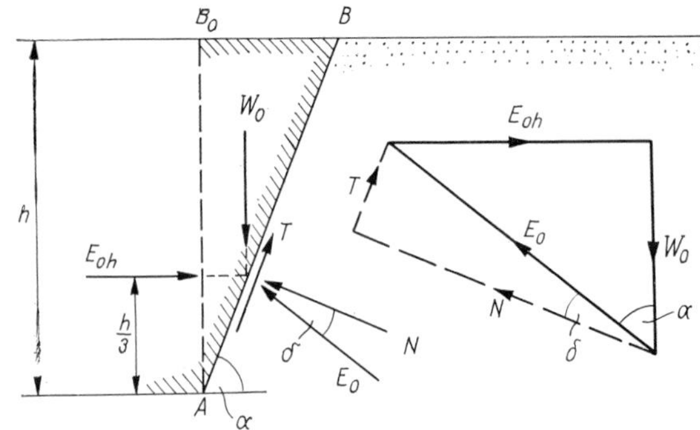

Fig. 365. Determination of the earth-pressure at rest acting on an oblique face

This result can be obtained by simple addition of the vectors representing the earth pressure on $\overline{AB_0}$ and the weight of the earth wedge AB_0B. The angle of direction is given by

$$\tan \delta = \frac{T}{N} = \frac{(1-K_0)\tan\alpha}{1+K_0\tan^2\alpha}. \quad (246)$$

This value has an upper limit, if

$$\tan\alpha = \tan\alpha_0 = \sqrt{\frac{1}{K_0}}. \quad (247)$$

We investigate now the case where there is ground water at a certain depth. The relationship

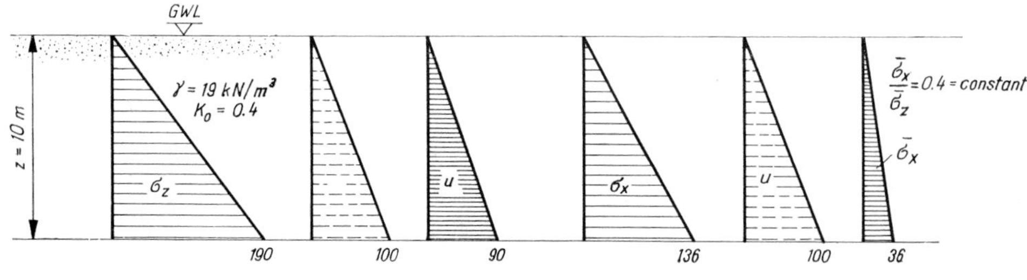

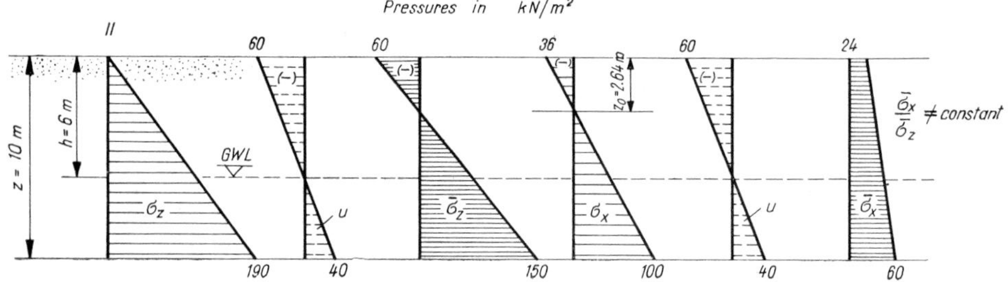

Fig. 366. Stress distribution in the at-rest condition in the presence of ground water

$\sigma_x = K_0 \bar{\sigma}_z$ is valid for effective stresses only, so we have to introduce the neutral stresses separately:

$$\sigma_z = \bar{\sigma}_z + u$$

and

$$\sigma_x = K_0\, \bar{\sigma}_z + u\,,$$

and, with the total stresses:

$$\sigma_x = K_0\, \sigma_z + (1 - K_0)\, u\,. \qquad (248)$$

The diagram of the total effective and neutral stresses is given in *Fig. 366*. Case I shows the stress distribution when the ground water is raised to the surface; in case II the ground water has been lowered to a depth h below the surface. If the soil remains saturated, the water is held by capillarity in the upper the layer with thickness h, the value of the total vertical stress does not change ($\Delta\sigma_z = 0$). The original value of the horizontal stress was

$$\sigma_x = K_0\, z\gamma + (1 - K_0)\, z\gamma_w$$

and, in case II

$$\Delta\sigma_x = (1 - K_0)\, \Delta u\,,$$

and the resulting total horizontal stress becomes:

$$\sigma_x + \Delta\sigma_x = K_0\, z\gamma + (1 - K_0)\, z\gamma_w - (1 - K_0)\, h\gamma_w\,.$$

Down to depth z_0 the total horizontal stress is negative. This depth (from Eq. $\sigma_x + \Delta\sigma_x = 0$) is equal to

$$z_0 = h\left[\frac{1}{1 + \dfrac{K_0}{1 - K_0}\dfrac{\gamma}{\gamma_w}}\right]. \qquad (249)$$

It can be seen from the stress distributions in Fig. 366 (Case II) that the ratio of the principal effective stresses is not a constant.

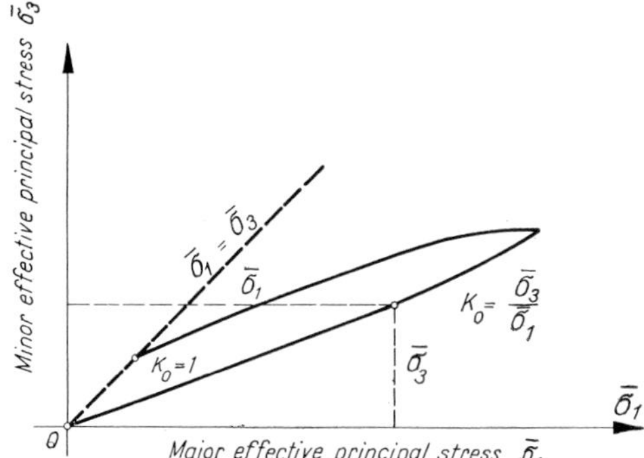

Fig. 367. Determination of the coefficient of the earth pressure at rest by triaxial compression test

In order to determine the earth pressure coefficient K_0, we use triaxial tests; the value of the horizontal stress will be adjusted in such a manner that the horizontal displacement be equal to zero. By plotting $\bar{\sigma}_3$ versus $\bar{\sigma}_1$, we obtain a straight line; the inclination furnishes the value K_0. A test result (after BISHOP and HENKEL, 1957) is given in *Fig. 367*. Removing the load we get another curve, which lies over this straight line: preloading greatly influences the coefficient of earth pressure at rest. At small stresses $K_0 = 1$, sometimes even greater. If we load the surface of a half space which is filled with sand, with a uniform load q, then we have vertical and horizontal stresses at a depth z which would correspond to the depth $z + \dfrac{q}{\gamma}$. After removing the load q, we have again the vertical stress $z\gamma$; however, the horizontal stress remains unchanged, if the limit value of elastic behaviour had been previously exceeded. If the soil is preloaded, it can happen that the horizontal stresses are greater than the vertical ones. The same effect occurs if we compact the soil by layers.

For the numerical value of the coefficient of earth pressure at rest, test results are available. According to TERZAGHI (1925), in loose sand $K_0 = 0.45 \div 0.50$; in dense sand $K_0 = 0.40 \div 0.45$. BERNATZIK (1947) gives, on the basis of very carefully conducted tests with sand cylinders, the values in *Table 32*. GERSEVANOV (1936) performed

Table 32. Coefficient of earth pressure at rest according to the test results of BERNATZIK (1947)

Soil state	Void ratio	K_0
Dense	0.60	0.49
Medium	0.70	0.52
Loose	0.88	0.64

tests with cohesive soils and suggested, for confined compression, on the basis of the equation $\partial\sigma_x/\partial z = K_0$, the following relationship

$$\sigma_x = K_0\, \sigma_z + K_1\,.$$

The constant depends on the initial stage of the compression; it is positive in a compacted or a preloaded sand, and negative in clays with negative pore pressures. A few test results are given in *Fig. 368*.

According to the tests of TSCHEBOTARIOFF (1948, 1951) and WELCH (1949) the value K_0, in a consolidated state of equilibrium (i.e. without pore water pressure) does not differ much from 0.5 either in sand or in clay.

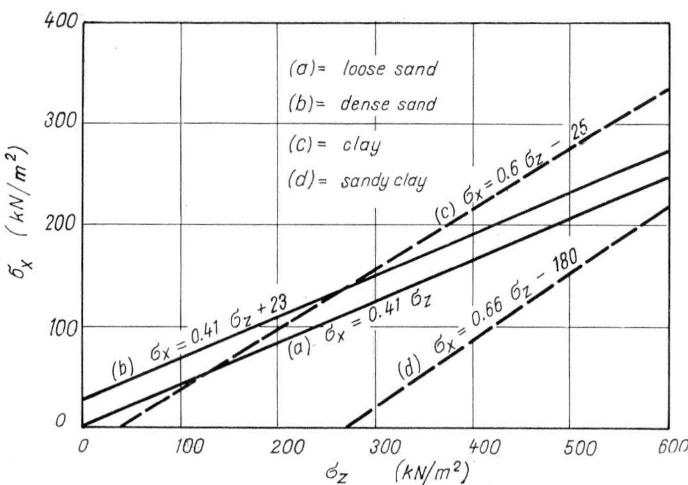

Fig. 368. Coefficient of the earth pressure at rest; tests made by GERSEVANOV

The test results obtained by BISHOP (1957, 1958) and SIMONS (1958) are given in *Table 33.*

Table 33. Coefficient of earth pressure at rest according to the test results of BISHOP (1957, 1958) and SIMONS (1958)

	w_L	w_p	I_p	Activity	K_0
Loose, saturated sand	—	—	—	—	0.46
Dense, saturated sand	—	—	—	—	0.36
Compacted residual clay	—	—	9.3	0.44	0.42
Compacted residual clay	—	—	31	1.55	0.66
Undisturbed, organic silty clay	74.0	28.6	45.4	1.2	0.57
Remoulded kaolin	61	38	23	0.32	0.66
Undisturbed marine clay, Oslo	37	21	16	0.21	0.48
Quick clay	34	24	10	0.18	0.52

According to HENKEL (1957) and BISHOP (1958) there is a definite relationship between the angle of internal friction (determined on the basis of effective stresses) and the coefficient of earth pressure at rest. This is explained by the fact that a certain part of the shearing strength will be mobilized if the soil is compressed in the vertical direction as was shown in connection with Fig. 338. The relationship is given in *Fig. 369;* the test results very well fit the theoretical formula derived by JÁKY (1944). According to this

$$K_0 = 1 - \sin \Phi . \qquad (250)$$

The derivation has been shown by KÉZDI (1962).

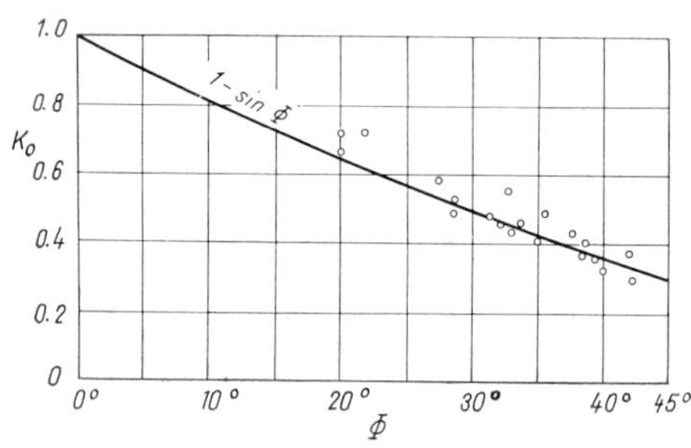

Fig. 369. Relation between the angle of internal friction and K_0 [test results and Eq. (250)]

9.3 Limit states of equilibrium in the half space

9.3.1 Horizontal surface; cohesionless material

The state of stress according to Fig. 364 will be maintained in the half space, as long as the earth remains in the "at rest" condition. We shall investigate the changes that occur in the half space upon a uniform expansion or compression. An expansion or compression will produce displacements, the shearing strength of the material will be mobilized until we reach the state of failure,

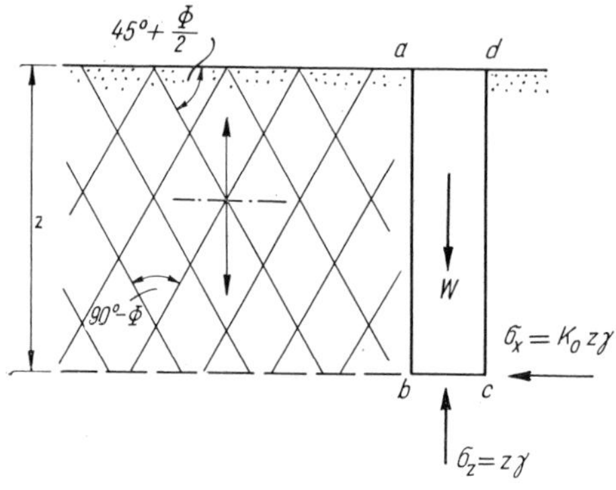

Fig. 370. Stresses in the semi-infinite half space

Fig. 371. Stresses in the at-rest and in the active state; Mohr's representation of stresses

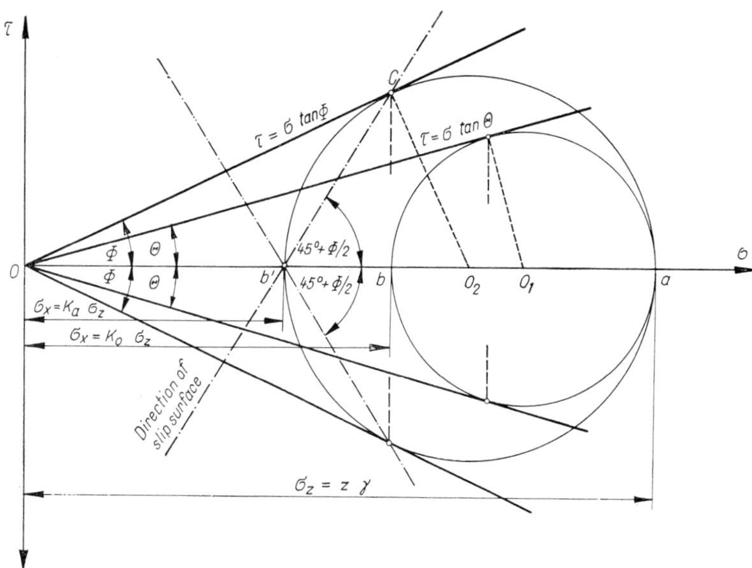

the limit equilibrium. From this point on, any differential increment in the deviator stress will produce a continuous deformation. The condition of failure is then fulfilled at every point of the half space.

This state of stress in the half space was given by RANKINE (1856). The half space is given in *Fig. 370*. We assume that the failure condition is

$$\tau = \sigma \tan \varphi ;$$

i.e. the medium is cohesionless. We have seen that the principal stresses in the "at rest" condition are:

$$\sigma_1 = \sigma_z = z\gamma ,$$

$$\sigma_2 = \sigma_3 = \sigma_x = \sigma_y = K_0\, z\gamma .$$

If we intend to change this state of stress, we have to allow an expansion or compression in the horizontal direction. Since the weight of the sand above a given horizontal plane will not change during these displacements, the vertical stress remains the same. However, the horizontal stress will be decreased if there is an expansion, and it will be increased if there is a compression. In the case of an expansion the lower limit of the horizontal stress will be given by the failure condition: when the *Mohr* circle of the stressed state will be tangent to the *Coulomb* line, failure occurs, and the described plastic state will be reached; the sand mass is in the *active Rankine state.*

The changes in the stressed state can be followed in *Fig. 371*, with the help of *Mohr*'s circle. In the at rest condition $\tau = \sigma \tan \Theta$; the failure condition is not yet fulfilled. When expansion starts, σ_x will decrease, σ_z remains the same, so the radius of the circle increases to the point where it will be tangent to the *Coulomb* line. In this state, the following relationship can be given:

$$\sin \varphi = \frac{\overline{CO_2}}{\overline{OO_2}} = \frac{\frac{1}{2}(\sigma_z - \sigma_x)}{\frac{1}{2}(\sigma_z + \sigma_x)} ,$$

hence

$$\frac{\sigma_x}{\sigma_z} = \frac{1 - \sin \varphi}{1 + \sin \varphi} = \tan^2 \left(45^\circ - \frac{\varphi}{2}\right)^* = K_a . \quad (251)$$

* The identity of the trigonometric expression can be proven from the figure. From triangles *adc* and *bcd* we get

$$\frac{\cos \Phi}{1 + \sin \Phi} = \tan \left(45^\circ - \frac{\Phi}{2}\right)$$

and

$$\frac{\cos \Phi}{1 - \sin \Phi} = \tan \left(45^\circ + \frac{\Phi}{2}\right) .$$

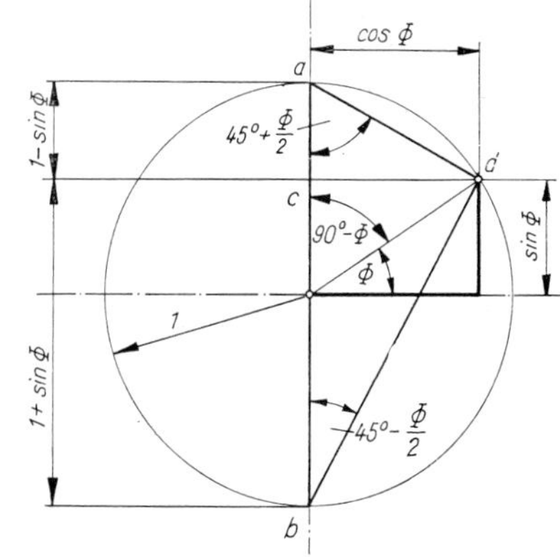

Since

$$\tan \left(45^\circ - \frac{\Phi}{2}\right) = \frac{1}{\tan \left(45^\circ + \frac{\Phi}{2}\right)}$$

dividing the equations:

$$\frac{1 - \sin \Phi}{1 + \sin \Phi} = \tan^2 \left(45^\circ - \frac{\Phi}{2}\right) .$$

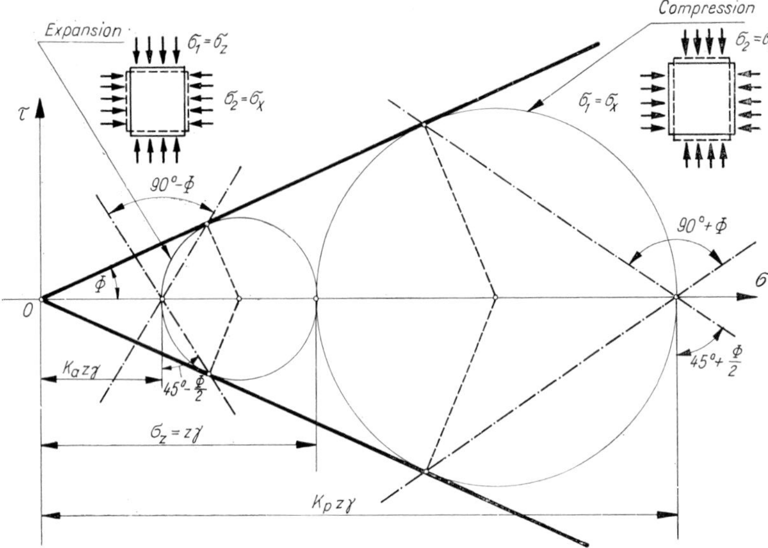

Fig. 372. Mohr's circles in active and passive Rankine states

The section $\overline{OA}$ represents the resulting stress acting on the slip surface; the direction of this plane is given by the line *b'C*. It makes an angle $45° + \frac{\varphi}{2}$ with the plane of the major principal stress, i.e. with the horizontal line (Fig. 370).

If there will be a uniform compression in the horizontal direction, the horizontal stress has to increase, and also the ratio $K = \sigma_x/\sigma_z$ will increase. The diameter of the stress circle becames smaller, then shrinks to zero (hydrostatic state of stress: $\sigma_1 = \sigma_2 = \sigma_3$); after that, the horizontal stress will be greater than the vertical one. The horizontal stress will increase until a failure state will be reached, i.e. the circle will be tangent to the *Coulomb* line. This is the passive *Rankine* state (*Fig. 372*). Now, the horizontal stress will be given by

$$\frac{\sigma_x}{\sigma_z} = \frac{1 + \sin \Phi}{1 - \sin \varphi} = \tan^2\left(45° + \frac{\varphi}{2}\right) = K_p. \quad (252)$$

The sliding surfaces make again an angle $45° + \varphi/2$ with the plane of the major principal stress, but now this is the vertical plane, so the angle with the horizontal plane will be equal to $45° - \varphi/2$.

The families of slip lines and the distributions of the horizontal stresses are given in *Fig. 373*.

Equation (251) gives the possibility to determine the earth pressure of dry sand, acting on a retaining wall, provided that the backfill has a horizontal surface and the back of the wall is vertical and completely smooth. If this wall tilts around its toe, there will be a uniform expansion in the backfill (*Fig. 374*). If the movement is sufficiently great to produce the plastic state in the backfill, sliding surfaces will develop and the stresses acting on the back of the wall can be calculated by Eq. (251). Due to the smoothness of the wall, only horizontal stresses will occur. The distribution of the stresses will be triangular (see Fig. 373); its area furnishes the earth pressure:

$$E_a^R = \int_0^h \sigma_x \, dz = K_a \frac{h^2 \gamma}{2}. \quad (253)$$

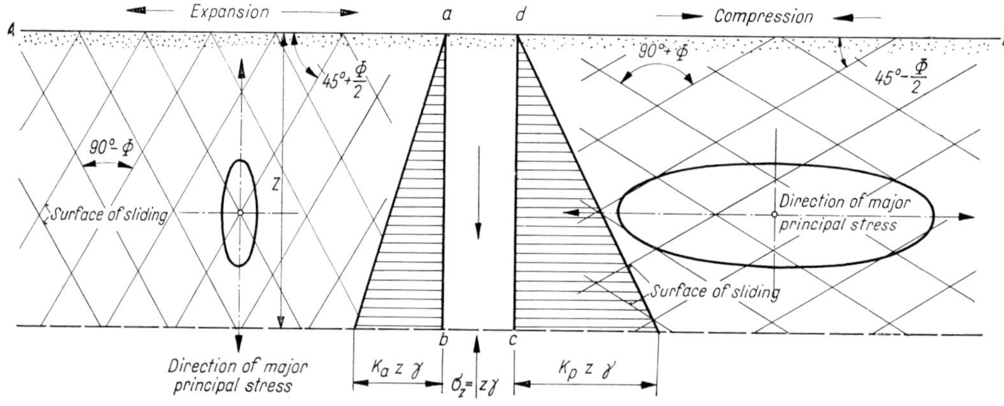

Fig. 373. The half-space in the Rankine states of stress

The point of application of the resulting force is in the lower third of the height h.

If the wall will be forced to tilt toward the backfill, a compression will be produced, and passive pressure develops. If the movement is sufficient to produce failure (see Fig. 359), we have the passive *Rankine* state in the backfill, and the resulting force (from the triangular stress distribution) will be given by

$$E_p^R = \int_0^h \sigma_x \, dz = K_p \frac{h^2 \gamma}{2} . \qquad (254)$$

Real retaining walls have rough backs so the above results do not strictly apply to them. Active earth pressure acting on rough surfaces will be smaller than the above value, while passive earth pressure will be greater, so this calculation gives values on the safe side. Besides, the calculation is very simple, therefore in practical work extensive use in made of the formulae (253) and (254). This approximation is justified for walls of minor importance.

If there is a uniform surcharge on the surface of the half space, then the vertical stresses acting on the lower $b - c$ surface of the *abcd* prism on Fig. 370 will increase:

$$\sigma_z = q + z\gamma = \gamma \left(\frac{q}{\gamma} + z \right) .$$

σ_z is a principal stress. The corresponding horizontal stress, in the limiting active state of stress is given by

$$\sigma_x = K_a \gamma \left(\frac{q}{\gamma} + z \right) , \qquad (255)$$

and in the passive case:

$$\sigma_x = K_p \gamma \left(\frac{q}{\gamma} + z \right) ; \qquad (256)$$

and the earth pressure:

$$E_a^R = \int_0^h \sigma_x \, dz = K_a \int_0^h (q + z\gamma) \, dz = K_a \left(qh + \frac{h^2 \gamma}{2} \right) =$$

$$= \frac{K_a h^2}{2} \left(\gamma + \frac{2q}{h} \right) = K_a \frac{h^2 \gamma'}{2} ,$$

where $\gamma' = \gamma + 2q/h$. The stress distribution is given in *Fig. 375a*; it is of trapezoidal shape and the point of application is located at the center of gravity, i.e.

$$m_0 = \frac{h}{3} \frac{h + 3\dfrac{q}{\gamma}}{h + 2\dfrac{q}{\gamma}} . \qquad (257)$$

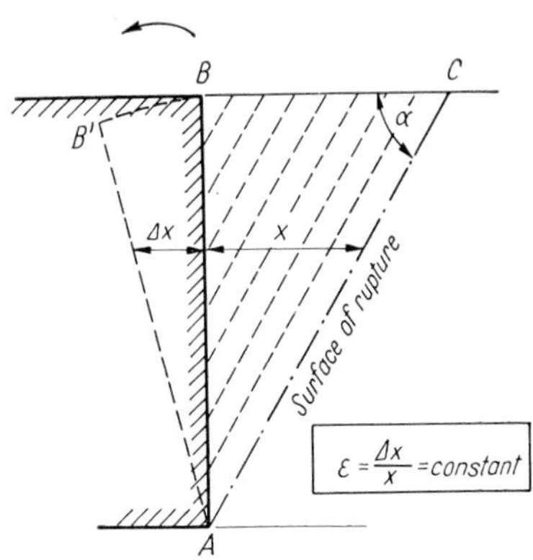

Fig. 374. Expansion in the backfill; tilting of the smooth retaining wall

In a layered soil, with different bulk densities and different angles of internal friction, an approximate stress distribution figure can be given as follows. In the uppermost layer Eq. (251) applies; for the second layer, we replace the thickness h_1 by an equivalent h_1', so that the vertical stress acting on bb_1 remains the same, but the bulk density of the equivalent layer is equal to γ_2. So, $h_1' = h_1 \gamma_1 / \gamma_2$ (see Fig. 375*b*), and the stress distribution in layer 2 can be drawn starting from point a_1. The vertical stress at depth $h_1 + h_2$ is equal to

$$\sigma_z = h_1 \gamma_1 + h_2 \gamma_2 = \gamma_2 \left(h_2 + h_1 \frac{\gamma_1}{\gamma_2} \right) = \gamma_2 (h_2 + h_1')$$

and the horizontal stress amounts to

$$\sigma_x = K_{a2} \gamma_2 (h_2 + h_1') .$$

The method can be generalized for several layers. The stress distribution figure is bounded by broken lines; the earth pressure is equal to the area of the same. The point of application is on the horizontal gravity line.

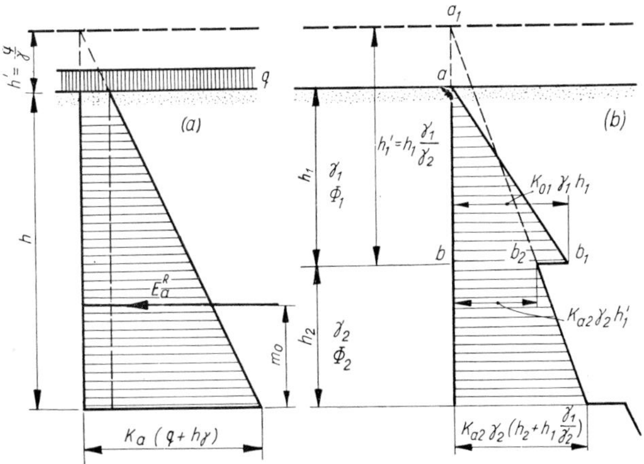

Fig. 375. Earth pressure in the Rankine state:
a — uniform surcharge on the backfill; *b* — layered backfill

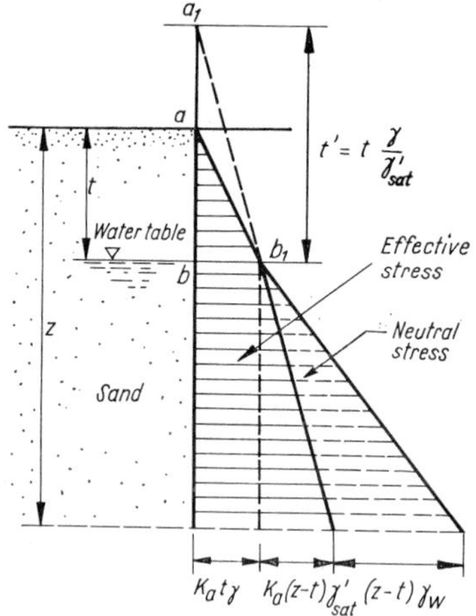

Fig. 376. Stresses in the Rankine state; presence of ground water

The same method can be used when there is ground water. The angle of internal friction of the sand will not be altered by the presence of water. Above the ground water level the stress distribution figure can be drawn; at a depth $z = t$, $\sigma_x = K_a t\gamma$. Below the water level the neutral stress amounts to

$$u = (z - t)\,\gamma_w$$

and the effective vertical stress is equal to

$$\bar{\sigma}_z = \gamma t + \gamma'(z - t)\,,$$

where γ' is the buoyant unit weight of the soil. The effective horizontal stress:

$$\bar{\sigma}_x = K_a[t\gamma + \gamma'(z - t)] = K_a\gamma'\left[z + \left(\frac{\gamma}{\gamma'} - 1\right)t\right]. \tag{258}$$

To construct the stress distribution figure, we use the same procedure as for layered soil; we calculate the equivalent thickness of the layer above the water level for buoyant bulk density; then we plot this from point b (*Fig. 376*). From point a_1, the line valid for the layer under water can be drawn.

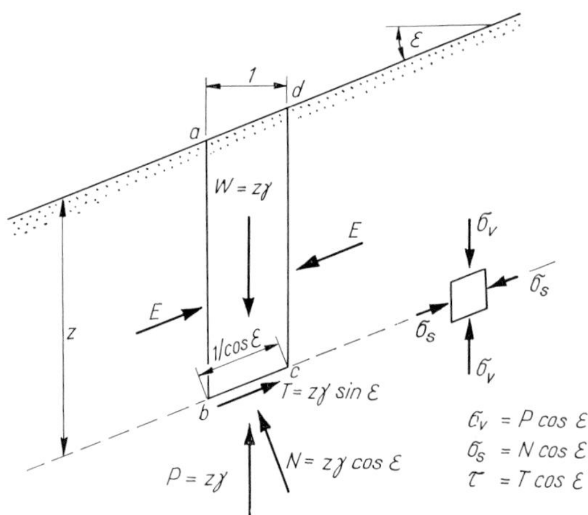

Fig. 377. Semi-infinite half space with oblique surface

9.3.2 Oblique surface; cohesionless material

We assume in the following that the surface of the half space makes an angle ε with the horizontal. The failure condition is fulfilled at every point.

This case is given in *Fig. 377*. At two points, b and c, at identical depth, there must be the same stressed state. Four forces act on the earth prism *abcd*, with a unit width: its own weight, the reaction on *bc* and the earth pressures on the sides. These forces are in equilibrium; therefore, the earth pressures are equal and they act along the same line. The lines of action of W and P are again identical. Stresses acting on an elementary quadrangle at depth z, cut out with four sides vertical and two sides parallel to the ground surface, are conjugate. This means that the resulting stress acting on the vertical side is parallel to the surface, and the resulting stress acting on the side parallel to the surface is vertical.

The vertical stress acting on *bc* can easily be given. The weight above $b - c$ is $1\,z\gamma$; the surface $bc = 1/\cos\varepsilon$, i.e.

$$\sigma_z = \frac{W}{A} = z\gamma\cos\varepsilon\,. \tag{259}$$

Resolving this stress into components which are normal and parallel to *bc*, we have

$$\left.\begin{aligned}\sigma &= \sigma_z\cos\varepsilon = z\gamma\cos^2\varepsilon\,,\\ \tau &= \sigma_z\sin\varepsilon = z\gamma\sin\varepsilon\cos\varepsilon\,.\end{aligned}\right\} \tag{260}$$

In *Fig. 378*, the failure condition is represented by lines through *0*, making an angle Φ with the σ axis. The state of stress on *bc*, at depth z can be given by a point Z, with the coordinates σ and τ according to Eq. (260). The resulting stress $\sigma_z = z\gamma\cos\varepsilon$ on *bc* makes an angle ε with the normal; the inclination of the line *0 2* is therefore also ε. ($\tau/\sigma = \tan\varepsilon$.) The Mohr circle of principal stresses has to go through point 2, and, since there is failure, it must be tangent to the Coulomb line. The center of the circle is located on the horizontal axis. In the active stressed state there is only one circle that fulfills these conditions; its center is at point C_1. The points where the circle cuts the horizontal axis give the values of the principal stresses σ_1 and σ_3.

In order to find the directions of the slip lines, we first determine the pole. Now, the σ and τ stresses pertaining to a given direction are known, so we have to make the construction shown in Section 7.2, in Fig. 243, in the opposite way. We draw a line through Z, parallel to the plane on

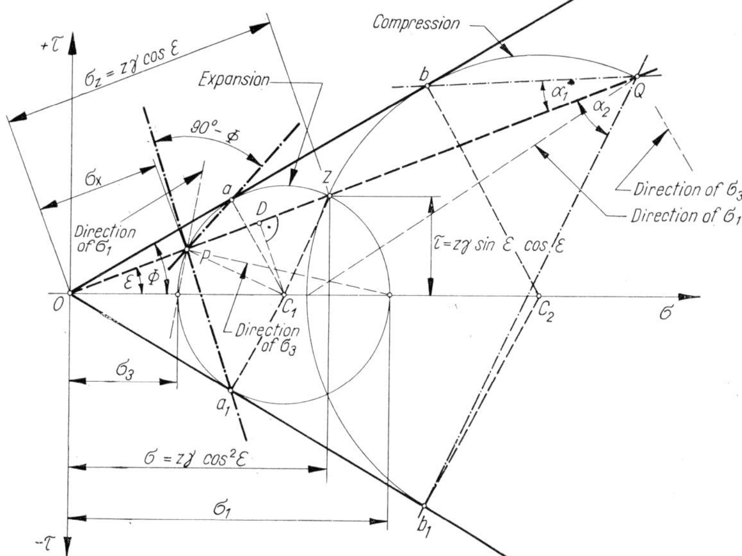

Fig. 378. Stresses in the half space bounded by sloping surface

which the stress is acting; this line makes an angle ε with the horizontal. The point where this line cuts the circle with the center C_1 is the pole P. One family of the sliding surfaces has the direction Pa, the other, Pa_1. The angle between them is $90° - \varphi$; they make an angle $45° - \varphi/2$ with the major principal direction.

The graphical procedure furnished the stressed state and also the slip line field. The same method can be applied to find the characteristics in the passive Rankine state. The corresponding stress circle runs again through point Z; the tangent points with the Coulomb's line are b and b_1. Based on the slip circle, we easily get the stresses and slip fields.

Figure 379 shows the slip line fields, the ellipses of the principal stresses and the directions of the trajectories, in the active as well as in the passive state.

If $\varepsilon = \Phi$, the first family of slip lines consists of vertical lines, the other family is parallel to the surface. The trajectories make an angle $45° + \Phi/2$ with the horizontal (*Fig. 380*).

The ratio of the conjugated stresses σ_v and σ_s (Fig. 377) can be determined also in the analytical way. From Fig. 378:

$$\frac{\sigma_s}{\sigma_v} = \frac{\overline{OP}}{\overline{OZ}} = \frac{\overline{OD} - \overline{DP}}{\overline{OD} + \overline{DZ}} \,.$$

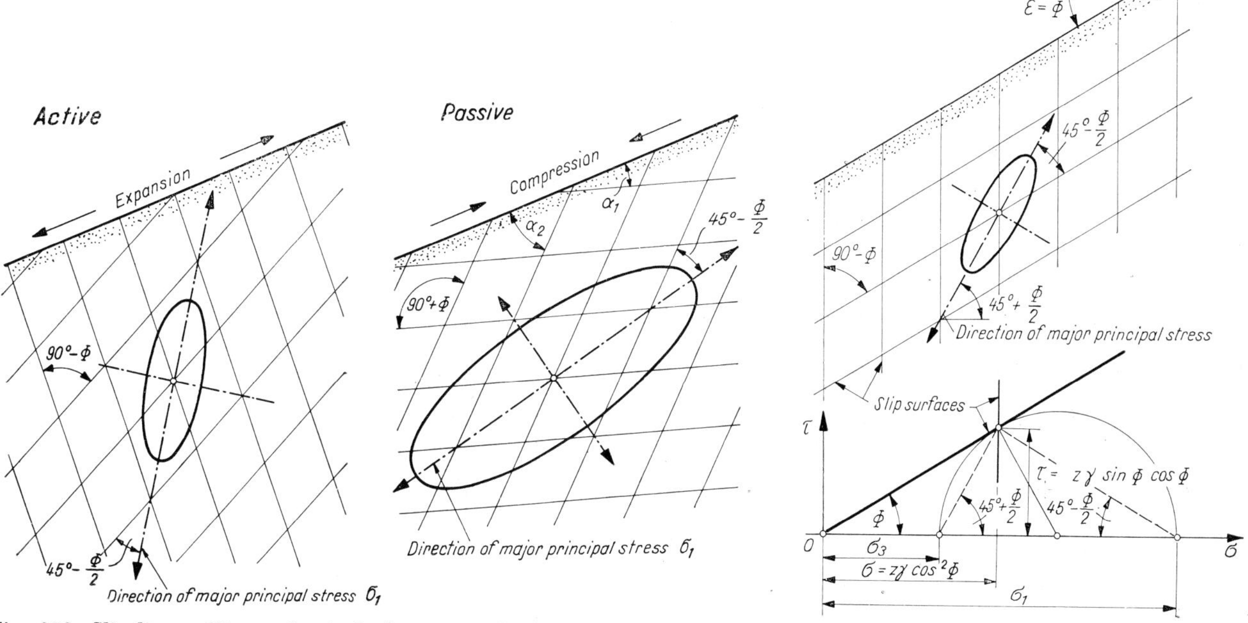

Fig. 379. Slip lines, ellipses of principal stresses, directions of trajectories in the half space bounded by sloping surface

Fig. 380. State of stress in the half-space bounded by a sloping surface, at $\varepsilon = \Phi$

It can be written:

$$\overline{OD} = \overline{OC}_1 \cos \varepsilon ,$$
$$\overline{DP} = \overline{DZ} = \sqrt{\overline{OC}^2 \sin^2 \Phi - \overline{DC}_1^2},$$
$$\overline{DC}_1 = \overline{OC} \sin \varepsilon .$$

Substituting into the above expression we get

$$\frac{\sigma_s}{\sigma_v} = \frac{\cos \varepsilon - \sqrt{\cos^2 \varepsilon - \cos^2 \Phi}}{\cos \varepsilon + \sqrt{\cos^2 \varepsilon - \cos^2 \Phi}} . \qquad (261)$$

JÁKY (1944) brought this formula into a much simpler form:

$$\frac{\cos \varepsilon - \sqrt{\cos^2 \varepsilon - \cos^2 \Phi}}{\cos \varepsilon + \sqrt{\cos^2 \varepsilon - \cos^2 \Phi}} = \frac{1 - \sqrt{1 - \cos^2 \nu}}{1 + \sqrt{1 - \cos^2 \nu}} =$$
$$= \tan^2 \left(45^\circ - \frac{\nu}{2}\right), \qquad (262)$$

where

$$\cos \nu = \frac{\cos \varphi}{\cos \varepsilon} .$$

The conjugated stresses are given by

$$\left.\begin{aligned} \sigma_v &= z\gamma \cos \varepsilon \\ \sigma_s &= z\gamma \tan^2 \left(45^\circ - \frac{\nu}{2}\right) \cos \varepsilon . \end{aligned}\right\} \qquad (263)$$

The stress distribution diagram is triangular, and the earth pressure acts in the lower third point, parallel to the surface. Its value is

$$E_a^R = \frac{1}{2} h^2 \gamma \tan^2 \left(45^\circ - \frac{\nu}{2}\right) \cos \varepsilon . \qquad (264)$$

In the passive *Rankine* state we get:

$$\frac{\sigma_s}{\sigma_v} = \frac{\overline{OQ}}{\overline{OZ}} = \frac{\overline{OZ}}{\overline{OP}} = \frac{1 + \sqrt{1 - \cos^2 \nu}}{1 - \sqrt{1 - \cos^2 \nu}} =$$
$$= \tan^2 \left(45^\circ + \frac{\nu}{2}\right) . \qquad (265)$$

The family of slip lines consists of straight lines; the direction can be determined as follows.

In a two-dimensional problem, if we know three coordinated stresses (σ_x, σ_z and τ_{xz}) in one point, the stresses acting on any plane making an angle α with the plane of σ_z, can also be calculated. Let us consider the coordinate system according to *Fig. 381a*. Then from Fig. 381*b*:

$$\left.\begin{aligned} \tau &= \frac{\sigma_z - \sigma_x}{2} \sin 2\alpha + \tau_{xz} \cos 2\alpha , \\ \sigma &= \frac{\sigma_z + \sigma_x}{2} + \frac{\sigma_z - \sigma_x}{2} \cos 2\alpha - \tau_{xz} \sin 2\alpha . \end{aligned}\right\} \qquad (266)$$

If the elementary surface with inclination α forms part of a sliding surface, then the circle of the principal stresses has to be tangent to *Coulomb*'s line. Therefore, it can be written:

$$\tau_{xz} = \tau \frac{\cos (2\alpha - \Phi)}{\cos \Phi} . \qquad (267)$$

Since $\sigma = \tau \cot \Phi$, we get after substitution:

$$\left.\begin{aligned} \sigma_z &= \tau \frac{1 + \sin \Phi \sin (2\alpha - \Phi)}{\sin \Phi \cos \Phi} , \\ \sigma_x &= \tau \frac{1 - \sin \Phi \sin (2\alpha - \Phi)}{\sin \Phi \cos \Phi} , \\ \tau_{xz} &= \tau \frac{\cos (2\alpha - \Phi)}{\cos \Phi} . \end{aligned}\right\} \qquad (268)$$

The conditions of equilibrium are

$$\left.\begin{aligned} \frac{\partial \sigma_z}{\partial z} + \frac{\partial \tau_{xz}}{\partial x} &= \gamma \cos \varepsilon , \\ \frac{\partial \sigma_x}{\partial x} + \frac{\partial \tau_{xz}}{\partial z} &= \gamma \sin \varepsilon . \end{aligned}\right\} \qquad (269)$$

In our case, the stress components are independent of x (half space), therefore

$$\frac{\partial \sigma_x}{\partial x} = \frac{\partial \tau_{xz}}{\partial x} = 0 .$$

So we have from Eq. (269)

$$\sigma_z = z\gamma \cos \varepsilon$$
$$\tau_{xz} = z\gamma \sin \varepsilon$$

and, making use of Eq. (268),

$$\frac{\tau_{xz}}{\sigma_z} = \tan \varepsilon = \frac{\cos (2\alpha - \Phi) \sin \Phi}{1 + \sin \Phi \sin (2\alpha - \Phi)} . \qquad (270)$$

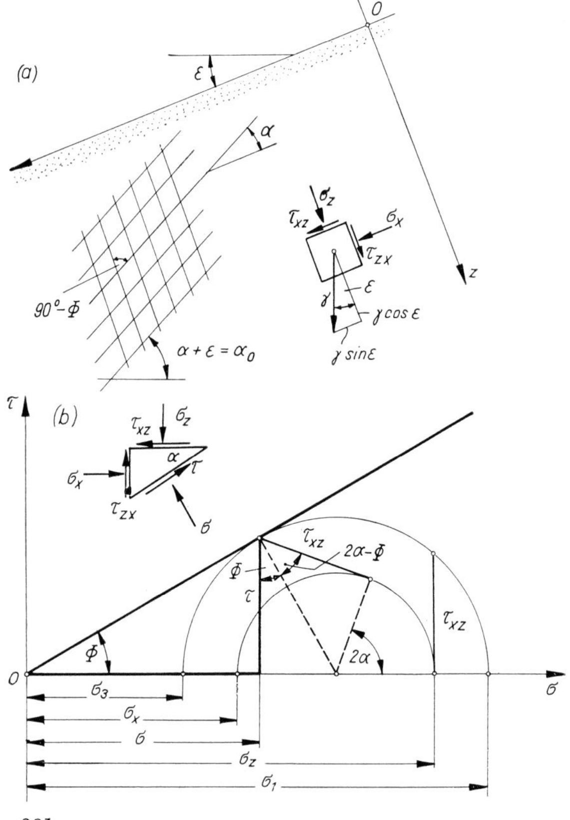

Fig. 381

a — Equilibrium in the semi-infinite half space; *b* — Mohr's circle of coordinated and of principal stresses

Fig. 382. Slope of the slip lines in the Rankine state

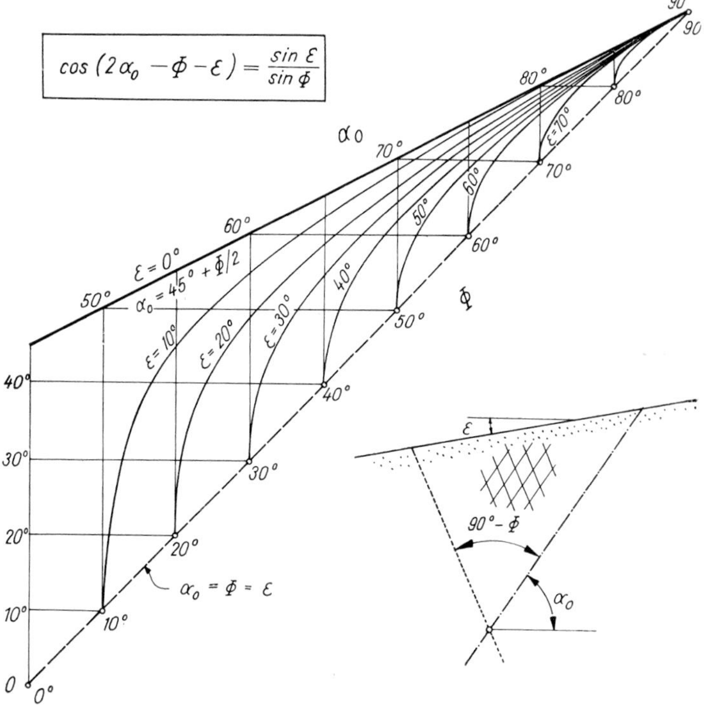

From Eq. (270):

$$\cos(2\alpha - \Phi)\sin\Phi\cos\varepsilon = \sin\varepsilon + \sin\Phi\sin(2\alpha - \Phi).$$

Taking $\alpha_0 = \alpha + \varepsilon$, i.e. the angle of the slip surface with the horizontal, we get:

$$\boxed{\cos(2\alpha_0 - \Phi - \varepsilon) = \frac{\sin\varepsilon}{\sin\Phi}} \tag{271}$$

The slip field consists of two families of slip lines; they make an angle of $(90° - \Phi)$ with each other (see Fig. 381). For example, if $\Phi = 35°$ and $\varepsilon = 25°$,

$$\alpha_0 = 51°15'.$$

The values of α_0 ∢ are given in *Fig. 382*.

The earth pressure in the *Rankine* state, acting on an oblique wall, can be determined with the help of the pole. We draw a line through the pole which runs parallel to the back of the wall; the coordinates of the point where this line cuts the Mohr circle of principal stresses, furnish the stresses σ and τ acting on the back of the wall. An example is given in *Fig. 383*.

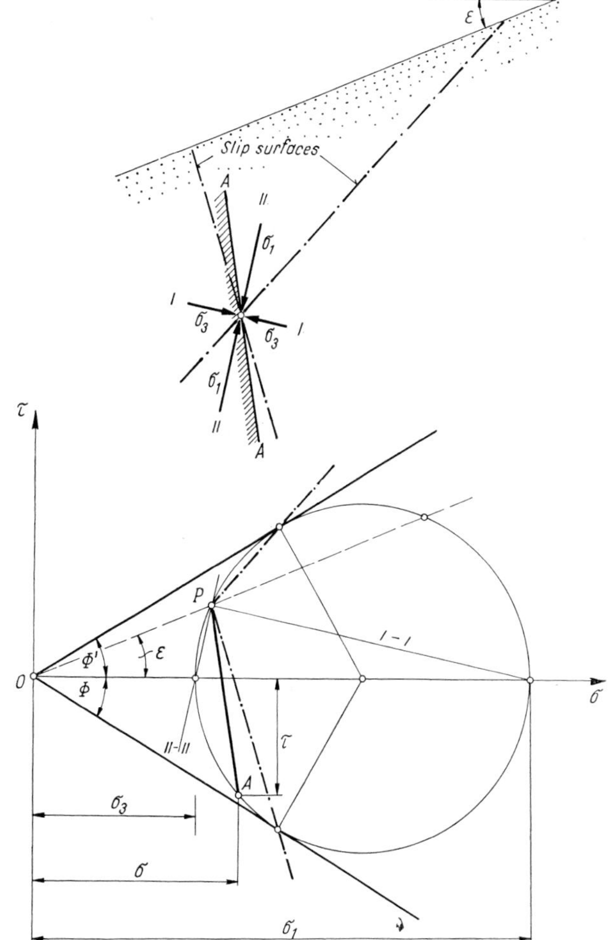

Fig. 383. Construction of slip line directions and of the stresses

9.3.3 Cohesive material

In cohesive soils, the failure condition is given by

$$\tau = \sigma\tan\Phi + c,$$

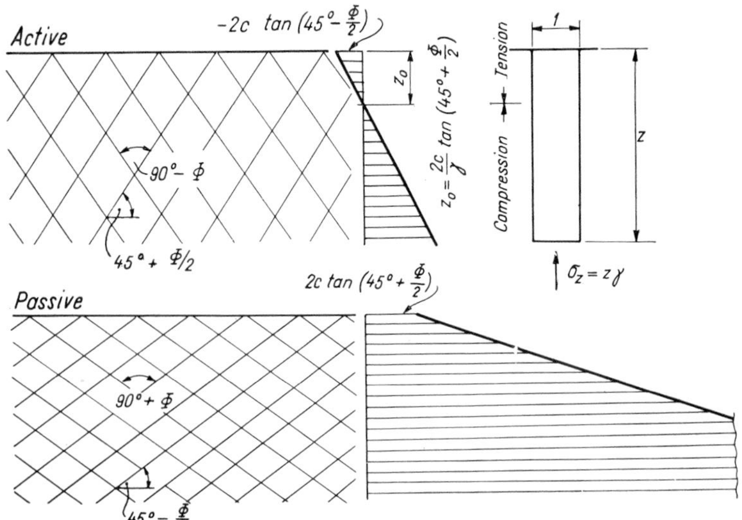

Fig. 384. Semi-infinite half space, horizontal boundary, cohesive material:
a — active case; *b* — passive case

the *Coulomb* line does not go through *O*. Therefore, if there will be a uniform expansion in the half space, filled with this material, we have to consider the fact, that the soil is capable of carrying tensile stresses.

We take a half space bounded by a horizontal surface (*Fig. 384*). The vertical stress at a depth 2, due to the soil's own weight, is again given by

$$\sigma_z = z\gamma .$$

We can draw the *Mohr* circle of the stressed state, for the case of active plastic failure (expansion); it will be tangent to the *Coulomb* line. If this circle cuts the horizontal axis in the negative part, the minor principal stress will be tension (*Fig. 385*).

For $z = 0$ and $\sigma_z = 0$, we get the circle of the unconfined tension; with increasing z we arrive at a state of stress, where $\sigma_x = 0$. This occurs at $z = z_0$; at this depth, the soil is in the state of unconfined compression: $\sigma_z = z_0\gamma$, $\sigma_x = 0$. It can be shown from Fig. 385 that

$$z_0 = \frac{2c}{\gamma} \tan\left(45° + \frac{\Phi}{2}\right) . \tag{272}$$

In this zone $0 \leq z < z_0$ the minor principal stress is tension. The angle of the slip lines with the horizontal is again, as in the case of cohesionless material, $\alpha = 45° + \Phi/2$.

If we produce a uniform compression in horizontal direction, no tension will occur, but in the case of failure, there will be an additional member in the expression of the horizontal stress, due to the cohesion of the material. The formulae for the

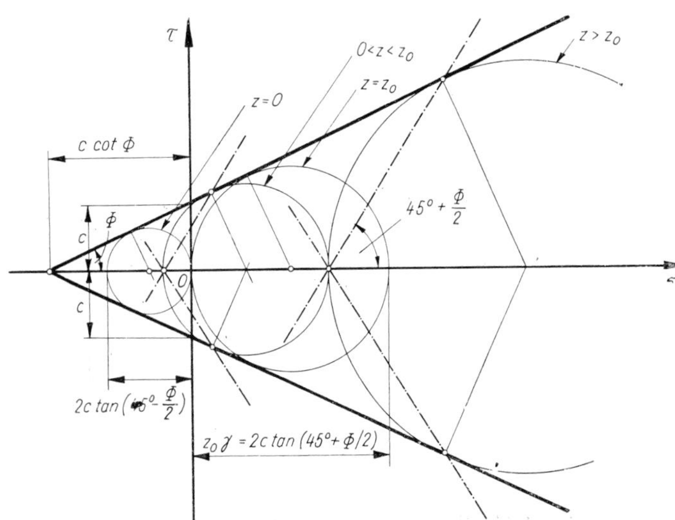

Fig. 385. Graphical representation of the state of stress

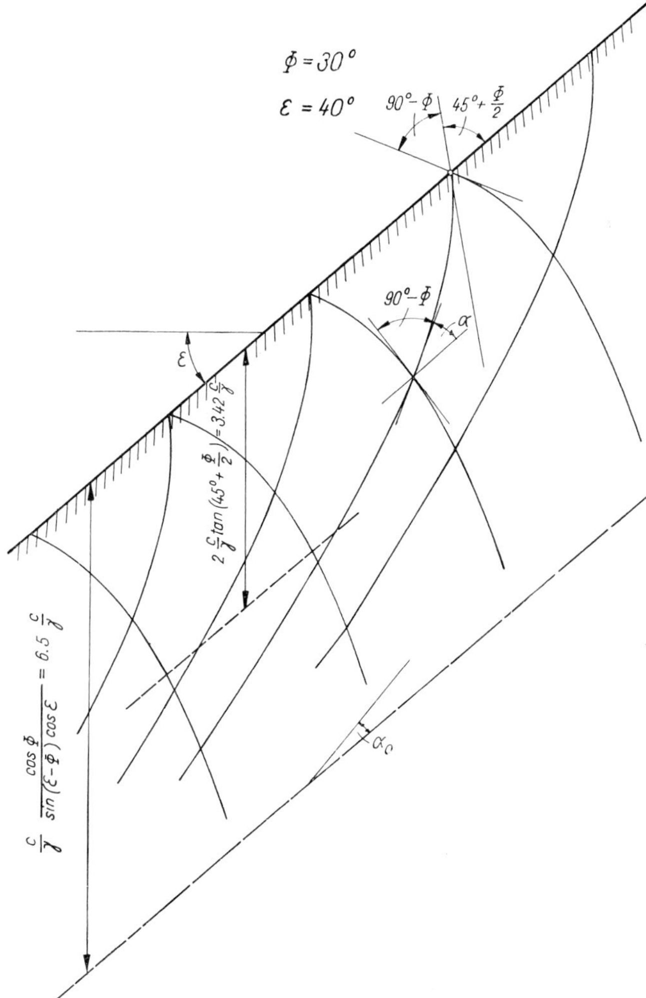

Fig. 386. Family of slip lines in the semi-infinite half space bounded by a sloping plane; cohesive material

stresses are:

$$\left.\begin{aligned}\sigma_z &= z\gamma \\ \sigma_x &= z\gamma \tan\left(45° \mp \frac{\Phi}{2}\right) \mp 2c \tan\left(45° \mp \frac{\Phi}{2}\right)\end{aligned}\right\} \quad (273)$$

and the resulting force acting on a vertical wall with height h:

$$E^R_{a,p} = \frac{h^2\gamma}{2} \tan^2\left(45° \mp \frac{\Phi}{2}\right) \mp 2ch \tan\left(45° \mp \frac{\Phi}{2}\right). \quad (274)$$

In Eqs (273) and (274) the upper sign gives the active, the lower sign the passive case.

In the case of an oblique surface, with $\varepsilon < \Phi$, we get the same critical depth, where $\sigma_s = 0$, but the sliding surfaces are curved, with the differential equation (see *Fig. 386*)

$$z = \frac{c}{\gamma} \frac{\cos \Phi \cos (2\alpha - \Phi)}{\sin \varepsilon - \sin \Phi \cos (2\alpha - \Phi + \varepsilon)}. \quad (275)$$

For the derivation see RÉSAL (1940), FRONTARD (1922) and KÉZDI (1962).

The conjugate stresses, σ_v and σ_s, can be determined either analytically or by graphical construction. Here we show the graphical construction for the case $\varepsilon < \Phi$ after JELINEK (1943).

We use the method shown for cohesionless material in Fig. 378. We draw first the *Coulomb* line of failure in the coordinate system (τ, σ) (*Fig. 387*). On the same plane, we draw from point O a straight line with angle ε. On the vertical axis, pointing downward, we measure the vertical depth. At a certain depth z, the vertical stress, acting on an element which is parallel to the surface, is equal to $\sigma_v = z\gamma \cos \varepsilon$; it is represented by the section $\overline{OP}$ in Fig. 387. We can draw two circles through P, both tangent to the failure line and having the centre on the horizontal axis. These are the Mohr circles of principal stresses, one for the active, and one for the passive case. The points P_1 and P_2, where the circles cut the line $\tau = \sigma \tan \varepsilon$, represent the end points of the vectors, σ_{s1} and σ_{s2}, i.e. they give the stresses parallel to the surface acting on vertical elements. If we draw the line Oa on the lower half of the figure, and making $\overline{AP'}$ equal to $\sigma_s = z\gamma \cos \varepsilon$ (i.e. $\overline{OP} = \overline{AP'}$), a vertical projection of the points P_1 and P_2 on $\overline{Oa}$, i.e. the points P'_1 and P'_2 give stress values σ_s at depth z, for the active and passive case. The principal stresses and their directions are also given in the figure.

For practical purposes it is convenient to use a length scale for the stresses and to plot σ/γ and

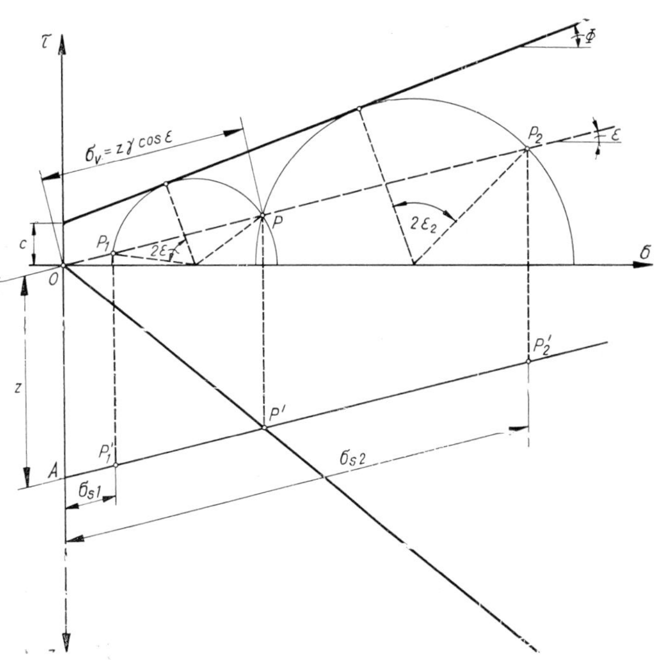

Fig. 387. Determination of the distribution of stresses by using Mohr's circle of stresses

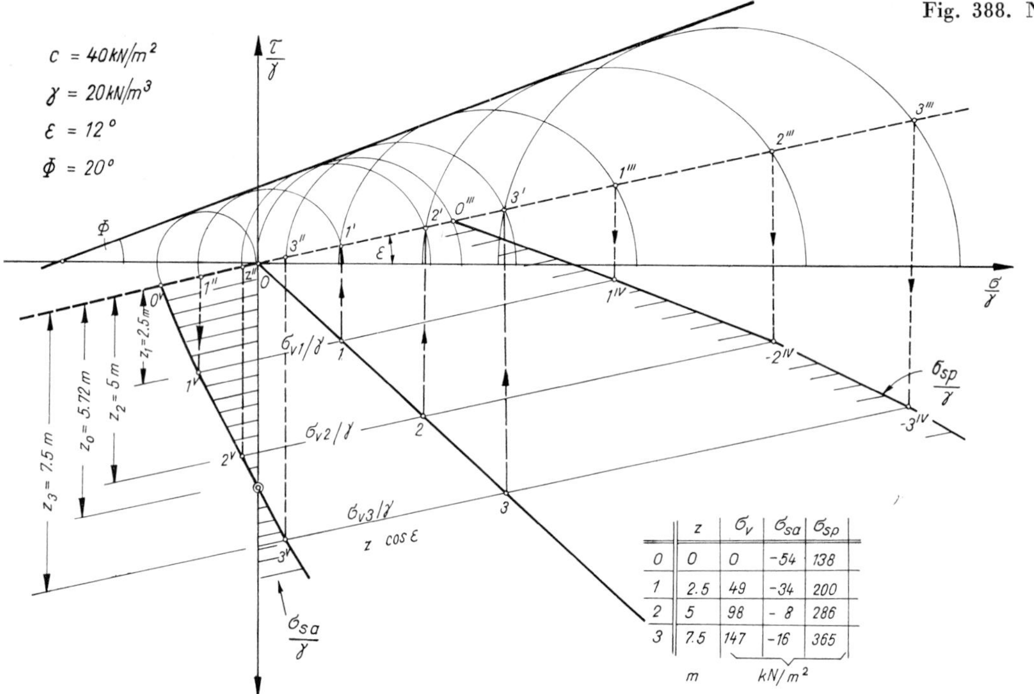

	z	σ_v	σ_{sa}	σ_{sp}
0	0	0	-54	138
1	2.5	49	-34	200
2	5	98	- 8	286
3	7.5	147	-16	365
	m	kN/m²		

Fig. 388. Numerical example

τ/γ values on the upper part of the figure. A numerical example is given in *Fig. 388*.

For the cases $\varepsilon = \Phi$ and $\varepsilon > \Phi$ see JELINEK (1947) and KÉZDI (1962).

9.4 Retaining wall problems

9.4.1 Introduction

If we have to form an earth slope steeper than it would stand safely by its own strength, then a retaining structure is required. If this structure would be taken away, one part of the backfill would come into movement. Consequently, the backfill exerts a force on the retaining structure; this force is the earth pressure.

The simplest type of structure of this kind is a gravity wall (*Fig. 389*). The wall has to be designed in such a manner that the earth pressure force (the resultant of the stresses acting along the back of the wall) and the weight of the wall give a resultant that can be transmitted safely to the underlying soil on the surface $\overline{AA_1}$.

The earth pressure will cause some displacements of the wall; an outward movement—tilting around the toe, horizontal displacement, or both—causes the backfill to expand. If the displacements are large enough, a plastic state may occur, or a slip line may develop and the earth wedge ABC reaches limit equilibrium. The condition of this limit equilibrium gives the possibility to determine the acting forces, particularly the earth pressure force, a knowledge of which is required to calculate the dimensions of the wall.

The displacement of the wall will mobilize further stresses in the earth mass. The surface $\overline{AA_1}$ will be acted upon by the resulting force R. If it amounts to the bearing capacity of the foundation, a base failure will occur, a slip surface will develop. When the bearing value of the foundation is high enough, then the wall moves more or less horizontally causing compression of the earth mass $A_1B_1C_1$ in front of the wall. This will cause passive earth pressure to develop on $\overline{A_1B_1}$; with a sufficient amount of movement, limit equilibrium—

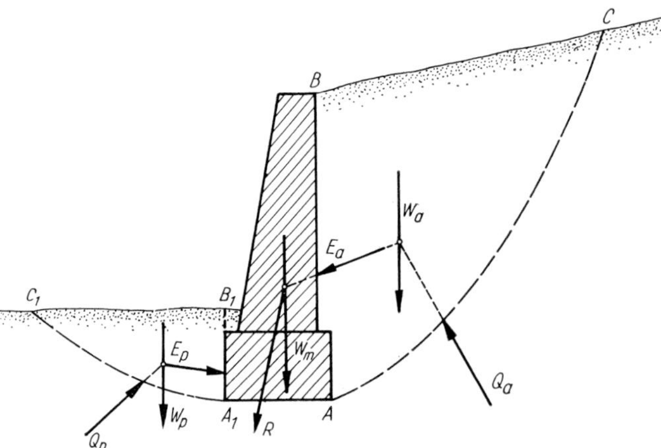

Fig. 389. Cross-section of a retaining wall; forces acting on the wall

passive case—will be reached and a slip line A_1C_1 develops.

In this chapter, we intend to show the determination of the limiting values of E_a and E_p. The problem of the bearing capacity will be discussed in Vol. 2.

9.4.2 Active earth pressure of dry sand on gravity wall

In Section 9.3 we discussed the problem of earth pressure in the half space. We mentioned that the final formulae furnishing earth pressure values, acting on vertical walls can be used for practical calculations of retaining walls, if the back of the wall is completely smooth and it tilts around its toe until failure occurs, due to uniform expansion. While the second condition is usually fulfilled in the case of a dry, cohesionless backfill and normal wall dimensions, the first condition is always a rough approximation. The back of real walls is never smooth and the movement of the wall always mobilizes frictional stresses. This will also influence the pattern of slip lines in the backfill.

Figure 390 illustrates this effect. If the wall moves outward and point A has been fixed (case Ia), theory and experience show that the sliding surface, after a sufficient amount of movement of the wall, starts with an upper straight section and joins the wall with a second curved section. The sand mass ABC moves downward, friction will be mobilized on both $\overline{AB}$ and $\overline{AC}$. The resulting force will make an angle δ with the normal of the wall. If we push the wall toward the backfill, its passive resistances have to be mobilized; if the weight of the wall is greater than the friction force mobilized on the back of the wall, the resulting force points downwards and makes an angle δ with the back of the wall (case IIa).

The wall friction can have an effect in that part of the backfill which adjoins the wall; by drawing a straight line from point B, having the inclination of the sliding surface in the Rankine state of stresses, we devide the backfill into two parts: above the line $\overline{BD}$ there is a Rankine state of stress, with two families of straight slip lines, making an angle $90° - \Phi$ and $90° + \Phi$, respectively. Below the line $\overline{BD}$ in the field ABD the slip lines are curved. The surface of the backfill, on section $\overline{BC}$, settles in the active case, heaves in the passive case.

If the wall, simultaneously with the outward tilting, moves downward because of a compression of the subsoil, the angle of direction changes to the opposite sign and the slip line becomes convex from above (case Ib). The same thing occurs in the passive case, when the wall is pushed toward the backfill and there is a force acting upwards (IIb). The exact mathematical expression of the sliding surface is unknown in the general case; there are some special cases where an exact solution is available. These are: weightless medium with internal friction and cohesion, weighty medium with cohesion. In the first case the slip line is a logarithmic spiral, in the second, a circle. These two and the straight line are today in use to solve the majority of earth pressure problems (*Fig. 391*). If the backfill is dry sand and we wish to determine the active pressure, the use of a straight slip line is general because the difference of the earth pressure so obtained and that obtained with a curved surface is very small. In other cases,

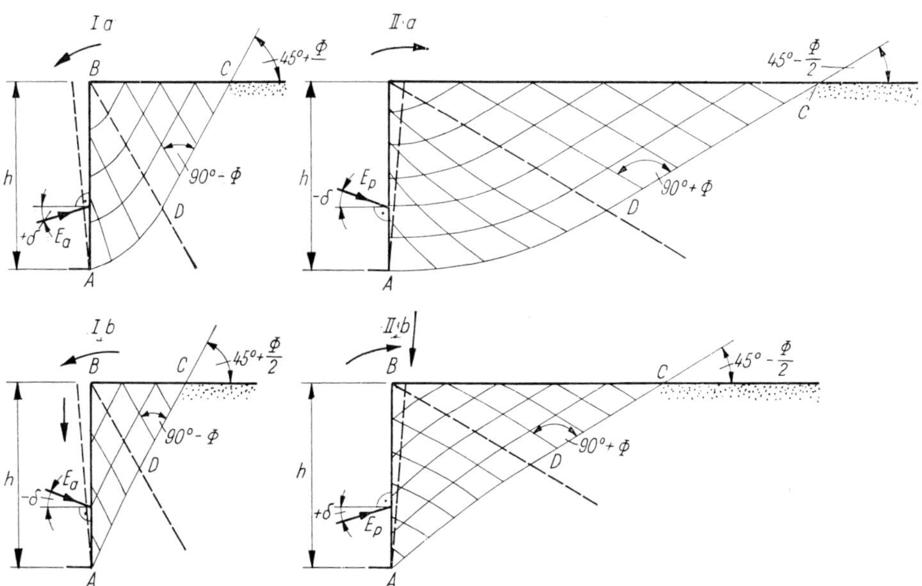

Fig. 390. Shear patterns (families of slip surfaces) in sand backfills behind rough vertical wall; failure state

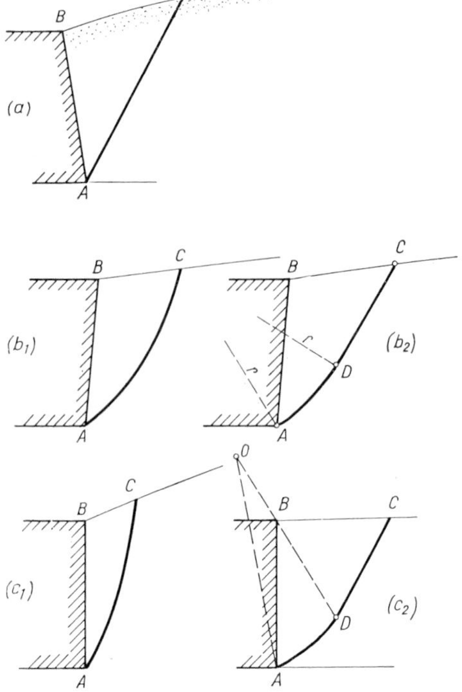

Fig. 391. Different forms of sliding surface; the three most usual assumptions

particularly in the passive state, either a curved or a composite surface is used, in accordance with the explanations given for Fig. 390.

The earliest theories (Feld, 1928) considered the earth pressure as a force that prevents the downward movement of a rigid sliding wedge on a smooth, oblique plane. The weight of the wedge was resolved into two components; one was normal to the sliding slope, the other parallel to it. So the earth pressure was equal and opposite to this parallel component. Later, for the sake of safety, the friction on both the wall and the sliding slope was neglected; the result was a hydrostatic pressure of a liquid with a density of γ.

The first theory founded on a scientific basis was developed by Ch. Coulomb. In a paper published in 1773, he investigated the stability of earth masses. He considered straight and curved surfaces and used the extremum principle to determine the earth pressure. *Figure 392* shows a facsimile of one of his drawings.

The assumptions of *Coulomb's* theory are the following (*Fig. 393*):

1. The slip surface is plane. This is true in the half space; it is an approximation for a backfill retained by a rough wall.

2. The wall is vertical, the surface of the backfill is horizontal, and there is no friction between wall and backfill.

The boundary conditions are thus identical with the *Rankine* state.

3. On the slip surface, in the moment of failure the friction is fully mobilized and the condition of failure

$$T = N \tan \Phi$$

is fulfilled.

4. Among the possible sliding surfaces $\overline{AD}$, the one furnishing the maximum value of earth pressure will be considered as final.

Considering these assumptions the solution is obtained easily. The weight of the wedge ABC will be in equilibrium with the reactions on $\overline{AB}$ and $\overline{AC}$; the first is horizontal (assumption 2) the second makes an angle with the normal to $\overline{AC}$ (assumption 3). From the triangle of the forces:

$$E_a^C = W \tan(\alpha - \Phi)\,.$$

The weight of the wedge is equal to

$$W = \frac{h^2\gamma}{2} \cot \alpha\,,$$

i.e.

$$E_a^C = \frac{h^2\gamma}{2} \cot \alpha \tan(\alpha - \Phi)\,. \qquad (276)$$

Making use of assumption 3, we have to determine the maximum value of E_a^C.

Solving the equation

$$\frac{dE_a^C}{d\alpha} = \frac{h^2\gamma}{2}\left[-\frac{\tan(\alpha - \Phi)}{\sin^2\alpha} + \frac{\cot\alpha}{\cos^2(\alpha - \Phi)}\right] = 0$$

for $\alpha \nless$, we obtain

$$\alpha = 45° + \frac{\Phi}{2}\,.$$

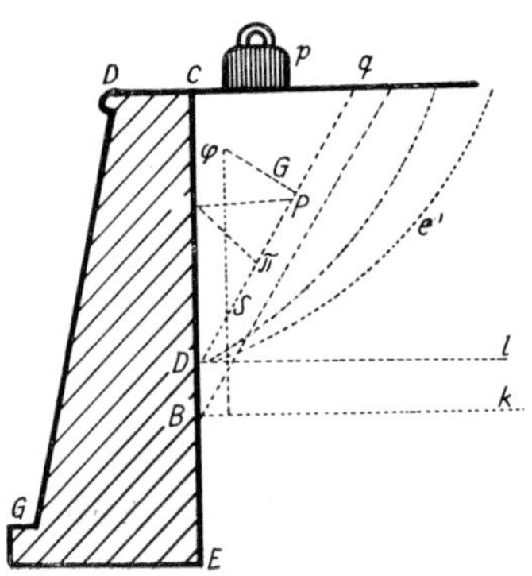

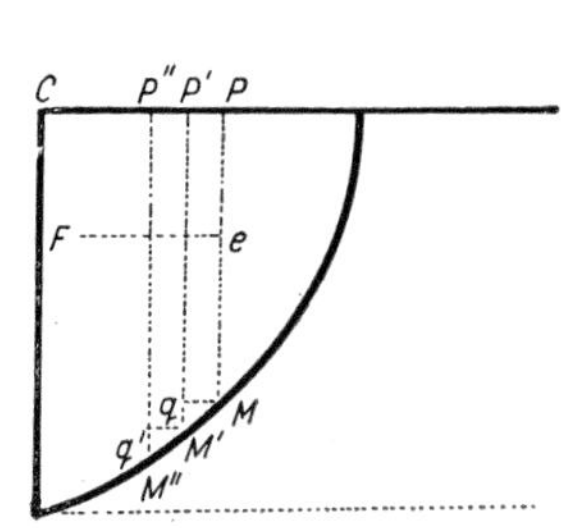

Fig. 392. Facsimile from the original paper of Coulomb

We have the same angle as in the Rankine state; this must be so because the boundary conditions are identical. The value of the earth pressure is also the same: substituting $\alpha \sphericalangle$ into Eq. (276) we get

$$E_a^C = \frac{h^2\gamma}{2}\tan^2\left(45° - \frac{\Phi}{2}\right). \tag{277}$$

However, there is a difference between the two theories: *Coulomb* assumed *one* sliding surface through point A; in the *Rankine* theory there were sliding surfaces through every point of the half space.

The point of application was obtained by assuming a triangular distribution of the earth pressure; thus, the resulting force acts in the lower third point of the wall.

Coulomb's theory was later generalized in order to cover more practical cases. The assumption of a plane sliding surface was retained, but the friction between wall and backfill was not neglected and other wall and backfill boundaries were considered (*Fig. 394*). The resulting earth pressure does not act perpendicular to the back of the wall, but makes an angle δ with the normal. This angle depends on the amount of displacement between the earth and the wall, but it cannot exceed the angle of surface friction there. We assume again that on the friction is fully mobilized, the reaction Q makes an angle Φ with the normal. The back of the wall makes an angle β with the horizontal, the surface an angle ε. Let the angle between the direction of the earth pressure and the vertical be ψ, the angle with the horizontal be ω.

The three forces, W, the weight of the wedge, Q, the reaction on the sliding surface and E_a, the earth pressure, form a closed force polygon. It can be written:

$$E_a = W\frac{\sin(\alpha - \Phi)}{\sin(\alpha - \Phi + \psi)} = f(\alpha, \psi). \tag{278}$$

The earth pressure force is a function of two variables: the angle α of the sliding surface and angle ψ, i.e. the angle of direction.

We mentioned that δ depends on the amount of displacement, this function is not known. There are some practical rules in use to assume a realistic value. Möller (1902) suggested to assume a value between $\frac{1}{3}\Phi$ and $\frac{2}{3}\Phi$. According to Müller-Breslau $\delta = 0.8\,\Phi$. Jáky gives a theoretical derivation (see later). If δ is known, the condition $\partial E_a/\partial\alpha = 0$ (4th assumption of *Coulomb)* gives an equation for α; substituting it into Eq. (278) we have

$$E_a = K_a\frac{z^2\gamma}{2}. \tag{279}$$

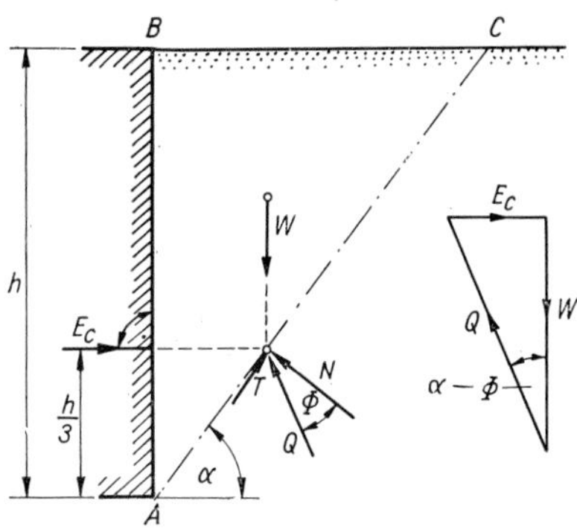

Fig. 393. Coulomb's earth pressure theory

K_a is the coefficient of active earth pressure. In the general case it is given by

$$K_a = \left[\frac{\sin(\beta - \Phi)}{\sqrt{\sin(\beta + \delta)}\sqrt{\dfrac{\sin(\beta + \Phi)\sin(\Phi - \varepsilon)}{\sin(\beta - \varepsilon)}}}\right]^2. \tag{280}$$

For $\beta = 90°$, $\delta = 0°$, $\zeta = 0°$ (*Coulomb's* case):

$$K_a^C = \left[\frac{\cos\Phi}{1 + \sin\Phi}\right]^2 = \tan^2\left(45° - \frac{\Phi}{2}\right). \tag{280a}$$

For $\beta = 90°$, $\varepsilon - \delta$ (*Rankine*)

$$K_a^R = \left[\frac{\cos\Phi\sqrt{\cos\varepsilon}}{\cos\varepsilon + \sqrt{\sin(\Phi + \varepsilon)\sin(\Phi - \varepsilon)}}\right]^2 =$$

$$= \tan^2\left(45° - \frac{\nu}{2}\right)\cos\varepsilon \tag{280b}$$

where

$$\cos\nu = \cos\Phi/\cos\varepsilon$$

and, finally, for $\delta = \Phi$

$$K_a = \left[\frac{\sin(\beta - \varepsilon)}{\sqrt{\sin(\beta + \Phi)} + \sqrt{\dfrac{\sin(\beta + \Phi)\sin(\Phi - \varepsilon)}{\sin(\beta - \varepsilon)}}}\right]^2. \tag{280c}$$

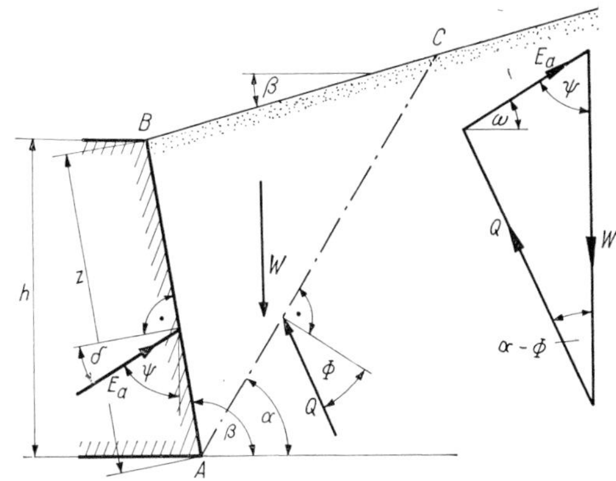

Fig. 394. General case; slanting back, sloping surface, rough wall

The above formulae are rather cumbersome, therefore, for practical problems, earth pressure tables and graphical methods are in use. Tables give the K_a values for plane back and plane and unloaded backfill surface; graphical methods are practical for irregular and/or loaded boundaries. In order to simplify the calculations we give, in *Table 34*, the most important coefficients, and, in *Table 35*, the coefficient of the horizontal component of the active earth pressure. The symbols are the same as in *Fig. 394*.

Table 34. Important coefficients for use in earth pressure calculations

$\Phi°$	$\tan \Phi$	$\tan (45° + \Phi/2)$	$\tan (45° - \Phi/2)$	$\tan^2 (45° + \Phi/2)$	$\tan^2 (45° - \Phi/2)$
0	0	1.000	1.000	1.000	1.000
1	0.017	1.018	0.983	1.036	0.966
2	0.035	1.036	0.966	1.073	0.933
3	0.052	1.054	0.949	1.111	0.901
4	0.070	1.072	0.933	1.149	0.870
5	0.087	1.091	0.916	1.190	0.839
6	0.105	1.111	0.900	1.234	0.810
7	0.123	1.130	0.885	1.277	0.783
8	0.140	1.150	0.869	1.322	0.755
9	0.158	1.171	0.854	1.371	0.729
10	0.176	1.192	0.839	1.420	0.704
11	0.194	1.213	0.824	1.472	0.680
12	0.213	1.235	0.810	1.525	0.656
13	0.231	1.257	0.795	1.580	0.633
14	0.249	1.280	0.781	1.638	0.610
15	0.268	1.303	0.767	1.698	0.589
16	0.287	1.327	0.754	1.761	0.568
17	0.306	1.351	0.740	1.826	0.548
18	0.325	1.376	0.727	1.894	0.528
19	0.344	1.402	0.713	1.965	0.509
20	0.364	1.428	0.700	2.040	0.490
21	0.384	1.455	0.687	2.117	0.472
22	0.404	1.483	0.675	2.198	0.455
23	0.424	1.511	0.662	2.283	0.438
24	0.445	1.540	0.649	2.371	0.422
25	0.466	1.570	0.637	2.464	0.406
26	0.488	1.600	0.625	2.561	0.390
27	0.510	1.632	0.613	2.676	0.376
28	0.532	1.664	0.601	2.770	0.361
29	0.554	1.698	0.583	2.882	0.347
30	0.577	1.732	0.577	3.000	0.333
31	0.601	1.767	0.566	3.124	0.320
32	0.625	1.804	0.554	3.255	0.307
33	0.649	1.842	0.543	3.392	0.295
34	0.675	1.881	0.532	3.537	0.283
35	0.700	1.921	0.521	3.690	0.271
36	0.727	1.963	0.510	3.852	0.260
37	0.754	2.006	0.499	4.023	0.249
38	0.781	2.050	0.488	4.204	0.238
39	0.810	2.097	0.477	4.395	0.228
40	0.839	2.145	0.466	4.599	0.217
41	0.869	2.194	0.456	4.815	0.208
42	0.900	2.246	0.445	5.045	0.198
43	0.933	2.300	0.435	5.289	0.189
44	0.966	2.356	0.424	5.550	0.180
45	1.000	2.414	0.414	5.828	0.172
46	1.036	2.475	0.404	6.126	0.163
47	1.072	2.539	0.394	6.445	0.155
48	1.111	2.605	0.384	6.786	0.147

Table 35. Coefficients of the horizontal components of the active earth pressure. Values of 1000 K_{ah}

$$E_{ah} = \frac{1}{2} K_{ah} \gamma H^2, \qquad E_a = E_{ah}/\sin (\beta + \delta)$$

$\tan \alpha$	Φ [°]	$\beta = 0$				$\beta = \Phi/2$				$\beta = \Phi$
		δ				δ				δ
		0	$\Phi/2$	$2/3\,\Phi$	Φ	0	$\Phi/2$	$2/3\,\Phi$	Φ	any value
	15	523	480	469	449	589	552	542	524	835
	20	417	378	367	348	482	446	435	418	763
	25	330	295	286	270	388	354	345	329	675
0.2	30	257	229	221	207	306	278	270	256	587
	35	198	175	169	158	237	214	208	195	496
	40	148	132	126	118	178	160	155	146	406
	45	108	96	93	86	129	116	102	95	320
	15	556	510	499	475	627	587	576	556	883
	20	456	409	397	376	526	485	473	453	822
	25	368	327	316	296	434	396	384	365	747
0.1	30	295	260	250	233	353	319	309	291	666
	35	234	250	196	181	282	253	245	228	580
	40	182	159	152	140	220	197	190	176	492
	45	139	121	116	106	168	149	140	133	405
	15	588	538	524	500	665	621	609	587	933
	20	490	440	426	401	569	523	510	486	883
	25	406	359	345	322	482	436	423	400	824
0	30	333	291	279	257	402	360	334	326	750
	35	271	235	224	205	330	293	283	262	672
	40	218	187	183	161	267	235	226	207	587
	45	172	148	145	125	210	185	177	160	500
	15	619	564	549	521	701	654	640	615	983
	20	525	469	453	424	612	561	545	518	948
	25	443	389	373	345	529	477	461	434	900
	30	372	321	306	280	452	402	387	359	839
−0.1	35	309	264	251	226	381	335	318	294	768
	40	254	216	204	180	316	275	263	237	689
	45	207	174	164	143	257	223	212	188	605
	15	648	588	571	541	737	684	669	642	1036
	20	559	495	477	444	654	596	579	548	1016
	25	479	416	398	365	576	516	498	465	982
	30	409	349	332	299	502	442	424	390	933
	35	347	292	275	244	432	376	360	323	872
−0.2	40	292	243	229	197	367	316	300	265	800
	45	243	200	186	157	307	262	247	213	720

The basis of the graphical methods is *Rebhann's theorem* (REBHANN, 1871). From the 4th assumption ($\partial E_a/\partial \alpha = 0$), and Eq. (278) we may write:

$$\frac{\partial E_a}{\partial \alpha} = \frac{\partial W}{\partial \alpha} \frac{\sin (\alpha - \Phi)}{\sin (\alpha - \varphi + \psi)} + $$

$$+ W \frac{\sin \psi}{\sin^2 (\alpha - \varphi + \psi)} = 0\,, \tag{281}$$

from the force polygon

$$W \frac{\sin \psi}{\sin (\alpha - \varphi + \psi)} = Q \tag{282}$$

and, from *Fig. 395*

$$\frac{\partial W}{\partial \alpha} = \frac{dW}{d\alpha} = -\frac{l^2\gamma}{2}. \qquad (283)$$

The negative sign is necessary, because if α increases, W decreases. Substituting (282) and (283) into Eq. (281) we obtain:

$$Q = \frac{l^2\gamma}{2} \sin(\alpha - \Phi). \qquad (284)$$

Now, let us assume that the angle α of the sliding surface which fulfills the extremum condition is known, and we draw the sliding surface in the cross-section of the backfill (*Fig. 396*). It cuts the surface in point C. From this point we draw a perpendicular to the line under angle Φ from A; and, from this, we measure the angle ω. The angle in D, will be equal to ψ. The direction of $\overline{CD}$ can also be obtained, if we measure, at point B, the angle $(\delta + \Phi)$ from the back of the wall.

The triangle $ACD_\triangle$ will be similar to the polygon of forces W, Q and E_a, because the angles are equal. Therefore, we may write:

$$\frac{Q}{l} = \frac{W}{w} = \frac{E}{e}.$$

Substituting Eq. (284):

$$\frac{l\gamma}{2} \sin(\alpha - \varphi) = \frac{W}{w}.$$

From Fig. 396:

$$l \sin(\alpha - \Phi) = p$$

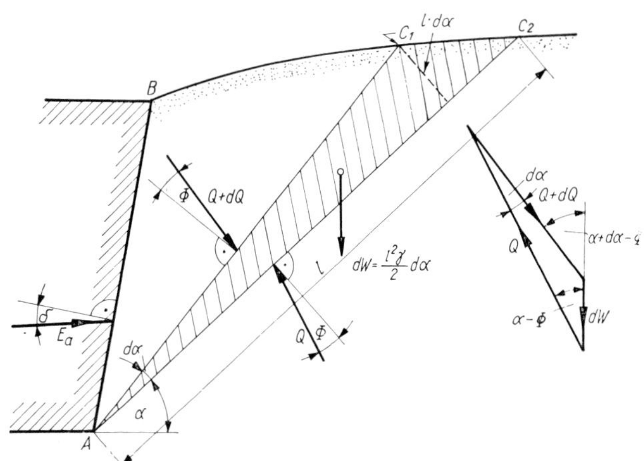

Fig. 395. Forces acting on the elementary prism

i.e.

$$\frac{p\gamma}{2} = \frac{W}{w}, \quad W = \frac{pw}{2}\gamma = aw. \qquad (285)$$

$pw/2$ is the area of the triangle $ACD_\triangle$; on the other hand, by definition, W = area $ABC_\triangle\,\gamma$. Therefore:

$$\text{area } ABC_\triangle = \text{area } ACD_\triangle. \qquad (286)$$

The sliding surface has such a position, that it divides the quadrangle $ABCD$ into two equal parts. This is *Rebhann's theorem*: it gives the possibility to determine the sliding surface. Once it has been found, the earth pressure can be obtained from the similarity of $ACD_\triangle$ and the force polygon:

$$E_a = ae = \frac{pe}{2}\gamma. \qquad (287)$$

p and e are measured from the figure.

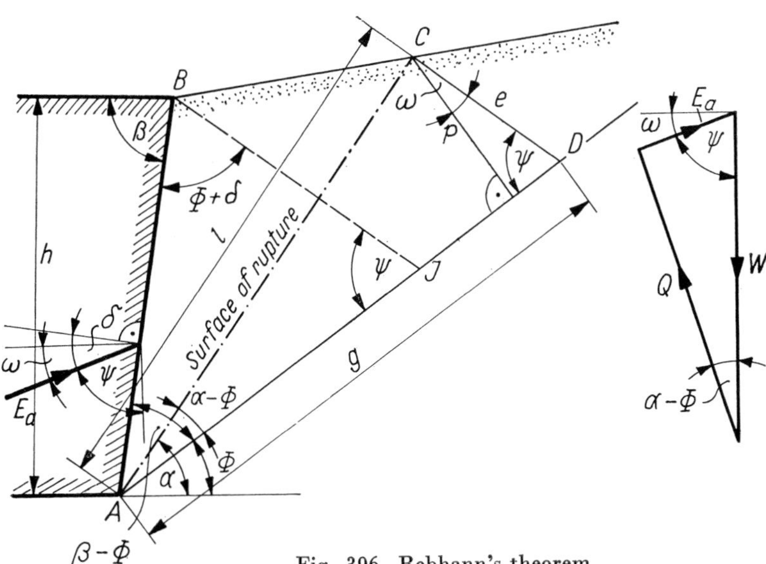

Fig. 396. Rebhann's theorem

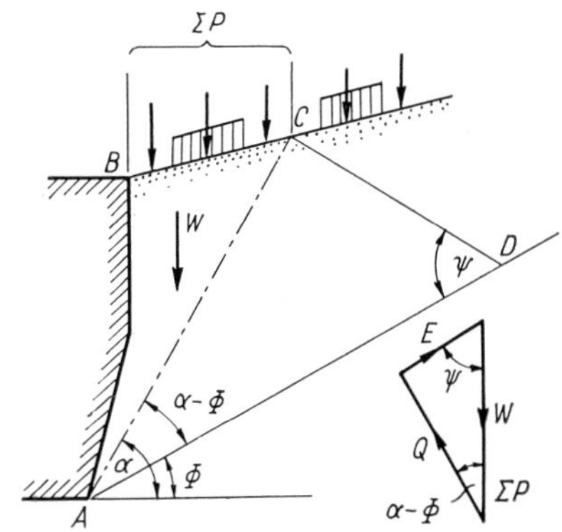

Fig. 397. The general form of Rebhann's theorem

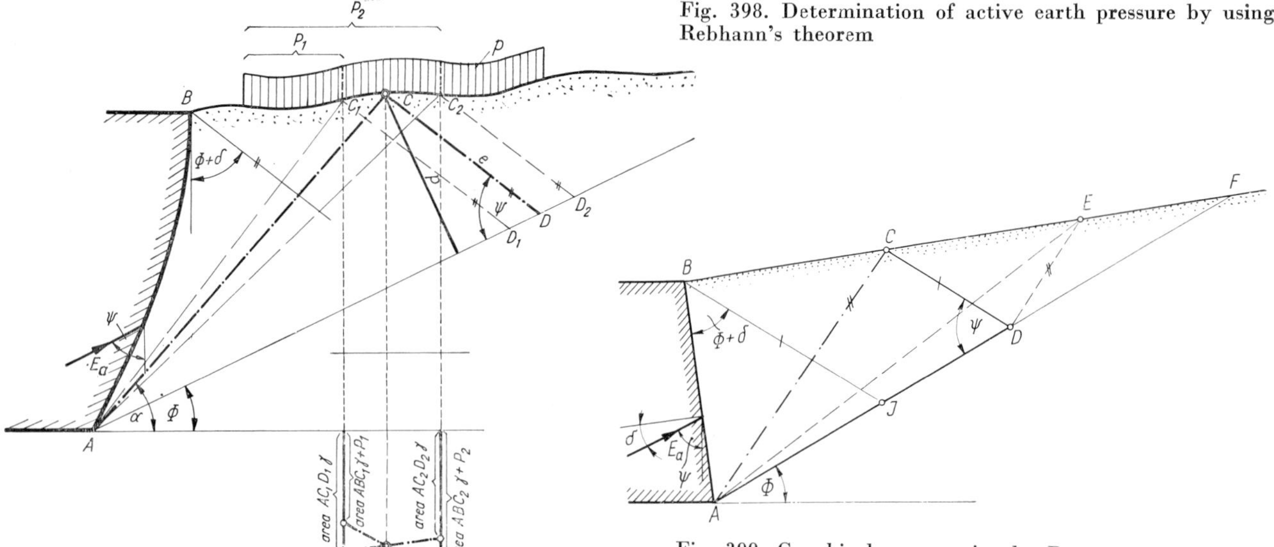

Fig. 398. Determination of active earth pressure by using Rebhann's theorem

Fig. 399. Graphical construction by PONCELET; basic principle

It can be proven that *Rebhann*'s theorem is valid for any boundaries and any loading, it can be written (*Fig. 397*):

$$W + \sum P = \text{area } ACD_{\triangle}\, \gamma\,. \tag{288}$$

A graphical interpolation method furnishes easily the sliding surface that meets this requirement; an example is given in *Fig. 398*.

To a given angle of direction and for plane and unloaded boundaries, the direct graphical construction introduced by PONCELET (1840) can be used.

The cross-section is given in *Fig. 399*. According to *Rebhann*'s theorem:

$$\text{area } ABC_{\triangle} = \text{area } ACD_{\triangle}\,.$$

We draw a line from point D, which runs parallel to the sliding surface; it cuts the surface at E. It can be seen, that

$$\text{area } ACD_{\triangle} = \text{area } ACE_{\triangle}\,.$$

It follows from the above equalities that

$$\overline{BC} = \overline{CE}\,,$$

therefore, the following ratios can be written:

$$\frac{\overline{AD}}{\overline{AF}} = \frac{\overline{CE}}{\overline{CF}} = \frac{\overline{BC}}{\overline{CF}} = \frac{\overline{DJ}}{\overline{DF}} = \frac{\overline{AD} - \overline{AJ}}{\overline{AF} - \overline{AD}}\,.$$

From the first and the last term of the series of equalities:

$$\overline{AD}^2 = \overline{AJ} \cdot \overline{AF},$$

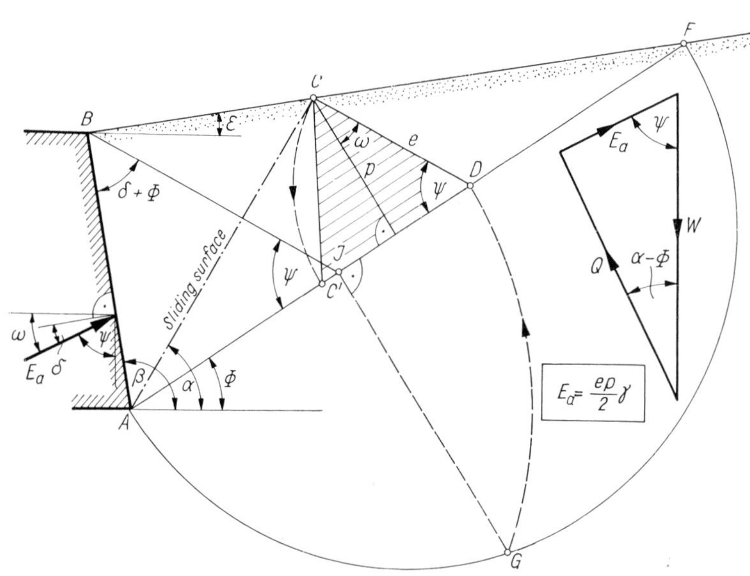

Fig. 400. Graphical construction by PONCELET

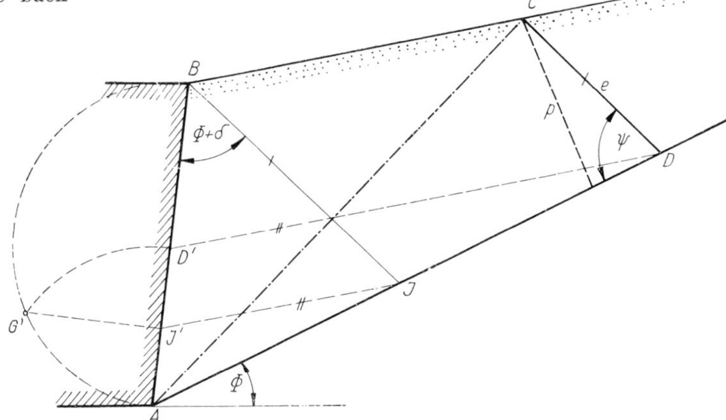

Fig. 401. Construction by using projection on the back of the wall

i.e. $\overline{AD}$ is the geometrical mean of $\overline{AJ}$ and $\overline{AF}$, so the point D can be constructed, and from that, the end point C of the sliding surface, if we trace the line $\overline{DC}$ which runs parallel to $\overline{BJ}$.

The procedure is shown in *Fig. 400*; the steps are the following:

1. Measure angle $(\delta + \Phi)$ at B, from $\overline{AB}$. This line of direction will cut the slope under Φ $(\overline{AF})$ at point J.
2. Draw a semi-circle over $\overline{AF}$ and project point J on it. (Point G.)
3. Draw a circle from the center A, with the radius AJ, which cuts the slope under Φ at D. Make $\overline{DC}$ parallel to $\overline{BJ}$; it gives the end point C of the sliding surface. The earth pressure can be calculated from Eq. (288).

By drawing a circle from point D with radius CD, we get the point $\overline{C}$. The area of the triangle $C\overline{C}'D_\triangle$ is proportional to the earth pressure ($E =$ = area $C\overline{C}'D_\triangle\,\gamma$).

If point F is too far, the graphical construction can be performed on the back of the wall, by projecting the respective points parallel to the surface (*Fig. 401*).

Other graphical solutions were given by CULLMANN, ENGESSER and others; these are discussed in the book "Erddrucktheorien" (KÉZDI, 1962). Here we give only one example of *Engesser's* method which clearly shows the application of the extremum methods. This is a trial-and-error method. *Figure 402* shows the cross section of the wall and of the backfill; they may be irregular. The direction of the earth pressure (the angle ψ) is given. We assume a sliding surface $\overline{AC}$ and determine the weight W_1 of the wedge ABC and construct the direction of the reaction Q_1 on the sliding surface. Q_1 makes an angle Φ with the normal. We draw the polygon of forces, then repeat the steps by taking other C points (C_2, C_3, C_4, C_5). The forces will be plotted in the same force polygon. The forces Q will have an envelope curve, and the point where the line of the force E cuts this curve will furnish the maximum value of the earth pressure which is sought. The curve e on Fig. 402 is the so-called *Engesser* curve.

These equations and graphical methods all require the knowledge of the angle of direction of the earth pressure (angle ψ or δ, respectively). The values given earlier in this chapter are just rough approximations. A definite value can be given if we extend the application of the extremum principle. This is justified by the earth pressure tests of TERZAGHI (1934 and 1936), which gave an insight into the pressure problems of dry, cohesionless sand. *Terzaghi* performed his tests with a rigid concrete wall (height 201 cm, length 420 cm, weight 130 kN), where he could measure the resulting force, its direction and point of application as well as the displacements. As long as the wall did not move, it was acted upon by the earth pressure at rest. The coefficient K_0 was equal to 0.42. A very slight tilting of the wall—the maxi-

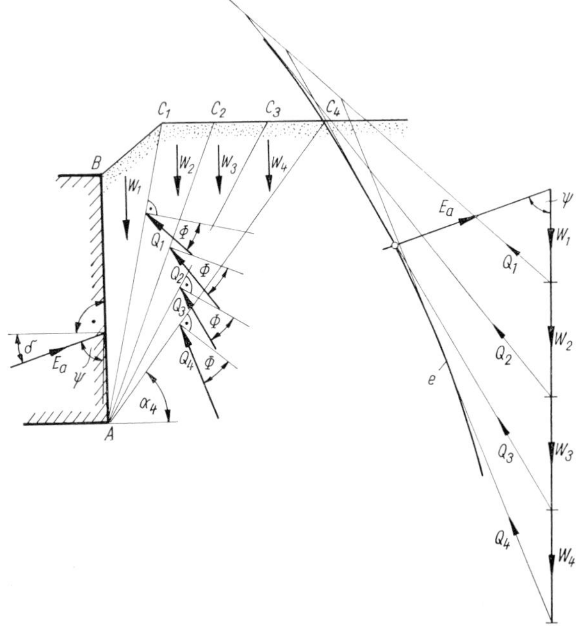

Fig. 402. Graphical construction according to ENGESSER

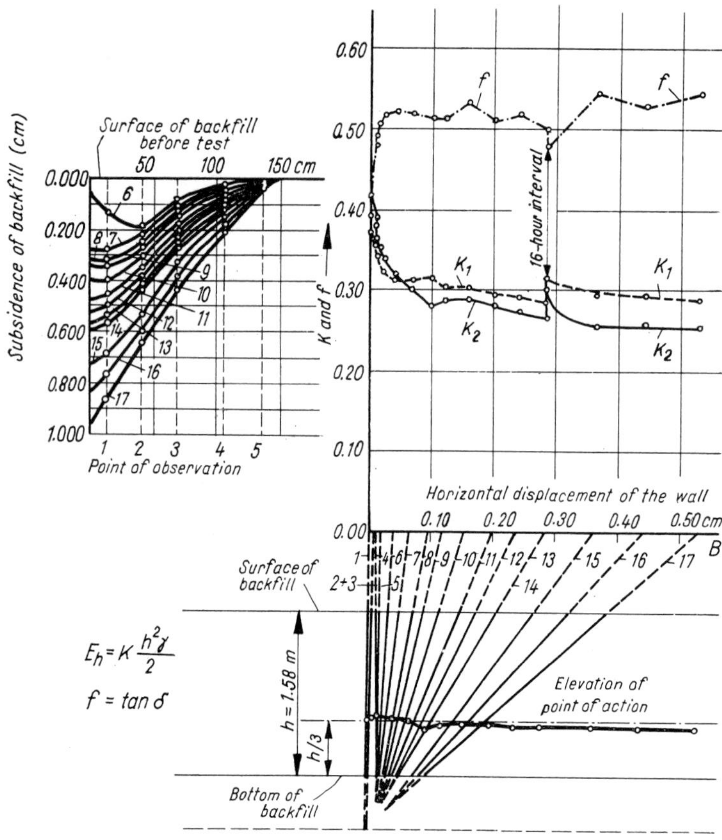

Fig. 403. Results of the earth pressure tests carried out by TERZAGHI (1934)

mum displacement amounted to 1/1000th of the height—caused a substantial decrease of the earth pressure which soon reached the extremum—minimum—value. The angle of direction increased from zero to a limiting value; the point of application was located somewhere near the lower third point of the wall height. An example of the test results is shown in *Fig. 403.*

As shown in these tests, the earth pressure quickly drops and tends toward an absolute minimum. Based on this fact, JÁKY (1946, 1948) applied the minimum principle to the function given in Eq. (278) and thus obtained an additional equation which can be used to determine the angle of direction. The function $E_a = f(\alpha, \delta)$ can be represented by a surface in the coordinate system with two variables; this surface is given, for assumed numerical values, in *Fig. 404.* It can be seen that the function has a maximum according to α and a minimum, according to δ. The extremum value sought is given by the saddle point of the surface (point P). The analytical condition of the extremum value is

$$\mathrm{d}E_a = \frac{\partial E_a}{\partial \alpha}\,\mathrm{d}\alpha + \frac{\partial E_a}{\partial \psi}\,\mathrm{d}\psi = 0. \qquad (289)$$

Equation (289) is fulfilled if

$$\text{(I)}\quad \frac{\partial E_a}{\partial \alpha} = 0\,; \quad \text{and} \quad \text{(II)}\quad \frac{\partial E_a}{\partial \psi} = 0\,. \qquad (290)$$

Equation (1) furnished *Rebhann's* theorem and served to determine the angle of the sliding surface. From Eq. (II), by differentiating Eq. (278) with respect to ψ we obtain:

$$\frac{\partial E_a}{\partial \psi} = \frac{W \sin(\alpha - \Phi)\cos(\alpha - \Phi + \psi)}{\sin^2(\alpha - \Phi + \psi)} = 0\,.$$

The equation will be fulfilled, if

$$\cos(\alpha - \Phi + \psi) = 0\,,$$

i.e.

$$\boxed{\alpha - \Phi + \psi = 90^\circ\,.} \qquad (291)$$

The second extremum condition will be fulfilled if the polygon of forces (E_a, Q, W) is a triangle, with a right angle between E_a and Q (*Fig. 405*). Since the triangle of forces is similar to triangle ACD, also the angle at C has to be 90°. The angle of direction is then given from $\psi = 180° - \beta + \delta$ to

$$\delta = 90° + \alpha - (\beta + \Phi)\,. \qquad (292)$$

Thus, the extremum value of E_a will be obtained, if we construct a sliding surface which fulfills the two requirements:

(*a*) area $ABC_\triangle$ = area $ACD_\triangle$;

(*b*) $ACD \sphericalangle = 90°$.

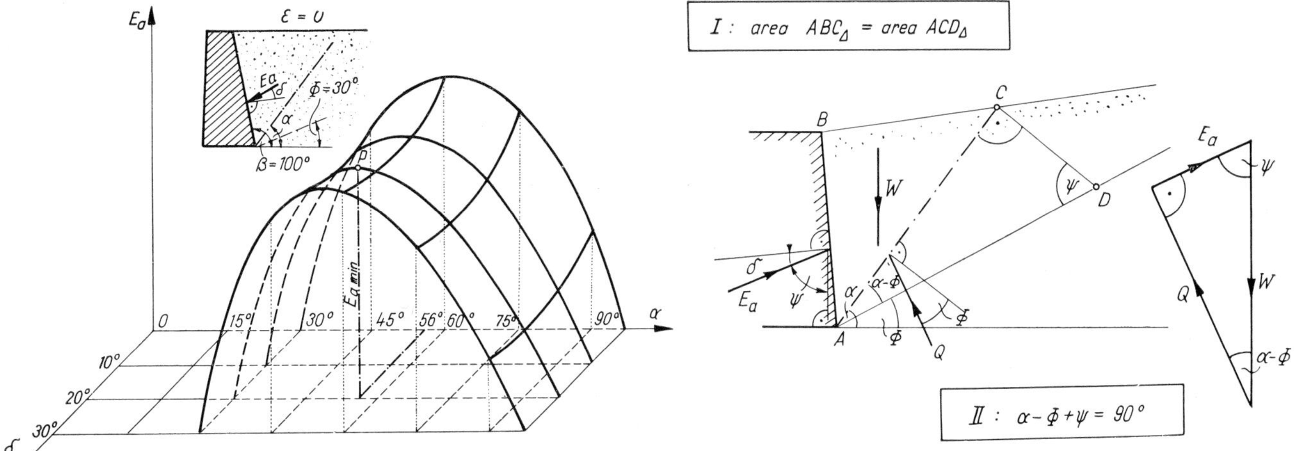

Fig. 404. Earth pressure values as functions of α and δ

Fig. 405. The extremum conditions

It can be shown that the analytical solution leads to an equation of the third degree, therefore, no direct graphical method can be given. A trial-and-error method was suggested by JÁKY (1946); this is shown in *Fig. 406*.

We assume a sliding surface $\overline{AC_1}$ and make $\overline{BE_1}$ normal to $\overline{AC_1}$. The same distance $\overline{BE_1}$ will be measured from point C_1; we get point D_1. If $\overline{AC_1}$ would fulfill both requirements (*a*) and (*b*), point D_1 would be located on the line under Φ from point A. This is not the case, so we take another surface $\overline{AC_2}$ and repeat the procedure. If points D_1 and D_2 are close together and they are on different sides of the natural slope, a linear interpolation is permitted and the straight line $\overline{D_1D_2}$ gives point D. A semi-circle over $\overline{AD}$ cuts the surface at C; this is the end point of the sliding surface. The earth pressure is given by $E_a = \frac{pe}{2}\gamma$; its angle of direction can be calculated from Eq. (262).

The point of application of the resulting earth pressure force can be obtained with the approximate method given in *Fig. 407*. We determine the center of gravity of the sliding wedge, and, from there, we draw a line parallel to the sliding surface; this cuts the back of the wall at the point of application. For straight boundary conditions, the point of application is located in the lower third point of the height of the wall.

Since the assumption of a plane sliding surface is an approximation in itself, no exact method to determine the point of application can be given.

9.4.3 Effect of surcharge on the earth pressure

When there is a surcharge on the backfill of a wall, the earth pressure can be determined by making use of the generalized *Rebhann* theorem. In some cases, if there is a regularity in the surcharge, simpler methods are available. In the following we investigate these cases.

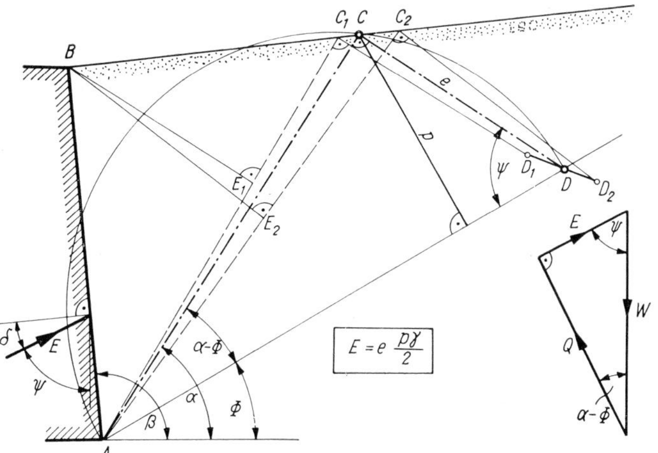

Fig. 406. Graphical construction to obtain the minimum value of earth pressure

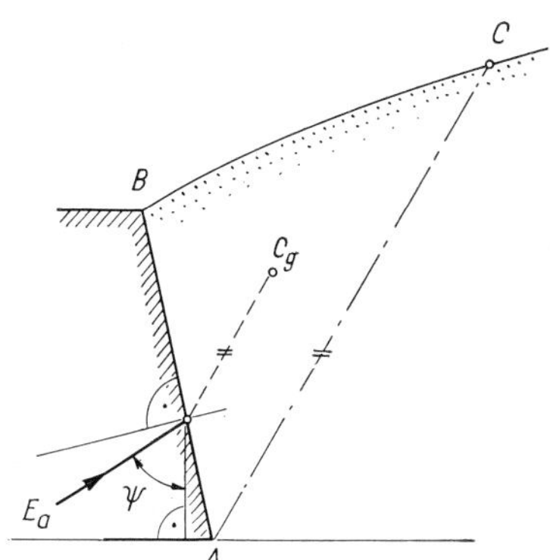

Fig. 407. Point of application of earth pressure

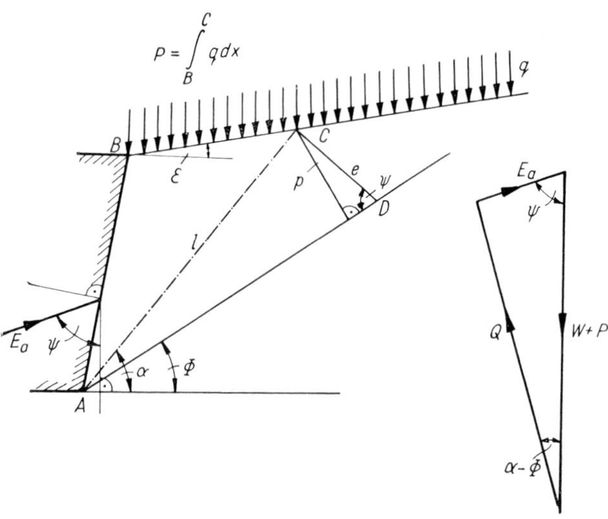

Fig. 408. Uniform surcharge on the surface of the backfill

We assume first that a uniformly distributed vertical load is acting on the surface of the backfill (*Fig. 408*). The equilibrium requires that

$$E_a = (W + P)\frac{\sin(\alpha - \Phi)}{\sin(\alpha - \Phi + \psi)},$$

where P is the sum of the load acting on the section BC. The extremum principle requires that $\partial E/\partial\alpha = 0$, i.e.

$$\frac{\partial E_a}{\partial \alpha} = \frac{\partial(W + P)}{\partial \alpha}\,\frac{\sin(\alpha - \Phi)}{\sin(\alpha - \varphi + \psi)} + (W + P)\frac{\sin\psi}{\sin^2(\alpha - \Phi + \psi)} = 0 .$$

Since

$$(W + P)\frac{\sin\psi}{\sin(\alpha - \Phi + \psi)} = Q$$

and

$$\frac{\partial(W + P)}{\partial\alpha} = \frac{dG}{d\alpha} + \frac{dP}{d\alpha}$$

$$Q = -\left(\frac{dW}{d\alpha} + \frac{dP}{d\alpha}\right)\sin(\alpha - \Phi).$$

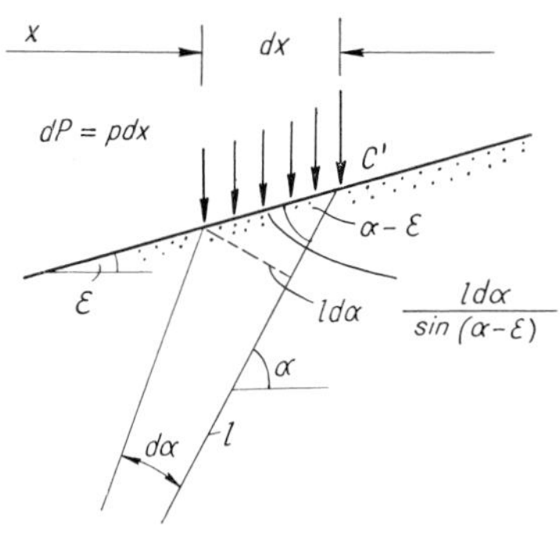

Fig. 409

Since

$$dW = -\frac{l^2\gamma}{2}\,d\alpha ; \quad dx = -\frac{l\,d\alpha}{\sin(\alpha - \varepsilon)}\cos\varepsilon$$

(see *Fig. 409*)

$$Q = \left[\frac{l^2\gamma}{2} + \frac{l\cos\varepsilon}{\sin(\alpha - \varepsilon)}\,p\right]\sin(\alpha - \Phi) = = \frac{l^2}{2}\left[\gamma + \frac{2\cos\varepsilon}{l\sin(\alpha - \varepsilon)}\,p\right]\sin(\alpha - \Phi) .$$

Introducing

$$\gamma' = \gamma + \frac{2\cos\varepsilon}{l\sin(\alpha - \varepsilon)}\,p \tag{293}$$

we have

$$Q = \frac{l^2\gamma'}{2}\sin(\alpha - \varphi) . \tag{294}$$

It can be proven, similarly to the case without surcharge, that

$$W + P = \text{area } ACD_{\triangle}\,\gamma' . \tag{295}$$

Rebhann's theorem is therefore valid; the earth pressure can be determined with the same graphical methods as without surcharge, the only difference being that γ' has to be used, instead of γ.

A further simplification comes if we express l from geometry (*Fig. 410*):

$$l = \frac{h\sin(\beta - \varepsilon)}{\sin\beta\sin(\alpha - \varepsilon)},$$

$$\gamma' = \gamma + \frac{2p}{h}\,\frac{\cos\varepsilon\sin\beta}{\sin(\beta - \varepsilon)} . \tag{296}$$

An analytical formula can also be given; from Eq. (279):

$$E_a^p = K_a\frac{z^2\gamma'}{2} = K_a\frac{h^2\gamma'}{2\sin^2\beta},$$

and the pressure at depth z:

$$e = \frac{dE_a^p}{dz} = K_a\frac{h\gamma}{\sin\beta} + K_a p\frac{\cos\varepsilon}{\sin(\beta - \varepsilon)} .$$

The stress distribution on the back of the wall is given in Fig. 410. If we lengthen the limiting line of the stress figure and make it a triangle, we may write:

$$h_e = \frac{p}{\gamma}\,\frac{\cos\varepsilon\sin\beta}{\sin(\beta - \varepsilon)} .$$

The earth pressure acting on the wall AB can therefore be calculated with the following formula:

$$E_{ap} = E_a\left(\frac{h + h_e}{h}\right)^2 - E_a\left(\frac{h_e}{h}\right)^2 = E_a\left(1 + 2\frac{h_e}{h}\right) \tag{297}$$

ribution of stresses on the back of

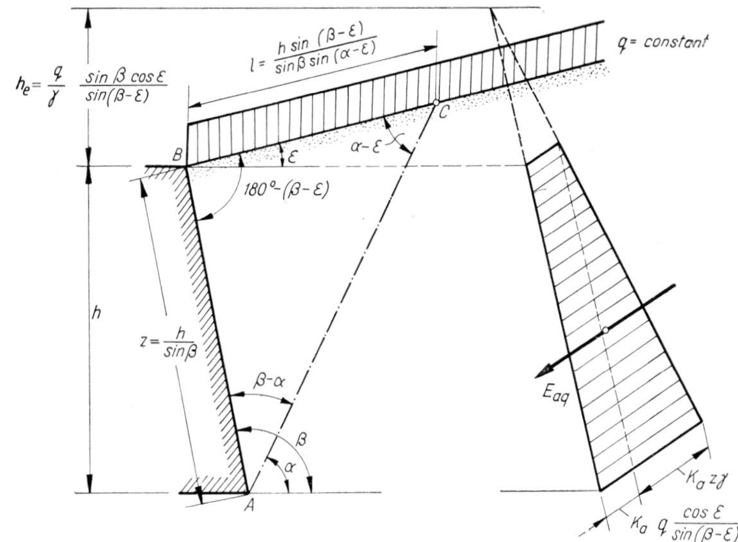

where E_a is the earth pressure acting on $\overline{AB}$ due to the soil's own weight. For a vertical wall and horizontal surface:

$$\gamma' = \gamma\frac{2p}{h};\ h_e = \frac{p}{\gamma};\ E_a^p = E_a\left(1 + \frac{2p}{\gamma h}\right) = K_a\frac{h^2\gamma'}{2}.$$

The point of application will be furnished by the line of gravity of the pressure distribution figure (Fig. 410).

If there is a *line load* acting on the surface of the backfill along a line which is parallel to the crest of the wall, the increase of the earth pressure will also depend on the distance of the load from the crest. Three sections can be distinguished. If we construct the sliding surface on the basis of the generalized *Rebhann*'s theorem, i.e. the plane surface $\overline{AC_1}$ which fulfills the condition that

$$\text{area } ABC_1\gamma + p = AC_1D\,\gamma$$

(see *Fig. 411*) we obtain the limit of the first section, $\overline{BC_1}$. For any position of the line load in section I, the sliding surface will be $\overline{AC_1}$; the value of the earth pressure is determined according to Fig. 411*b*. On the other hand, if we construct the sliding surface to the own weight, without taking the line load into consideration, the end point of this (C_2) will give the limit between sections II and III. A line load acting in section III has no influence on the earth pressure; so the latter can be determined on the basis of the own weight. In section II (C_1C_2) the sliding surface runs to the point of application of the line load. For these cases, the earth pressure can be calculated by the methods shown in Fig. 411.

The point of application of the earth pressure can be obtained by the following approximate methods. If the load acts in section II, we draw a line from its point of application C, which is parallel

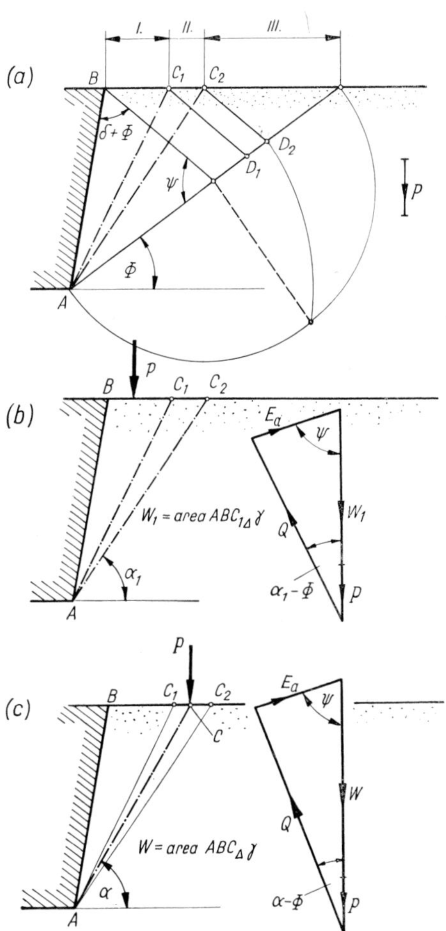

Fig. 411. Finding the boundaries of the sections on the surface of the backfill

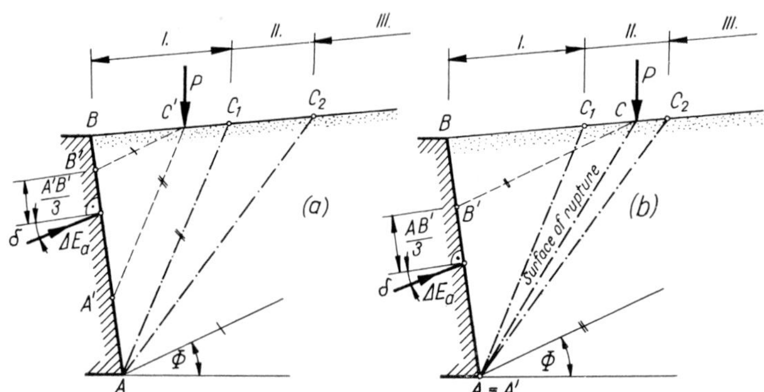

Fig. 412. Point of application of ea
the case of a line load

to the line under the angle Φ from point A (*Fig. 412*). This line cuts the back of the wall at point B. The ΔE value—which is the difference between the earth pressure due to the own weight plus the line load and the earth pressure due to the own weight only—acts then at a point which lies at a distance $\overline{AB'}/3$ from point B'. If the load acts in section I, we draw two lines from its point of application C: one making an angle Φ with the horizontal, and another, making an angle α_1, which runs parallel to the sliding surface $\overline{AC_1}$. The force ΔE acts in the upper third point of the section $\overline{A'B'}$. The point of application of the resulting earth pressure will be obtained by writing the moment equilibrium with respect to A.

If the backfill consists of several different layers, with different angles of friction, we can calculate the earth pressure by making use of the following approximation.

This case is shown in *Fig. 413*. First we determine the earth pressure due to the upper layer. In order to calculate the earth pressure acting on the lower layer we substitute the upper layer with a uniform loading of an intensity $q = h_1\gamma_1$, and, neglecting the shearing forces acting on the limit surface and in vertical planes as well, we determine the earth pressure to the height h_2. The resulting force will be obtained by vectorial summation. If we calculate the coefficients K_{a1} and K_{a2} with Φ_1 and Φ_2 respectively, the stress distribution figure can also be plotted (Fig. 413).

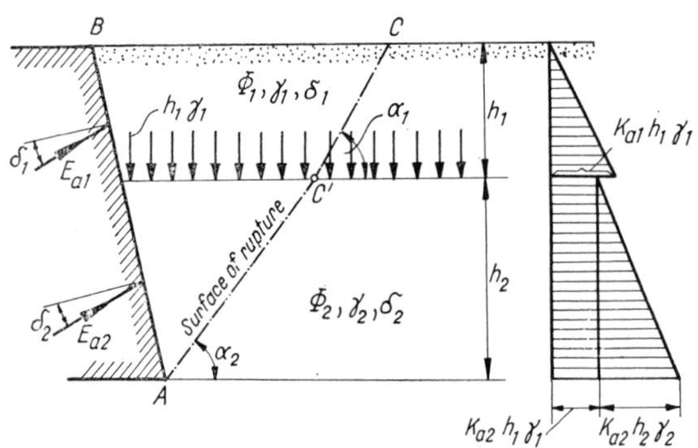

Fig. 413. The backfill consists of different layers

9.4.4 Use of curved slip surfaces

In Section 9.4.2, in connection with Fig. 390, we demonstrated the shape of the earth mass, which comes into the plastic state if the wall tilts around its toe. Experiments and theoretical investigations established the fact that the sliding surface was composed of a curved part, in the lower section, and of a plane part which corresponds to the *Rankine* state. The exact equation of the curved part has not yet been found—it is likely that the respective differential equations cannot be solved in a closed form—therefore we use approximate methods in practice. We replace the real curve by some simple one, which can be easily applied in graphical constructions. Two types of slip curves are mainly in use: the circle and the logarithmic spiral. The first one is easily drawn: in the case of the second, the radius vector and the tangent make the same angle at every point of the curve.

For retaining wall problems, it is advisable to determine the wall friction angle by direct shear tests. For this purpose we prepare a rectangular piece out of the wall material and we put it into the upper frame of the direct shear machine. On performing the shear test, we get the wall friction angle. Assuming—in the case of a tilting wall—a linear distribution of the stresses along the back of the wall; the point of application lies in the lower third point of the height. Now, we have to determine a sliding surface, composed of a curved and a straight part, which gives the maximum value of the earth pressure.

The construction is shown in *Fig. 414*, for a horizontal backfill surface.

First, we determine that part of the backfill, where the wall friction does not exert any effect. In this field, the active Rankine state will occur, since there is a uniform expansion, due to the tilting of the wall. The boundary line of this field will be obtained if we draw a line from point B of the wall which makes an angle $45 + \Phi/2$ with the horizontal. There will be a point D on this line where the straight and the curved part of the sliding surface meet. Above point D, the sliding surface is plane, with an inclination of $45° + \Phi/2$. We

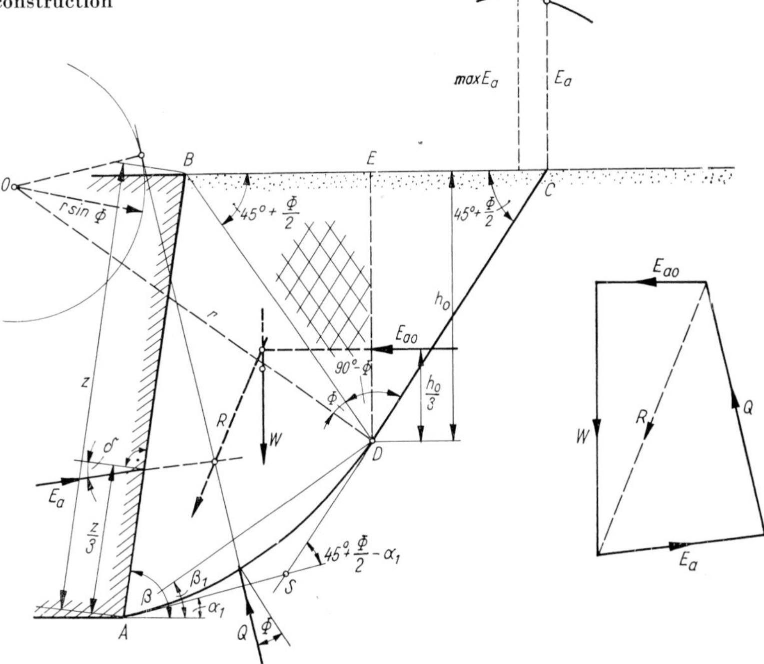

Fig. 414. Active earth pressure; graphical construction using composite surface of sliding

take an arbitrary angle α_1, which is the inclination of the tangent to the sliding surface at point A. In the case of a circle, the angle β_1 can be determined from the triangle ASD. It can be written

$$\beta_1 - \alpha_1 = \frac{45^\circ + \dfrac{\Phi}{2} - \alpha_1}{2}$$

i.e.

$$\beta_1 = \frac{45^\circ + \dfrac{\Phi}{2} + \alpha_1}{2}. \tag{298}$$

Knowing β_1, point D can be determined. The center of the circle can be obtained by drawing perpendiculars to the tangents of it.

Now, after constructing the sliding surface, we carry out the equilibrium investigation. We draw the vertical line $\overline{DE}$ at point D. This surface lies entirely in the active Rankine field, so we can get the earth pressure E_{a0} acting on it by using Eq. (250). Value, direction and point of application are thus known. Since the effect of the prism DEC is represented by the earth pressure E_{a0}, we have to deal with the equilibrium of the earth prism with the cross section $ABED$ only.

The acting forces are the following: the weight of the prism, W, the earth pressure E_{a0} on the plane $\overline{DE}$, the resulting force Q of the stresses acting on the sliding surface, finally, the earth pressure E_a, the point of application of which and direction are known; its magnitude has to be determined. E_{a0} is known, the magnitude of W is given by the surface $ABED$ multiplied by the bulk density, and the line of action passes through the center of gravity of the prism $ABED$. So, E_{a0} and W can be put together to give the resulting force R.

We have to determine the direction of Q. The surface is a sliding surface, therefore, the resulting force $dQ = q\,ds$ on every elementary part ds (see *Fig. 415*) makes an angle Φ with the normal to the surface; the components τ and σ fulfill the *Coulomb* relation $\tau = \sigma \tan \Phi$. The line of action of dQ is therefore tangent to a circle drawn around point 0, having radius $r \sin \Phi$. This circle is called the friction circle. The former statement is true for every elementary arch of the sliding surface, thus, we may assume that the resulting force Q is also tangent to the friction circle. As we shall see in the chapter dealing with

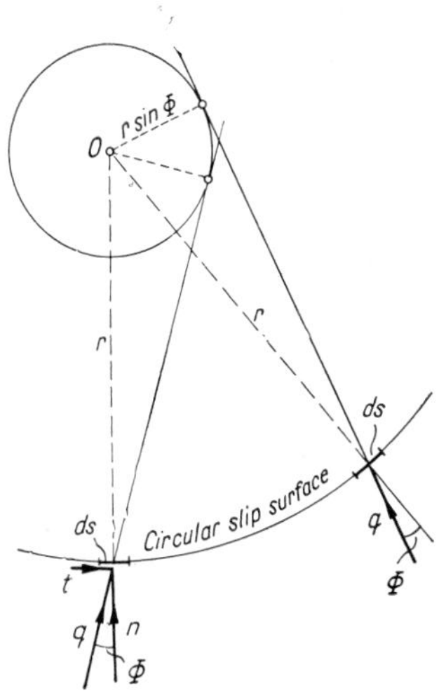

Fig. 415. Direction of stresses acting on the circular sliding surface

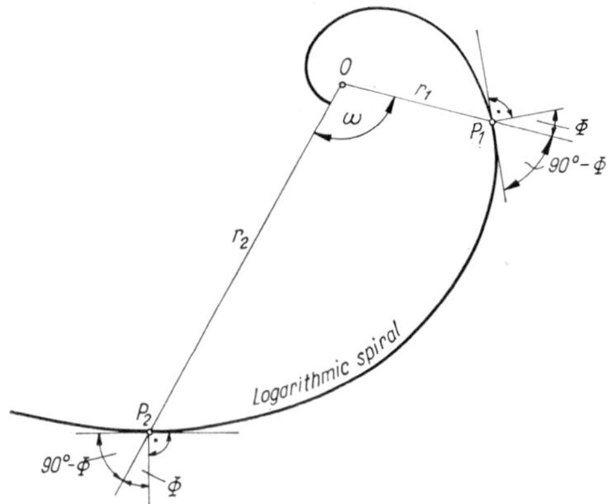

Fig. 416. Characteristics of a logarithmic spiral

the stability of slopes, this is an approximation only; however, it gives an earth pressure that is on the safe side.

Based on the above considerations, it is possible to resolve the force R into the forces E_a and Q; Q goes through the point of intersection of E_a and R and it is tangent to the friction circle. The polygon of forces can thus be completed to give the force E_a. Assuming another value for α_1 and repeating the construction, we obtain different earth pressure values; if we plot E_a, by selecting a suitable scale, above each point C, which is the uppermost point of the sliding surface, we are able to draw a curve, the maximum of which furnishes the earth pressure value sought. This maximum value will be considered as final for the direction δ. By performing the construction with several different values of $\delta \sphericalangle$, we see that there is a minimum value of the individual maxima pertaining to a given $\delta \sphericalangle$, so we have the same situation; the final sliding surface must fulfill the conditions

$$\frac{\partial E}{\partial \alpha_1} = 0 \quad \text{and} \quad \frac{\partial E}{\partial \delta} = 0\,,$$

respectively. Usually, we do not determine this earth pressure, but assume a suitable angle of direction $\delta \sphericalangle$. The best way is to take the result of the experiment described above.

The construction can be carried out also by using a logarithmic spiral for the curved part of the sliding surface.

The equation of a logarithmic spiral can be written, based on the properties shown in *Fig. 416*, in the following form:

$$r_2 = r_1 e^{\omega \tan \Phi}\,. \tag{299}$$

Therefore, if we assume a logarithmic spiral for the sliding surface, the resulting stress, acting on an elementary arch, goes through the center 0 of the spiral; also the resultant of elementary forces dQ goes through this point. By making use of this property, the construction can be easily carried out.

The construction is shown in *Fig. 417*. Because it has to be repeated several times to find the maximum value of the earth pressure, it is convenient to cut a sheet of carboard paper accord-

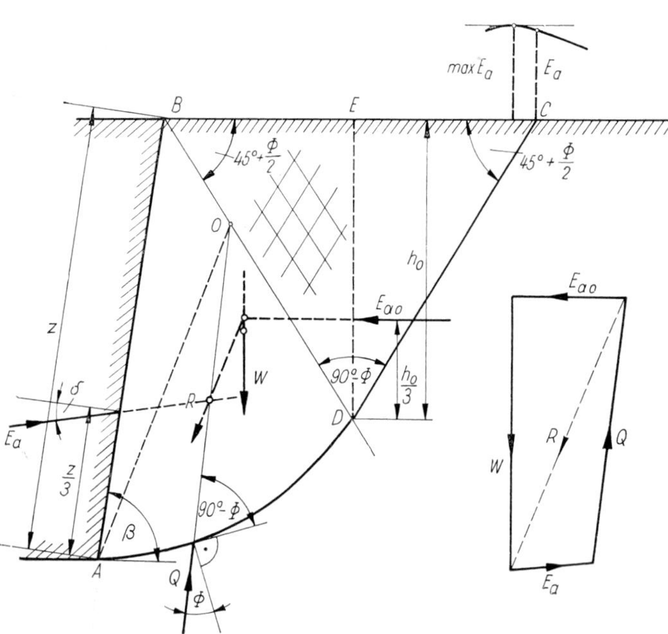

Fig. 417. Active earth pressure; graphical construction using logarithmic spiral for the curved part of the sliding surface

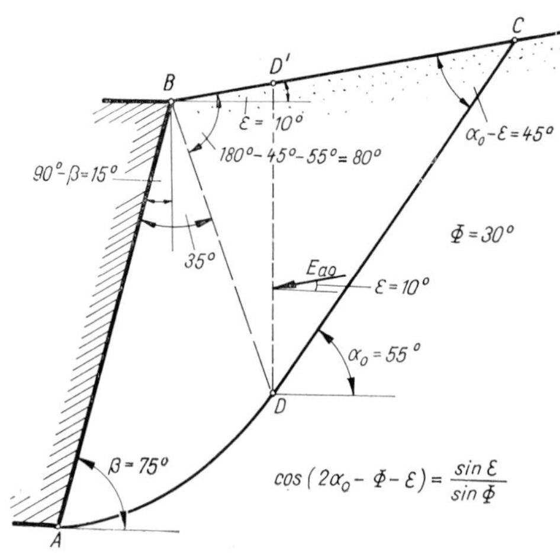

Fig. 418. Sliding surface for sloping backfill

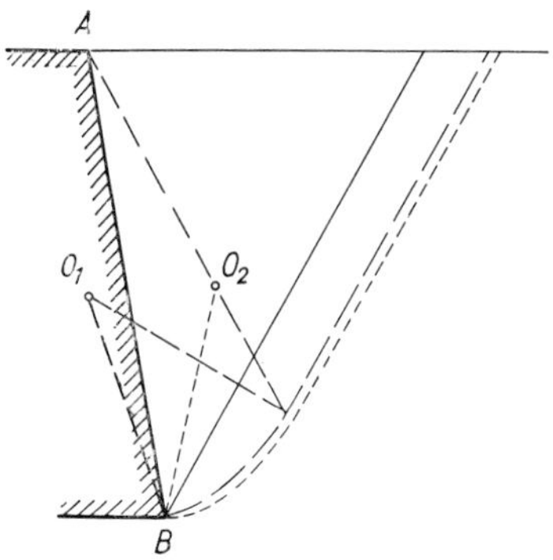

Fig. 419. Variant slip surfaces in the numerical example

ing to the logarithmic spiral for the given value of Φ. First, we draw again the straight line $\overline{BD}$, making an angle $45° + \Phi/2$ with the horizontal. The spiral cuts this straight line at a certain point D, and has to be tangent to the straight section $\overline{CD}$ of the sliding surface. It also goes through point A. So, we take an arbitrary point for the center of the spiral on the line $\overline{BD}$, and put the point 0 of the cardboard spiral on it. By turning the cardboard we easily find the location where the spiral goes through point A; the point of intersection with $\overline{BD}$ furnishes point D. After having constructed the sliding surface, the procedure is essentially the same as in the case treated above. The resultant R goes through the point of intersection of E_{a0} and W, the force R cuts the line of action of the force E_a; this point will

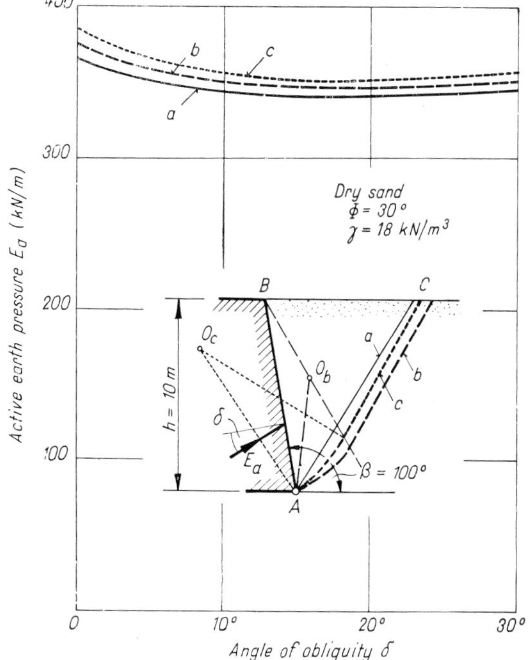

Fig. 420. Active earth pressure for different sliding surfaces versus angle of wall friction:
a — plane rupture surface; *b* — logarithmic spiral; *c* — circular arc

be connected with the center of the spiral. This is the line of action of Q. Resolving the force R into Q and E_a, as shown in the figure, we get the magnitude of E_a. Assuming several points 0, and drawing the approximate spirals, we get different E_a values, the maximum of which furnishes the final earth pressure sought.

It should be mentioned that the procedure is essentially the same also for the case when the terrain is not horizontal, except that we have to determine the angle of inclination α_0 of the plane section of the sliding surface; accordingly, by using the formula (269), the straight line $\overline{BD}$ makes an angle $90° - \Phi$ with this line. The earth pressure E_0 can be calculated by Eq. (262). *Figure 418* shows a sliding surface for a numerical example.

It is clear from the preceding considerations that the determination of the earth pressure with curved surface of sliding is much more time-consuming; it requires much more construction work than the method using plane surface of sliding. Therefore, the question arises whether the use of curved sliding surfaces is absolutely necessary, and whether the differences in the E values are so great that either economical or safety reasons call for the use of the latter. Besides, we have to consider that, due to the errors in the laboratory testing of the soils, and to the inherent heterogeneities of the material, the values Φ and γ are not exact; therefore, it is not worth-while to use extremely refined calculation methods.

The results of numerical investigations for the retaining wall given in *Fig. 419* are shown in *Fig. 420*. Here, earth pressure values for a reclining retaining wall ($\beta = 100°$) having a height of $h = 10.0$ m are given. The physical characteristics of the soil are: $\gamma = 18$ kN/m³; $\Phi = 30°$.

The earth pressure is given as a function of the angle of direction, for three different types of sliding surface: plane; plane + circle; plane + logarithmic spiral. It can be clearly seen that the differences between these forces are very small; the values obtained with composite sliding surfaces are slightly greater; however, the differences amount to only $2 \div 3\%$. It can be concluded that, for all practical purposes, the active earth pressure of cohesionless soil may be calculated by assuming a plane surface of sliding. This is not the case, however, with passive earth pressure.

9.5 Pressure of cohesive soils

9.5.1 General considerations

The treatment of earth pressure problems for soils without cohesion has shown that theoretical investigations and experimental results are available for this case; graphical constructions and numerical methods serve to calculate the earth pressure. Earth retaining structures can be designed for such soils safely and economically. This is not the case for cohesive soil, for the following reasons. First, the earth pressure depends heavily on the shear strength of the soil, and—as we emphasized in Ch. 7—the exact determination of the shear strength of clays has not been worked out yet in Soil Mechanics. Second, the amount of deformation that is required to produce the active state in an earth mass, cannot be defined unequivocally; it is not possible to give the exact relation unequivocally; it is not possible to give relation between stress and deformation for cohesive soils, as we did for dry sands (see for example, *Terzaghi's* tests, Fig. 401). If a sliding surface occurs in the cohesive earth mass, and the shear strength along it remains constant, then we can count on the constancy of the earth pressure as well. However, if the shear stresses in the earth mass exceed the fundamental shearing strength of the soil, then the shearing stress will be constant only if there is a continuous slow deformation; the wall tilts slowly but continuously outwards.

If this movement cannot take place, and therefore no shear deformations can occur on the potential sliding surface, then the shear stresses decrease and, at the same time, the earth pressure increases in order to maintain the equilibrium.

Retaining walls for cohesive earth masses have to be designed on the basis of the active earth pressure only, if the continuous, slow deformations may take place without damaging the wall or some adjacent structures, or, if the wall is not so rigid or not so rigidly connected to other structures that the deformation required to maintain the smaller active earth pressure cannot take place. If we design the wall to withstand the active earth pressure, then we have to take the occurrence of such deformations into consideration.

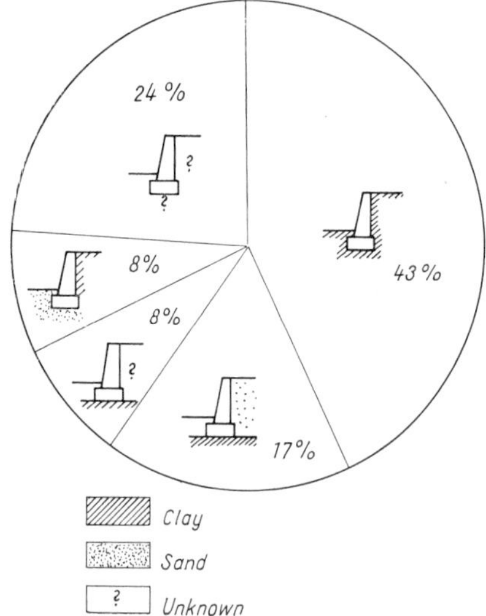

Fig. 421. Statistical data on damaged retaining walls

This was not the case with cohesionless soils, since there was a unique relationship between deformation and shear stress.

If no deformations are allowable, the wall has to be designed to withstand the earth pressure at rest. Then, the distribution of the horizontal stresses is hydrostatic, and their value may reach 0.9 times the vertical overburden pressure; the coefficient of earth pressure is thus very close to unity. Even a wall designed to withstand this earth pressure will perform appreciable movements if the backfill is exposed to atmospheric effects. In such circumstances the temperature and the moisture content of the upper layers will be subjected to seasonal changes, and shrinkage and swelling may occur. It is possible that the earth pressure will be even greater than the earth pressure at rest. This is the case when the moisture content of the backfill increases and the backfill swells. The developing swelling pressure may be even greater than the passive one; its magnitude cannot be determined by earth pressure theories. In order to avoid the occurrence of swelling pressure, care should be taken that good drainage of surface water from the area behind the crest of the wall is provided; we have to prevent the entrance of water into the backfill. Also the backfill itself has to be drained with French drains.

The importance of the above considerations is emphasized by some statistical data collected by IRELAND (1964). He investigated many walls that suffered damages; the data have shown (see *Fig. 421*) that 68% of these walls was founded on clay; 51% (8 + 43) had clay backfill. If we consider further that also some "clay" cases were present in the 24% where neither the type of foundation soil nor the backfill was known, we have to conclude that retaining walls founded on clay or having a clay backfill are much less safe than generally assumed. For practical purposes it can be concluded that retaining walls have to be designed on the basis of the fundamental shearing strength of the clay, which amounts to 40 to 50% of the failure strength. Furthermore, it is advisable to avoid the use of clay as a backfill.

9.5.2 Active earth pressure of cohesive backfill

The magnitude of the earth pressure acting on a vertical plane in the semi-infinite half space in the *Rankine* state was given by Eq. (217). This value is identical with the result furnished by *Coulomb*'s theory taking a plane sliding surface, if we assume that the cohesion as a part of the shearing resistance is evenly distributed on that sliding surface $\overline{AC}$ (see *Fig. 422*). According to these results, the distribution of the earth pressure can be represented by a double triangle; there is tension in the upper part of the earth mass and there is no pressure acting if the height of the wall us smaller than

$$h_0 \leq \frac{4c}{\gamma} \tan\left(45° + \frac{\Phi}{2}\right). \qquad (300)$$

This height is called the cohesive height; a vertical earth wall will be stable without any support if its height is smaller than this value. If $h > h_0$, a trapezoidal part of the pressure diagram will produce the earth pressure; the resulting force is horizontal and its line of action is located at the height of

$$m_0 = \frac{h}{3} - \frac{h_0}{6h}(h + h_0)\,. \tag{301}$$

The application of these results to earth pressure problems, and to tilting retaining walls cannot, however, be accepted. As the upper part of the backfill is in a state of tension, vertical fissures will soon occur, resulting in a reduction in the cohesive to h_0'. As we shall see in Vol. 2, this height is equal to $\frac{2}{3}h_0$. In the lower section of the back of the wall, adhesion and friction will be acting, thus, the earth pressure will make an angle δ with the horizontal. Considering these facts, the simplest way to determine the earth pressure due to cohesive backfill is the following.

We assume that the fissures reach depth h_0' and this upper part of the earth mass acts as a uniformly distributed load on the having the height $h - h_0'$ (*Fig. 423*). A plane surface of sliding will be considered; then, the following forces act: C, the weight of the earth mass $ABCD$, the forces due to friction and adhesion, respectively, on the surface $\overline{AD}$; these are Q and $K = cl$; finally, the earth pressure E_a, and the adhesion $A = a(h - h_0')$ on the section $\overline{AB'}$. An arbitrarily selected sliding plane reaches the depth of the potential fissures at point D_1. We investigate the equilibrium of the earth mass ABC_1D_1. The weight W is given by area $ABC_1D_1 \cdot \gamma$; K acts parallel to $\overline{AD_1}$, and its magnitude equals cl, so it can be plotted in the polygon of forces. The same is true

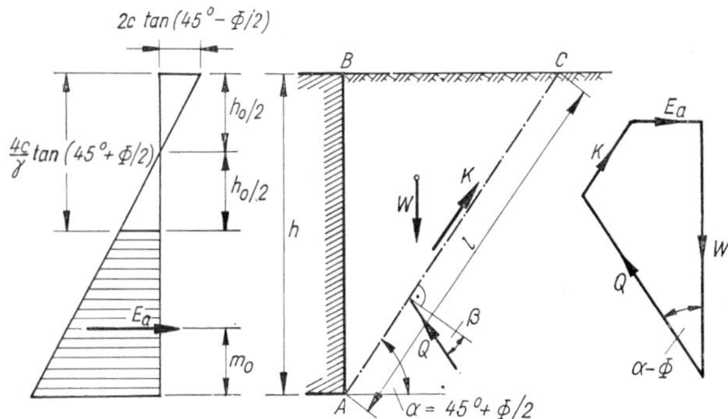

Fig. 422. Coulomb's earth pressure theory for cohesive soil

for $A = a(h - h_0')$; the resulting force R has to be resolved into E and Q, whose directions are known. Repeating the graphical construction for several different sliding surfaces and plotting the obtained earth pressure values as shown in Fig. 422, the maximum value can be obtained.

If surface drainage is not perfect or if the surface of the clay backfill is not covered with a layer of sand or topsoil, we have to consider the possibility that the fissures become filled with water during heavy rains. In this case, a horizontal water pressure will act on the wall of the fissure. This force has to be included into the investigation of the equilibrium.

In order to determine the point of application of the earth pressure, the shape of the pressure diagram has to be known. There are pressures on the back of the wall below depth h_0' only, because in the upper part the tensions and pressures are in equilibrium. Assuming the zero point of the pressure diagram to be located at depth $h_0/2$ (point *3* in Fig. 422), and drawing an arbitrary line through this point (line *3 — 2*), we get the

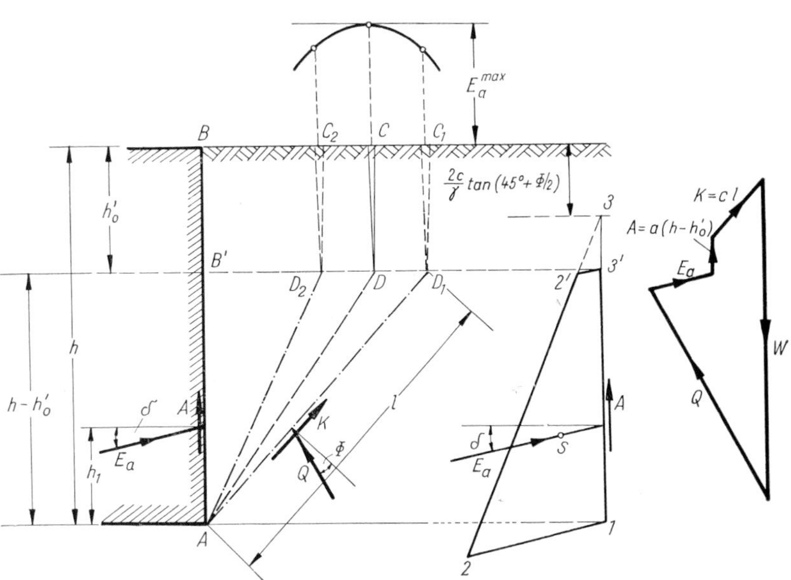

Fig. 423. Determination of earth pressure for cohesive backfill

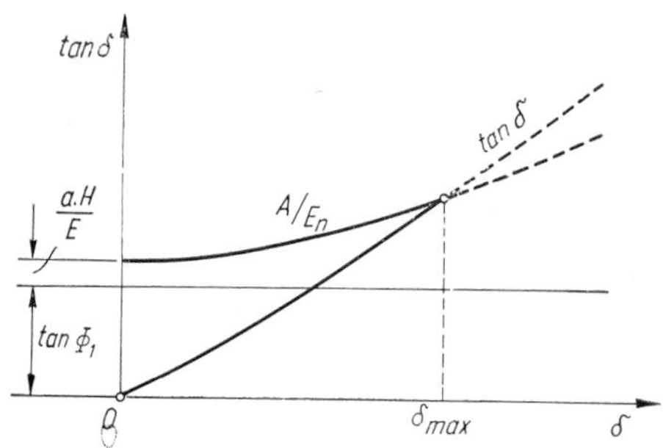

Fig. 424. The upper limit of the angle of wall friction if adhesion is acting

triangle *1 2 3*. We cut from the upper part the triangle *3 2' 3'*, to depth h_0'; the remaining part *1 3'2'2* represents the pressure diagram on an arbitrary scale. Through the center of gravity of this diagram we draw a line parallel to the direction of the earth pressure; it cuts the line *1—3* at the height of the point of application.

Some authors are of the opinion that, for the sake of safety, it is advisable to assume that the soil is incapable of carrying tensions. If we do so, the pressure diagram will be represented by the triangle *1 2 3*. However, in the course of the investigation we considered the possibility of tension cracks; besides, the earth wall of height h_0' is always stable, and does not exert pressures on the back of the wall; therefore, the method of calculation here presented is safe enough to be acceptable. This is not the case, however, for soft plastic clays, because their fundamental shearing strength is very low and plastic deformations will occur. Thus, no gap will exist between wall and earth mass, and, if further development of the plastic deformation will be prevented by the wall itself, also the upper part of the wall will get pressures.

The angle of direction of the resultant of the forces of adhesion and earth pressure has some limitations. If φ is the angle of friction between wall and earth, and A is the adhesion force, then the normal and the tangential component of the earth pressures are related by the following equation:

$$E_t = E_n \tan \varphi + A = E_n \tan \delta$$

i.e.

$$\tan \delta = \tan \varphi + \frac{ah}{E_n}. \tag{302}$$

Plotting Eq. (302), we get the maximum possible value for δ (*Fig. 424*).

If we wish to assume a curved surface of sliding, we may proceed similarly to the case of noncohesive backfill. In the following, we point out the differences only. We take a circular surface of sliding (*Fig. 425*).

Investigating the equilibrium of forces, we have to consider the cohesion as well. We take an elementary part ds of the curved section of the sliding surface; the elementary cohesive force $dK = c\,ds$ (*Fig. 426*) acts there. According to this equation, the total arch represents, on a certain scale, the polygon of the elementary cohesive forces; therefore, the resulting force runs parallel to chord $\overline{AB}$. Its magnitude is equal to $K = cl$. In order to determine the distance z measured from the center of the sliding surface, we write down the sum of the moments of the elementary forces and equate it to the moment of the resulting force K. For instance

$$\int_0^L r\, c\, ds = K z\,.$$

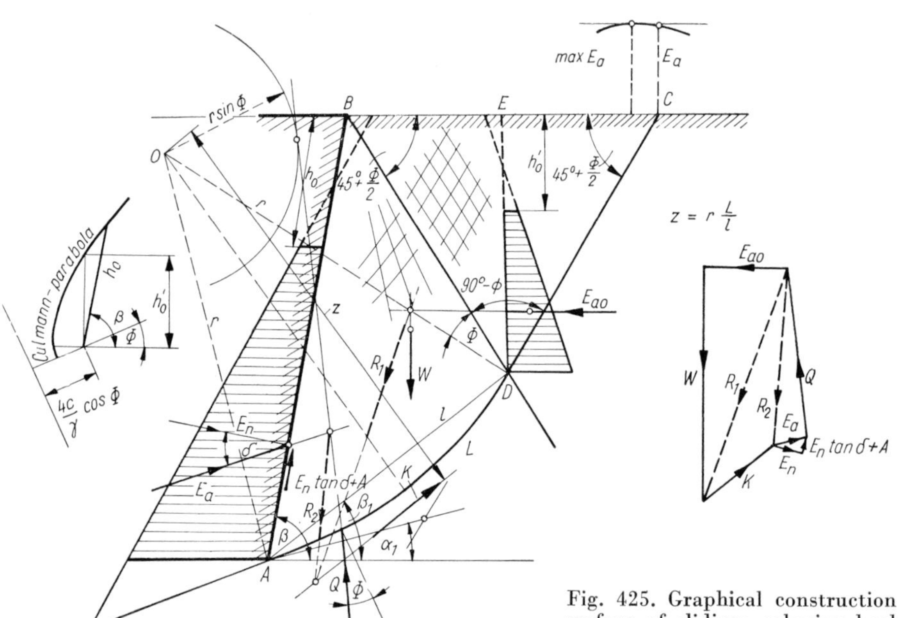

Fig. 425. Graphical construction of active earth pressure; circular surface of sliding; cohesive backfill

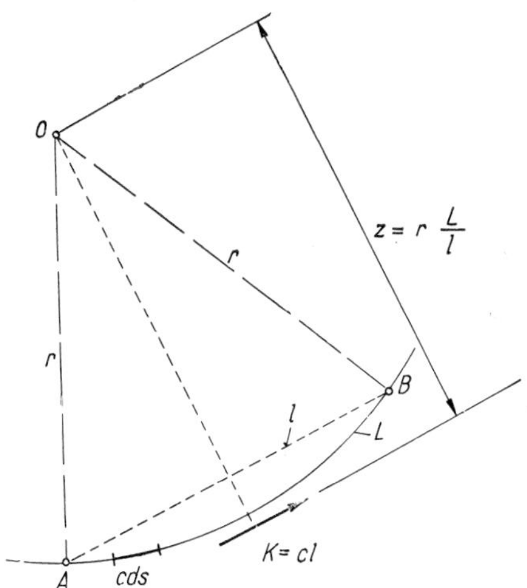

Fig. 426. Magnitude, direction and point of application of cohesive force

Substituting $K = cl$ and integrating:

$$r\,c\,L = c\,l\,z$$

and

$$z = r\frac{L}{l}. \tag{303}$$

Thus, the force K became completely known. The pressure distribution diagram of the force E_{a0} is the trapezoidal figure which is well known from *Coulomb*'s theory. To determine the point of application of the earth pressure E_a, which acts on the slanting back $\overline{AB}$ of the wall, we have to determine the height h_1 up to which the earth mass remains stable without any support. This can be given by using *Culmann*'s parabola (see Vol. 2). The zero point of the pressure diagram is located at depth $h_1/2$; through this, we draw an arbitrary line, it represents, on a certain scale, the limiting line of the pressure diagram. Now, the center of gravity of this diagram furnishes the point of application. Otherwise, the procedure requires the adding and resolving of forces, the details of which can be seen in Fig. 424.

9.6 Passive earth pressure

In a somewhat broader sense, passive earth pressure means the resistance of a certain earth mass to sliding or failure when horizontal forces are acting. This horizontal force will be produced for instance by the foundation of a retaining wall, the buried part of a sheet pile wall, or the abutment of an arch bridge; but also vertical forces, acting on earth masses, may cause horizontal stresses in the vicinity. Passive earth pressure is thus acting, when due to the presence of a wall or another construction, eventually an earth mass is pressed against a neighbouring earth mass. Its magnitude is always equal to the acting forces and in the case that this force overcomes the internal resistances of the soil, a sliding, a failure state will occur; an earth mass begins to move on a sliding surface (*Fig. 427*), and the outer force cannot be increased further.

According to the preceding definition the stability of every earth-retaining structure and the bearing capacity of every shallow foundation are determined by the magnitude of a certain passive earth pressure, thus its determination is of great importance.

We investigated, in Ch. 9, the plastic equilibrium of the semi-infinite half space, and determined the state of stress also in the case of a uniform compression, i.e. in the *passive* state. These results are identical with the results furnished by *Coulomb*'s theory for the passive case, since the boundary conditions are the same. Following *Coulomb*'s basic principles, the researchers assumed a plane sliding surface to determine the passive earth pressure. As we shall see, this approximate assumption furnishes earth pressure values with errors that are never on the safe side. However, for cases where the wall friction angle is small, the sliding surface is just slightly curved and the error is acceptable. For greater values of δ the assumption of a plane sliding surface leads to erroneous results, and a curved sliding surface has to be used.

If we assume a plane sliding surface, the investigation leads to the following results.

We have to determine the force which has to be applied to a retaining wall in order to produce failure in the adjoining earth mass by rotating the wall around its toe, toward the earth mass. Assuming a plane surface of sliding, failure occurs when an earth wedge (*ABC* in *Fig. 428*) moves upward and the frictional forces are fully mobilized on the sides $\overline{AB}$ and $\overline{AC}$. The direction of the movement determines the direction of the frictional resistances which always act in the opposite direction.

The position of the critical sliding surface can be found by using the same procedure as in the active case; however, the different direction of the frictional forces has to be considered. For a given value of the wall friction angle δ, the critical sliding surface will be that which furnishes a minimum for the passive earth pressure; which is in contrast with the active case where we deter-

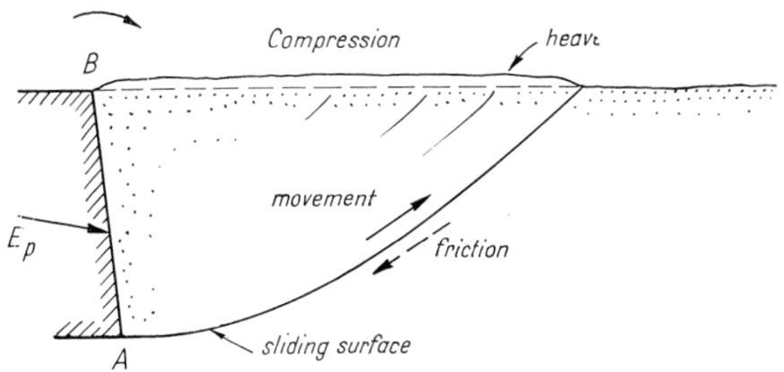

Fig. 427. Retaining wall tilting toward the backfill; mobilization of passive earth pressure

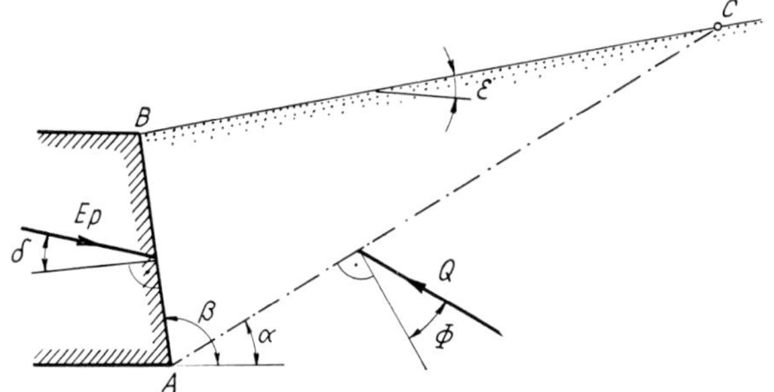

Fig. 428. Passive earth pressure; plane surface of sliding

mined the maximum value. The critical sliding surface makes a much smaller angle with the horizontal and its magnitude is several times greater than the active earth pressure.

An analytical formula may be obtained if we change the signs of the appropriate variable in the formulae for the active pressure, i.e. if we write $-\Phi$ and $-\delta$ instead of $+\Phi$ and $+\delta$. So, in the case represented in Fig. 427, we obtain the following:

$$E_p = K_p \frac{z^2 \gamma}{2}, \tag{304}$$

wherein

$$K_p = \left[\frac{\sin(\beta + \Phi)}{\sqrt{\sin(\beta + \delta)} - \sqrt{\frac{\sin(\delta + \varphi)\sin(\Phi + \varepsilon)}{\sin(\beta - \varepsilon)}}} \right]^2. \tag{304a}$$

For $\beta = 90°$, $\delta = 0$ and $\varepsilon = 0$ we get *Coulomb*'s formula:

$$K_p = \tan^2\left(45° + \frac{\Phi}{2}\right). \tag{252}$$

The angle of wall friction δ may sometimes reach its maximum value, i.e. the limiting angle of friction φ between the earth mass and the back of the wall; it is, however, in general smaller than that. For the sake of safety it is advisable not to take values greater than $\Phi/3$ when working with a plane surface of sliding. For a given value of δ, the passive earth pressure can be determined by making use of ***Poncelet***'s graphical construction. This is similar to the method used for the active case. We have to consider, again, the change of the sign only, i.e. the appropriate angles have to be plotted on the opposite side of the normals. The graphical construction is shown in *Fig. 429*.

If we assume curved sliding surfaces, experience shows that in the cases where a mutual displacement occurs between earth mass and wall, so that considerable friction will be mobilized along the back of the wall, it will take the shape shown in *Fig. 430a*. The upper part is more or less near plane, but the lower part is definitely curved. The curved part is concave from above, if the wall rotates and moves downward; it is convex, when the wall moves upward. In practice, we generally deal with case IIa shown in Fig. 390.

Figure 431 shows the determination of the passive earth pressure for cohesionless backfill, using—for the lower part—a circular surface of sliding.

We draw first the sliding surface for the Rankine state, from point B; it makes an angle $45° - \Phi/2$ with the horizontal. We assume arbitrarily point D; this is the end point of the curved part of the sliding surface. Drawing line $\overline{AD}$, we measure the angle β_1, and calculate $\alpha_1 = 45° - \dfrac{\Phi}{2} - 2\beta_1$. Thus, it is possible to construct the circle. We investigate the equilibrium of the mass $ABED$. In the field BDC, the passive Rankine state prevails; therefore, the force acting on the vertical plane $\overline{ED}$ can be calculated. ($E_{0p} = \dfrac{1}{2}\, h_0^2 \gamma K_p^R$.) W is the weight of the mass $ABED$; its line of action goes through the center of gravity of this area. The resultant of W and E_0 is R. We assume now, by judging the degree of mobilization of frictional resistance along AB, the angle δ; the

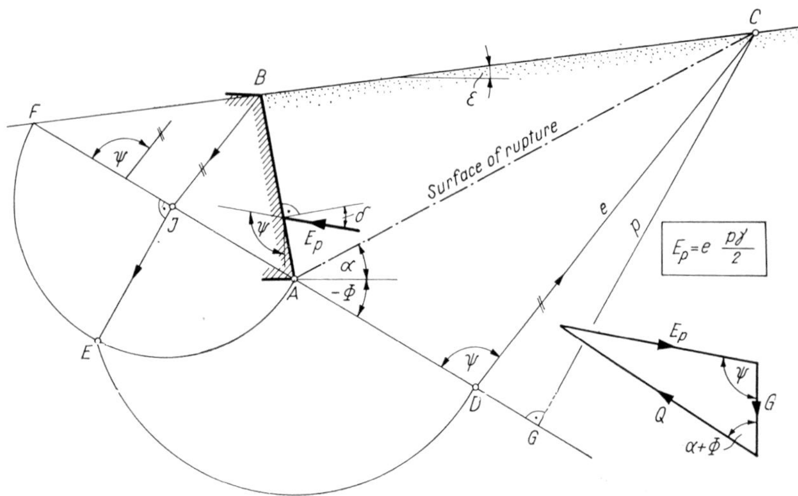

Fig. 429. Passive earth pressure; Poncelet's graphic construction

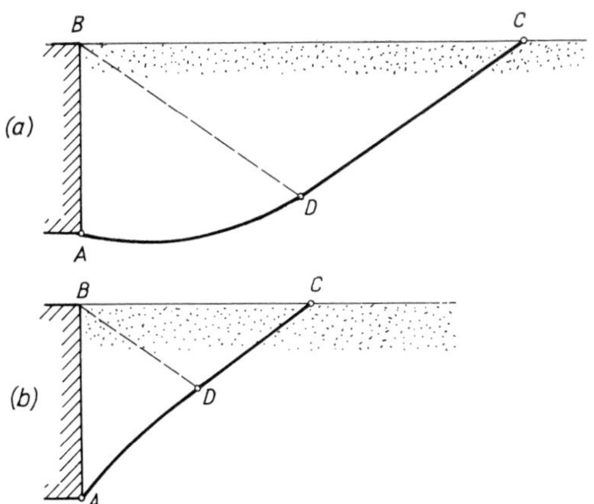

Fig. 430. Development of sliding surfaces in passive state: *a* — rotation about the toe with upward movement of the sliding block; *b* — rotation about the toe and an additional upward movement of the wall

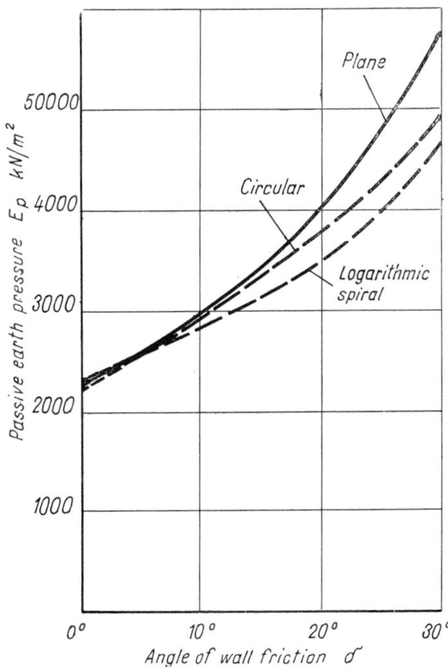

Fig. 432. Passive earth pressure for different forms of surface of sliding versus angle of wall friction

point of application is at the lower third point. E_p and Q being the components of R, we determine the point of intersection of E_p and R; through this point goes the force Q, which is the reaction on the sliding surface, AD. Since the friction here is fully mobilized, Q is tangent to the friction circle around O, with radius $r \sin \Phi$. So, the line of action of Q can be drawn (care has to be taken in selecting the correct side of the circle; the friction acts always against the movement) and R resolved into the components Q and E_p.

After having determined E_p, we select another point D and repeat the construction; the extremum (minimum) value of E_p will be furnished by interpolation.

If we wish to use a logarithmic spiral for the curved part of the surface of sliding, we can again make use of the cardboard spiral which was described in connection with the active earth pressure. The other steps of the construction are identical with the above.

If we determine the passive earth pressure of sand ($\Phi = 30°$) acting on a wall of 10 m height, and we use plane, circular and logarithmic spiral sliding surfaces, and we plot the results versus the angle of wall friction, we obtain the diagram

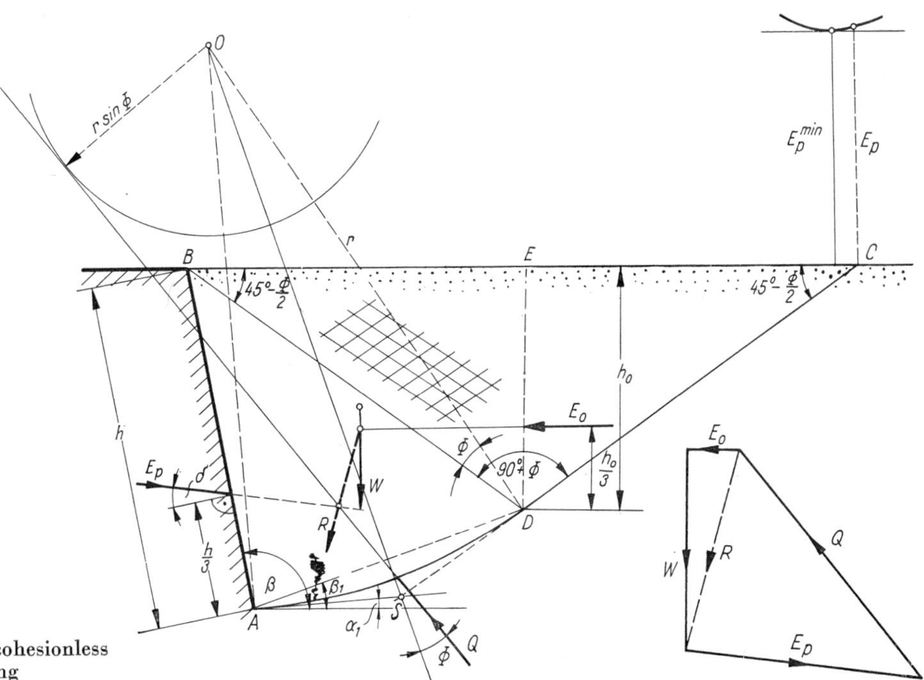

Fig. 431. Passive earth pressure; cohesionless backfill, circular surface of sliding

in *Fig. 432*, from which the following general conclusions can be drawn.

1. The passive earth pressure increases with increasing wall friction angle; it has no extremum value with respect to $\delta \sphericalangle$. The rate of increase is rather high.

2. If $\delta \leq 1/3\, \Phi$, the difference between the values obtained with plane and with curved sliding surface, respectively, is very small; in these cases we may use a plane sliding surface which gives a much simpler construction.

3. If $\delta > \Phi/3$, the earth pressure obtained with curved sliding surface is significantly smaller, the difference may amount to $15 \div 20\%$. In these cases, it would be dangerous to assume a plane, since the real surface will be curved, so, this has to be considered.

4. With regard to the magnitude of the earth pressure, it is irrelevant whether a circle or a logarithmic spiral will be used.

If the surface of the backfill is not horizontal, the same procedure can be used, only the inclination of Rankine's plane has to be determined accordingly.

For cohesive soils, it is advisable to use composite sliding surfaces. The passive earth pressure can be given as the sum of two terms; the first comes from friction, i.e. from the own weight of the earth mass, the second, from cohesion. In the *Rankine* state of stresses, the passive earth pressure is given by

$$E_{0p} = \frac{h^2\gamma}{2} \tan^2\left(45^\circ + \frac{\Phi}{2}\right) + 2ch \tan\left(45^\circ + \frac{\Phi}{2}\right).$$

The distribution of stresses is shown in *Fig. 433*. It can be seen that the part due to friction gives a triangular stress distribution; therefore, the resulting horizontal force acts at the lower third point of the height *h;* the second part, which is

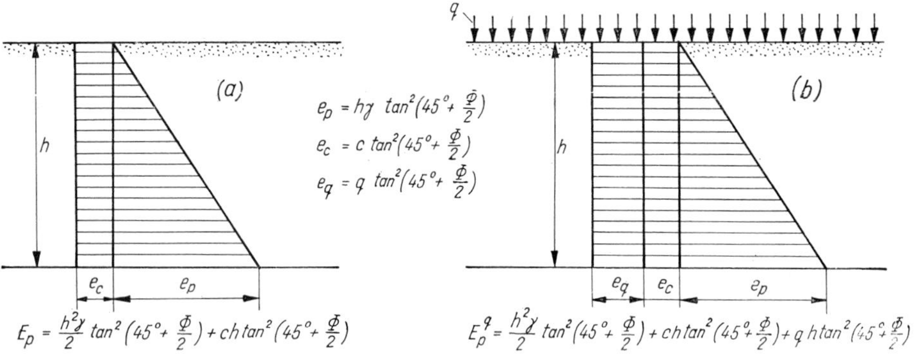

Fig. 433. Stress distribution in the passive Rankine state, in cohesive soil

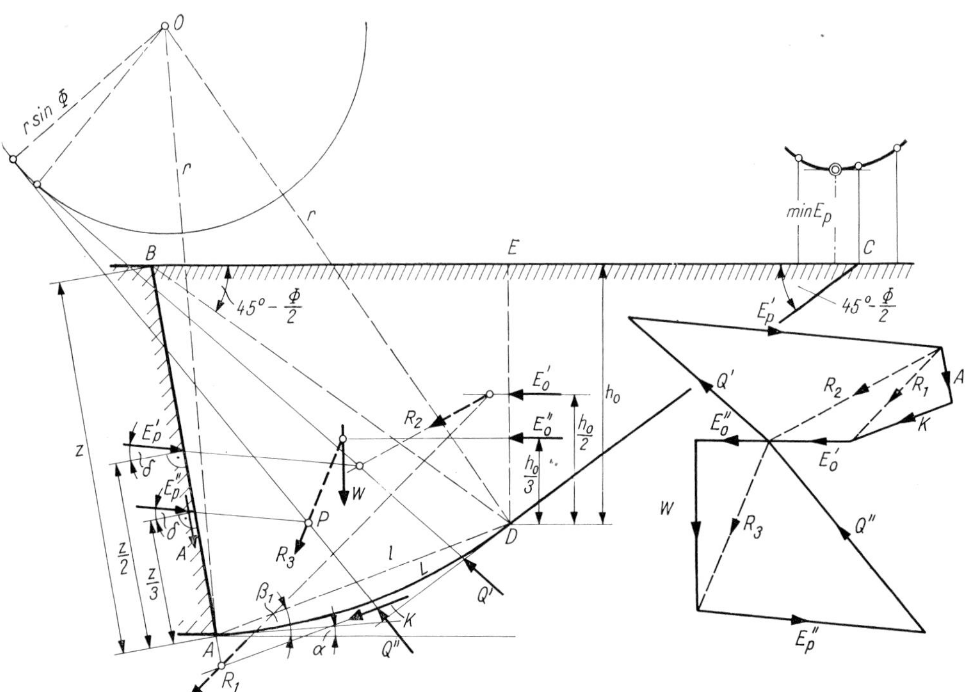

Fig. 434. Graphical construction to determine the passive earth pressure, for cohesive backfill

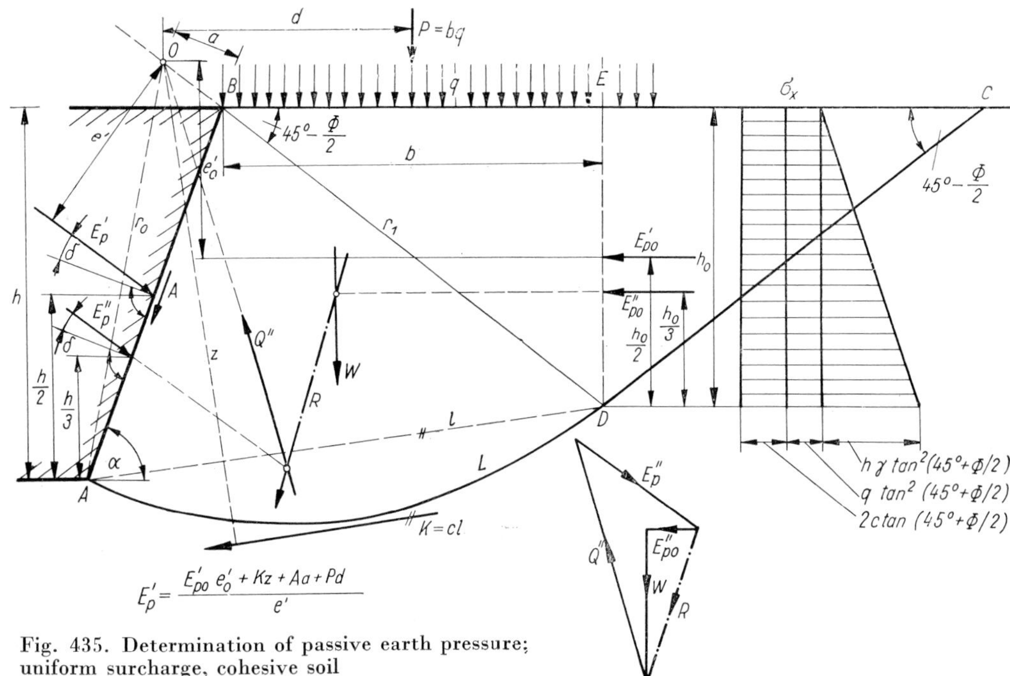

Fig. 435. Determination of passive earth pressure; uniform surcharge, cohesive soil

due to cohesion, is uniformly distributed. Its resultant acts at mid-height. We assume that the resulting passive earth pressure force on the back $\overline{AB}$ of the wall, acting at the wall friction angle $\delta\sphericalangle$, has a similar distribution. With this assumption, it is possible to arrange the construction so that we determine the magnitude of both parts separately. The resulting force will be obtained by vectorial addition.

We show the graphical construction for a circular surface of sliding.

The shearing resistance of the soil is characterized by the relationship $\tau_s = \sigma \tan \Phi + c$. Along the back of the wall the distribution of the shear stresses is governed by the equation

$$\tau_1 = \delta_1 \tan \delta + a\,.$$

Herein δ is the wall friction angle and a the specific value of the adhesion between soil and wall.

The sliding surface has again two parts: a circular section and a plane section. The latter makes an angle $45° - \Phi/2$ with the horizontal (*Fig. 434*). We draw, from point B, the straight line with the inclination of $45° - \Phi/2$; the joining point of the two sections will be located on this line. We assume this arbitrarily; we measure the angle β_1, of the chord, and calculate the angle of the tangent at point A. Now, the circle can be constructed.

We determine the resulting passive earth pressure, as mentioned above, in two parts. First, we assume a cohesive but weightless material ($\gamma = 0$); then, the frictional part is equal to zero. The passive earth pressure E_p', which is due to the cohesion, will act at mid-height of the wall, at an angle δ. We calculate the earth pressure from cohesion on the vertical plane $\overline{DE}$:

$$E_0' = 2ch_0 \tan (45° + \Phi/2);$$

it acts horizontally at mid-height of h_0. The resultant force of the cohesion acting along the circular arch $\widehat{AD}$ is given by

$$K = cl\,,$$

its line of action is located at a distance z_c from point 0:

$$z_c = r\frac{L}{l}\,.$$

Herein, r is the radius of the circle, $L = \overline{AD}$ and $1 = \overline{AD}$ the length of the arch and of the chord, respectively. There is an adhesion acting along the back of the wall: its magnitude is given by $az = A$. Adding vectorially the forces K and A, we get R_1, and the sum of R_1 and E_0 gives R_2. Now, we have to resolve R_2 into E_p' and Q'. The former acts at the mid-point of the $\overline{AB}$ wall, at an angle $\delta\sphericalangle$, the latter passes through the point of intersection of E_p' and R_2 and is tangent to the circle with radius $r \sin \Phi$ around 0. Thus, the polygon of forces furnishes the magnitude of E_p'.

In order to obtain E_p'' we assume $c = 0$ and bulk a density γ. The force E_0'' acting on the plane $\overline{DE}$ is given by

$$E_0'' = \frac{h_0^2\gamma}{2} \tan^2\left(45° + \frac{\Phi}{2}\right);$$

Table 36. Earth pressure coefficients (SOKOLOVSKI)

	$\Phi°$	10			20			30			40		
	$\delta°$	0	5	10	0	10	20	0	15	30	0	20	40
70°	K_{an}	0.58	0.54	0.51	0.35	0.31	0.28	0.20	0.17	0.15	0.11	0.09	0.07
	K_{at}	0.00	0.05	0.07	0.00	0.05	0.10	0.00	0.04	0.09	0.00	0.03	0.06
	K_{pn}	1.53	1.61	1.80	2.53	3.26	3.79	4.42	7.13	9.31	8.34	18.3	29.9
	K_{pt}	0.00	0.15	0.32	0.00	0.57	1.38	0.00	1.91	5.37	0.00	6.67	25.1
90°	K_{an}	1.20	0.66	0.64	0.49	0.44	0.41	0.33	0.29	0.27	0.22	0.19	0.17
	K_{at}	0.00	0.06	0.11	0.00	0.08	0.15	0.00	0.08	0.15	0.00	0.07	0.14
	K_{pn}	1.42	1.55	1.63	2.04	2.51	2.86	3.00	4.46	5.67	4.60	9.10	14.0
	K_{pt}	0.00	0.14	0.29	0.00	0.44	1.04	0.00	1.20	3.27	0.00	3.31	11.7
110°	K_{an}	0.73	0.70	0.69	0.57	0.53	0.51	0.46	0.42	0.39	0.35	0.32	0.29
	K_{at}	0.00	0.06	0.12	0.00	0.09	0.19	0.00	0.11	0.22	0.00	0.12	0.24
	K_{pn}	1.18	1.28	1.33	1.51	1.80	2.00	1.90	2.70	3.29	2.50	4.41	6.30
	K_{pt}	0.00	0.11	0.24	0.00	0.32	0.73	0.00	0.22	1.90	0.00	1.61	5.29

it acts in horizontal direction, in the lower third point of the height. The procedure is further identical with that used for cohesionless soil; it can be followed in the figure. The resulting force will be given by $\overline{E}_p = \overline{E}'_p + \overline{E}''_p$.

The construction has to be repeated for several sliding surfaces; the final value of E_p is the minimum of the forces thus obtained. The point of application is located between the mid-point and the lower third point of the height.

Finally, *Fig. 435* shows the graphical construction for the case when there is a uniformly distributed load on the surface of the backfill and the soil has cohesion.

Figure 436 gives a diagram for the quick determination of the coefficient of earth pressure both for the active and the passive case; for horizontal surface and vertical wall ($\varepsilon = 0°$; $\beta = 90°$), for different values of Φ and δ. The numerical correspond to $K = 2E/h^2\gamma$. The diagram can be conveniently used for quick orientation.

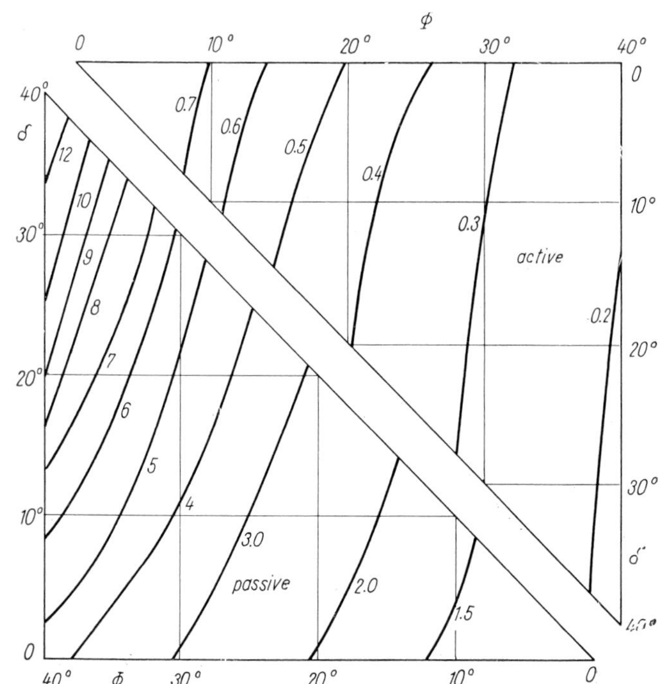

Fig. 436

Fig. 437

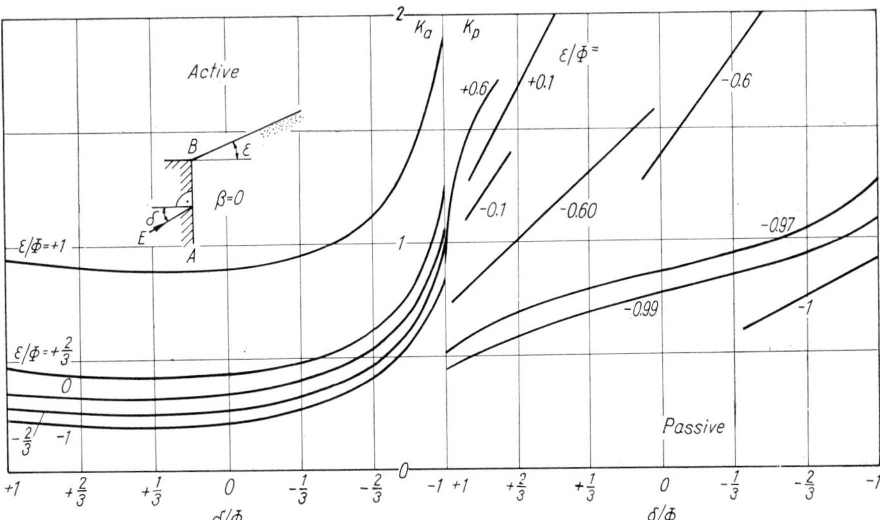

The diagram in *Fig. 437* takes also a sloping backfill into consideration. Here composite sliding surfaces were assumed and again the values $K = 2E/h^2\gamma$ are given as functions of δ/Φ. The ratio ε/Φ serves as a parameter; there are also curves for negative ε/Φ values, i.e. for backward slopes. The back of the wall is vertical; the angle of internal friction is equal to 30° which is assumed in many practical calculations.

The diagram furnishes coefficients for a single value of Φ only; therefore, it serves primarily the purpose of visualization: it gives a good picture of the effect of the different variables. There are coefficients given for + and — values of the wall friction angles, both in the active and the passive case. Positive values represent cases according to the sketch given in the figure, negative ones the opposite cases. Direction and magnitude of the wall friction angle depend finally on the direction and magnitude of the displacement between wall and backfill. The friction acts always in opposite direction; the degree of mobilization is determined by the amount of movement.

Recommended reading

Bölling, W. H. (1972): Bodenmechanik der Stützbauwerke, Strassen und Flugpisten. Anwendungsbeispiele und Aufgaben. Springer-Verlag, Wien, New York.

Brinch Hansen, J. (1953): Earth Pressure Calculation. Teknisk Forlag, Kopenhagen.

Caquot, A.–Kérisel, J. (1956): Traité de mécanique des sols. Gauthier Villars, Paris.

Harr, M. E. (1966): Foundations of Theoretical Soil Mechanics. McGraw-Hill Book Co., New York.

Jáky, J. (1938): Die klassische Erddrucktheorie mit besonderer Rücksicht auf die Stützwandbewegung. Abh. Int. Vereinigung f. Brückenbau u. Hochbau Nr. 5. Zürich.

Kézdi, Á.: (1962) Erddrucktheorien. Springer Verlag, Berlin–Göttingen–Heidelberg.

Kézdi, Á. (1973): 5th Chapter in "Foundation Engineering Handbook", edited by Winterkorn, H. F. and Fang, C. S. Van Nostrand Co., New York, N. Y.

Krey, H.–Ehrenberg (1936): Erddruck, Erdwiderstand und Tragfähigkeit des Baugrundes. W. Ernst, Berlin, 5. Aufl.

Proceedings, Brussels Conference 58 on Earth Pressure Problems. Bruxelles, 1958.

Proceedings, Conference on Pore Pressure and Suction in Soils. London, Butterworths, 1961.

Soil Mechanics Lecture Series: Design of Structures to Resist Earth Pressures. Soil Mech. Found. Div., Illinois Section ASCF. Chicago, Ill. 1964.

Terzaghi, K.: (1943): Theoretical Soil Mechanics. J. Wiley & Sons, New York.

Vereinigung Schweizerischer Strassenfachmänner: Stützmauern Grundlagen zur Berechnung und Konstruktion. Bemessungstabellen. Zürich, 1966.

Author index

Subject index

Contents of volume 2
Soil mechanics of earthworks, foundations and highway engineering

Contents of volume 3
Soil testing in the laboratory and in the field

Contents of Vol. 4
Application of soil mechanics in practice. Examples and case histories